T0180216

# Springer Series in Reliability Engineering

**Series editor**

Hoang Pham, Piscataway, NJ, USA

More information about this series at http://www.springer.com/series/6917

Prabhakar V. Varde · Michael G. Pecht

# Risk-Based Engineering

An Integrated Approach to Complex
Systems—Special Reference to Nuclear Plants

 Springer

Prabhakar V. Varde
Research Reactor Services Division,
   Homi Bhabha National Institute
Bhabha Atomic Research Centre
Mumbai
India

Michael G. Pecht
Center for Advanced Life Cycle
   Engineering
University of Maryland
College Park, MA
USA

ISSN 1614-7839                    ISSN 2196-999X    (electronic)
Springer Series in Reliability Engineering
ISBN 978-981-13-4327-8          ISBN 978-981-13-0090-5    (eBook)
https://doi.org/10.1007/978-981-13-0090-5

Printed on acid-free paper

This Springer imprint is published by the registered company Springer Nature Singapore Pte Ltd. part of Springer Nature
The registered company address is: 152 Beach Road, #21-01/04 Gateway East, Singapore 189721, Singapore

# Preface

This book has been written for the students, researchers, and professionals in the area of reliability and risk modeling who are looking for a single reference that could help them solve problems related to practical applications with risk as the bottom line. Worldwide, there is an increasing trend for application of risk-informed approach that uses available risk insights, in support of design, operation, and regulation of complex system.

This book proposes an integrated risk-based engineering approach where the basic premise is that deterministic and probabilistic aspects are integral to any problem space, and for holistic solution, these two components need to be treated suitably for effective modeling and analysis of safety cases. For example, traditionally, the safety community has been depending on the deterministic risk insights for design, operations, and regulations employing conservative approach that addresses the lack of knowledge or data. With the advent of risk assessment techniques in general and probabilistic risk assessment in particular, it has become possible to quantify the reliability and safety indicators, like failure frequency, system unavailability, and core damage frequency. Hence, this work proposes an integrated risk-based engineering (IRBE) approach which employs qualitative insights from deterministic and quantitative insights from probabilistic risk assessment to address the safety aspects in design, operation, and maintenance.

The more popular approach in this context is risk-informed approach which is targeted to a complex issue of decision-making. In risk-informed approach to decision-making, particularly in support of regulatory decisions, the insight from PRA forms one input along with other insights, be it design, operation, and regulatory stimulation that relate to available margins, provision of defense in depth, etc., in support of decision-making. Therefore, this book has a separate Chap. 14 on risk-informed decisions.

The chapters in this book deal with individual topics that support the implementation of integrated risk-based engineering as a subject. There are host of topics which are related or support risk-based decisions; however, in this book, a conscious decision has been taken in respect of level of details that need to be covered in individual chapter and aspects that are more relevant and help built the subject in an effective manner.

Chapter 1 introduces the integrated risk-based engineering—the subject of this book. This chapter explains the premise; that is, deterministic and probabilistic aspects are integral to any engineering issue, and integrated risk-based approach seeks to apply them toward reducing uncertainty in results of the analysis. The increasing application of probabilistic approach in many fields, like structural engineering, thermal hydraulics in general and passive systems in particular, nuclear physics, damage and degradation modeling, and surveillance and monitoring through implementation of prognostics and health management to reduce uncertainty in predictions, are some of the areas where extensive development work is being performed. As such, the literature shows many applications of risk-based approach in complex systems. It can be argued that this book discusses a new approach which might at the outset appear like risk informed; however, it looks at the subject in a different way where deterministic and probabilistic aspects work hand in hand and not as separate subject.

Chapter 2 provides a brief overview of risk characterization that is directly relevant to interpretation of results of risk analysis for engineering systems. The risk characterization deals with two aspects: The first one is the presentation of results of reliability studies at system level and risk metrics at plant level, e.g., core damage frequency which is a quantified statement of level of safety of the plant. Risk metrics is also used in support of application of probabilistic risk assessment where risk prioritization is performed considering likelihood and consequence associated with the items under considerations. The second aspect is the presentation of uncertainty, which also forms part of risk characterization, and this chapter just touches upon this aspect.

Chapter 3 provides the fundamentals on probabilistic/reliability modeling. This chapter begins with the discussion on bathtub curve, then introduces major reliability indicators, provides insights into and application of probability distributions, and further provides models and methods for data interpretation and analysis. This chapter is a prerequisite to system-level modeling.

Chapter 4 presents the models and methods for system-level reliability or risk prediction. In this chapter, the current practices in reliability and risk modeling, reliability block diagram, failure mode and effect analysis, and fault tree and event tree analyses are presented. However, these methods are static in nature and there is an increasing interest in the development and application of dynamic methods. In this context, the dynamic tools, like Markov modeling, Petri net, and dynamic fault tree and event tree, have been introduced in this chapter.

Chapter 5 on life testing discusses the requirements and application of life-testing approach in support of risk modeling. The basic requirements are to have an approach where data from field experience are not available, particularly for new components, and this is where the accelerated life-testing approach provides insights into the competing failure mechanism as also reliability or life prediction. Accelerated life-testing methods and the related models have been discussed in this chapter. Life prediction is an essential component of physics of failure and prognostics and health management for component and systems. In this context, this chapter, apart from life and failure rate prediction, is a prerequisite for Chaps. 12 and 13.

Chapter 6 on probabilistic risk assessment is central to IRBE. The available literature shows that probabilistic approaches are increasingly being used in design of engineering systems. In this chapter, the major objectives are to develop a risk model of a complex system that enables (a) quantified statement of risk or conversely speaking safety, (b) propagation of uncertainty from component to system further to plant level, and (c) integration of human factor with the plant model. In IRBE approach, the role of PRA is to provide the quantified estimate of risk and uncertainty associated with results of the analysis. In this chapter, a conscious decision is taken to deal with Level 1 PRA in detail, while Level 2 and Level 3 PRA are kept to introductory level as major aspects of risk are addressed using system-level analysis using statement of core damage frequency for the plant.

Chapter 7 provides an overview of risk-based design methodology to complex engineering systems. Risk-based design at varying levels of details is being employed to many engineering systems. Major elements of risk-based design have been discussed. In the approach proposed, the complete design approach has been divided into two parts—higher-level modeling, which deals with plant-level modeling employing PRA framework and lower-level modeling, which requires probabilistic structural analysis methods. Supporting tools and methods, applicable codes and standards, and a case study have been discussed in brief.

Fatigue is one of the major contributors to mechanical failure and requires modeling and analysis. There are relatively large uncertainties associated with model and parameters in conventional fracture analysis methods. In this context, Chap. 8 introduces probabilistic fracture mechanics as part of fracture risk assessment. Characterization of uncertainty is central to any advanced safety methodology and this is where probabilistic tools and methods can provide a framework to quantify uncertainty primary to address the issues related to application and use of factor of safety.

Uncertainty characterization is a vital component of integrated risk-based engineering. Chapter 9 provides an overview of uncertainty modeling approaches. This chapter establishes relevance of uncertainty with risk-based approaches, particularly the aspects related to decision under uncertainty. A brief overview of codes, guides, etc. has been provided.

Probabilistic risk assessment methods provide a way to integrate human factor in risk modeling, and in this context, human factor or human reliability is a critical aspect of risk evaluation. Chapter 10 provides an overview of the available methods, research work being done, and a generic methodology for integrating human factor in probabilistic risk assessment. Most of the available approaches lack considerations of a robust human model that can link, apart from cognitive aspect, the consciousness and conscience phenomenon that governs human behavior, particularly in emergency conditions and security environment. This chapter introduces a consciousness, cognition, consciousness, and brain ($C^3B$ or CQB)-based framework. In this approach, direct measurement of parameters for assessing physiological and psychological stresses is proposed apart from the traditional performance shaping factors. A human reliability model has been proposed based on the research work performed with field data and simulator environment.

The electronic/electrical controls and logic in a complex system require, apart from normal risk modeling approaches, treatment of subject considering the specific characteristics and the constraints posed by these systems. Chapter 11 deals with the reliability modeling of digital systems. There are challenges in digital system modeling, like software reliability, digital being new technology, non-availability of hardware failure data, and non-availability of a generally accepted methodology. In this context, this chapter provides the state of the art in the digital system modeling and presents a simplified approach for digital system modeling, where the available taxonomy along with conservative approach attempts to provide a solution.

Traditional approach to electronic component reliability employs handbook or statistical methodology. Handbook approach is being discouraged by the experts as the modifying factors may not account for the stresses actually experienced by the component in field environment. The physics-of-failure-based approach is based on the science-based degradation models for life and reliability prediction. Chapter 12 on physics of failure provides a brief overview of this methodology as this approach has shown potential to be improved over of traditional handbook approach which is being discouraged by the experts in the field. This chapter introduces various failure mechanisms and available physics-of-failure software.

Traditionally, the deterministic approach depends on the periodic surveillance and maintenance programs to ensure availability of the process- and safety-related components and systems. There is an increasing trend in employing condition monitoring methods for predicting failure trends. However, these methods lack instant of failure prediction with acceptable level of uncertainty. Chapter 13 provides an overview and potential for application of prognostics and health management (PHM) to electronics and control systems. The PHM approach provides tools and methods to predict the prognostic distance that facilitates application of management of failures in a real-time environment.

Decision-making is a complex process where, apart from technical rationales, a judgment is required to comprehend the uncertainty in respect of decision alternatives. Apart from this, decision makers are required to evaluate the long-term and short-term consequences. Also, the ethical aspects form an important aspect of decision-making. Chapter 14 provides research work on the application of integrated risk-based engineering to risk-informed decisions. This chapter attempts to present the role of PRA, its limitation, and the role of humans and deliberations in decision-making.

Chapter 15 provides the R&D on the application of risk-based/risk-informed approach to real-time challenges, like life extension, test interval optimization, and risk monitor. The case studies presented deal with the R&D efforts in developing risk-based/risk-informed decisions. However, the available literature shows that risk-based in-service inspection, deployment of risk monitor for nuclear power plants, test interval, and allowable outage time evaluation as part of technical specification optimization, etc. are some of the major applications of risk-based applications.

This book, as the author feels, has been able to develop a new subject, i.e., integrated risk-based engineering which can be taught in university-level courses, and at the same time, risk professional can apply it to address the real-time issues.

It goes without saying that for any project support and encouragement forms a major driving force. We thank Dr. S. Basu, Chairman, Atomic Energy Commission, India, for providing the necessary support and encouragement for this project. Mr. K. N. Vyas, Director, BARC, Mumbai, has unconditionally supported this work and that provided us the needed driving force for this work. We sincerely appreciate the help provided by Mr. S. Bhattacharya, Associate Director, Reactor Group, BARC, Mumbai.

We thank Dr. Michael Osterman, Research Scientist, and Mr. Guru Prasad Pandian, CALCE, University of Maryland, for providing important modification and changes in Chap. 12 on physics of failure. The authors of this book would like to thank Dr. Myeongsu Kang, CALCE, University of Maryland, for providing valuable input and updates on the original work in respect of Chap. 13 on prognostics and health management. We also, in all sincerity, appreciate the editorial support from Ms. Cheryl Wurzbacher, Staff at CALCE, University of Maryland. We thank Mr. Arihant Jain, Scientist at BARC, in providing review comments and support in document editing and management. Support from N. S. Joshi, Vivek Mishra, and other colleagues at BARC and CALCE is also appreciated.

The authors are aware that like any other work this is not a perfect piece of work, and by this token, the readers will have comments. We sincerely welcome review comments on this book in general and specific comments on chapters, and we assure you that these reviews will help us update/enrich the work.

Mumbai, India                                    Prabhakar V. Varde
College Park, USA                                   Michael G. Pecht

# Contents

# About the Authors

**Prof. Prabhakar V. Varde** is an expert in the field of application of reliability and probabilistic risk assessment to nuclear plants and is currently working as Head of the Research Reactor Services Division and Senior Professor at Homi Bhabha National Institute, Bhabha Atomic Research Centre, Mumbai, India, where he also serves in advisory and administrative capacities in Atomic Energy Regulatory Board (AERB), India, and the Homi Bhabha National Institute, India. He is the Founder and President of the Society for Reliability and Safety (SRESA) and is one of the chief editors for its international journal—*Life Cycle Reliability and Safety Engineering*. He completed his B.E. (Mech) from Government Engineering College, Rewa, in 1983 and joined BARC, Mumbai, in 1984, where he worked as a Shift Engineer in the Reactor Operations Division until 1995. In 1996, he received his Ph.D. in Reliability Engineering from the Indian Institute of Technology, Bombay, Mumbai, following which he worked as a Postdoctoral Fellow at the Korea Atomic Energy Research Institute, South Korea, and a Visiting Professor at the Center for Advanced Life Cycle Engineering (CALCE) at the University of Maryland, USA.

Professor Varde is also a consultant/specialist/Indian expert for many international organizations, including OECD/NEA (WGRISK), Paris; International Atomic Energy Agency, Vienna; University of Maryland, USA; Korea Atomic Energy Research Institute, South Korea. Based on his R & D work, he has published over 200 publications in journals and conferences, including 11 conference proceedings books.

**Prof. Michael G. Pecht** is a world-renowned expert in strategic planning, design, test, and risk assessment of electronics and information systems. Professor Pecht has a B.S. in Physics, an M.S. in Electrical Engineering, and an M.S. and Ph.D. in Engineering Mechanics from the University of Wisconsin at Madison. He is a Professional Engineer, an IEEE Fellow, an ASME Fellow, an SAE Fellow, and an IMAPS Fellow. He is the Editor-in-Chief of IEEE Access and served as Chief Editor of the IEEE Transactions on Reliability for nine years and Chief Editor for Microelectronics Reliability for sixteen years. He has also served for three US

National Academy of Science studies, two US Congressional investigations in automotive safety, and as an expert to the US Food and Drug Administration (FDA). He is the Founder and Director of Center for Advanced Life Cycle Engineering (CALCE) at the University of Maryland, which is funded by over 150 of the world's leading electronics companies at more than US$6M/year. The CALCE received the NSF Innovation Award in 2009 and the National Defense Industries Association Award. Professor Pecht is currently a Chair Professor in Mechanical Engineering and a Professor in Applied Mathematics, Statistics, and Scientific Computation at the University of Maryland. He has written more than twenty books on product reliability, development, use, and supply chain management. He has also written a series of books of the electronics industry in China, Korea, Japan, and India. He has written over 700 technical articles and has 8 patents. In 2015, he was awarded the IEEE Components, Packaging, and Manufacturing Award for visionary leadership in the development of physics-of-failure-based and prognostics-based approaches to electronic packaging reliability. He was also awarded the Chinese Academy of Sciences President's International Fellowship. In 2013, he was awarded the University of Wisconsin–Madison's College of Engineering Distinguished Achievement Award. In 2011, he received the University of Maryland's Innovation Award for his new concepts in risk management. In 2010, he received the IEEE Exceptional Technical Achievement Award for his innovations in the area of prognostics and systems health management. In 2008, he was awarded the highest reliability honor, the IEEE Reliability Society's Lifetime Achievement Award.

# Chapter 1
# Introduction

*When you can measure what you are speaking about and express it in numbers, you know something about it.*

Lord Kelvin

## 1.1 Introduction

The traditional approach to design, operation, and regulation for complex engineering systems in general and nuclear systems, in particular, has been deterministic in nature where safety principles that mainly include, defense in depth, fail-safe criteria, redundancy, and diversity form the basic framework. Given the competitive market conditions in this era of globalization, nuclear systems have to have a strategy where it meets the market demand effectively without comprising safety. The major goal here is to realistically understand the available engineering margins and associated uncertainties to meet safety and availability goals and make these systems more sustainable.

Even though deterministic approach is time-tested and worked well for all these years, it has some limitations. It is conservative and prescriptive in nature and does not provide a measurable parameter for safety and reliability of engineering systems. Experience and research have shown that the deterministic approach where defense in depth is integral to this approach, while reasonably assures safety, often leads to expensive systems and technologies that the society and market would not be able to afford. Further studies have shown that while some designs and regulations based on conservative approaches appear to reduce risk of complex engineering systems, e.g., nuclear plants, this may come at exorbitant cost and still may not guaranty safety [1]. With advances in technology, for example, advances in computational techniques, improved understanding of materials, simulation methods availability of data and information, there is increasing interest in application of best estimate and further risk-informed approach [2]. In spite of these

P. V. Varde and M. G. Pecht, *Risk-Based Engineering*, Springer Series in Reliability Engineering, https://doi.org/10.1007/978-981-13-0090-5_1

developments, the approach to address uncertainty, by and large remains conservative, i.e., based on application of relatively large safety factor. The considerations of safety factors provided a way to compensate for a lack of knowledge and data; but often make the systems more complex, costly, and unsustainable. The requirement 15 of IAEA Safety Standard entitled safety assessment for facilities and activities states that both deterministic and probabilistic approaches should be used in safety demonstration [3]. Further, the integrated risk-informed approach employing deterministic and probabilistic framework is an established tool in support of decision-making [4].

Probabilistic risk assessment (PRA) or probabilistic safety assessment (PSA) and deterministic approach along with comprehensive considerations of uncertainty and human factor provide the basic framework for a holistic risk-based approach. Apart from PRA, other methods, such as hazard and operability analysis (HAZOP) and failure mode and effect analysis (FMEA), which are often considered as qualitative risk evaluation methods, are also used to derive risk insights in support of design and operation evaluation in the chemical and process industry.

## 1.2   Historical Perspective on Probabilistic Risk Assessment and Risk-Based Applications

Former's paper in 1967 [5] provided a new approach for site selection. However, Rasmussen's [6] comprehensive and landmark PRA study, known as the WASH-1400 safety study, and later the German risk assessment performed for European Union nuclear plants [7], laid the foundation for the risk-based/ risk-informed approaches.

The evidence over six decades shows that nuclear power is a safe means of generating electricity; however, the three major nuclear plant accidents—Three Mile Island, USA, in 1979; Chernobyl, Russian Federation, in 1986; and Fukushima Dai-Ichi, Japan, in 2011 [8]—raised some questions that mainly include the capability or tolerance of plant and systems for human and institutional failures, considerations related to combined events, multi-unit site issues, gap areas in utility regulatory relations, and public communications particularly during emergency conditions.

After the Three Mile Island accident, role of PRA was recognized as a tool for improved and systematic understanding of nuclear plant design and operational safety issues. It became clear that it is not possible to base safety issues on a few selected accident scenarios, referred to as maximum credible accidents in the traditional deterministic approach, and it was found to be necessary to incorporate all the modes of individual component failures to construct a system model that facilitates quantitative prediction of all levels of safety.

The literature shows that over 200 Level 1 PRA studies have been performed for nuclear power plants up to 2011 [9]. In fact the current literature highlights the fact that for most of the NPPs, PRA studies have been completed. Further, there is growing interest in applying the PRA process to chemical plants [10]. The publication of USNRC Guide 1.200 [11] has encouraged application of PRA studies in support of risk-informed decisions. Successful development and deployment of risk monitors in many advanced countries and development of risk-based in-service inspection programs and risk-based maintenance management programs are just a few examples of applications of Level 1 PRA for addressing risk.

Some of the areas where the deterministic and probabilistic approaches work together at component level are reliability/risk-based approach to design a component, physics-of-failure approach to electronic system risk modeling, probabilistic fracture mechanics for mechanical component failure risk, damage modeling, software reliability modeling, where apart from estimate of component failure probability, the uncertainty characterization also forms part of the results. There is increasing interest in applying the risk-based approach [12] to shipbuilding [13], off-shore drilling [14, 15], aircraft design [16], space systems [17], chemical and process systems [18], marine systems [19], flood risk mitigation and management [20, 21], and the banking and financial sector [22]. Most of these applications employ risk analysis tools, be it probabilistic risk assessment, failure mode effect analysis, event tree or fault tree models, to assess and manage risks. Basically these applications focus on ways to address management of hazard in question to effectively address safety while achieving performance objectives.

For regulatory applications, the integrated risk-based approach can be applied at two levels. First, it can improve inspection procedures that form an integral part of regulatory review by prioritizing activities and enable optimization of resources in support of regulatory inspections, and second, performance of review based on risk-based approach, with stipulated deterministic and probabilistic goals and criteria, organizational and management framework for principles, etc.

## 1.3  Integrated Risk-Based Engineering Approach

There are concerns about the risks associated with system failure on the one hand and its impact on reliability and system availability on the other. Unfortunately, this information, particularly in terms of quantified estimates, is not available from the deterministic approach. Nevertheless, the deterministic approach has always provided intuitive and qualitative insights into risk and reliability of systems. In fact, the traditional deterministic approach to decision-making has also been risk-informed in nature as the "qualitative notion of risk" forms an integral part of decision-making, even in traditional approach to decisions. The only difference is that the current definition of the risk-informed approach uses quantified estimates of risk obtained from probabilistic risk assessment as input, along with the traditional deterministic considerations, for decision-making

Therefore, it can be argued that the qualitative and quantitative risk insights available from deterministic and probabilistic approach, respectively, provide a robust foundation for an integrated risk-based engineering approach. The deterministic approach provides basic principles, goals, and criteria, while probabilistic approach provides a rational and integrated methodology for system modeling toward arriving at quantified estimates of risk and reliability. These estimates are based on the system process configuration, engineering details, understanding of root causes of failures, and component reliability data.

The conventional definition of risk-based approach is where decisions are based on the insights provided by probabilistic safety assessment only. The other definition of risk-based approach is "an approach where the decisions are based on risk assessment," without going into the argument for probabilistic or deterministic. The conventional notion of a risk-based approach essentially means decisions are based on inputs and insights from risk assessment.

Here we define integrated risk-based engineering (IRBE) approach as "an approach where the deterministic and probabilistic methodology is employed in an integrated manner toward reducing uncertainty in arriving at the solution. The considerations of human reliability and quality assurance form a part of IRBE framework. The provision for prognostics and surveillance forms an integral part of solution metrics."

There is no explicit and clearly demarcated boundary between deterministic and probabilistic considerations for engineering systems, as these two aspects are overlapping and to some extent integrate into any engineering problem. Figure 1.1 depicts the role of deterministic and probabilistic approach in IRBE. As can be

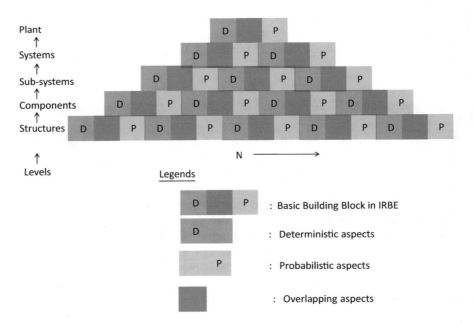

**Fig. 1.1** Role of deterministic and probabilistic elements in IRBE

seen, the basic building blocks that make this approach have both deterministic as well as probabilistic elements and have overlapping functions at all the levels, viz. structural level to component and through subsystem to system and finally plant level for hardware and software systems.

Requirement of PRA stems from, apart from requirements-related quantified statement of safety, the characterization of uncertainty such that there is an improved understanding of safety margins. On the other hand, for example, the fundamental scientific formulations and models, design criteria, design rules, and failure criteria are derived from the deterministic approach and probabilistic approach starts with the availability of these information.

The probabilistic approach in IRBE provides an integrated model of a system using sound and well-understood reliability engineering tools that further allows assessment of the performance of individual components and their impact on plant safety. It also provides a quantified statement of safety of a system (as core damage frequency for the case of nuclear plants) and at lower level the system unavailability, as well as a quantitative evaluation of both aleatory and epistemic uncertainties. Finally, the approach integrates human performance with the integrated model of a system. IRBE also provides an integrated framework for validation of results against stipulated goals and criteria and provision for monitoring and feedback for continued assurance for safety in support of any design, operation, and regulatory review. The major premise of IRBE is that deterministic and probabilistic approaches together have the capability to provide holistic solutions based on risk considerations in the present context of global and competitive market scenario.

## 1.4  Factor of Safety and Uncertainty

In a deterministic approach, the factor of safety is based on conservative assumptions that served to account for uncertainty in data and model with an intent to ensure safety margins. The factor of safety F is defined as:

$$F = \frac{S_m}{s_m} \tag{1.1}$$

where $s_m$ and $S_m$ are mean stress and strength. Often factor of safety has relatively large value. This approach worked well when the data on mechanics of failure and advanced technologies and computational systems were not available. With the advent of technology, availability of material databases, and best estimate codes, it was possible to quantitatively characterize material properties in terms of not only material stress (s) and strength as point value of $s_m$ and $S_m$ but also the uncertainty associated with these properties as shown in Fig. 1.2. Without these distributions, the ratio of $s_m$ and $S_m$ used to be relatively large. With the availability of mean value of $s_m$ and $S_m$ and associated standard deviation, $\sigma_s$ and $\sigma_S$, it became easier to

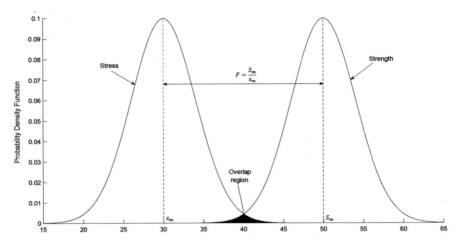

**Fig. 1.2** Stress–strength representation—a perspective on safety margin

provide effective and efficient designs. This formulation also provided an effective mechanism to evaluate failure probability or a statement of safety or reliability.

Figure 1.2 highlights the importance of uncertainty assessment in optimizing the design stress and strength toward removing unnecessary conservatism. Stress and strength are both assumed to follow a normal distribution with respective means as $s_m$ and $S_m$, respectively. The overlapping shaded area indicates the probability of failure region. Probability of failure, i.e., probability that stress is greater than strength, is given as:

$$F_{pr} = \int\limits_{-\infty}^{\infty} f_s(s) \left\{ \int\limits_{-\infty}^{s} f_S(S)\mathrm{d}S \right\} \mathrm{d}s \qquad (1.2)$$

where $f_s(s)$ and $f_S(S)$ are probability density function for stress and strength, respectively. Design objective is to reduce the probability of failure to an acceptable value and at the same time to remove unnecessary conservatism.

Increasing need for highly efficient, cost-effective, and reliable systems is pushing toward probabilistic design approach. Probabilistic design approach addresses the probability of failure directly and hence gives better design estimates. It helps in taking the conservative design approach from deterministic approach toward more realistic design. This, as part of IRBE, allows the designer to prepare optimum design that caters to safety requirements along with being cost-effective and efficient. Quantitative notion of risk and uncertainties reduces the excessive conservatism from the design.

## 1.5 Basic Framework for Integrated Risk-Based Engineering

Figure 1.3 shows the basic IRBE framework. The first step in IRBE is identification and formulation of requirement specifications. Depending on the type of application, e.g., system design, change evaluation in support of plant operation, or regulatory review, the requirements need to be formulated. The next step is identification of the deterministic and probabilistic components of the problem or issue on hand and detailed analysis. A deterministic approach, for example, a thermal-hydraulic analysis to assess passive system reliability involves characterizing uncertainty in the deterministic variables through one of the available simulation approaches, like Monte Carlo simulation of the governing equation to arrive at the final estimate of the parameter with uncertainty bounds. Similarly, modeling the frequency of pipe rupture failure might require a probabilistic fracture mechanics approach, wherein the structural inputs and crack initiation and growth parameters are treated as a random variable. The reliability parameters for hardware, software, and human error are derived either from a generic or plant-specific sources along with associated uncertainty that forms the input for the PRA modeling. A PRA approach is required to create an integrated model to estimate the failure probability of the system. The PRA model requires a fault tree or event tree approach depending on the nature of problem.

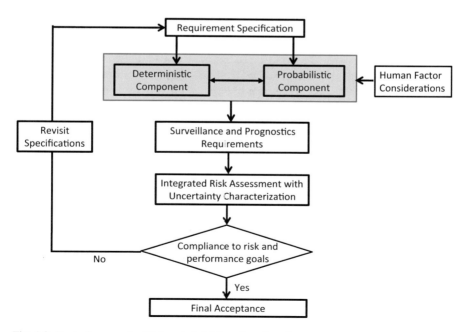

**Fig. 1.3** Basic framework of integrated risk-based engineering

Human factor considerations form part of deterministic as well as probabilistic evaluation. While for a given design, deterministic considerations might provide information on type of human–machine interface, time required for a given action, plant procedures, like technical specifications and emergency procedures; the probabilistic approach provides an effective framework for integrating human performance in the risk model of the plant.

Further, the IRBE procedure requires input, on plant surveillance provisions which include testing frequency, online monitoring, condition monitoring, and prognostic capability at plant, system, and component levels. This information is required to evaluate the plant's capability to identify the failure in advanced such that corrective actions can be initiated. The surveillance and monitoring capability also plays critical role after the change has been affected to get feedback on subject system performance.

Integrated risk assessment is an iterative process where all the assumptions and uncertainty bounds at component, system, and plant levels are evaluated toward ensuring that the change is complying with the risk and performance goals, so that the subject change can be accepted. The results obtained from the integrated model are subjected to validation with applicable deterministic and probabilistic performance criteria, achievable goals, or regulatory stipulations. In case the evaluation is not meeting the set goals and criteria, the complete process is revisited till acceptable solution is obtained.

## 1.6   Major Elements of Integrated Risk-Based Engineering

This section deals with the typical procedural steps in IRBE to introduce and demonstrate the concepts in this approach. These steps are indicative only and not exhaustive, and depending on the type of application, it may be required to add or remove few steps. For example, the following steps are typical for implementing a change in specifications in the plant.

1. Define the problem,
2. Define the objective and scope of the analysis,
3. Identify major assumptions,
4. Identify system boundary and limitations/constraints,
5. Identify quality attributes for specific tasks or applicable code/standard,
6. Identify initiating events and set of input variables/data,
7. Identify safety provisions or engineering safety features,
8. Gather design and operational/expected performance data and information,
9. Perform deterministic and probabilistic modeling analysis,
10. Create an integrated (preferably dynamic) model for effective sensitivity analysis,
11. Ensure that human performance aspects are integrated in the model,

12. Estimate uncertainty in input variables,
13. Simulation and modeling/analysis and sensitivity analysis,
14. Compare the set of probabilistic and deterministic established goals and criteria in conjunction with insights on uncertainty,
15. Provide independent and/or regulatory review,
16. Ensure regulatory compliance and oversight,
17. Implement change, retrofit, or modification,
18. Follow-up and monitor trends through PHM procedures,
19. Document and record.

Even though the steps in IRBE have been listed in chronological order, the nature of the complete implementation involves many parallel activities with recursive tasks. The problem in hand should be formulated in a crisp and clear manner so that the objective and scope function can be defined in a clear and unambiguous manner.

Many steps in IRBE are governed by the nature of issue in hand; e.g., the problem in hand deals with new design, in support of existing plant operations (change in configurations or operating policy) or in support of regulatory review. Hence, nature of evaluation and resources required will change depending on scope and objective of the analysis.

The deterministic analysis provides failure criteria as an input for probabilistic analysis. For example, for a loss of coolant accident scenario the coolability criteria and accordingly the injection flow requirements are derived from deterministic analysis. In IRBE, the specific requirement is to provide apart from the point value of the parameter, like emergency flow, the uncertainty bounds of parameter (like minimum and maximum flow) also. In fact, the probabilistic model is developed based on the given plant configurations that satisfy the objective and performance function of the plant. The probabilistic goal could deal with complying with system-level unavailability ($1 \times 10^{-3}$/demand) and change in core damage frequency (CDF) goal (change in CDF <1% of the reference value). The acceptable radiation level criteria may require considerations of as low as reasonably achievable (ALARA) principles.

The analysis should reflect the assumptions, constraints, limitation of analysis tools and methods, data, etc., and validation of the same by performing the sensitivity analysis. As mentioned, uncertainty modeling, be it deterministic or probabilistic aspect, is an integral part of IRBE. Similarly, the system boundary should be clearly marked, and in case, there is some interfacing connection with other system then this needs considerations as part of the analysis.

The quality of a deterministic and probabilistic analysis is critical and has direct bearing on the results of the analysis. For example, only validated best estimate code should be used in the analysis. The analysis should take note of the code and standards followed for designing these systems. In case, any part of system is not meeting the standards then provision of supporting additional safety criteria should be demonstrated in the analysis. In case, a change is affecting plant technical specification then compliance should be checked. It should be ensured that human error events which form part of the analysis are covered in plant emergency operating procedures. If certain human actions are not considered in available plant emergency operating procedures, then the recommendation of the analysis should reflect this requirement.

Similarly, quality assurance for PRA is also critical, particularly when a real-time application is being developed. Subjecting various procedural and logical steps to establish risk assessment quality attributes is vital to get the required confidence in decision-making. The available framework; for implementation of a quality checklist for each of the procedural elements, including initiating event selection, system modeling, human reliability analysis, common cause failure analysis; include international standards/documents such as the IAEA-TECDOC-1101 on Quality assurance in PRA [23], IAEA-TECDOC-1804 on quality attributes for Level 1 PRA [24] and ASME/ANS standards [25]. Apart from this, national guide and standards also enable characterization of quality of the applications.

The plant system description, like any other analysis, forms an important element of IRBE. This includes systems descriptions, associated drawings, system modes of operations, description plant/system logic, and expected behavior of the system in various modes like normal operation, transient, and emergency conditions.

The deterministic principles and criteria provide the basic framework for ensuring safety. The role of the probabilistic approach should be to consolidate but not dilute these principles. For example, if the deterministic principles provide for a redundant system to cater to certain safety functions, the probabilistic approach should provide the probabilistic evaluation and should provide quantitative statement reliability or availability of these redundant provisions.

Other important deterministic or design/operation and maintenance aspects that need to be considered for evaluation include: (a) safety code or plant technical specification applicability, (b) assessment of consequences, e.g., in terms of leak size, containment scenario, and radioactivity release and, hence, adequacy of emergency provision in respect of on-site and off-site emergencies, (c) considerations of in-service inspection insights in respect of structural integrity assessment, (d) credit to be given for condition monitoring or prognostics provisions and status indications, (e) assurance against human performance and training-related aspects, (f) requirements related to the test override provision, (g) demonstration of built-in system capacity and plans and procedures to assess plant's coping capability for a certain scenario, and (h) requirements of plans and provisions for regulatory review and oversight.

The objective of data collection and analysis is to generate reliability estimates for hardware components and common cause failures. The data should facilitate simulation with a range of test intervals to aid optimization. For example, steady-state unavailability data or data on demand failure probability do not allow dynamic modeling. Also, test interval optimization requires dynamic equations with respect to incremental changes in time intervals, and hence, use of steady-state values is not appropriate. Most of the components involved in the safety systems are modeled as standby-tested components. Accordingly, the standby-tested model is used to determine component failure probability [26].

The insights obtained from deterministic analysis in the form of available system redundancy, diversity, cooling criteria, or system failure criteria form the inputs to the probabilistic methods for risk assessment. Keeping in view the complexity of the problem and the problem definition, the Level 1 PRA model of the plant should be used. There are two major aspects of PRA model simulation—event tree analysis and fault tree analysis. Event tree modeling is performed to generate accident sequences associated with given initiating event and plant response in terms of success or failure of actuation of safety systems or human action. The fault tree analysis is performed for arriving at safety system failure probability.

Judgment or decision-making should not be based on point estimates. It is advisable that uncertainty analysis should be performed for real-time applications. Uncertainty analysis accounts for randomness in the variables, lack of knowledge, or data or inadequacy of the model employed for probabilistic modeling. Uncertainty analysis provides the upper and lower bounds of the estimates along with the median and mean values of the parameter. However, for this illustration, uncertainty and sensitivity analyses have not been performed.

Sensitivity analysis is performed to evaluate the impact of various assumptions on the overall results of PRA. Apart from this, sensitivity analysis is also performed considering the range of uncertainty bounds for selected components/human actions, for which the uncertainty is relatively high.

Technical review is performed at two levels, viz. peer review and regulatory review. The data and assumptions that form input for the analysis are important aspects of this review. The peer review is performed by experts dealing in the respective areas, but not the individual or group involved in the analysis. The role of the probabilistic expert is to assess the model and data accuracy and overall representation of the system. The design and operation and maintenance aspects are checked and verified, keeping in view the overall objective of the change. It is also verified that the fundamental safety aspects such as defense in depth are not diluted, and therefore, no compromise in redundancy or diversity level takes place. It is also ensured that the provision of fail-safe criteria is not compromised.

Finally, this review includes aspects such as assessment and implication of system unavailability and CDF estimates and its comparison with the target goals. A checklist procedure is used for ensuring that all the safety issues have been addressed. For example, the major metrics that form the inputs for decision-making could include, say for a given LOCA scenario, which is as follows:

1. For ensuring that the provision of defense in depth is not diluted following is typical checkpoints:

   (a) Provision of redundancy meets the design intent.
   (b) Provision of diversity meets the design intent.
   (c) Fail-safe criteria are maintained and meet the design intent.

(d)  System success criteria for applicable postulated scenario are met.
(e)  Provision exists to detect latent critical failures, or the likelihood of this event is low.
(f)  Alternative provisions still provide the same backup as before the change.
(g)  The results of in-service inspection and condition monitoring have not shown any symptom of significant degradation.
(h)  No new issues in respect of human actions and procedures.
(i)  The provision of surveillance and prognostics are commensurate with the change.

2.  Probabilistic criteria

(a)  The data used for the analysis are from plant-specific source
(b)  The increase in system unavailability, e.g., <5%.
(c)  The increase in CDF is, e.g., <1.0%.
(d)  Sensitivity analysis has not shown any significant impact of any single assumptions.
(e)  Uncertainty analysis bound for the system unavailability and CDF is well within the acceptable range.
(f)  PRA attributes analysis provides adequate confidence in PRA model, data, assumptions, and uncertainty in final results.

After obtaining the regulatory clearance, it is management's responsibility to ensure the recommended changes are implemented. This means modifying the technical specifications, changing system procedures and schedules, and training the staff, wherever required. Management should ensure regulatory compliance in words and "spirit." Generating plans and procedures for follow-up and trend monitoring is an important part of a change implementation program. The follow-up program should ensure that adequate provision exists for prognostics and health management or a condition monitoring program for critical functions in the system. Finally, the activities related to documentation and records should comply with various communication protocols.

# References

1. M. Modarres, *Risk Analysis in Engineering, NW* (CRC Press, USA, 2006)
2. International Atomic Energy Agency, *Deterministic Safety Analysis for Nuclear Power Plants, IAEA Specific Safety Guide No. SSG-2* (IAEA, Vienna, 2009)
3. International Atomic Energy Agency, *Safety Assessment for Facilities and Activities, General Safety Requirements Part 4, No. GSR Part 4* (IAEA, Vienna, 2009)
4. International Atomic Energy Agency, *A Framework for an Integrated Risk Informed Decision Making Process, INSAG-25* (IAEA, Vienna, 2011)
5. F. Farmer, *Siting Criteria—A New Approach* (Containment and Siting of Nuclear Power Plants, IAEA, Vienna, 1967)

6. U.S. Nuclear Regulatory Commission, Reactor Safety Study—WASH-1400, NUREG 75/104, 1975
7. A. Birkhofer, German Risk Study for Nuclear Power Plants, IAEA Bulletin **22**(5/6) (1979)
8. World Nuclear Organization, http://www.world-nuclear.org/information-library/safety-and-security/safety-of-plants/safety-of-nuclear-power-reactors.aspx. Accessed 9 Feb 2018
9. International Atomic Energy Agency, *Applications of Probabilistic Safety Assessment (PSA) for Nuclear Power Plants* (IAEA, Vienna, 2001)
10. R.R. Fullwood, *Probabilistic Safety Assessment in the Chemical and Nuclear Industries* (Butterworth-Heinemann, Woburn, MA, 1988)
11. U.S. Nuclear Regulatory Commission, An Approach for Determining the Technical Adequacy of Probabilistic Risk Assessment Results for Risk-Informed Activities, Revision 2, Regulatory Guide, Office of Nuclear Regulatory Research Regulatory Guide 1.200, March 2009
12. DNV, M/s DNV Website and relevant document, https://rules.dnvgl.com/docs/pdf/DNV/codes/docs/2012-04/Oss-304.pdf. Accessed 26 Jan 2018
13. Papanikolaou and Apostolos, Eds., Risk-Based Ship Design: Methods, Tools and Applications, Springer
14. Risk-based Casing Design, http://petrowiki.spe.org/
15. S. Lee, B. Chu, D. Chang, Risk-based design of dolly assembly control system of drilling top drive. Int. J. Precis. Eng. Manuf. **15**(2), 331–337 (2014)
16. J. Gruenwald, Risk-Based Structural Design: Designing for Future Aircraft, AE 440, Individual Technical Report, October 9, (2008)
17. I.Y. Turner, F. Barrientos, L. Meshkat, Towards Risk Based Design For NASA's Missions, NASA Ames Research Center & Jet Propulsion Laboratory
18. Center for Chemical Process Safety, Guidelines for Risk Based Process Safety, WILEY, April 10, 2007
19. A risk-based approach to cumulative effect assessments for marine management. Sci. Total Environ. **612**, 1132–1140 (2018)
20. R. Kellagher, P. Sayers, C. Counsell, Developing a risk-based approach to urban flood analysis, in *11th International Conference on Urban Drainage*, Edinburgh, Scotland, UK, 2008
21. A. Roser, R. Voge, P. Kirshan, A risk-based approach to flood management decisions in a nonstationary world. Water Resour. Res. **50**(3), 1928–1942 (2014)
22. OECD, OECD, Guidance for Risk-based approach: The banking sector, Financial Action Task Force (FATF), www.fatf-gafi.org
23. International Atomic Energy Agency, *A Framework for Quality Assurance Program for PSA, IAEA-TECDOC-1101* (IAEA, Vienna, 1999)
24. International Atomic Energy Agency, *Attribute of Full Scope Level 1 Probabilistic Safety Assessment (PSA) for Applications in Nuclear Power Plants, IAEA-TECDOC-1804* (IAEA, Vienna, 2016)
25. ASME/ANS RA-S Standard for Level 1/Large Early Release Frequency PRA for NPP Applications, ASME, 2008
26. International Atomic Energy Agency, *Procedure for Conducting Level 1 PSA for NPPs, Safety Series No. 50-P-4* (IAEA, Vienna, 1992)

# Chapter 2
# Risk Characterization

> *The biggest risk is not taking any risk ....In a world that changing really quickly, the only strategy that is guaranteed to fail is not taking risks.*
>
> Mark Zuckerberg, Rebornealist

## 2.1 Background

As one of the critical steps in risk assessment, risk characterization is yet to be fully developed. Risk characterization involves preparing a document that provides the main results of the analysis, conclusions, and recommendations. The document accounts for the assumptions made for the analysis, the uncertainty or variability in the data and model, the limitations and strengths of the analysis, and the quality level addressed during each step of the analysis. The document plays a vital role in the risk assessment process because it accounts for any subjectivity or technical gap that can occur while communicating the results of the risk analysis is to the stakeholder, be it the regulator or the plant risk manager. The goal is to produce a document that lays out the requirements of the risk-based application from the point of view of a utility/plant manager or a regulator.

The available literature shows that risk characterization has become a standard practice in environmental protection, either as one of the steps in risk analysis or as an independent study [1]. Specific examples of risk characterization in the nuclear sector, mostly related to environmental evaluation and are as follows: a World Health Organization (WHO) health risk assessment of the 2011 Fukushima Dai-ichi accident in Japan, that includes radiological characterization of shutdown nuclear reactors for decommissioning; a WHO health risk assessment after the Fukushima Dai-ichi accident in 2011 in Japan [2]; an environmental risk characterization work plan at the Yankee Nuclear Power Station [3]; and risk characterization of the potential consequences of an armed terrorist ground attack [4]. However, these studies were carried out not as a standard practice and as part of risk assessment, but to address issues in specific contexts.

© Springer Nature Singapore Pte Ltd. 2018
P. V. Varde and M. G. Pecht, *Risk-Based Engineering*, Springer Series in Reliability Engineering, https://doi.org/10.1007/978-981-13-0090-5_2

Most of these studies deal with nuclear environmental consequences, be it radiation or chemical pollution. The studies support the view that risk characterization should be one of the steps in risk assessment, particularly when dealing with integrated risk-based engineering applications to address real-time scenarios.

Hence, this chapter deals with development of a formal framework for risk characterization by adopting the available knowledge base and experience available in other sectors such as environmental protection. Even though the PRA approach provides a well-developed framework and methodology for development of Level 1, Level 2, and Level 3 PRAs, risk characterization is not formally carried out as an integral part of risk assessment.

For nuclear power plants, many countries have conducted Level 1 PRA, some conduct Level 2 PRA, but only a few countries conduct all three levels of PRA. Another noteworthy observation is that most of the risk-based/risk-informed studies use knowledge developed in Level 1 PRA to address real-time issues related to system safety improvements while there are only limited examples of utilization of insights from Level 2 and Level 3 for emergency preparedness.

In this book, there is a motivation to develop a risk characterization procedure that considers results and insights obtained from a Level 1 PRA framework. Accordingly, keeping in view requirements of IRBE, this chapter discusses the definition of risk and various aspects associated with risk characterization.

## 2.2  Definition of "Risk"

The definition of "risk" as given in the IAEA glossary [5] can best be adopted and modified for qualitative as well as quantitative formulation of risk as follows:

*Qualitative measure of risk* is expressed as the probability of a specified health effect occurring in a person or group because of exposure to radiation. The health effect(s) in question must be stated—for example, the risks of fatal cancer, serious hereditary effects, or overall radiation detriment—as there is no generally accepted "default."

Quantitative measure of risk $R$, expressed as the product of the probability that exposure will occur and the probability that the exposure, if it occurs, will cause the specified health effect:

$$R = \sum_i p_i C_i \qquad (2.1)$$

where $p_i$ is the probability of occurrence of scenario or event sequence $i$ and $C_i$ is a measure of the consequence of that scenario or event sequence.

Typical consequence measures $C_i$ for nuclear plant system, plant and public level include core damage scenario, magnitude of release, or exposure of individual or population, respectively. If the number of scenarios or event sequences is large, the summation is replaced by an integral. Methods for treating uncertainty in the

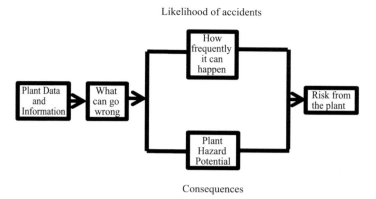

Likelihood of accidents

**Fig. 2.1**  Basic elements of risk analysis

values of $p_i$ and $C_i$, and whether such uncertainty is represented as an element of risk itself or as uncertainty in estimates of risk, vary.

Figure 2.1 shows the basic elements of risk analysis, which involves asking the following questions: What can go wrong? How frequently can it—(the failure) happen, and what are the hazard potentials for a set of failures in the plant? The same philosophy is depicted graphically in Fig. 2.1. The consequences can be characterized as either short term or spontaneous (e.g., the number of fatalities or injuries) or long term (e.g., cancer risk or any other ailments referred to as the "stochastic effect"). The consequences can also be defined as either loss of property or societal effects or any other economy as loss of working hours.

The following factors can be inferred from the IRBE point of view:

- The risk-based approach employs mathematical notions of risk where the quantified estimates of probability and consequences are used to estimate risk.
- One basic assumption that goes into building a mathematical risk model is "rare event approximation" and this should be considered suitably in risk characterization.
- The list of postulated initiating events along with their frequency and uncertainty bounds should form part of the risk characterization.
- The procedure followed for uncertainty characterization of hardware failure, human error, software components, should be part of the results of the analysis.
- Consequences such as core damage and system unavailability should be defined in a clear and related to deterministic parameters, e.g., physical quantities like maximum fuel/clad temperature, core cooling criteria, fuel damage conditions.
- The results of sensitivity analysis and the justification/basis for why certain items have been selected or omitted for sensitivity analysis should form part of the risk characterization.

## 2.3  Risk Characterization

In the context of nuclear plants and complex engineering systems such as process plants and industrial systems, risk characterization essentially consists of (a) a brief description of the plant/systems/applications along with the context that required the risk characterization; (b) scope and objective of the risk analysis; (c) major assumptions that may have potential impacts on the results of the analysis; (d) salient features of the approach/methodology used for risk modeling; (e) description of the major initiating events or hazards considered in the analysis; (f) comparisons of the results of the analysis with other sources of risk; (g) discussion of results, not only point estimates but more specifically, uncertainty bounds or variability; (h) sensitivity analysis for the key assumptions and parameters to demonstrate the robustness of the model and methods; and (i) documentation and presentation in the format that is suitable to the stakeholders, viz. regulatory body, public communications, other government bodies. This document is often prepared as an executive summary to a comprehensive document on risk assessment.

Is there a difference between "risk characterization" and "risk assessment"? The answer is "yes." Risk assessment is a process that deals with the selection of initiating events for the analysis, detailed system modeling, event sequences propagation and analysis, data analysis, simulation review, human reliability and common cause failure analysis, uncertainty and sensitivity analysis, and documentation. Risk assessment is iterative in nature and essentially deals with simulation at the component, system, and plant level to ensure that the analysis reflects the way the actual plant was designed, built, and operated. This process culminated in an often comprehensive and voluminous document called a risk assessment document, which is self-sufficient when it comes to peer or regulatory review. For example, a typical Level 1 PRA document is compiled in more than one volume, each volume comprising a couple of hundred pages.

The risk characterization document is basically a summary report that highlights the salient features, insights, and recommendations of the study keeping complexity of the studies away in main documents such that issues related to risk communication to plant management and regulatory bodies are carried forward in an effective manner. It creates a bridge between the risk analyst and risk manager or regulator and addresses those aspects relevant to the case under review. Therefore, risk characterization is an essential part of risk assessment because it only deals with concise information that serves the purposes of the decision maker. The risk characterization document may be around 50–100 pages as compared to main documents which might runs, as mentioned earlier, into more than one volume while each volume in term may run into more than one hundred pages.

In their 1996 report "Understanding Risk, NAS [6]" defines risk characterization as:

> ... a synthesis and summary of information about a potentially hazardous situation that addresses the needs and interests of decision makers and of interested and affected parties. Risk characterization is a prelude to decision-making and depends on an iterative, analytic-deliberative process.

They go on to refer risk characterization as "the process of organizing, evaluating, and communicating information about the nature, strength of evidence, and the likelihood of adverse health or ecological effects from particular exposures."

Even though the above definition covers the essence of risk characterization, it is more attuned to the task of environmental/ecological hazard characterization. The definition of risk characterization, for engineering systems in general and nuclear plants in particular, based on the foregoing discussions can be proposed as follows:

Comprehension of information on various potential levels of risks should include, along with assumptions, uncertainty bounds, data, facts, and figures that are the results of a systematic risk analysis that deals with sources of hazards, considers the engineering safety provisions in the plant, and presentation of the insights and results; considering interests of the stakeholders.

In the context of PRA, Level 1 risk characterization includes, apart from the statement of various categories of core damages and associated uncertainties, the major assumptions, discussion of plant strengths and weaknesses, and sensitivity analysis of various assumptions. When risk characterization is done at the Level 2 PSA, the statement of release frequencies will form the major element (in place of core damage frequency, CDF, as in Level 1) along with various potential exposure pathways. Note that the other elements of risk characterization as in Level 1 PRA (e.g., assumptions, treatment of uncertainty for weather and source term parameters, discussions of salient modeling aspects, and sensitivity analysis) also form part of risk characterization in Level 2 PSA. Similarly, risk characterization at Level 3 includes the dose/response curve, discussions of uncertainty, and sensitivity analysis, and finally a statement of risk for scenarios considered relevant to the requirements of the study.

The Environmental Protection Agency handbook on risk characterization [7] is a vital guidance document that discusses the associated aspects of risk characterization (from the environmental protection perspective) in detail along with case studies. In the following section, the relevant aspects, such as (a) the risk characterization policy, (b) major elements of risk characterization, and (c) the role of people and organizations in risk characterization, as applicable to the IRBE process, are either reproduced or are suitably adopted from this handbook.

## 2.3.1  Risk Characterization Policy and Principles

The risk characterization policy states that "A risk characterization should be prepared in a manner that is clear, transparent, reasonable and consistent with the regulatory or organization safety principles and other risk characterizations of similar scope cleared earlier by the regulating agency." The attributes transparency, clarity, consistency and reasonableness become the principles of risk characterization. However, these principles are expressed in natural language, and therefore vague and imprecise, how well are these attributes met? The EPA handbook provides detail on how to address these attributes by using a well-defined compilation of qualitative attributes or criteria as a checklist, where one case uses the ranking system to characterize the lower-level attributes.

## 2.3.2  Major Elements of Risk Characterization

Keeping in view the preceding discussions, it is prudent at this point to list what the elements of risk characterization should be. The number generated as part of risk analysis is often emphasized as risk characterization. However, a number should not be generated without providing the basis as to how it was generated, the associated uncertainty, the context of the analysis, the strength and weakness associated with the assessment, and reference to earlier study on similar context or analysis should also form part of risk characterization. This information should be conveyed to the decision makers to appreciate the number. It may be noted that the list presented below is not exhaustive and may require additional features to be included in the risk characterization for an application. Considering the nature of IRBE applications, the major elements of risk characterization are as follows:

- The *objective and scope* of risk assessment should be defined clearly and explicitly.
- Key information that has a direct impact on the results of the analysis should be highlighted.
- Contribution of various initiating events to the net CDF and large early release frequency along with its uncertainty bound forms part of the risk metrics.
- The main assumptions of the analysis which the analyst thinks may be sensitive to the results of the analysis should be listed and discussed.
- The *context* in which the risk characterization is required should be spelt out.
- The risk assessment methodology/approach should be discussed in brief.
- The complexity of the risk assessment should be kept out such that it helps the risk manager, or decision maker to appreciate the results of the analysis.
- The codes and guides used for the analysis should be documented. The verification and validation of the codes should be brought out in the characterization.

- Suitable comparison with similar options or other forms of risk coming from other socio-economic aspects should be included. The goal is to have improved perceptions of the risk in the subject area.
- The main results in the risk matrix should be presented such that various factors affecting the risk are communicated in an effective manner.
- The exposure pathways should be included (e.g., when characterization is based on the results of Level 3 PRA).
- The methodology used for human factors assessment and common cause failure and the limitations of these approaches should be brought out clearly.
- The results of the uncertainty analysis should form part of the results.
- A detailed sensitivity assessment should be performed and included to demonstrate that the risk analysis has been validated for the considered assumptions.
- Contributions to the total risk from various hazards or initiating events and important measure for various safety and mitigation system should be listed.
- The risk characterization should be performed keeping in view the stakeholders' requirements, viz. regulators, utilities, the public, or other vendors.
- Comparisons with the set goals and criteria, qualitative or quantitative, should form part of the risk characterization.
- Aspects related to residual risk or cliff-edge effect should be discussed.
- The design/selection of risk metrics should be consistent and should conform to the requirements of the application.

The basic idea is to generate an executive summary document which brings out the essence of the risk analysis that forms the channel of communication with the stakeholders. The objective of risk characterization is not to present the entire risk evaluation but to present the key findings and conclusions. Further, the handbook can be referred to for details of specific elements [7].

## 2.3.3  Roles of People and Organizations

*Risk analyst*: The risk analysis team has a team leader, preferably a risk analysis expert and experts in various areas, such as operations, maintenance, human factor modeling, thermal hydraulics, and structural analysis. The risk analysis team has primary responsibility for conducting and commissioning the risk analysis, where risk characterization forms one of the modules. The risk analysis team should have a communication channel with the plant or facility in general and with the risk manager in particular. Even though the major activities are analysis, modeling, and simulation, the deliberation should be part of the activities. The documentation and submission of the analysis document for peer review is the responsibility of the risk analysts. The selection of a peer review agency should be such that peer review team comprised of independent experts who have adequate experience on similar reviews.

*Peer review*: As a policy, any risk analysis and its characterization should be peer-reviewed. The peer review agency/expert should be independent in the sense that they should be selected from different organizations so that an independent review can be maintained. Peer review is vital for scientific credibility of the analysis.

*Regulator*: The regulatory authority is the decision-making body at national level with power to give final decision. Risk characterization followed by peer reviewers' comments is crucial for the deliberative process of regulatory decision-making.

*Risk manager*: In a plant or facility, the risk manager is the final authority who implements the changes/modifications. The risk characterization document should ensure that the final user of this information, along with regulatory rules, is the risk manager. The risk manager should have a good understanding of spirit of the recommendations generated by risk characterization, particularly its safety implications in short as well as long term, on the subject systems and the interface of the systems with other system and overall operational aspects of the plant. Further, the risk manager is responsible for periodic monitoring of the effects of change and for providing feedback to the regulators, if required.

The risk characterization policy and associated guides and standard should be adhered to throughout the process of decision-making.

## 2.4   Risk Assessment Techniques

Even though the basic ingredients of risk characterizations remain the same, the presentation of the results and the overall matrix for the presentation of the results change depending on the methods used for risk assessment. Hence, it is important to discuss, at the outset, various methods for risk assessment that can form part of the risk-based approach. Even though some of these methods have been discussed in detail in Chap. 4 in this book, a brief reference is made here in respect of risk characterization.

### 2.4.1   Failure Mode Effect Analysis (FMEA)

Failure mode and effect analysis (FMEA) is a qualitative approach to risk assessment involving the use of standard table formats from available standards/codes. FMEA caters to the aims and objective of the subject analysis. The major features of the FMEA include component identifications, failure modes, effects of a given failure mode on the system, compensating identification (e.g., alarms or symptoms) provisions in the plant or system, and categorization of likelihood for each failure and associated mechanism(s) for risk assessment. Even though this analysis is qualitative in nature, there are some cases in the literature where limited-scope quantification has been used by providing the component failure probabilities/

frequency either by grouping a range of probabilities or by incorporating the failure rates or frequencies for individual components. This is a simple approach where individual components along with various failure modes are inducted and examined for effects at various levels, viz. component level, system level, and plant level.

This approach as an independent tool has limited applicability for complex engineering systems, such as nuclear plants, process plants, and industrial systems, because it does not include analysis of complex interactions between components and human actions, including common cause failure (CCF). However, FMEA is performed to identify unsafe failure modes of components that can cause a system to fail in an unsafe mode. In this approach, risk characterization is done by prioritizing components to identify those components that may have: (a) consequences, such as loss of plant availability, economic loss, safety (loss of life or property); (b) effects at the local/system/plant level; (c) low/high and very high frequency; and (d) maintenance options, such as repair/replacement. These aspects are weighted against the recovery provisions in the plant.

## 2.4.2 Hazard and Operability (HAZOP) Analysis

HAZOP analysis is a very common risk assessment tool in chemical and process plants. The plant is divided into various zones demarked by either physical or functional boundaries depending on the objective and scope of the analysis. Like FMECA, this analysis is also documented in a tabular matrix. The use of guide words such as "increasing," "decreasing," "low," "high," "flow," and "temperature," forms the significant element of HAZOP. An event tree approach is employed, where propagation of an accident scenario forms part of the HAZOP analysis. The safety provision and human actions are used as header elements for event tree modeling. Assessment of likelihood and consequences forms the major analysis part of HAZOP. Important scenarios are identified using a risk matrix comprised of categorizing the events based on likelihood, which are characterized as low, medium, high, or very high. Similarly, the consequences are also characterized as low, medium, high, or catastrophic. Qualification of these consequences is governed by factors such as loss of production, extent of damage to the plant (recovery possibilities), injuries, death, and long-/short-term effects.

## 2.4.3 Probabilistic Risk Assessment (PRA)

This technique is generally employed for integrated risk assessment of nuclear plants and space and aviation systems. There are applications of PRA for chemical as well as process plants. For the case of nuclear plants, the approach comprises identification and analysis of potential accident sequences for estimating the reactor "core damage frequency," "large early release frequencies," and "risk to members of the public." These estimates comprise a major risk matrix when performed at Level 1, 2, and 3 PRA, respectively. Generally, for the major part of the analysis, an event tree approach and fault tree approach are employed for system-level modeling, while event tree analysis is used for accident sequence modeling. Application of the Markov model has been noted in many PRA studies for analysis of complex scenarios. The various categories of core damage frequencies for various states, such as full-power operations, shutdown, and low-power operations, and various categories of events, such as external and internal, form the major matrix for risk characterization at the plant level. At the system level, the initiating event frequencies and safety system unavailability form the second lower level of the matrix. Apart from this, the assessment of uncertainties and sensitivity analysis also forms a major part of the risk characterization. Mapping of the contributions of the systems to the core damage frequency and further to large early release frequencies and finally to risk to the members of the public, form an integral part of risk characterization. PRA is discussed in detail in Chap. 6.

### 2.4.4   Quantitative Risk Assessment

The approach adopted for quantitative risk assessment is similar to PRA, but for certain aspects which are governed by the application. In fact, PRA framework has been extensively employed for modeling of mission-critical systems like space and aviation systems. The emphasis in this technique is on mission reliability evaluation whereas consideration of the human factor, onboard electronics, and structural reliability considering thermal and mechanical fatigue loads forms an integral part of risk assessment.

### 2.4.5   Other Risk Assessment Approaches

The methodology and scope of risk assessment differ based on the application requirements. For example, the risk assessment methods in healthcare systems may be more focused on drugs that need to be evaluated before they are certified for clinical trials or launched. Here, data collected from the experiments may become the central part of the risk assessment. Similarly, the risk assessment approaches in finance may focus on the bottom-line decisions. Here, evaluation of risk factors and various options becomes part of the mechanism for risk assessment.

However, keeping in view the scope of this book, PRA methodology will be discussed in detail in Chap. 6.

## 2.5   Risk Metrics

The outcome of risk assessment is either qualitative or quantitative estimates of risk metrics. We will focus our discussion considering Level 1 PRA. In this regard, the results of risk assessment consist of the following four major components: likelihood of the core damage frequency; definition of the core damage; uncertainty bounds of the core damage frequency; and sensitivity analysis of the major assumptions and uncertainty bounds.

Presenting these results is an art. As discussed earlier, merely providing the quantified estimates is not adequate, unless the estimates are presented with the background data and information (i.e., the assumptions, scope, and strengths/weaknesses of the analysis), required for interpretation of results. It is crucial to discuss the number of iterations used in Monte Carlo simulation to determine the distribution of the CDF. For example, it helps to provide the CDF results as shown in Fig. 2.2. This figure shows the CDF on the x-axis and frequency samples corresponding to a given CDF on the y-axis. This presentation gives holistic information about the central tendency (mean and median) and the number of points that are populated for this simulation.

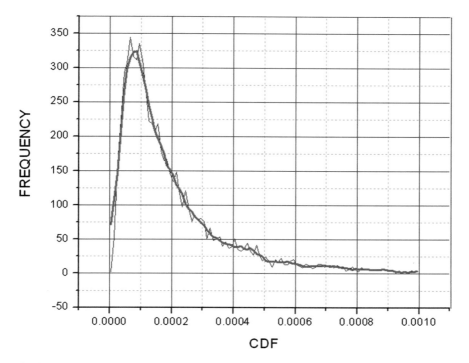

**Fig. 2.2**  Core damage frequency representation

PRA model simulation involves evaluation of more than one core damage states considering various categories of accident sequences. Considering the IRBE application, it is important to explain which categories of core damage states constitute final CDF of the plant.

What constitutes core damage should be defined? There is no unique definition of core damage because each plant type has its own definition. At this stage, let us conclude that this is a complex subject. However, there are broad definitions of core damage that may be general, reactor-specific, or specific to parameter degradation conditions. For example, one broad definition of core damage by the US Nuclear Regulatory Commission (USNRC) is "an expression of the likelihood that, given the way a reactor is designed and operated, an accident could cause the fuel in the reactor to be damaged [8]." PRA engineers might use a conservative definition where core damage comprised of those accident scenarios that do not terminate in safe conditions.

Some among the many conditions/parameters that forms input to the definition of core damage may include, extent of (a) reactor core uncovery (i.e., fuel portion in the core exposed to steam environment), (b) fuel/clad temperature increasing more than the predetermined value, (c) clad oxidation, (d) metal water reaction, (e) structural damage channel ballooning, and (f) en-mass fuel damage conditions; that has potential for leading to release of radioactivity into the coolant system. The PRA engineer works out the definition of core damage based on the type of plant and on the measurable parameters, direct and/or indirect. However, it is important to understand that the definition of core damage is a subjective aspect of PRA and the discussion here is only academic and not exhaustive but indicative.

Asensitivity analysis is a useful tool to evaluate the impact of various assumptions in the analysis. Often the risk-based application involves estimation at an individual system level only, without involving calculation at the plant level (i.e., without considering estimation of CDF). In such cases, the risk metric is system "unavailability" and not CDF. To demonstrate this, let us consider a case of risk-based evaluation, where the problem deals with optimizing the test interval for a safety system, for example, the emergency core cooling system (ECCS). The case involves evaluation of unavailability to obtain the surveillance test interval for the minimum unavailability such that it meets the regulatory guideline on ensuring an unavailability goal of $\sim 1 \times 10^{-3}$/demand. This optimization shows the sensitivity of the unavailability estimate for various surveillance test intervals. The fault tree simulation for various test intervals results in the optimization shown in Fig. 2.3. As can be seen, the test interval of 2000 h meets the regulatory stipulation on availability criteria.

So far, we have discussed that CDF and unavailability form important risk metrics for IRBE applications. We discussed the importance of sensitivity and uncertainty analysis as part of the risk metrics. Often the list of parameters to be included is exhaustive and depends upon the applications and the expectations of the analyst from the assessment; hence, every assessment will have to work out its own set of metrics.

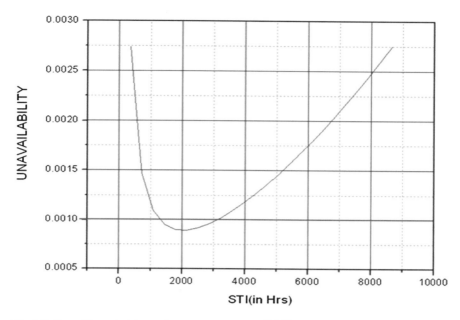

**Fig. 2.3** Surveillance test interval versus system unavailability

| Likelihood | Consequences | | | |
|---|---|---|---|---|
| | Catastrophic | Major | Low | Minor |
| Highly likely | | | | |
| Likely | | | | |
| Moderate | | | | |
| Rare | | | | |
| Unlikely | | | | |

High ▮  Medium ▢  Low ▢  Negligible ▢

**Fig. 2.4** Qualitative risk metrics

The most common form of risk matrices, particularly those used in risk-based in-service inspection, that can be used in many other applications is shown in Figs. 2.4 and 2.5. Although there are no universal or well-accepted norms to draw the matrix, generally the rows represent the likelihood, whereas the columns represent the consequences. The choice of number of rows and columns rests with the analyst and is governed by the spectrum of categories that need to be used for risk communication.

Examples of the qualitative and quantitative matrixes normally used in risk-based in-service inspection are shown in Figs. 2.4 and 2.5. For example, in

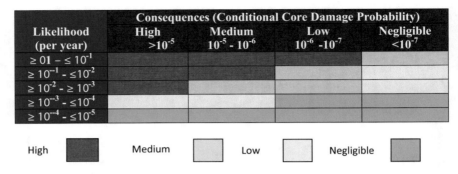

| Likelihood (per year) | Consequences (Conditional Core Damage Probability) | | | |
|---|---|---|---|---|
| | High >$10^{-5}$ | Medium $10^{-5}$ - $10^{-6}$ | Low $10^{-6}$ -$10^{-7}$ | Negligible <$10^{-7}$ |
| ≥ 01 – ≤ $10^{-1}$ | | | | |
| ≥ $10^{-1}$ - ≤$10^{-2}$ | | | | |
| ≥ $10^{-2}$ - ≥ $10^{-3}$ | | | | |
| ≥ $10^{-3}$ - ≤$10^{-4}$ | | | | |
| ≥ $10^{-4}$ - ≤$10^{-5}$ | | | | |

High [ ]   Medium [ ]   Low [ ]   Negligible [ ]

**Fig. 2.5** Quantitative risk metrics

these two figures, the metrics have been designed by considering four categories of consequences and five categories of likelihood to characterize the risk. Risk metrics also changes depending on whether the analysis is qualitative or quantitative. The traditional in-service inspection program is built around the qualitative notion of risk and consequences.

The qualitative likelihood of failure is generated based on materials, stress intensity factor, number of welds, length, primary and secondary stresses experienced by a given pipe segment. These likelihoods are further categorized into, say, highly likely, likely, moderate, rare, and unlikely. Further, the consequences are characterized, for example, into catastrophic, major, low, and minor based on attributes such as size, location (e.g., inlet of the core or outlet of the core), pressure, or flow).

Finally, a risk metrics is generated as shown in Fig. 2.5. The color coding is used to indicate risk category, e.g., red (high), medium (orange), low (yellow), and negligible (green). Inspection planning in terms of periodicity and intensity is carried out based on these metrics.

The procedure for developing the quantitative risk metrics is also same as for qualitative risk metrics, except that in these metrics, quantitative estimates of piping failure frequency are generated and categorized into five categories, as shown in Fig. 2.5.

Risk-based in-service inspection helps focus on those areas that are critical to safety and result in optimization of resources. Apart from this, pie charts, bar charts, and regular graphs are often used to characterize the risk contributors. In this section, only a few examples of risk metrics have been discussed, however, risk metrics will change based on the definition of the problem.

The other risk metrics are importance measures, including risk achievement worth, risk reduction worth, Fussel-Vesely importance measures, and other measures of importance, quantitative health objectives (acceptable level of numerical criteria for risk to member of public), large early release frequency (major results of level 2 PRA) and risk statement (major results of Level 3 PRA).

# References

1. National Research Council (NRC), *Risk Assessment in the Federal Government: Managing the Process* (National Academy Press, Washington, D.C., 1983)
2. International Atomic Energy Agency, *Radiological Characterization of Shut Down Nuclear Reactors for Decommissioning Purposes, TECHNICAL REPORTS SERIES No. 389* (IAEA, Vienna, 1998)
3. Environmental Risk Characterization Work Plan Yankee Nuclear Power Station Rowe, Yankee Atomic Electric Company, Massachusetts, U.S, 2006
4. D. True, D. Leaver, E. Fenstermacher, J. Gaertner, *Risk Characterization of the Potential Consequences of an Armed Terrorist Ground Attack on a U.S. Nuclear Power Plant* (Electric Power Research Institute, Washington, D.C., 2003)
5. International Atomic Energy Agency, *IAEA Safety Glossary Terminology Used in Nuclear Safety and Radiation Protection* (IAEA, Vienna, 2007)
6. N. R. C. National Academy of Sciences, Understanding Risk, NAS, USA, 1996
7. J.R. Fowle, K.L. Dearfield, *Risk Characterization Handbook: Science Policy Council Handbook* (EPA 100-B-00-002, U.S. Environmental Protection Agency, Washington, D.C., 2000)
8. USNRC, *USNRC Glossary* (United States Nuclear Regulatory Commission, Washington D.C.)

# Chapter 3
# Probabilistic Approach to Reliability Engineering

> *But to us, probability is the very guide of life.*
> Joseph Butler, Picturequotes.com

## 3.1 Introduction

Traditionally, the approach to reliability assessment is probabilistic approach, and it is this approach that is key to assessing risk. This approach works well in risk modeling; however, the limitation of this approach is that even though it predicts the probability of failure, it is not capable of predicting the instant of failure. When it comes to predicting the failure in advance, the techniques like physics-of-failure or data-driven-based prognostic techniques are used such that incipient failure can be detected in time. Accordingly, the prognostics and health management approach is employed for remaining life estimation in the field conditions. This aspect has been discussed in Chaps. 12 and 13 on Physics-of-Failure and Prognostics and health management.

For risk characterization, the probabilistic approach provides the needed tools and methods [1]. This chapter deals with the probabilistic aspects of reliability and risk modeling for complex engineering systems. The bottom line in the IRBE approach is to, on one hand, evaluate the risk employing the probabilistic approach while predicting failure in advance in field conditions employing prognostics and health management approach and thereby develop a holistic system to reduce the consequences of failure or manage the failure such that safety and cost implications are addressed effectively. The real power of IRBE lies in its capability to generate quantified estimates of system performance and safety along with other important parameters that characterize criticality and uncertainty. In this respect, the fundamentals and concepts of reliability engineering and associated/supporting topics such as statistical analysis, probability distributions, confidence interval estimations, and data elicitation techniques form the basic framework for utilizing the raw performance data for reliability estimates. The PRA or quantitative risk assessment that forms one of the pillars of the IRBE application is one of the major applications of reliability engineering discipline. It is not possible to realize a representative

© Springer Nature Singapore Pte Ltd. 2018
P. V. Varde and M. G. Pecht, *Risk-Based Engineering*, Springer Series in Reliability Engineering, https://doi.org/10.1007/978-981-13-0090-5_3

PRA model of the plant without implementing the reliability engineering concept in perspective.

This chapter presents the fundamental principles as applicable to the probabilistic approach to reliability engineering.


## 3.2   Life Characteristics: The Bathtub Curve

Before we go into the details of the probabilistic approach to reliability, we need to understand the generic life cycle representation of any class of engineering component. As shown in Fig. 3.1, this representation is referred to as the bathtub curve; the shape of this curve resembles a typical bathtub. The curve can be divided into three regions, viz., region I depicts the decreasing hazard rate and is referred to as the infant mortality period; region II, where the hazard rate is constant, is referred to as the constant hazard rate or useful period region; and region III and the right portion of the curve is characterized by an increasing failure rate that represents aging-related hazard failures. However, in practice, there are very few components or systems that actually follow this curve.

To better appreciate this representation, consider the example of a certain group of electrically driven pumps that have similar specifications and are installed in a complex engineering setup or plant. When the pumps were commissioned and put into operation, operations and maintenance staff often faced initial problems due to a higher failure/hazard rate; however, the failure rate shows the decreasing trend and enters a region where it tends to be constant. This region of decreasing failure rate is characterized as the infant mortality or initial failure rate region. After a few months or nearly one year of operation, the failure rate stabilizes and becomes constant, as shown in the region on the bathtub curve called the useful life period. However, as the pump sets becomes older, aging-related issues start cropping up. The failures start increasing and in spite of the increasing rate of repair and maintenance, the failure rate keeps increasing until it becomes difficult to keep the

**Fig. 3.1** Bathtub curve representing typical life cycle of a component or system

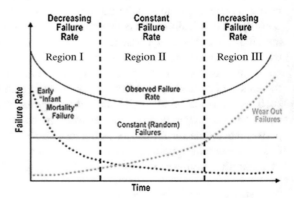

pump set operating. This is the point when we say that the pump sets have become old and need replacement because considerable time and resources are required to keep these pump sets in operation. This is the plant's perspective on the life cycle usage of pump sets.

Consider, however, the pump manufacturer's perspective. The experience of the pump manufacturer that sells 100 pump sets during a year is that most of the complaints are received during the initial few months of the supply. The manufacturer's records also show that as time passes, the rate of complaints concerning pump performance drops and then becomes fairly constant for over a period of 10–50 years.

The initial experience of the supplier is characterized by the early failure rate for which the supplier's quality control department is also responsible to some extent. As the experience shows, the initial failure rate is mostly caused by manufacturing-related faults but also installation and commissioning faults at the user or customer facility. However, once these faults are overcome, then the component enters the useful life period where the routine maintenance activities are enough to keep the failure rate at the lowest and most constant level. However, once aging of the component starts, even repair/replacement of subcomponents does not help and the component is rendered not viable from a safety as well as economy perspective.

During the early period in the life of a component, the decreasing trend is generally related to manufacturing or quality-related issue as shown in region I. The region II, i.e., the middle portion of the curve, is characterized by random failures and represented by constant failure rate, while region III depicts increasing failures due to aging phenomenon on the bathtub curve. However, the bathtub curve is one of the most common and most popular expressions of component life cycle, in actual practice the life curve could be anything as discussed below.

When new components are tested thoroughly by the vendors (e.g., testing pumps for all the postulated conditions and thermal-hydraulic parameters and transients), the initial failure trend in region I of the bathtub curve will not occur when the pump set is integrated with the rest of the plant. Right from the beginning, the trend could be represented by the constant failure rate model.

Similarly, it is quite possible that for certain components, the constant failure rate may not occur. It could be characterized by a small gradient showing either an increasing or a decreasing failure rate. A small gradient in the increasing trend, which has been observed by many agencies dealing in reliability-centered maintenance, generally happens when wear-related damage starts cropping up during what should have been the useful period. This situation requires urgent revision and re-engineering of the component material selection and design. Similarly, there are instances when the trend in region II shows a small decreasing trend. This could be attributed to continuous learning on operation- and maintenance-related aspects of the components or the implementation of a reliability growth program at system and component level.

The existing international practice requires the replacement of the component at the first sign of aging. This is required as there are perennial safety implications or availability issues at system or plant level due to individual component recurrent failures in spite of enhanced repair and maintenance. Hence, the components are replaced by new ones when the component enter region III, i.e., aging region.

It will be interesting to argue the application of bathtub curve for at plant level or at system level. Often it can be observed that at higher level, i.e., at system or plant level, the bathtub curve might represent the system/plant life cycle performance. For example, the availability/capacity factor for systems as also at plant level may be limited by initial or early failures while it attains a stable or constant availability after some period as characterized in region II. Later as the plant or system ages, after couple of 10 years, plant availability/capacity decreases due to increased system failures, such that even extended repair and maintenance is not able to arrest the increasing failure trends of components and subsystems, and behavior at plant level could be depicted by region III.

## 3.3   Probability Theory: Main Concepts

The concept of probability forms the core of reliability and risk engineering. Dealing with random events and thereby predicting the likelihood of an undesired event requires that the nature of the raw data is properly interpreted so that predictions are correct. This requires understanding of the terms "reliability," "availability," and "failure distributions"—the fundamentals for probabilistic modeling.

### 3.3.1   Reliability

*Reliability is defined as the probability of failure-free operation of the component, system, or service for the intended period of time under stated conditions.* Assume the time $t$ represents the intended mission time of the successful operation of a component, and $c_1$, $c_2$, and $c_3$ represent the stated condition. Then, the reliability $R(t)$ for the mission $t$ is mathematically given by:

$$R(t) = \Pr(T > t | c_1, c_2, \text{and } c_3 \ldots) \qquad (3.1)$$

where $T$ is a random variable for the time to failure of the components put under test, and Pr is the abbreviation for the term "probability." The vertical line | is read as "given that" in mathematical parlance. The above expression is read as the reliability of an entity to perform satisfactorily up to time $t$ is equal to the probability that the component will perform its intended function from time $t = 0$ when it was started until time $t$, the mission time of the entity given the conditions $c_1$, $c_2$, and $c_3$.... These conditions could be related to maintenance of temperature under certain limits, maintenance of humidity within a range, or a maintenance practice such as maintenance of the oil level in a pump bearing casing above the minimum limit. Hence, reliability is a subjective issue.

The axioms of reliability are:

- Reliability of a component at time $t = 0$ is 1.0
- Reliability of a component for time $t = \infty$ is 0.0
- Reliability of a component lies between 0 and 1
- If $F(t)$ is the probability that the component will fail within time $t$, then

$$R(t) + F(t) = 1;$$

$F(t)$ is also called the cumulative distribution function.

### 3.3.2 Derivation of Reliability Function from the First Principle

A population of components comprising total $N$ components was subjected to testing in parallel. After time $t$, the test was terminated. The test results were as follows.

Out of $N$ components, $n_f(t)$ denoted the number of component failed at time $t$ while remaining component $n_s(t)$ number of component were found to be successfully operating until time $t$ when the test was terminated. The formulation for reliability is as follows [2]:

$$R(t) = \frac{n_s(t)}{N}$$

$$R(t) = \frac{N - n_f(t)}{N}$$

$$R(t) = \frac{N - n_f(t)}{N}$$

Differentiating both sides with respect to $t$ results in

$$\frac{dR(t)}{dt} = -\frac{1}{N}\frac{dn_f(t)}{dt}$$

or

$$\frac{dn_f(t)}{dt} = -N\frac{dR(t)}{dt}$$

Dividing both sides by $N_s(t)$ results in

$$\frac{1}{n_s(t)}\frac{dn_f(t)}{dt} = -\frac{N}{n_s(t)}\frac{dR(t)}{dt}$$

The term on left-hand side defines the "instantaneous failure rate" and is also called the "hazard rate." The term $dn_f(t)/dt$ defines the failure rate of the component in a small interval between $t$ and $\Delta t$. Suppose that total $n_s(t)$ components are surviving at time $t$. Accordingly, when $dn_f(t)/dt$ is divided by $n_s(t)$, the surviving components at time $t$ which are available for further testing, we get the instantaneous failure rate or simply the failure rate or hazard rate, which are conventionally denoted as $\lambda(t)$.

Since

$$\frac{n_s(t)}{N} = R(t)$$

The above equation can be written as

$$\lambda(t) = -\frac{1}{R(t)}\frac{dR(t)}{dt} \tag{3.2}$$

or

$$\lambda(t)dt = -\frac{dR(t)}{R(t)}$$

For getting the failure characterization over the life of a component up to time $t$, we integrate

$$\int_0^t \lambda(t)dt = -\int_0^t \frac{1}{R(t)}dR(t)$$

Further simplification yields

$$\ln R(t) = -\int_0^t \lambda(t)dt$$

$$R(t) = e^{-\int_0^t \lambda(t)dt} \tag{3.3}$$

Equation 3.3 provides the general equation for reliability of a component or system for which the failure rate is characterized by $\lambda(t)$, and the mission time of operation is defined as $t$. For a mission time of $t = 0$, i.e., at the time when the component has not even started to perform its intended function, we find from the above equation that reliability $R(t) = 1$ (on the right-hand side of the equation when we put $t = 0$, we get $e^0 = 1$). Similarly, for a hypothetical mission when $t = \infty$, we get reliability as zero. So we conclude as follows:

$$R(0) = 1$$

$$R(\infty) = 0$$

For the case when the failure rate is constant, as it is the case for exponential distribution (i.e., $\lambda(t) = $ constant), the right-hand term in reliability equitation can be given as follows:

$$-\int_0^t \lambda(t)\mathrm{d}t = -\lambda t$$

Hence, for the case of exponential distribution, the reliability equation is given as

$$R(t) = e^{-\lambda t} \tag{3.4}$$

Equation 3.4 provides an equation of reliability when the component failure rate follows exponential distribution. Most of the discussions subsequently in this book will be based on this equation. This is called the single-parameter distribution formulation for exponential distribution.

We extend the discussion further to understand the relation between reliability $R(t)$ and cumulative failure probability $F(t)$ as follows:

We know now that

$$R(t) = n_s(t)/N$$

Similarly,

$$F(t) = n_f(t)/N$$

From equations above, we have

$$R(t) + F(t) = n_s(t)/N + n_f(t)/N$$

However, $n_s(t) + n_f(t) = N$;
Therefore,

$$R(t) + F(t) = 1 \tag{3.5}$$

For the case of exponential distribution

$$R(t) = e^{-\lambda t}$$

Hence,

$$F(t) = 1 - e^{-\lambda t} \tag{3.6}$$

This shows that at any given time, the sum of the probability of component success and the probability of component failure is equal to unity. In respect of any individual component, with a well-defined point and rigid failure criteria, the component can acquire either a failed state or an operating state. The argument for failure criteria is introduced because there could be other states of the component apart from completely failed or completely successful, such as partially failed or partially successful. It happens in the engineering scenario that the component keeps performing at a lower capacity than the rated capacity (e.g., a diesel generator, which does not deliver the required power but keep operating at lower capacity). Similarly, a valve in a cooling system does not acquire either fully open or fully closed status but gets stuck at a partially open status from a fully closed status. It may be doing its function. This partial opening may sometime be very vital for avoiding extreme emergency conditions, or in some situations it may not be effective.

The point being made here is that a clear definition of the failure mode of the component is crucial for correct representation of the system state in mathematical expressions.

### 3.3.3  Reliability Characteristics

Even though the reliability expression provides complete information about the probability of a component meeting its mission objective, in practice, other expressions like mean time to failure (MTTF), mean time between failure (MTBF), and availability expressions are also used to describe reliability of the systems, structure, and components. For example, MTTF is often used to describe expected life of the non-repairable components, like batteries, lighting and optoelectronics devices, categories of relays and switches which are non-repairable. MTBF expression is often used to indicate in place of reliability expression for repairable components, like pumps, valves, diesel generators, transformers, electronic boards. The following section will provide a brief expression about these reliability parameters.

#### 3.3.3.1  Mean Time to Failure (MTTF)

The traditional and conventional (but generally of little value in risk assessment) parameter that characterizes reliability of a component is mean time to failure (MTTF). This term is associated with non-repairable or mission-oriented components. As the term signifies, the mean or average time taken to failure is called

MTTF. This estimate refers to the average time to failure derived from the test where a finite sample was subjected to test under a similar environment.

Suppose $n$ components are put to test at time $t = 0$ and their failure times were recorded as $t1$, $t2$, $t3$ … $tn$ for components 1, 2, 3, … and $n$, respectively.

$$\text{MTTF} = \frac{t1 + t2 + t3 \dots tn}{n} \tag{3.7}$$

If the accumulated time to failure for all the $n$ components is given by $T$, then $t1 + t2 + t3 \dots tn = T$. The average failure rate $\lambda$ of the component can be given as

$$\lambda = \frac{\text{total number of failure}(n)}{\text{accumulated time to failure of } n \text{ components}} \tag{3.8}$$

$$\lambda = \frac{n}{(t1 + t2 + t3 \dots + tn)} \tag{3.9}$$

It can be easily deduced that

$$\text{MTTF} = 1/\lambda \tag{3.10}$$

Hence, the inverse of the failure rate is commonly used to estimate the MTTF of the components and vice versa.

### 3.3.3.2 Reliability

As seen in previous Sect. 3.3.2 and Eq. 3.4, the parameter $R(t)$ is used to describe the expectation of failure-free operation of non-repairable components for a certain mission $t$ for a given sets of conditions. The most simple and popular way of defining the reliability of a component is through exponential distribution as follows:

$$R(t) = \exp(-\lambda t) \tag{3.11}$$

### 3.3.3.3 Availability

Availability is defined as the probability of a repairable component to be available when the demand is placed. Steady-state availability of a component having failure rate $\lambda$ and repair rate $\mu$, due to random causes, is defined as:

$$\text{Availability}(A) = \frac{\text{Uptime}}{\text{Uptime} + \text{Downtime}} = \frac{\text{MTTF}}{\text{MTTF} + \text{MTTR}} \tag{3.12}$$

$$A = \frac{\frac{1}{\lambda}}{\frac{1}{\lambda} + \frac{1}{\mu}}$$

$$A = \frac{\mu}{\mu + \lambda} \tag{3.13}$$

Generally, unavailability $U$ is used in practice to give failure probability of a component not to come on demand due to random causes as

Further,

$$\text{Unavailability}(U) = 1 - \text{Availability} = 1 - \frac{\mu}{\mu + \lambda}$$

$$U = \frac{\lambda}{\mu + \lambda} \tag{3.14}$$

The unavailability function is used in fault tree analysis for characterizing the failure probability of components and subsystems of a safety system or safety support systems. Safety systems are required to be actuated or started on a demand. For example, diesel generators, as part of emergency supply systems, remain standby during normal condition and get signal to start automatically on failure of normal grid (or Class IV) power to the station. For standby components, system unavailability estimates are used for characterizing basic component or subsystem reliability in system fault trees.

## 3.4   Probability Distribution Functions

Let us discuss the role of various statistical distributions in reliability engineering. It is important to recognize that probabilistic modeling attempts to estimate the mean value and uncertainty bound associated with a random variable. This random variable could be the time to failure of a component, frequency of operation of an electronic device (e.g., a switch), or a design variable (e.g., pressure, temperature, or physical dimension such as the diameter or height of a storage tank). Having recognized that these are random variables and uncertainty in estimate is inherent to these estimates, it is required to find ways to represent the data trends and characterize these trends using mathematical parameters. Why these parameters are required? These parameters and graphical representations of the data enable us to better understand the variation in the parameter trends and thereby quantify the otherwise qualitative information. Quantification is vital for comparing, for example, the performance of two similar components, prioritizing a list of components based on quantified estimates, and assessing the criticality of components for safety applications. Hence, assessing the correct distribution to represent the data forms an important element of quantification.

There are two types of events—discrete events and continuous events. Based on the type of data being modeled, the distributions have been divided into two categories, viz., discrete distributions and continuous distributions. The discrete events are characterized by nature of their occurrence, which is sort of intermittent, such as the failure of a component when it is subjected to demand. The generic example of rolling of a die is a discrete event because on a time scale, there is no continuity between two successive events. For example, a diesel generator in a process industry or nuclear power plant starts when a demand is placed to start the set. Similarly, operation of a switch on demand is also modeled using discrete distribution. The unit of failure probability of discrete events is defined in per unit demand.

When the operation of a component is characterized as a continuum over a period of time, a like number of failures over an accumulated period of time is expressed in failure per unit time, and the events are referred to as a continuous event. These events are modeled using continuous distribution.

The Poisson and binomial distributions are examples of discrete distributions. The normal, log-normal, exponential, Weibull, and gamma distributions are examples of continuous distributions. The following section deals with various distributions generally used in the probabilistic approach to reliability assessment.

## 3.4.1   Continuous Distribution Function

Even though there are host of distributions to model engineering component reliability, this section will describe only those continuous distributions which find application in PRA modeling.

### 3.4.1.1   Normal Distribution

The normal distribution was invented by De Moivre in 1773, and later, Gauss used the normal curve to analyze astronomical data in 1809. Hence, this distribution is often referred to as Gaussian distribution.

Many of the statistical observations not only in the area of science and engineering but also in the areas of social studies, economics, business, or management studies extensively use normal distribution. This is because the data or observations performed in these areas exhibit the normality in the sense that the observations are all concentrated around a mean value with a spread around this mean that describes the variance around the mean. This distribution is characterized by its bell-shaped representation as shown in Fig. 3.2. This distribution is characterized by two parameters: mean ($\mu$) and standard deviation ($\sigma$). The empirical model that gives the probability distribution function (pdf) of the normal distribution is as follows:

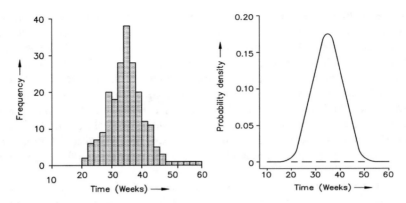

**Fig. 3.2** Normal probability distribution functions

$$f(t) = \frac{1}{\sigma\sqrt{2\pi}} \exp\left[-\frac{1}{2}\left(\frac{t-\mu}{\sigma}\right)^2\right], -\infty < t > \infty \tag{3.15}$$

The cumulative distribution is obtained by integrating both sides of equation of $F(t)$ from $-\infty$ to $t$ as follows:

$$F(t) = \frac{1}{\sigma\sqrt{2\pi}} \int_{-\infty}^{t} \exp\left[-\frac{1}{2}\left(\frac{y-\mu}{\sigma}\right)^2\right] dy \tag{3.16}$$

where $y$ is a dummy variable for $t$.

As can be seen, the solution of Eq. 3.16 for $F(t)$ can be performed only by numerical integration of the normal distribution to a standard normal distribution by putting

$$\frac{y-\mu}{\sigma} = z$$

where, $z$ is the standard normal parameter.

$$y - \mu = \sigma z$$

Differentiating both sides w.r.t. $y$, we get

$$\sigma \frac{dz}{dy} = 1$$

$$\frac{dz}{dy} = \frac{1}{\sigma}$$

$$dy = \sigma dz$$

Substituting the value of $(y - \mu)/\sigma$ by $v$ and $dy$ in equation of the normal distribution model will give a standard normal distribution as follows:

$$F(t) = \frac{1}{\sqrt{2\pi}} \int_{-\infty}^{z} \exp\left(-\frac{z^2}{2}\right) dz \tag{3.17}$$

The normal practice in reliability engineering is to represent

$$\frac{1}{\sqrt{2\pi}} \exp\left(-\frac{z^2}{2}\right) = \Phi(z)$$

The equation of standard normal cumulative distribution function (CDF) becomes

$$F(t) = \int_{-\infty}^{z} \Phi(z) dz \tag{3.18}$$

Now we have seen that the computation has become simpler by normalization with $z$ having characteristics of standard normal distribution with mean 0 and standard deviation as 1. The pdf and CDF of standard normal deviation are shown in Figs. 3.3 and 3.4.

Since it is easy to transform normal distribution parameters into standard normal as shown above, computation of CDF becomes easier by employing standard normal table given in Annexure.

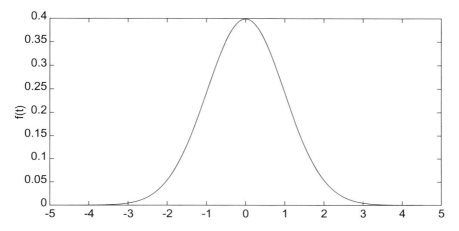

**Fig. 3.3** Probability distribution function (pdf) of standard normal distribution

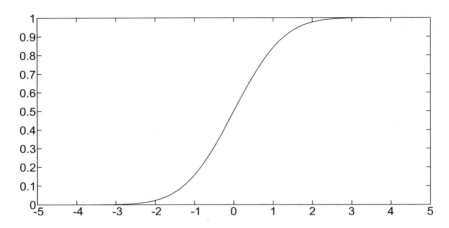

**Fig. 3.4** Standard normal distribution cumulative distribution function (CDF)

### 3.4.1.2   Log-Normal Distribution

If the logarithms of the random variable, say, time $T$, follows normal distribution, then the variable $T$ is considered to follow log-normal distribution. Log-normal distribution is extensively used to represent failure data from a wide range of sources. For example, conclusions drawn from the set of data derived from different sources for a component are generally modeled assuming log-normal distribution. This distribution is also used to model life test data. This distribution finds wider application for uncertainty. This is also a two-parameter distribution, where $\mu$ and $\sigma$ represent the logarithmic mean and variance of the distribution. Unlike normal distribution, this distribution is generally skewed toward the right. Log-normal distribution is defined as follows:

$$f(t) = \frac{1}{\sigma\sqrt{2\pi t}} \exp -\frac{1}{2}\left(\frac{\ln t - \mu}{\sigma}\right)^2 \quad \text{for } t > 0 \tag{3.19}$$

Figure 3.5 shows the log-normal distribution for different values of $\mu$ and $\sigma$.

### 3.4.1.3   Exponential Distribution

Exponential distribution is the most widely used distribution in reliability engineering owing to its simplicity and attributes such as its constant failure rate, which makes it suitable for applications to components in its useful period. This distribution does not have memory. For example, failure probability of two components, one that has operated for a period $t$ and another that has just been put in service at time $t$, will be same for period $t + \Delta t$. In short, this distribution does not capture aging. At every epoch of time, the component is treated as new.

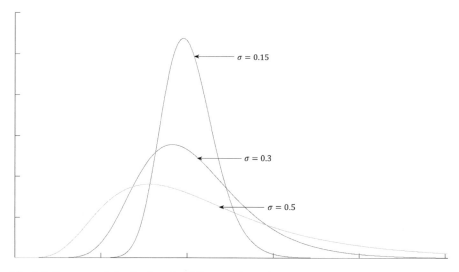

**Fig. 3.5** Log-normal distributions for different values of $\sigma$

This is a single-parameter distribution. The pdf of the exponential distribution is given as

$$f(t) = \lambda \exp(-\lambda t) \tag{3.20}$$

where $\lambda$ is the parameter of the distribution and $t$ as the mission time.

The CDF for exponential distribution can be given as

$$F(t) = \int_0^t f(t)\mathrm{d}t \tag{3.21}$$

$$F(t) = \int_0^t \lambda \exp(-\lambda t)\mathrm{d}t \tag{3.22}$$

$$F(t) = 1 - \exp(-\lambda t) \tag{3.23}$$

Since

$$R(t) = (1 - F(t)) \tag{3.24}$$

Therefore,

$$R(t) = \exp(-\lambda t)$$

It can be shown that the hazard rate for exponential distribution is constant $= \lambda$.

$$h(t) = \frac{f(t)}{R(t)} \qquad\qquad (3.25)$$

$$h(t) = \frac{\lambda \exp(-\lambda t)}{\exp(-\lambda t)} = \lambda = \text{A constant} \qquad\qquad (3.26)$$

In real life, the constant failure rate characteristics of this distribution also are suitable for modeling of the maintained system where the periodic maintenance activities make the constant failure rate assumption more close to real-life situations. Let us take a component on which periodic maintenance has just been performed and has been brought back to its new condition. As the time $t$ passes after the maintenance, the component degrades to an acceptable limit before new maintenance action is performed. This cycle repeats, and it can be seen that over the life cycle of the component, the average failure rate represents the constant failure characteristics as shown in Fig. 3.6. Here, the assumption of constant failure rate to most of the real-time models using exponential distribution appears to be justified.

From the above discussions, we should be able to draw some conclusions regarding the areas where this distribution can be used to model real-life problems, bearing in mind that this distribution is an applicable scenario where the constant failure rate is the characteristics of the system. For example, exponential distribution can be used for an engineering component that has been extensively tested by either the vendor or in the plant in situ, where we expect that initial failures due to manufacturing- or quality-related problems have been overcome. Similarly, the aging-related failure, such as wear out and extensive corrosion, should not be expected to degrade the component in a significant manner. However, minor wear out and aging of subcomponents is the order of the day and can be corrected by routine maintenance or testing schedules. These components can also be modeled by exponential distribution. Exponential distribution should not be used for cases where even the routine maintenance is not able to arrest the frequent failures of the components involving wear out as the main failure mechanism. This increase in failure frequency shows that the component has crossed the constant failure rate domain, and these types of scenarios can be modeled using other time-to-failure distribution, such as Weibull distribution.

There are many situations when the failure data on components or systems are available over a period of time. Consider off-site power supply failure events in a nuclear plant. In this hypothetical case, the events have been recorded over the past

**Fig. 3.6** Average reliability concepts for assumption of constant failure rate model during useful life (0 to $T$) of the component

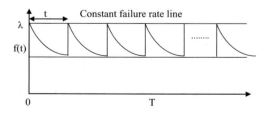

10 years (see Table 3.1). This data set is typically comprised of the time between successive failures. In such cases, it easier to know the applicable distribution and exponential graphs are available. If the data are arranged in ascending order, which means the point where the time to failure is minimum is ranked first, the time to the next higher failure is the second point, and likewise the time to highest failure forms the final point in the series. After organizing the data in ascending order, it is plotted on an exponential graph. If the data fall on a straight line, it is assumed that the data follow exponential distribution. The graphical method, even though approximate in nature, provides a good insight into the nature and trends of data.

Table 3.1 shows the data arranged in ascending order. The exponential graph is plotted in Fig. 3.7. It can be seen that the straight line appears to represent the data well. There are mathematical techniques called regression analysis, which will be covered in the subsequent analysis, that show how the line that represents the data can be estimated. However, a visual judgment is often used to assess the applicability of a distribution for a data set.

**Table 3.1** Off-site power failure events in a nuclear plant recorded over a period of 20 years

| Power failure data | Data arranged in ascending order | Rank | Median rank $(i - 0.4)/(N + 0.3)$ |
|---|---|---|---|
| 198 | 5 | 1 | 0.034313725 |
| 112 | 16 | 2 | 0.083333333 |
| 27 | 16 | 3 | 0.132352941 |
| 20 | 19 | 4 | 0.181372549 |
| 27 | 20 | 5 | 0.230392157 |
| 50 | 24 | 6 | 0.279411765 |
| 38 | 27 | 7 | 0.328431373 |
| 77 | 27 | 8 | 0.37745098 |
| 19 | 30 | 9 | 0.426470588 |
| 105 | 38 | 10 | 0.475490196 |
| 24 | 45 | 11 | 0.524509804 |
| 107 | 50 | 12 | 0.573529412 |
| 63 | 52 | 13 | 0.62254902 |
| 30 | 63 | 14 | 0.671568627 |
| 16 | 77 | 15 | 0.720588235 |
| 158 | 105 | 16 | 0.769607843 |
| 5 | 107 | 17 | 0.818627451 |
| 16 | 112 | 18 | 0.867647059 |
| 45 | 158 | 19 | 0.916666667 |
| 52 | 198 | 20 | 0.965686275 |

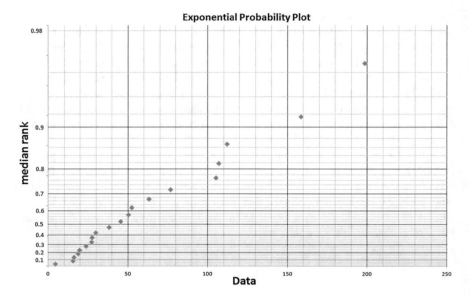

**Fig. 3.7**  Plot of power supply failure data on exponential graph

### 3.4.1.4  Weibull Distribution

The Weibull distribution is the most widely used distribution in the area of life prediction of components. The reason for its popularity is its flexibility to model data following other distributions such as exponential or normal. In fact, Weibull analysis has developed into a separate branch for prediction of life and failure data.

This is a continuous type of three-parameter distribution. Suppose $x$ is a continuous random variable, then the pdf of the distribution is given as follows:

$$f(x) = \frac{\beta}{\alpha - \theta} \left( \frac{x - \theta}{\alpha - \theta} \right)^{\beta - 1} \mathrm{Exp}\left[ -\left( \frac{x - \theta}{\alpha - \theta} \right)^{\beta} \right] \qquad (3.27)$$

where $\beta$ is the slope or shape parameter, $\theta$ is the location parameter, and $\alpha$ is the characteristic value or the characteristic life when dealing with life data of the component. The shape parameter (i.e., $\beta$) determines the distribution features, e.g., if in Eq. 3.27, when $\beta = 1$, the distribution reduces to exponential distribution, similarly when $\beta = 3$–$4$, then the Weibull distribution exhibits the property of normal distribution (Fig. 3.8).

The parameter $\theta$ accounts for the expected minimum value or life (for the life data) of the component. For example, if the analysis involves life prediction of a mechanical bearing and the minimum expected life of the bearing is 1000 h, then $\theta = 1000$ h fixes the location parameter of the distribution. The parameter $\alpha$

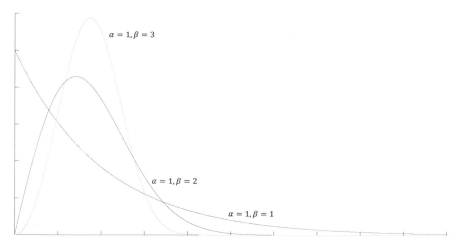

**Fig. 3.8**  Density function of Weibull distribution (when $\theta = 0$)

characteristic value determines the life of 63.2% of the total components in the Weibull distribution.

For most of the situations, it is possible to assume the minimum, i.e., the scale parameter $\theta$, to be zero. Accordingly, the Weibull distribution reduces to a two-parameter distribution with pdf as

$$f(x) = \frac{\beta}{\alpha} \left(\frac{x}{\alpha}\right)^{\beta-1} \mathrm{Exp}\left[-\left(\frac{x}{\alpha}\right)^{\beta}\right] \tag{3.28}$$

The cumulative distribution function (CDF) for Weibull distribution can be derived as follows:

$$F(x) = \int_{0}^{x} f(x)\mathrm{d}x \tag{3.29}$$

$$F(x) = \int_{0}^{x} \frac{\beta}{\alpha} \left(\frac{x}{\alpha}\right)^{\beta-1} \mathrm{Exp}\left[-\left(\frac{x}{\alpha}\right)^{\beta}\right] \mathrm{d}x\ldots \tag{3.30}$$

or

$$\left(\frac{x}{\alpha}\right)^{\beta} = y$$

$$\beta \left(\frac{x}{\alpha}\right)^{\beta-1} \frac{1}{\alpha} \mathrm{d}x = \mathrm{d}y$$

Rearranging the equation

$$\frac{\beta}{\alpha}\left(\frac{x}{\alpha}\right)^{\beta-1} dx = dy$$

$$F(x) = \int_0^y e^{-y} dy$$

$$F(x) = [-e^{-y}]_0^y$$

$$F(x) = 1 - e^{-y}$$

$$F(x) = 1 - \exp\left[-\left(\frac{x}{\alpha}\right)^{\beta}\right] \tag{3.31}$$

This is the equation for the CDF of the Weibull distribution. As mentioned earlier, for $\beta = 1$, the CDF equation reduces to the CDF of exponential distribution. In the above equation, when $x = \alpha$, it follows:

$$F(x) = 1 - \exp\left[-\left(\frac{\alpha}{\alpha}\right)^{\beta}\right]$$

$$F(x) = 1 - e^{-1}$$

$$F(x) = 1 - \frac{1}{e} = 1 - \frac{1}{2.718} = 0.632 \tag{3.32}$$

$$F(x) = 63.2\%$$

Hence, it can be inferred that for the Weibull distribution, $\alpha$ represents the life in units of time (year or hours, etc.) when 63.2% of the components under consideration would have failed. This observation is of great significance for understanding and interpreting the information from the Weibull plot.


### 3.4.2  Discrete Distributions

Even though there are many discrete distributions, this section discusses only the three most extensively used distributions in reliability modeling—the binomial distribution, Poisson distribution, and $F$-distribution.

### 3.4.2.1 Binomial Distribution

Consider an experiment comprised of $n$ trials where the outcome is discrete and characterized by either "success" or "failure." Suppose the number of success event is $x$ with probability $p$. This also implies that number of failure events is $n - x$ having probability $1 - p$ denoted by $q$. Then the pdf for binomial distribution can be described as:

$$P[x; n, p] = f(x) = \binom{n}{x} p^x q^{n-x} \text{ for, } x = 0, 1, 2 \ldots n \qquad (3.33)$$

The above model estimates the probability of $x$-out-of-$n$ success when the probability of success if given as $p$. The following example provides better clarity:

In a nuclear plant, there is one shutdown device to shutdown the reactor on demand. For a given demand, the device either actuates successfully or fails to actuate. The probability of success is denoted by p, and probability of failure is denoted by $q$. Mathematically, it can be represented as [3]:

| S | F |
|---|---|
| p | q |

$$p + q = 1$$

If there are two shutdown devices, then the four outcomes are:

| SS | SF | FS | FF |
|---|---|---|---|
| pp | pq | qp | qq |

$$p^2 + 2pq + q^2 = (p + q)^2$$

Further, with three devices there will be eight outcomes as follows:

| SSS | SFS | SSF | SFF | FSF | FSS | FFS | FFF |
|---|---|---|---|---|---|---|---|
| ppp | pqp | ppq | pqq | qpq | qpp | qqp | qqq |

$$p^3 + 3p^2q + 3pq^2 + q^3 = (p + q)^3$$

Accordingly, if there are $n$ shutdown devices, the probability of obtaining $n$, $n - 1$, $n - 3$, $n - 4$, ... 3, 2, 1, 0 successful actuation can be estimated by the binomial expression as follows:

$$(p+q)^n = p^n + c_1^n p^{n-1} q + c_2^n p^{n-2} q^2 \ldots + c_r^n p^{n-r} q^r + \ldots q^n$$

Hence, probability of the $x$ successes from $n$ trials can be written as

$$P[x; n, p] = f(x) = \binom{n}{x} p^x q^{n-x} \tag{3.34}$$

which is nothing but the equation for binomial distribution where

$$x = 0, 1, 2, 3 \ldots n$$
$$n = 1, 2, 3 \ldots$$
$$0 \leq p \geq 1, \text{and } q = 1 - p$$

Here, $n$ and $p$ are the parameters of the binomial distribution.

### 3.4.2.2  Poisson Distribution

The Poisson distribution finds wider application when dealing with discrete events over a given interval of interest. It is a natural choice for the analyst when the probability associated with a discrete event is very low. Take the case of a piping analysis when the analysts are trying to model the defect density or, to be precise, pitting corrosion-related defective sites in the primary coolant system pipelines in a nuclear plant. Other examples are the number of defective units produced or the number of accidents in certain types of process or industrial plants.

Let the random variable $X$ represent the number of independent events of interest; it could be the number of defects or accidents in a given time interval. The Poisson probability distribution has one parameter $\lambda$, which is constant and always positive. The pdf for $x$ occurrences is given as:

$$p(X = x) = \frac{(\lambda)^x \exp(-\lambda)}{x!} x = 0, 1, 2, 3 \ldots n \tag{3.35}$$

Here, $\lambda = E(X)$ is the "expectation of $X$" or in more general form mean number of occurrence per unit of population space. For example, it could be failures per unit time or number of corrosion defects observed in a nuclear plant pipeline per feet. This distribution is used to characterize the corrosion-related defects in a pipeline network in a process or industrial system.

### 3.4.3  Joint Probability and Marginal Distribution

The previous section dealt with one random variable; however, there are many practical situations when it is required to model the function with two or more random variables. Such situations require a joint probability density function. Suppose there are two continuous random variables $x$ and $y$, then the joint probability density function is $f(x, y)$. If the two random variables $X$ and $Y$ are discrete, then the joint pdf is written as $\Pr(x, y)$.

The joint continuous pdf $f(x, y)$ satisfies the following conditions:

(a) $f(x, y) \geq 0$;    $-\infty < x, y > \infty$ and
(b) $\int\limits_{-\infty}^{\infty} \int\limits_{-\infty}^{\infty} f(x, y)dxdy = 1$.

Similarly, the joint discrete pdf $\Pr(x, y)$ should satisfy the following conditions:

(a) $\Pr(x, y) \geq 0$ for all the values of $x$ and $y$ and
(b) $\sum\limits_{x} \sum\limits_{y} \Pr(x, y) = 1$

In a joint pdf $f(x, y)$ when pdf of $X$—a discrete random variable—is obtained by summing $f(x, y)$ for the entire range of $Y$, then this function is known as a marginal density function of $X$, $g(x)$ and is written as:

$$g(x) = \sum_{y} f(x, y)$$

Also, the marginal density function of $Y$ denoted as $h(y)$ is written as:

$$h(y) = \sum_{x} f(x, y)$$

For the case when $X$ and $Y$ are continuous random variables, then $g(x)$ and $h(y)$ take the form:

$$g(x) = \int\limits_{y=-\infty}^{\infty} f(x, y)dy$$

$$h(y) = \int\limits_{x=-\infty}^{\infty} f(x, y)dx$$

We extend this argument for conditional probability for joint pdf for both the cases of discrete and continuous random variable as follows:

$$P(y/X = x) = f(y/x) = \frac{f(x,y)}{g(x)} \tag{3.36}$$

and

$$P(x/Y = y) = f(x/y) = \frac{f(x,y)}{h(y)} \tag{3.37}$$

Further, the probability that the discrete random variable $X$ occupies $a$ value between a and $b$ while the random variable $Y = y$

$$P(a < X < b/Y = y) = \sum_{x=a}^{b} f(x/y) \tag{3.38}$$

Similarly, when $X$ and $Y$ are continuous random variable, then

$$P(a < X < b/Y = y) = \int_{x=a}^{b} f(x/y) \mathrm{d}x \tag{3.39}$$

These cases of two random variables can be extended to more than two random variables, and such cases are known as multivariate distributions.

### 3.4.4 Determining Applicable Distribution

We have seen in earlier sections that it is required to identify the data type, viz., discrete or continuous, and accordingly work for assessing the type of distribution. For example, for continuous random variables, the distribution could be normal, log-normal, or Weibull, whereas for discrete random variables based on the nature of problem, the binomial, Poisson, or any other distribution could be used. The question is, given a data set, how to select the most appropriate distribution because these models are used for prediction of reliability and safety parameters. Hence, the choice of distribution should represent the data set appropriately. Accordingly, the distribution should be able to first reduce the error in the point estimate and then to assess the uncertainty bounds for the estimates in the most appropriate manner.

Three approaches are commonly used to select an applicable distribution for a given problem or data set:

(a) *Based on the nature of data*: As discussed in the earlier section, the general practices for selecting the distribution by traditional heuristics are as follows: (i) for a production line involved in manufacturing of shafts, the variation in diameter and length of the shaft is characterized using the normal distribution,

(ii) Weibull and exponential distribution are used for life prediction in general and accelerated testing of components, (iii) log-normal distribution is used to model data coming from independent sources and where the scatter in data is rather wide, (iv) binomial distribution is used for cases where the probability of failure of a desired number of parts from the given population is to be estimated for, e.g., the probability of 3 out of a total of 7 shutdown devices failure, (v) Poisson distribution is used to analyze events that are randomly distributed in the sample space, e.g., distribution of defects in coolant pipelines and number of faults/incidents occurring in a plant. However, it may be noted that even though these heuristics serve a good purpose when the scope of the problem is rather wide, the available resources and the input information prohibit a detailed data analysis to determine applicable distribution.

(b) *Probability plotting*: Probability plots are available for various distributions, such as exponential, Weibull, normal, and log-normal distributions. Even though this method is considered an approximate method, it provides reasonable rationales for determining the applicability of a distribution for a data set. One limitation of this method is that a visual judgment of how well the line drawn through the data point represents the data set sometimes is not sufficient to confirm the adequacy of the distribution in question. However, statistical methods, such as (i) regression analysis where the value of coefficient of correlation $\sim 0.9$ or better confirms the fit of the line for the given data set or (ii) assessment using the $\chi^2$ (chi-square) method where the value of $\chi^2 \sim 17$ or better confirms the fit for the data set, are some of the methods that can be used to ensure that the line drawn through the data set represents the data points adequately.

(c) *Analytical technique*: The Weibull distribution has become one of the most sought-after distribution for understanding the trends of a data set and thereby provides a sound rationale for finding the applicability of a distribution for a given data set. The slope parameter $\beta$ provides reasonable approximations. For example, if $\beta = 1$, then the data set follows exponential distribution, whereas if $\beta$ is between 3 and 4, then the data set is considered to follow normal distribution. There are a few other distributions that provide information about the applicability of a particular distribution. This method is considered to be more accurate compared to probability plotting.

One distribution does not necessarily represent an entire data set in the given population. Many times it has been found that one distribution represents only a segment of the data. Here, it is required to assess the suitable distribution for the remaining segment of the line or data points. Note that there are many more continuous and discrete distributions given in the literature that have not been discussed. Keeping in view the requirements of the subject area (i.e., risk-based engineering), this section focuses on only those distributions that are used for data analysis. For other distributions, any book on statistical analysis or reliability engineering may be referred to.

## 3.5  Statistical Estimation of Failure Rate

Probability plotting provides an effective procedure for approximating a distribution; however, analytical techniques are required to evaluate the parameters and associated confidence interval, if required. Further, how well the distribution presents the data also needs to be analyzed by performing either goodness of fit or regression, depending upon the requirements. Hence, the following section is dedicated to parameter evaluation and deals with single- and two-parameter cases, such as exponential distribution for a single parameter and normal and Weibull distributions for two parameters.

### *3.5.1  Point Estimate*

Obtaining the point estimate of a given random variable is the key to component and system reliability modeling. Determining the probability distribution through a graphical approach only provides a basis for an approximation to a distribution; however, point estimation is a prerequisite to systems structures and components SSC reliability modeling.

In point estimation, the parameter of a distribution is estimated based on the sample data obtained from reliability tests or field observations. Let us assume that we are interested in determining the distribution function $f(X, \theta)$, where $X$ represents a random variable and $\theta$ a parameter of the distribution, for example, average failure rate $\lambda$. The random variable $X$ can have values $x_1, x_2, x_3, x_4 \ldots x_n$;

If the $\theta$ is a true value or ideal value of the parameter, then the estimate obtained from the single-valued function, $h(x_1, x_2, x_3, x_4 \ldots x_n)$, will be a statistical estimate for $\theta$. Hence, we can say that this estimate will be expectation or $E[h(x_1, x_2, x_3, x_4 \ldots x_n)]$. By this argument, we can expect a bias between the actual value of $\theta$ and the expectation or statistical estimate of $\theta$. Hence, parameter $\theta$ is called the estimator for $\theta$ and the numerical estimate obtained is called the estimate of $\theta$.

It is very important to understand this concept because the point estimate forms the fundamental building block in risk-based engineering. In a nutshell, this can be understood as follows, the estimates obtained from various sets of test data are random and the estimate based on these data might represent only central tendency (mean or median) parameter value of the failure rate. However, as the size of the population increases, the estimated value tends to move closer to the actual value.

That is the reason the value estimated from a population is estimator $\hat{\theta}$ of the parameter $\theta$. Consider the scenario where there is no bias, the value of the estimator is the same as the parameter, i.e.,

$$\overset{\wedge}{\theta} = E[h(x_1, x_2, x_3, x_4 \ldots x_n)] = \theta. \tag{3.40}$$

Bias $b$ is deduced as

$$\overset{\wedge}{\theta} = E[h(x_1, x_2, x_3, x_4 \ldots x_n)] - \theta = b \tag{3.41}$$

The estimator should be such that it tends to satisfy the following condition:

$$\underset{n \to \infty}{\mathrm{Lim}} P[|h(x_1, x_2, x_3, x_4 \ldots x_n) - \theta| < b] = 1 \tag{3.42}$$

The practical interpretation of the above discussion is that, not all situations require a decision in respect of what is the accuracy required for an estimator to represent the "real parameter." If the analysts feel that the random variables have all the needed information in the data set that forms the input, then the estimator could be considered an adequate representation of the estimate. This is called the condition of sufficiency.

However, there are two commonly used procedures for statistical estimation, viz., method of moment and maximum likelihood estimation, which are discussed in the following section.

### 3.5.1.1 Method of Moment

Method of moment is the most simple and elegant approach to arrive at point estimation, particularly with fewer properties. This method is used when the application of other methods such as the maximum likelihood method, discussed in the next section, becomes complex in terms of calculations. The mean and variance of a continuous random variable characterize the expected value of $X$ and expected value of $(X - \mu)^2$ and are the point estimate that describes properties of population. Suppose there are $n$ observation $(x)$ for a random variable $X$ as $x_1, x_2, x_3, \ldots, x_n$; then the expectation of the sample mean and variance is written as follows:

$$\bar{x} = \frac{1}{n} \sum_{i=1}^{n} x_i \tag{3.43}$$

$$S^2 = \frac{1}{n} \sum_{i=1}^{n} (x_i - \bar{x})^2 \tag{3.44}$$

where the $\bar{x}$ and $S^2$ can be used as the point estimate for distribution mean, $\mu$ and variance $\sigma^2$ or standard deviation $\sigma$. Since $S^2$ uses the estimate of $\bar{x}$, this estimate is biased and this bias can be removed by multiplying the $S^2$ term by $\frac{n}{n-1}$ as follows:

$$S^2 = \frac{1}{n-1} \sum_{i=1}^{n} (x_i - \bar{x})^2 \tag{3.45}$$

As can be seen, the mean and variance are the first and second moments of the distribution.

### 3.5.1.2  Maximum Likelihood Estimate

The maximum likelihood estimate (MLE) is one of the most commonly used point estimation procedures because it involves estimation of the parameter from the pdf of a given random variable [4]. Apart from this, this estimating procedure has all the desirable properties.

Consider a test where all the observations on a random variable $X$ are coming from $n$ independent source as $x_1, x_2, x_3, \ldots x_n$. This information can be characterized by pdf $f(x_i, \theta)$, where $\theta$ is the parameter of the distribution. If the likelihood function is written as

$$L(x_1, x_2, x_3 \ldots x_n, \theta) = \prod_{i=1}^{n} f(x_i, \theta) \tag{3.46}$$

the MLE method requires determination of an estimator $\overset{\wedge}{\theta}$.
Such that

$$L(x, \hat{\theta}) \geq L(x_i, \theta)$$

For maximizing this function, we differentiate the likelihood function continuously such that at one point the gradient is zero as follows:

$$\frac{\partial}{\partial \theta} L(x_i, \theta) = 0$$

Keeping in view the requirement of mathematical treatment, often it is required to consider the logarithm of likelihood function as follows:

$$\frac{\partial \log L(x_i, \theta)}{\partial \theta} = 0$$

It can be shown that MLE is consistent, possibly biased, sufficient, asymptotically normal, and efficient.

*Example*: Finding the MLE for exponential distribution dealing with a random observation on time to failure has been produced as sample of observation $t_1, t_2, t_3, \ldots t_n$; the parameter to be estimated is failure rate $\lambda$. Find the MLE for $\lambda$.

The likelihood function is written as the joint probability density function of $f(t_i, \lambda)$

$$L(t_i, \lambda) = \prod_{i=1}^{n} \lambda \exp(-\lambda t_i) \qquad (3.47)$$

$$L(t_i, \lambda) = \lambda^n \exp\left(-\lambda \sum_{i=1}^{n} t_i\right) \qquad (3.48)$$

Taking natural log on both sides, we have

$$\ln L = n(\ln \lambda) - \lambda \sum_{i=1}^{n} t_i$$

Differentiating both sides of the equation for maximizing for $\lambda$, i.e.,

$$\frac{\partial \ln L}{\partial \lambda} = \frac{n}{\lambda} - \sum_{i=1}^{n} t_i = 0$$

$$\hat{\lambda} = \frac{n}{\sum_{i=1}^{n} t_i} \qquad (3.49)$$

To check that this estimate of $\lambda$, i.e., $\hat{\lambda}$, is indeed maximized, we go for second-order derivative as follows:

$$\frac{\partial^2 \ln L}{\partial \lambda^2} = -\frac{n}{\lambda^2} < 0 \qquad (3.50)$$

Hence, it is clear the $\hat{\lambda}$ is MLE.

### 3.5.1.3 Bayesian Estimator

The Bayesian methodology for parameter estimation is comprised of estimation of the posterior probability using the prior information available for the parameter in question and evidence. In the PRA approach, this methodology is used for updating the prior generic data on the failure rate with the evidence, that is, the information available from plant sources to give the updated posterior parameter [5]. This approach provides an effective mechanism for failure rate estimation when only limited data are available from the plant-specific source. For better visualization, Fig. 3.9 depicts the update mechanism.

Let $g(\lambda)$ is the joint prior probability distribution function of $\lambda$;

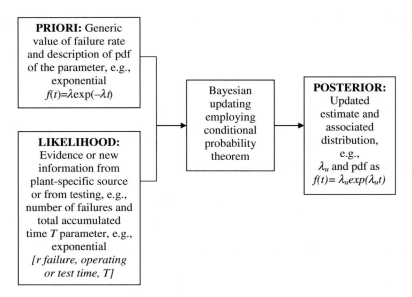

**Fig. 3.9** Depiction of Bayesian updating procedure in probabilistic risk assessment (PRA)

$f(t|\lambda)$ is the conditional probability of random variable $t$ given $\lambda$;
$f(t,\lambda)$ is the joint probability of random variable $t$ and $\lambda$;
Hence,

$f(t|\lambda) = \prod\limits_{i=1}^{n} f(t_i|\lambda)$ is the joint conditional probability function of $t$ given $\lambda$

$f(t)$ is the marginal density function of random variable $t$;
if $f(\lambda|t)$ = joint posterior probability distribution of $\lambda$ given $t$,

then accordingly to Bayes theorem, the posteriori and the priori along with evidence are related as follows:

$$g(\lambda|t) = \frac{\prod\limits_{i=1}^{n} f(t_i|\lambda)g(\lambda)}{f(t)} \qquad (3.51)$$

or

$$g(\lambda|t) = \frac{f(t|\lambda)g(\lambda)}{f(t)} \qquad (3.52)$$

$g(\lambda|t)$ is the posterior point estimator or update for the $\lambda$.

## 3.5.2   *Confidence Interval Estimation*

The failure rate data, and more importantly its applicability or rather its accuracy, is crucial to the quality of results in a PRA study. The failure rate data are generated by two methods. The first method is to perform a life test on a sample size of n components where the estimate of accumulated time of the population along with number of failures observed provides the estimate of average failure rate. The second method, which is more common when adequate operating experience on the subject component is available, is to use the plant records to estimate the accumulated time on the component and the number of failures recorded during the data collection. This input is used to estimate the average failure rate. However, the accuracy of these estimates depends upon many factors, such as the adequacy of the length of the data collection period, the definition of the failure criteria, the definition of the period that is considered as the operating period as the plant records adequacy is another factor that need to be considered, and the assumption of an applicable distribution (i.e., often the exponential distribution is considered in PRA), which remains a subject of debate.

However, the confidence limit or interval estimation provides a way out of this situation, in a way. This approach enables estimation of the upper and lower confidence bounds. For example, the upper bound with more than 90% confidence in the estimate so obtained can address the estimation uncertainty that forms part of the average value. Apart from point estimation confidence, interval estimation should be part of PRA data modeling such that the uncertainty in the data and the model is accounted for or presented while making a decision.

We will explain the procedure of confidence interval estimation using the life test approach. There are two methods of testing that can predict the component failure rate.

*Type I test*: The test is terminated after a predetermined time. The average failure rate is obtained by dividing the number of failures ($r$) observed during the test by the total accumulated time, $T$, i.e.,

$$\bar{\lambda} = \frac{r}{T}$$

*Type II test*: The test is terminated when a predetermined number of failures are observed. The formula for estimating the average failure rate is the same; however, the results might be different. Further, there are other aspects, such as tests with replacement and without replacement. The estimate of the average failure rates might change based on the type of test and other factors. Hence, the confidence interval estimation forms part of many reliability studies.

It has been demonstrated that for the case of exponential distribution with failure rate $\lambda$, the following expression has a chi-square ($\chi^2$) distribution. For the case of Type I and Type II tests, the applicable degree of freedom is $2n + 2$, and $2n$, respectively.

$$\frac{2n\lambda}{\bar{\lambda}} = 2\lambda T$$

where $n$ is the number of failures observed (either from the life test or plant data); $T$ is the accumulated time (either from the life test or plant data); $\lambda$ is the actual failure rate; and $\bar{\lambda}$ is the estimated average failure rate from the life test or plant data.

It was shown that it is possible to estimate the probability of having an actual failure rate $\lambda$ within a confidence bound. This interval is referred to as "confidence interval." The confidence interval was estimated by Epstein, as follows [6];

$$\Pr\left[\frac{\chi^2_{\frac{\alpha}{2}}(2n)}{2T} \le \lambda \le \frac{\chi^2_{\left(1-\frac{\alpha}{2}\right)}(2n)}{2T}\right] = 1 - \alpha \qquad (3.53)$$

Here, if the confidence interval is 90%, i.e., $1 - \alpha = 0.9$, so $\alpha = 0.1$ is the probability that the failure rate will not be a subset of the 90% confidence. This formulation provides the lower and upper confidence limits and is called "double-sided confidence bounds."

Often in confidence estimation, a one-sided confidence, i.e., either upper or lower, is required. The one-sided confidence bound is given as follows:

$$\Pr\left[0 \le \lambda \le \frac{\chi^2_{1-\alpha}(2n)}{2T}\right] = 1 - \alpha \qquad (3.54)$$

In any real-life situation, the limit is put on one side only, such as the highest value of failure rate, the lowest mean time-to-failure (MTTF) value, the highest value of stress, or the lowest thickness that serves the purpose in risk assessment. Therefore, in most of the situations, the one-sided confidence interval is more relevant than the two-sided confidence interval. Similarly, the confidence bound for other distributions for the failure rate as well as for MTTF can also be evaluated.

## 3.6   Goodness-of-Fit Test

In earlier sections in this chapter, we assume a distribution to represent a parameter. Furthermore, we also use approximate approaches such as probability graph plotting to figure out how well the data are represented by a chosen distribution. This term, "how well," will be further evaluated in the next section on regression analysis, but in this section, we will discuss an analytical approach to determine how well the chosen probability distribution represents a data set.

Before we go over to the method of two important methods for evaluating goodness of fit for data in terms of a distribution, we discuss the approach to formulate criteria for accepting/rejecting the hypothesis that a data set "represents a data set." Consider that the reliability parameter, such as the failure rate and MTTF, is represented by a common notation $\theta$.

The procedure involves formulation of a *null hypothesis* ($H_0$) and an *alternate hypothesis* ($H_1$) as follows:

$H_0$: $\theta = \theta_0$; i.e., when the data come from a specified distribution
$H_1$: $\theta \neq \theta_0$; i.e., when the data do not come from a specified distribution

Furthermore, it involves evaluating the statistic based on the available observed data and the estimates obtained from the specified distribution and comparing the results of this evaluation with the critical value obtained from the table, of either chi-square ($\chi^2$) or Kalmogorov–Smirnov (K–S), for example. If the statistic so obtained is less than the critical value obtained from the table, the null hypothesis, i.e., $H_0$: $\theta = \theta_0$, is accepted. Otherwise, the alternate hypothesis is accepted.

As mentioned above, there are two common methods of checking goodness of fit —the chi-square and K–S tests.

### 3.6.1 Chi-Square Test

In this test method, as the name suggests, the test statistic is developed assuming approximation of $\chi^2$ distribution. Even though both continuous as well as discrete data analysis can be done, there are certain guidelines for deploying the $\chi^2$ test, namely the sample size should be large, the test is only applicable for censored data, and the data should be split into non-overlapping classes/intervals, usually 5 or more. The $\chi^2$ test statistic is formulated as follows:

$$\chi^2 = \sum_{i=1}^{k} \frac{(O_i - E_i)^2}{E_i} \tag{3.55}$$

where $O_i$ is the observed failure data from the plant or test records;
$E_i$ is the estimated data from the considered distribution;
$k$ is the number of classes/intervals, $E = np_i$;
$n$ is the sample size; and $p_i$ is the probability of a failure in a given interval.

*Example*: In a nuclear plant, the data for off-site power failures per year were collected. In one of the PRA studies, there was an assumption that the number of power failures per year follows the Poisson distribution. The analysts wanted to confirm the validity of this assumption. The table shows the number of failures per year, $f$, and followed by observed frequency. Evaluate the adequacy of the assumption of Poisson distribution by employing the chi-square test.

| Failures per year, $x$ | Observed frequency, $O_i$ | Expected frequency, $E_i$ | $\chi^2$-statistic, $[(O_i - E_i)^2 / E_i]$ |
|---|---|---|---|
| 0 | 5 | 9.65 | 2.24 |
| 1 | 10 | 9.65 | 0.012 |
| 2 | 7 | 4.86 | 0.942 |
| 3 | 2 | 1.62 | 0.09 |
| 4 | 2 | 0.40 | 6.4 |
| 5 | 0 | 0.08 | 0.08 |
| 6 | 0 | 0.01 | 0.01 |
| Total | 27 | 26.27 | 9.77 |

*Solution*:

Estimation for the expected frequency is performed as follows:

The exponential and Poisson distributions are single-rate parameter distributions. The parameter $\rho$ is estimated as follows:

$$\rho = \frac{N}{T}$$

$N$: Number of power failures = 21
$T$: Total number of years = 27

$$\rho = 0.78/\text{year}$$

The probability of power failure ranging from 1 to 6/year, from the Poisson distribution, is estimated as follows:

$$\Pr(X = x_i) = \frac{\rho e^{-\rho}}{x_i} \tag{3.56}$$

The value of expected frequency is given by $np_i$, where $n$ is the total number of power failures and $p_i$ is the probability estimated from the Poisson distribution for 0–6 failures in a year. The expected frequency column contains $np_i$ estimates for all the $i_s$ from 1 to 6.

The $\chi^2$ statistic is obtained in the last column. The summation is $W = 9.77$.

The rejection region

$$R \geq \chi^2_{1-\alpha}(k - m - 1);$$

where $k - m - 1$ is the degree of freedom; $k$ is number of classes = 7; $m$ is the number of parameters = 1; hence, the degree of freedom is 5; and if the significance level $\alpha = 10\%$, i.e., 0.1; i.e., $1 - \alpha = 0.9$;

From $\chi^2$ table (Annexure), the value of $R = \chi^2_{(0.90, 5)} = 9.23$;

$$W = 9.77 \quad \text{and} \quad R = 0.923;$$

Since $W > R$, the hypothesis that the data follow the exponential distribution is rejected. Hence, null hypothesis, $H_0$ that the data follow the Poisson distribution is rejected and the alternate hypothesis $H_1$ that the data do not follow the Poisson distribution is accepted.

### 3.6.2 Kolmogorov–Smirnov Test

The Kolmogorov–Smirnov (K–S) test involves ranking of the observation in ascending order and unlike $\chi^2$, which does not require creating classes/intervals. However, this test is similar to the $\chi^2$ test, in the sense that it involves comparison of the assumed or hypothesized and empirical estimation, CDF in this case. We will consider normal distribution to illustrate the K–S test.

The hypotheses for the K–S test are as follows:

$H_0$ = Failure times are specified distribution, i.e., normal
$H_1$ = Failure times are not normal

The test statistics are $D_n = \max\{D_1, D_2\}$

$$D_1 = \max_{1 \leq i \leq n} \left[ \phi\left(\frac{t_i - \bar{t}}{s}\right) - \frac{i - 1}{n} \right] \tag{3.57}$$

$$D_2 = \max_{1 \leq i \leq n} \left[ \frac{i}{n} - \phi\left(\frac{t_i - \bar{t}}{s}\right) \right] \tag{3.58}$$

where $\bar{t} = \sum_{i=1}^{n} \frac{t_i}{n}$ and $S^2 = \frac{\sum_{i=1}^{n} (t_i - \bar{t})^2}{n - 1}$.

Acceptance criteria: if $D_n < D_{\text{crit}}$, accept $H_0$; if $D_n \geq D_{\text{crit}}$, accept $H_1$.

The value of $D_{\text{crit}}$ can be found for given sample size $n$ and significance $\alpha$, such as $\chi^2$, can be found from the K–S table in Annexure.

## 3.7   Regression Analysis

In the section on graph plotting, we introduced an approach to determine or approximate a distribution or model for a given set of data. The limitation of this approach is it provides only an approximation of a considered distribution model. Traditional analysis requires a manual method to draw a line through the data points and thus has the potential of introducing an error of judgment for fitting the line to the given data points. Of course, software programs are now used to determine a fit. However, the issue remains—how well does an assumed model represent the data point?

The regression analysis enables analysts to have regression parameters that provide confidence in the relationship between the response variable and the explanatory variables ($x_i$) by estimating regression coefficient [7]. The regression coefficient ($\beta_i$) also signifies the impact of the respective explanatory variables to the overall response variable ($y$).

The least squares analysis (LSA) approach is commonly used for regression analysis. LSA is a vital tool that can help reduce uncertainty in PRA that is associated with data, model, and assumptions.

The regression equation performs regression analysis as follows:

$$y = \beta_0 + \beta_1 X_1 + \beta_2 X_2 \ldots \beta_n X_n + \varepsilon \tag{3.59}$$

The response variable $y$ is a dependent variable, and $X_i$ represents the random independent variable. $\varepsilon$ is the random error term. Note that this regression is being modeled considering stochastic characteristics of the problem. Therefore, when regression analysis is used for modeling a physics-based equation, apart from random error terms, the deterministic part of the equation, i.e., the physical parameters, will also be applicable.

Linear regression is performed to evaluate the equation of a line having a form $y = mx + c$ that represents the data and the $R^2$, also called the coefficient of correlation, which shows how well this line represents the data. The value of $R^2$ ranges from 0 to 1. If the distribution fits the observed data values perfectly, $R^2$ is 1. Normally, a value of above around 0.9 is considered acceptable. The complete procedure will be illustrated by an example of regression analysis.

*Example*: Consider some data available in form of values of dependent variable ($y$) corresponding to some independent variable ($x$) given in Table 3.2.

Our first aim is to fit a straight line in the given data. Straight line is given by:

$$y = mx + c \tag{3.60}$$

where $m$ is the slope and $c$ is the intercept. $m$ and $c$ are given by:

**Table 3.2** Data for regression analysis

| S. No. | $x$ | $y$ | $xy$ | $x^2$ | $y^2$ | $\hat{y}$ | $(\hat{y} - \hat{y})^2$ | $(y - \hat{y})^2$ |
|---|---|---|---|---|---|---|---|---|
| 1 | 2 | 3.49 | 6.98 | 4 | 12.19 | −43.22 | 19298.07 | 8502.29 |
| 2 | 25 | 8.70 | 217.53 | 625 | 75.71 | −12.90 | 11794.72 | 7568.72 |
| 3 | 34 | 14.20 | 482.70 | 1156 | 201.56 | −1.04 | 9358.92 | 6642.65 |
| 4 | 44 | 20.01 | 880.56 | 1936 | 400.50 | 12.14 | 6982.52 | 5728.50 |
| 5 | 54 | 26.19 | 1414.12 | 2916 | 685.78 | 25.32 | 4953.56 | 4831.93 |
| 6 | 62 | 32.77 | 2031.66 | 3844 | 1073.79 | 35.86 | 3580.53 | 3960.28 |
| 7 | 66 | 39.81 | 2627.72 | 4356 | 1585.15 | 41.13 | 2977.39 | 3123.20 |
| 8 | 86 | 47.39 | 4075.82 | 7396 | 2246.12 | 67.49 | 795.55 | 2333.49 |
| 9 | 86 | 55.59 | 4781.14 | 7396 | 3090.76 | 67.49 | 795.55 | 1608.40 |
| 10 | 105 | 64.53 | 6775.56 | 11,025 | 4164.01 | 92.54 | 10.01 | 971.59 |
| 11 | 113 | 74.34 | 8400.52 | 12,769 | 5526.57 | 103.08 | 54.47 | 456.19 |
| 12 | 117 | 85.22 | 9970.88 | 13,689 | 7262.65 | 108.35 | 160.09 | 109.80 |
| 13 | 120 | 97.43 | 11691.77 | 14,400 | 9492.89 | 112.31 | 275.78 | 3.00 |
| 14 | 122 | 111.34 | 13583.81 | 14,884 | 12397.21 | 114.94 | 370.27 | 244.71 |
| 15 | 140 | 127.51 | 17850.96 | 19,600 | 16258.00 | 138.67 | 1846.12 | 1011.71 |
| 16 | 150 | 146.80 | 22019.59 | 22,500 | 21549.43 | 151.85 | 3152.43 | 2610.98 |
| 17 | 178 | 170.72 | 30388.20 | 31,684 | 29145.39 | 188.75 | 8658.40 | 5628.11 |
| 18 | 191 | 202.23 | 38625.61 | 36,481 | 40896.29 | 205.88 | 12140.62 | 11348.39 |
| 19 | 203 | 248.49 | 50443.61 | 41,209 | 61747.61 | 221.70 | 15876.13 | 23345.14 |
| 20 | 250 | 337.22 | 84305.25 | 62,500 | 113717.99 | 283.65 | 35323.89 | 58332.63 |
| **Sum** | **2148** | **1914** | **310573.98** | **310,370** | **331529.60** | **1913.98** | 138405.01 | 148361.71 |

$$m = \frac{n \sum xy - \sum x \sum y}{n \sum x^2 - \left(\sum x\right)^2}$$

$$c = \frac{\sum y \sum x^2 - \sum x \sum xy}{n \sum x^2 - \left(\sum x\right)^2}$$

where $n$ is the number of data points available. For given data, we get

$$m = 1.318$$

$$c = -45.854$$

Hence, the straight line that can be fitted into the given data is:

$$\hat{y} = 1.318x - 45.854$$

Next, we check if the given data points closely follow the fitted line. This will give us the goodness of fit. We will find the $R^2$ value to determine the goodness of fit. $R^2$ is given as:

$$R^2 = \frac{\sum(\hat{y} - \bar{y})^2}{\sum(y - \bar{y})^2}$$

(3.61)

where $\bar{y}$ is the mean value of $y$.

We get for the given data:

$$R^2 = 0.9329$$

The value of $R^2$ indicates that the data closely follow the fitted straight line. The given data and the fitted straight line are shown in Fig. 3.10.

To ensure that the regression analysis effectively predicts the parameters, one has to look at certain heuristics that need to be considered while performing regression analysis, otherwise the results could be inadequate, incorrect, or even misleading. Hence, the following need to be checked:

(a) The correctness and applicability of the data point should be checked and should reflect the motivation and purpose of the analysis.

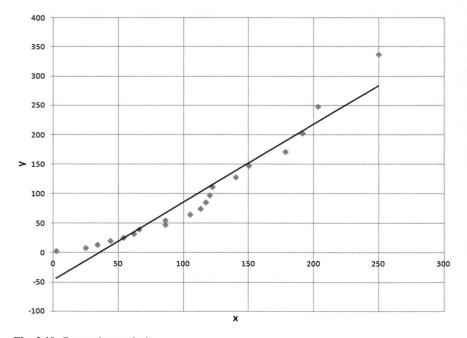

**Fig. 3.10** Regression analysis

(b) Only bench marked software is used for the analysis.
(c) In case the manual procedure is performed for regression analysis then the calculations are verified and validated.
(d) The final decision on the data is based on the estimates of regression coefficient, and this should be reflected in the confidence that need to be posed on our decision.

# References

1. M. Modarres, *Risk Analysis in Engineering: Techniques, Tools and Trends* (CRC Press, Boca Raton, FL, USA, 2006)
2. K. Agarwal, *Reliability Engineering* (Kluwer Academic Publishers, 1993)
3. A.B. Rao, *Quantitative Techniques in Business*, 2nd edn. (Jaico Publishing House, 2004)
4. K.B. Mishra, *Reliability Analysis and Prediction—A Methodology Oriented Treatment* (Elsevier, 1992)
5. M. Modarres, *Reliability and Risk* (Marcel Dekker, USA, 1993)
6. B. Epistein, Trucanted life test in the exponential case. Ann. Math. Stat. **25**, 255–264 (1954)
7. C. Lipson, N.J. Sheth, *Statistical Design and Analysis of Engineering Experiments* (McGraw-Hill Kogakusha Ltd., 1973)

# Chapter 4
# System Reliability Modeling

> *Success consists of going from failure to failure without loss of enthusiasm.*
>
> Winston Churchill, Goalcast

## 4.1 Background and Overview

In Chap. 3, we have seen the approach and methodology for component reliability modeling. However, the aim of any reliability analysis is to model a system and complete a plant in an integrated manner so that decisions can be made using insights available at the system level and lower down at the component level. Even at the component level, reliability modeling may require an understanding of various operational states and associated failure modes. The analysis may require assessment of the probability of a component in more than one state or condition. For example, analysis of an electronic module may require, apart from the module's operating state and failed state, the probability of detection and location. Reliability modeling may require modeling for contribution of repair to arrive at statement of system unavailability. Hence, characterization of the reliability of basic components or subsystems depends on the objective function and aim of the analysis.

Some well-established techniques are available for reliability modeling. Some of these techniques are routinely employed in risk analysis, while other techniques can be considered as advanced techniques that are still in development stages and are being used for specific applications, in a limited sense. Keeping in view this aspect, the scope of this chapter covers the following techniques:

- Reliability block diagram,
- Failure mode and effect analysis,
- Fault tree analysis,
- Event tree analysis,
- Markov analysis,

© Springer Nature Singapore Pte Ltd. 2018
P. V. Varde and M. G. Pecht, *Risk-Based Engineering*, Springer Series in Reliability Engineering, https://doi.org/10.1007/978-981-13-0090-5_4

- Dynamic fault tree,
- Dynamic event tree,
- Binary decision diagram.

Reliability block diagram (RBD) is the most commonly used modeling technique, where the system can be represented as series and parallel configurations. RBD representation often represents the natural components as a physical interrelationship and thereby provides effective understanding of the system. Failure mode and effect analysis or failure mode effect and criticality analysis are primarily employed for developing an intimate understanding of a relatively simple system where emphasis is on examining the individual component failure mode, its local effects, and effect on the system. This approach lacks integration capability, which can be achieved through fault tree and event tree methods. The static fault tree and event tree approaches are an integral part of PRA studies. Among the techniques listed above, the top six methods are conventional techniques. Markov modeling is basically used to model complex scenarios including dynamic modeling as part of a static fault tree and also as an independent dynamic modeling tool. For example, in traditional probabilistic safety assessment or quantitative risk assessment methods, the static fault tree and event tree methods are more commonly used. The modeling approaches for dynamic fault tree and dynamic event tree are the subjects of current research; however, there are increasing applications of these approaches, as part of PRA and analysis of specific cases to model real-time scenarios.

## 4.2   Reliability Block Diagram

In a reliability block diagram (RBD), each component of the system is represented by a block. The interrelationship between two components is represented by an arrow or simply a line. These diagrams are characterized by one input and one output node. Reliability block diagrams often represent the physical relation between the components of the system. This approach provides one of the simplest methodologies for system reliability modeling. Most of the commercial software available in the market provides the environment for system reliability modeling using reliability block diagrams.

Any given system can be modeled by considering three configurations: (a) series configuration, (b) parallel configuration, or (c) mixed or complex configurations (mixture of series and parallel or dependent configuration).

### 4.2.1   Series System

In a series system configuration, there is one input to the system and one output. The components are connected in series as per their functional requirements. Interruption in input and output connection results in system failure.

**Fig. 4.1** Valve system
comprising two valves
physically connected in series

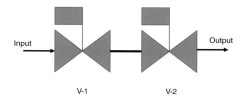

For example, consider a system consisting of two motor operated valves, V-1 and V-2, connected as shown in Fig. 4.1.

One of the key steps of system reliability modeling is defining the success or, conversely, failure criteria of the system. Since we are dealing with reliability modeling, we will define the success criteria for the system. Let us assume this valve system is part of a coolant injection system into the core. For the success of the system, the two valves should be in the open position to facilitate the injection of coolant. So the success criteria require that these two valves should be in open position. Conversely, failure of any one valve will result in failure of the above valve system.

Furthermore, we assume that the connected pipeline has negligible contribution to the valve system reliability, and then the reliability block diagram for the above system will be as shown in Fig. 4.2.

For the above system, the system reliability $R_s(t)$ for a given mission $t$ is given as:

$R_s(t)$ = reliability of valve V-1 and reliability of valve V-2 for a given mission time $t$. Let us assume the reliability of valve V-1 is denoted by $R_1(t)$ and valve V-2 is denoted by $R_2(t)$. Then the expression for $R_s(t)$ is given as:

$$R_s(t) = R_1(t) \times R_2(t) \tag{4.1}$$

If we assume that the exponential distribution adequately represents component reliability, and then the expression for system reliability can be given as:

$$R_s(t) = \exp(-\lambda_1 t) \times \exp(-\lambda_2 t) \tag{4.2}$$

$$R_s(t) = \exp[-(\lambda_1 + \lambda_2)]t \tag{4.3}$$

$$R_s(t) = \exp(-\lambda_s t) \tag{4.4}$$

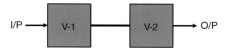

**Fig. 4.2** Reliability block diagram of valve system (success criteria: open status of V-1 and V-2 at any given time)

**Fig. 4.3** Representation of a
two-valve system (success
criteria is the closed status of
both or at least one of the two
valves)

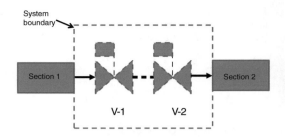

In the above equation, $\lambda_1$ and $\lambda_2$ are failure rate per unit time for components V-1
and V-2. As we see from this equation that

$$\lambda_1 + \lambda_2 = \lambda_s \tag{4.5}$$

where $\lambda_s$ is the system failure rate (failure per unit time). Hence from the above, we
conclude that for series system configuration the system failure is nothing but the
addition of the failure rate for individual components.

Let us change the failure criteria and see how the RBD looks different than the
series system as discussed in the previous case. Consider a system where the
function of the above valve system (comprising two valves V-1 and V-2 connected
physically in series) is to provide the isolation between two sections, sections 1 and
2, as shown in Fig. 4.3.

### 4.2.2   Parallel Configurations

Redundant components are modeled using the parallel configuration, which also
consists of one input and one output. In case the operating component fails, the next
component connected in parallel caters to the system's functional requirements. The
system keeps working until the last component fails.

The same example discussed in Sect. 4.2.1 will require representation in RBD as
a parallel system if the failure criterion is changed. Suppose the functional criteria
require isolation of the supply of water from input to output, then closure of any
valve, out of the two, can meet the function. Thus, there is redundancy in the
system. If one valve fails, then the other valve can isolate the supply. If the second
valve also fails, then the system fails.

The case discussed here deals with valve system modeling, and the objective is
to isolate section 1 from section 2 using the same valve system (note the system
boundary) discussed in the previous example. The RBD for the above system
considering the success criteria as isolation of section 1 from section 2 is shown in
Fig. 4.4.

Although the physical configuration of the system resembles series systems
configuration, the RBD for this case shows valve V-1 and valve V-2 in parallel

**Fig. 4.4** Reliability block
diagram for valve system
(success criteria: even one
valve results into system
success)

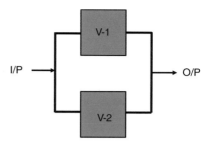

configuration. Considering the same notation for the previous valve systems, the
reliability equation for the above system can be written as follows:

$R_s(t)$ = Successful functioning of Valve 1 or Valve V2 or V1   and   V2

Hence,

$$R_s(t) = R_1(t) + R_2(t) + R_1(t)R_2(t) \tag{4.6}$$

$$R_s(t) = \exp(-\lambda_1 t) + \exp(-\lambda_2 t) + \exp(-\lambda_1 t)\exp(-\lambda_2 t) \tag{4.7}$$

As can be seen for giving system reliability, the reliability terms for the two
valves not only gets added up but also is further improved by the product term
comprising the reliability of valves V-1 and V-2.

From the above we can conclude that the RBD provides an elegant and simple
approach for system modeling where the components are represented by blocks and
the relationship or physical connection between two or more components is represented by a simple line. These diagrams are characterized by one input and one
output connection. Further details about the complex representation of the RBD
will be covered in the section on system representation.

### 4.2.3   Complex Configurations

Complex   system   modeling   deals   with   (a)   series-parallel   configurations,
(b) k-out-of-N (failure), (c) dependent failure, particularly, common cause failure,
and (d) dynamic aspects of system failure. These aspects will be discussed in other
sections on fault tree event tree analysis, etc.

## 4.3  Failure Mode and Effects Analysis

Failure mode and effect analysis (FMEA) can be defined as a systematic activity under an identified framework to analyze component failure and its effect at a local as well as system level. The goal of FMEA is to identify actions to eliminate or reduce the consequences of failure and create documentation in a systematic manner such that this information can be used to improve overall reliability and safety of the plant. FMEA is the simplest and most widely used approach for qualitative assessment/analysis; however, this approach finds a limited application where quantified estimates of component or subsystem reliability are used to identify and prioritize the safety significance of components. Traditionally, electrical and electronic system components are subjected to FMEA or a further developed version of FMEA referred to as failure mode effect and criticality analysis (FMECA).

The objective of this analysis is to investigate system failure by considering the effect of failure modes of individual components that comprise the system. FMEA is useful especially during the conceptual or design stages because it provides invaluable insights into the contributions of individual component failure modes to system failure. This component could be an electrical circuit, an electronic board, an individual component on a board, and in the case of an electrical or mechanical system, an electrical breaker, battery, diesel generator system, pump, valve, or piping segment. The FMEA is a bottom-up approach that is part of the safety system in a process or nuclear plant, or, for that matter, an aviation system. Putting it simply, FMEA consists of selecting a component, analyzing its effect on the system, and documenting the finding in a standard FMEA table.

There are many international standards [1–5] available that are employed for FMEA of processes, products, and even complex systems. However, keeping in view the requirements of IRBE, this book proposes minor deviations from the approach or table provided in the existing standards/literature. Certain attributes, such as components and system quality level, system and further component boundaries, identification of precursors, and quantified estimates of frequency of failures, should form part of FMEA/FMECA. This book recommends that FMEA/FMECA should form a prerequisite to the fault tree analysis such that important and applicable failure modes are systematically included and documented with their rationales (for inclusion) in the analysis. The major steps involved in FMEA are as follows:

1. Identify the system, its functional description, and system boundary.
2. Identify the components, their types, their major specifications, and the component boundary.
3. Develop the component coding using the existing system such that the code contains information not only about the component but also about the system it belongs to, generic category of the component, and applicable failure mode. Record the quality standards used to design and fabricate the system. Also, record the reference environment in which the system is supposed to operate.

4. Characterize the input/output to and from the interfacing or connecting systems.
5. Define the state of the system, such as operating or standby, for which the analysis is to be performed. Define the failure criteria of the system/component. There can be more than one failure criterion. For example, some failures may take the system to a safe state while in other cases it may go to an unsafe state, which has potential for serious consequences. Some failures may result in momentary interruptions. The definition of failure can be revisited if the analysis requires so. If require characterize the input/output to and from the interfacing or connecting systems.

- Develop a broad framework for screening the component to identify the candidate components for detailed FMEA. The failure criteria, a general feel of the likelihood of failure, and the failure mode could be used as the guiding factors for screening.
- Identify the applicable failure modes for each component. This step also looks for any potential or precursor to a failure.
- Investigate the effect of each failure at the local level and the system level.
- Identify and record the mechanism/provision that enables identification of the failure.
- Identify the condition and any control action that enables the system to recover or just a change of state that prevents the system from failing.
- Record the likelihood and severity of the failure. If the analysis is being performed at the qualitative level, then attributes such as improbable, low, medium, and high, or the classification given in a standard can be followed. It may be noted that, for assigning the quantified estimate, the value can be used either from a plant-specific source or from a generic source (from an available data source in the literature).
- Recommend a system modification or a change in the operational/maintenance procedure and action taken and remarks.

Figure 4.5 shows an FMEA format that has been designed to support safety and reliability analysis that leads to not only identification of the dominant component failure modes, but also provides a lead for detailed root cause analysis. This FMEA format is in line with the standard FMEA format available in the literature, with minor modifications that fulfill the safety and reliability requirements in complex systems.

In case the criticality of the failure must be identified, the same FMEA gets extended as FMECA. It may be noted that FMECA provides additional details about the criticality of the failure along with the information available in an FMEA. The ratings of the three parameters, i.e., "occurrence frequency," "severity," and "detection probability" are obtained from Tables 4.1, 4.2, and 4.3 [5].

One of the most widely used methods for criticality scoring, which is presented in this document, is an evaluation of the risk priority number (RPN). The RPN is a product of "occurrence or frequency of failure," "severity rating," and "detection probability." Mathematically it is written as:

**Failure Mode and Effect Analysis**

Company: BARC, Mumbai                                                                Analyst: P. V. Varde

System being analyzed: 24 VDC power supply; failure criteria: Module fail to deliver 24DC power supply within the given range

Operating environment: Temperature/Humidity/Radiation Level;   safety classification: Class 1A

Mode of operation: (On-line or Standby): On-line (with redundancy)

System boundary: DC power supply module, connector and cable and its termination;   Quality attributes of the system: Class 1A classification;

Objective/purpose of the FMEA/design/operation/safety analysis/other:reliability analysis (recurrent failure)

| Component code | Component type and broad specification | Component system boundary | Quality attribute | Failure modes &causes | Local effects | Effect on the system | Frequency of occurrence | Detection ratings | Precursor to the failure | Severity Rating | Corrective actions | Remark |
|---|---|---|---|---|---|---|---|---|---|---|---|---|
| 1 | 2 | 3 | 4 | 5 | 6 | 7 | 8 | 9 | 10 | 13 | 12 | 14 |
| RPS-PS-CAP | Aluminium electrolytic capacitor, 100µF, 24V | Capacitor, solder joint and cooling arrangement | MIL screened | Power supply failure | Increased harmonics | Power supply module erratic operations | Moderate (rating 4) | High (rating 3) | Increased in harmonics across the capacitor | Very low (rating 4) | Capacitors are replaced in advance, root cause analysis | - |
| | | | | | | | | | | | | |
| | | | | | | | | | | | | |
| | | | | | | | | | | | | |
| | | | | | | | | | | | | |

**Fig. 4.5**  Failure mode and effect analysis format for reliability and safety analysis in IRBE

**Table 4.1** Occurrence frequency rating [5]

| Rating | Estimate/expected frequency (probability): quantified value | Occurrence likelihood: qualitative attribute |
|---|---|---|
| 10 | 1 in 2 (0.5) | Very high (failure is almost inevitable) |
| 9 | 1 in 3 (0.33) | |
| 8 | 1 in 8 (0.125) | High (frequently repeated failure) |
| 7 | 1 in 20 (0.05) | |
| 6 | 1 in 80 ($1.25 \times 10^{-2}$) | Moderate (occasional failures) |
| 5 | 1 in 400 ($2.5 \times 10^{-3}$) | |
| 4 | 1 in 2000 ($5.0 \times 10^{-4}$) | |
| 3 | 1 in 15,000 ($6.7 \times 10^{-5}$) | Low (rare failures) |
| 2 | 1 in 150,000 ($6.7 \times 10^{-6}$) | |
| 1 | <1 in 150,000 ($6.7 \times 10^{-6}$) | Unlikely failures |

**Table 4.2** Severity rating [5]

| Rating | Effect | Criteria |
|---|---|---|
| 10 | Hazardous | Safety-related failure mode causing non-compliance with governmental regulations without warning |
| 9 | Serious | Safety-related failure mode causing non-compliance with governmental regulations with warning |
| 8 | Very high | Failure modes resulting in loss of primary system/vehicle/component performance/customer satisfaction |
| 7 | High | Failure modes resulting in reduced system/component/vehicle performance and customer satisfaction |
| 6 | Moderate | Failure mode resulting in loss of function by comfort/convenience/system/component |
| 5 | Low | Failure modes resulting in reduced level of performance of comfort/convenience/system/components |
| 4 | Very low | Failure modes resulting in loss of fit and finish, squeak, and rattle function |
| 3 | Minor | Failure modes resulting in partial loss of fit and finish, squeak, and rattle function |
| 2 | Very minor | Failure modes resulting in minor loss of fit and finish, squeak, and rattle function |
| 1 | None | No effect |

$$RPN = frequency * severity * detection \qquad (4.8)$$

From Tables 4.1, 4.2, and 4.3, respective ratings of frequency/probability, severity, and detection chances are obtained to arrive at the estimate of RPN. For example, from the FMEA table, we can obtain these ratings and estimate the RPN for capacitor failure as follows:

**Table 4.3** Detection rating based on design control criteria (or defense in depth) implementation [5]

| Rating | Detection | Criteria |
|---|---|---|
| 10 | Uncertain | Design control (defense in depth) provision will not and/or cannot detect a potential cause/mechanism and subsequent failure mode |
| 9 | Very unlikely | Very unlikely that design control (defense in depth) provision will detect a potential cause/mechanism and subsequent failure mode |
| 8 | Unlikely | Very unlikely that design control (defense in depth) provision will detect a potential cause/mechanism and subsequent failure mode |
| 7 | Very low | Very low chance that design control (defense in depth) provision will detect a potential cause/mechanism and subsequent failure mode |
| 6 | Low | Low chance that design control (defense in depth) provision will detect a potential cause/mechanism and subsequent failure mode |
| 5 | Moderate | Moderate chance that design control (defense in depth) provision will detect a potential cause/mechanism and subsequent failure mode |
| 4 | Moderately high | Moderate high chance that design control (defense in depth) provision will detect a potential cause/mechanism and subsequent failure mode |
| 3 | High | High chance that design control (defense in depth) provision will detect a potential cause/mechanism and subsequent failure mode |
| 2 | Very high | Very high chance that design control (defense in depth) provision will detect a potential cause/mechanism and subsequent failure mode |
| 1 | Very minor | Failure modes resulting in minor loss of fit and finish, squeak, and rattle function |
| 0 | Almost certain | The design control (defense in depth) provision will detect a potential cause/mechanism and subsequent failure mode |

$$RPN = 4 * 4 * 3 = 48$$

Using the value of RPN obtained from each component that makes the 24 VDC power supply module, we can prioritize the components. This information is vital to organize and optimize the design and, later during operational phase, the maintenance policy.

There are other extensions of FMEA and one of them is failure modes, mechanisms, and effects analysis (FMMEA). This approach is used in support of root cause analysis in general and physics-of-failure analysis in particular, where understanding the applicable mode as well as the failure mechanism is critical. In this approach, the major thrust is to perform experiments or accelerated tests to understand the failure mechanism and use the feedback to reduce or eliminate the effect of this failure mechanism, often called the dominating mechanism, by changing either the design, the materials, or the operational aspects of the equipment.

Even though the FMEA family provides one of the most widely used approaches for system reliability analysis, this approach has limitations because it operates on an individual component and analyzes its effect on the system. It does not take into account the effects of components, human actions, and software interactions on system failure. The results of the FMEA are qualitative in nature when it comes to giving an integrated statement of system reliability or availability. Hence, in an integrated risk-based approach, FMEA is used primarily to identify those components that have potential to take the system to "unsafe failure." Further, FMEA is used to analyze the modes and mechanisms that induce common cause failure in a deterministic sense, or for root cause analysis.

However, the fault tree and event tree analyses form the major elements of probabilistic risk assessment [6]. For complex redundancy modeling, where repair and test coverage, apart from operations, also form part of the analysis, Markov modeling is used when the analysis requires modeling for further in-depth analysis, dynamic modeling where component or system has more than one states and requires modeling for repair, or test aspects, etc., that cannot be captured by static fault tree analysis.

## 4.4 Fault Tree Analysis

The fault tree analysis approach was developed by Bell Labs in 1962 and can be considered to have changed the way risk and reliability modeling is performed. Thereby, fault tree analysis offers a powerful tool for system modeling. Fault tree modeling is a graphical and top-down approach that is deductive in nature. The definition of the top event is normally an undesired state of the system, viz. "failure of braking system" in an automobile or, in the context of nuclear plant, "failure of emergency core cooling system" are some typical examples. After having defined the top undesired event, one can think of the causes or failures in the subsystems that can lead to the top event. The lower level, in this case the second level of events, is called intermediate events, which are connected to the higher-level events, in this case the top event, by the logic gate. Say, for example, the brake failure can be caused by many hardware system failures, such as a brake shoe failure, hydraulic failure, brake link failure, hinge pin failure, or an event that could also be a human failure. Further, these intermittent events can be developed in a similar way when going from a top event to an intermediate event, until we reach the basic component failures. A basic component cannot be broken down further into subcomponents, or the analysis decides that the level of detail required for a subsystem is adequate such that it need not be broken down further into lower-level components which constitute the subsystem.

Now we understand that the "event" (top and intermediate and basic events) includes the connecting linkages and the logic gates (e.g., OR and AND) that form the minimum basic elements of a fault tree diagram. To better understand and appreciate this discussion, the fault tree analysis approach is discussed further in the following section:

## 4.4.1   Basic Entities in a Fault Tree

Table 4.4 shows the basic elements of a fault tree, their usual nomenclature, and the logical operation that the respective elements perform on the input and the output that it produces as a result of these logical operations.

**Table 4.4** Description of basic entities involved in making a fault tree diagram

| Name of entity | Graphical representation of entity | Logical operations | Remarks |
|---|---|---|---|
| Top event | | Contains a description of the undesired event, its coding, and associated results as quantified estimates | The top event is a starting point of the fault tree development and represents termination of the analysis by depicting the results of the analysis. It is a unique entity, and one fault tree can have only one top event |
| Linkages | ———————— | Depicts logical connections in the fault tree diagram | It is represented by a line that connects top or intermediate events as well as connects intermediate events with logic gates until the development leads to basic events |
| OR gate | $A+B$ $A$ $B$ | Performs OR operations, i.e., "summation" on the input, and produces the result of the summation | For example, the operations of the OR gate are depicted by using two inputs $A$ and $B$ and the result is $A + B$, i.e., addition of two qualitative inputs $A$ and $B$. The quantification process of the OR gate produces the addition of events $A$ and $B$ probability (the rare event approximation forms a basic assumption here, which will be discussed later) |

(continued)

**Table 4.4**  (continued)

| Name of entity | Graphical representation of entity | Logical operations | Remarks |
|---|---|---|---|
| AND gate | A•B <br><br> A          B | AND produces product operations on input | For example, the AND gate operations have been depicted by considering two inputs $A$ and $B$, and the result is shown by $A \cdot B$, i.e., $A$ multiplied by $B$. The use of quantified input for $A$ and $B$, i.e., probability of $A$ as $p_a$ probability of $B$ as $p_b$, will produce an output as $p_a \cdot p_b$ |
| Intermediate event | Input to higher gate <br><br> Lower gate output | No logical operations are performed in the intermediate events. It simply provides a description of the event, including the code of the intermediate event and associated estimates (during the quantification process of the fault tree) | These events provide information on intermediate or contributing state/ factors for the higher event of the system |
| Basic event | Input to logicgate | The basic event is a fundamental component, human action, etc., of the fault tree and forms input to logical gate as shown by the arrow | Represented by a circle, having only output but no input as it represents a basic component in the system which cannot or need not be broken down into further lower-level constituting parts. Normally, the components like electrical motor, pump, pipe segments, and electromagnetic relay, an indicating bulb, check valve, electrical breaker. are examples of basic event |
| Transfer-in gate | | Only a transfer event does not perform logical operations. It transfers output of one subfault tree to the | For better representation and space optimization, the fault trees are developed on different |

**Table 4.4** (continued)

| Name of entity | Graphical representation of entity | Logical operations | Remarks |
|---|---|---|---|
| | | next higher-level fault tree | pages and integrated using transfer gates |
| Transfer-out gate | | Does not perform a logical operation but transfers the output to the next subfault tree | It shows connection of the higher-level fault tree to a lower or subfault tree. This gate along with transfer-in gate shows the connection between higher- and lower-level fault trees |
| Undeveloped event | | Merely represents that a certain event, which could be a subsystem, component, or function, has not been developed due to a variety of reasons (e.g., no information on the event, analysts decided to just provide reference and not develop it due to no consequences, etc.) | Represented by a diamond symbol. It is depicted on the fault tree to indicate that the fault tree could have been more complete if this event had been developed or to indicate this dimension while making an assumption to this regard |
| House or external event | | Plays a role in logical operations which need to be set high or low depending on what assumption or external input the analyst wants to set for a given fault tree simulation | Can be used to see the impact on the result of the analysis when certain assumptions are validated by setting an input at high or low |
| Exclusive OR gate | A          C*          B | The logic operation results in an output only if exactly one of the inputs occurs. For example, if $C$ is that one event, then the output will occur only if the $C$ is high. On the contrary, if only $A$ or $B$ is high, then the output will not occur as it happens in the OR gate. $A$ is high, and then it requires the $C$ condition to be high, likewise with $B$ and $C$ high case | This is a special case of the OR gate. Use of OR gate is more common than the exclusive gate |

(continued)

**Table 4.4**   (continued)

| Name of entity | Graphical representation of entity | Logical operations | Remarks |
|---|---|---|---|
| Priority-AND gate | A•B•C  <br> A   C   B | The output occurs if the input occurs in sequence as set by the analyst. For example, the output $A.B.$ $C$ occurs if the events $A$, $B$, and $C$ follow the chronological sequence $A$ then $B$ and then $C$ | This gate is rarely used. However, in a dynamic fault tree, this gate can be used to model an evolving scenario over a period of time |
| INHIBIT gate |  | The output occurs if the condition event is satisfied for an input event | – |

The above sets of gates, events, and transfer symbols are adequate to develop fault trees. The most used entities for fault tree development are top event, OR gate, AND gate, basic event, intermediate event, undeveloped event, transfer-in, transfer-out, and linkages. As can be seen in the house events, exclusive OR, priority AND, and INHIBIT gates are used to impose logic conditions and are rarely used.

## 4.4.2   Fault Tree Analysis: General Considerations

It may be noted that fault tree development has more to do with intimate understanding of the system design and operation along with logical modeling as discussed in the above section. Fault tree analysis is performed at two levels, qualitative and quantitative. First, the qualitative fault is developed and checked whether it represents the as-built and as-operated system considering the top undesired state, i.e., the top undesired event. Boolean logic is used to analyze a fault tree. The qualitative analysis results in a list of minimal cut sets (MCS). A minimal cut set is a minimum number of basic event combinations that together (combination of components) or alone (single-order cut set) result in system failure or a top undesired event.

Before we discuss the analysis of the fault tree, let us have clarity on what we mean by cut set and minimal cut set (MCS) as this concept is vital for fault tree analysis. Let us look at the following definitions:

Cut set: The single or combination of components that result in system failure is called a cut set. When the cut set comprises a single failure event or component failure, it is called as ingle-order cut set; when the cut set comprises two components that lead to system failure, it is called a second-order cut set; when the cut set

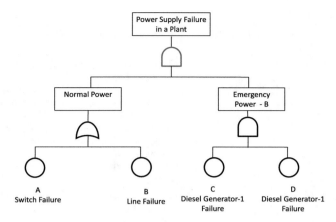

**Fig. 4.6** Illustration of a fault tree analysis for deriving the minimal cut sets

contains three events or components, it is called a third-order cut set and so on. Further, a cut set is called a minimal cut set when the minimum number of components in the set results in system failure. Further explanation and understanding of the minimal cut set are obtained through the following fault tree analysis illustration in Fig. 4.6.

Figure 4.6 shows that the power supply to the plant (e.g., an industrial or nuclear plant) fails when the normal power fails, and during this demand, the emergency power supply also fails. The normal power supply failure can be caused by either a switch failure or failures of the grid or line itself. During this demand, the alternative power source is emergency captive power supplied by two diesel generators. If the two diesel generators also do not start and meet the power supply requirements, then it will lead to the top event, i.e., "Power Supply Failure in Plant," represented by $T$.

By applying the usual logic operations for AND and OR gate, the equation for the power failure fault tree is as

$$T = (A + B).(C.D) \qquad (4.9)$$

or

$$T = A.C.D + B.C.D \qquad (4.10)$$

The above equation shows that the plant has adequate redundancy in terms of alternate sources of failure. Since each cut has three components, it can be seen that (a) no single-component failure results in power supply failure to the plant (i.e., there is no single-order minimal cut set in the equation), (b) even a combination of two component failures does not result in a power supply failure in the plant(i.e., there is no second-order minimal cut set in this equation), (c) only a combination of three component failures leads to power supply failure (i.e., there are two three-order cut sets that result in power supply failure in the plant, and they are *A.C. D* and *B.C.D*). Hence, it can be inferred that there is adequate redundancy at the component level, and if the reliability of each component is good, the system can be said to be robust and well designed. This is the value of the minimal cut sets—they provide information about the reliability and robustness of the design.

It may be noted that the above example depicts a qualitative analysis where it provides information about the mode in which the power supply to the plant can fail. However, the basic purpose of the fault tree analysis is to derive estimation of failure probability associated with the top event, in this case, power supply failure to the plant.

### 4.4.3 Quantitative Analysis

This section provides a simplified illustration for obtaining the quantified output of system failure frequency when the input data are available from generic sources as follows:

- Failure probability of power supply switch ($A$) = $4.2 \times 10^{-4}$/demand;
- Line or grid failure frequency—Obtained from plant-specific data ($B$) = $2.0 \times 10^{-3}$ demand;
- Diesel generator failure probability ($C$ & $D$) = $4.0 \times 10^{-3}$/demand.

By using the following equation:

$$T = (A+B).(C.D.) \tag{4.11}$$

Therefore,

$$T = \left(4.2 \times 10^{-4} + 2.0 \times 10^{-3}\right).\left(4.0 \times 10^{-3}.4.0 \times 10^{-3}\right)$$
$$T = 2.42 \times 10^{-3}.1.6 \times 10^{-5}$$
$$T = -3.87 \times 10^{-8} \text{ per demand}$$

The above calculations are only for illustration of a process for quantitative analysis of the fault tree. For a real-time scenario, common cause failure should be considered for redundant systems. Also, the failure data for an operational system such as power supply line failure frequency should be considered in place of failure probability as used in this example. Hence, this result has been obtained considering only (a) independent failure of components, for example, diesel generators are redundant components and one common cause is adequate to knock out both the units; hence, the above estimates are optimistic, and (b) the normal grid supplies continuous supply and its failure rates strictly should be defined as failures per unit of time, as failure/year.

Let us consider a simplified example of an emergency core cooling system (ECCS) in a nuclear plant to demonstrate the complete and detailed analysis of the fault tree.

Example 4.1: A demand is generated automatically to start and operate the ECCS in a nuclear plant consequent to confirmation of a loss of coolant accident (LOCA). The control signal gives a command to open the injection valves and start the emergency cooling pumps. The pumps take suction from the water accumulator having adequate inventory by injecting water into the reactor core and mitigate the consequence of the LOCA event. The ECCS and its interface with the primary coolant system are shown in Fig. 4.7. The ECCS operation is assumed to be successful if the system injects water and caters to the core cooling requirements for the given LOCA scenario.

As can be seen there are two redundant trains to inject water from its respective accumulators. Each train comprises an accumulator, a pump, a rupture disk, and two redundant injection valves. The two trains, upon sensing LOCA, start simultaneously and inject water into the primary circuit at cold lag.

The fault tree for this system is shown in Fig. 4.8.

Note that in Chap. 1, we manually estimated the failure probability without any considerations for systematic minimal cut-set analysis. Hence, the results were approximate. In this example, we will perform a qualitative fault tree analysis whereby we will develop the equation for the top event $T$. Later in step 2 a quantification of the fault tree will be performed.

$$T = I + C$$

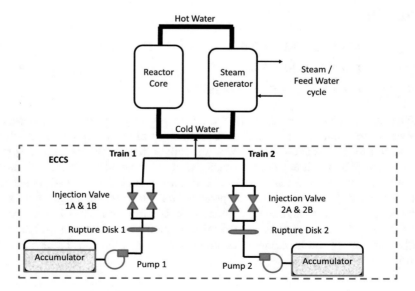

**Fig. 4.7** Simplified emergency core cooling system for a nuclear plant

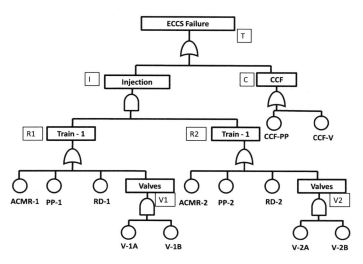

**Fig. 4.8** Fault tree for simplified emergency core cooling system in a nuclear plant

$$I = R1.R2 \quad \& \quad C = CCFPP.CCFV$$

$$R1 = ACMR1 + PP1 + RD1 + V1$$
$$R1 = ACMR1 + PP1 + RD1 + (V1A.V1B)$$
$$R1 = ACMR1 + PP1 + RD1 + V1A.V1B$$

Similarly, we derive an equation for R2 as follows:

$$R2 = ACMR2 + PP2 + RD2 + V2A.V2B$$

$$I = (ACMR1 + PP1 + RD1 + V1A.V1B).(ACMR2 + PP2 + RD2 + V2A.V2B)$$

$$T = (ACMR1 + PP1 + RD1 + V1A.V1B).(ACMR2 + PP2 + RD2 + V2A.V2B) + CCFPP + CCFV$$

$$T = ACMR1.ACMR2 + PP1.ACMR2 + RD1.ACMR2 + V1A.V1B.ACMR2 + ACMR1.PP2$$
$$+ PP1.PP2 + RD1.PP2.V1A.V1B.PP2 + ACRM1.RD2 + PP1.RD2 + RD1.RD2 + V1A.V1B.RD2$$
$$+ ACMR1.V2A.V2B + PP1.V2A.V2B + RD1.V2A.V2B + V1A.V1B.V2A.V2B + CCFPP + CCFV$$

As can be seen, this procedure provides the list of cut sets that cannot be reduced by applying the Boolean logic, viz. $A + A = A$ or $A.A = A$ or any other laws of Boolean reduction. Hence, this equation can be said to represent the list of minimal cut sets. What does it mean? This equation is final and can be used for estimating the top event failure probability for event $T$.

For estimating the top event probability, we need the failure probability of basic components. Table 4.5 lists the failure probability of the ECCS components. It may

**Table 4.5** Basic component failure probability

| S. No. | Component/event description | Generic component code | Failure probability (per demand) |
|--------|---------------------------|------------------------|----------------------------------|
| 1 | ECCS accumulator | ACMR | $1 \times 10^{-6}$ |
| 2 | Injection pump | PP | $3 \times 10^{-3}$ |
| 3 | Injection valves | V | $2.8 \times 10^{-3}$ |
| 4 | Rupture disk | RD | $2.2 \times 10^{-3}$ |
| 5 | Common cause failure—pumps | CCFPP | $3 \times 10^{-3}$ |
| 6 | Common cause failure—valves | CCFV | $2.8 \times 10^{-4}$ |

be noted that to simplify the calculation, the failure probability of the components in Trains 1 and 2 has been assumed to be the same.

Certain observations/assumptions in respect of the above analysis are as follows:

- A beta factor of 0.1 has been used to determine the common cause failures (CCFs) of the pumps and valves. This means the CCF failure probability is 10% of the independent component failure probability.
- The human contribution to the ECCS has not been considered, assuming that no human intervention is required for injection of the ECCS.
- It has been assumed that the ECCS will operate successfully after it got activated on demand. No considerations are given for successful continued operation of the ECCS pumps for a given mission time.
- Certain components such as piping and instrumentation have been assumed to remain in a safe state for the given mission.

These aspects, such as CCF, mission-related modeling, and different categories of models, such as mission models, standby-tested, and repairable models, will be discussed in Chap. 6 on PSA.

## 4.5   Event Tree Analysis

Event tree methodology, like fault tree analysis, is a graphical approach; however, unlike fault tree analysis this approach is inductive in nature. This methodology provides an elegant representation of a plant's response to an initiating event. Engineering systems in general and process systems and nuclear plants in particular have multiple levels of safety provisions that are activated automatically in response to a predetermined set of initiating events and take the plant to a safe state. Hence, event tree analysis provides a mechanism to propagate the postulated scenario by inducing plant safety function response to a given initiating event. The probabilistic risk assessment (PRA) procedure involves accident sequence analysis which is complex and requires computational algorithms to analysis often involving large number of event trees. The result of the event tree analysis is a list of accident sequences which are categorized as safe or unsafe or more than one category that provide the statement of safety of the plant.

The first step in constructing an event tree deals with identifying the initiating event and the applicable set of safety functions required to mitigate the undesired consequences. The safety function might include a safety system or a safety support system, a human action as a safety function, or even a recovery mechanism. The safety function forms the header elements in the event tree, which are arranged in a chronological order (i.e., the order in which they are designed to be activated or the logical order in which they are considered to depict the safety philosophy of the plant). The initiating event is inducted at the left-hand side of the event tree diagram.

Consider the event tree shown for loss of off-site power (LOOP) in a nuclear plant. LOOP is an anticipated operational occurrence in nuclear plants. The event is expected to occur with a frequency of $\sim 1.0$/year in the life of a nuclear plant. Unlike LOCA, which occurs with a frequency of $\sim 10^{-4}$/year, has the potential for high consequences and is not expected to occur in the lifetime of the plant, and LOOP is considered a high-frequency but low-consequence event. Of course, design safety provisions, apart from reactor trip and automatic starting of emergency power supply systems, exist in the plant to cater to emergency power supply for essential loads in the plant and to ensure that the likelihood of station blackout is reduced or avoided.

Let us look at the simplified event tree, meant to illustrate the event tree approach as part of probabilistic risk modeling, in Fig. 4.9. For the second step in constructing an event tree, we will draw the event tree diagram for a LOOP event.

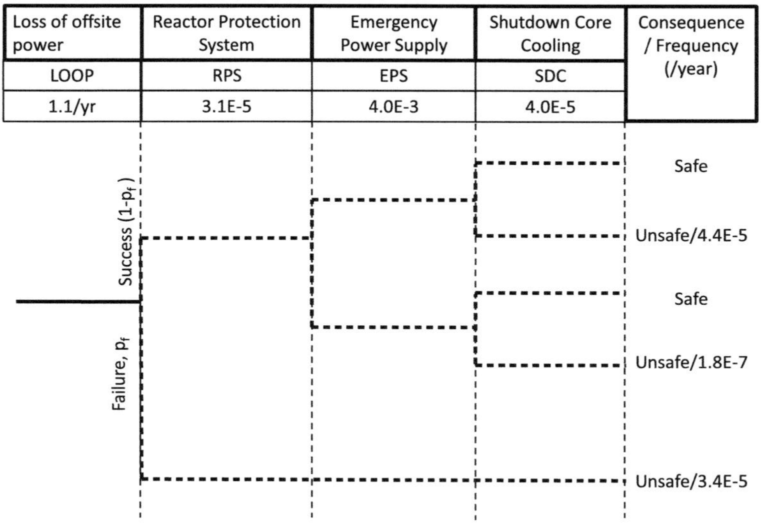

| Loss of offsite power | Reactor Protection System | Emergency Power Supply | Shutdown Core Cooling | Consequence / Frequency (/year) |
|---|---|---|---|---|
| LOOP | RPS | EPS | SDC | |
| 1.1/yr | 3.1E-5 | 4.0E-3 | 4.0E-5 | |

Success $(1-p_f)$

Failure, $p_f$

Safe

Unsafe/4.4E-5

Safe

Unsafe/1.8E-7

Unsafe/3.4E-5

**Fig. 4.9** Simplified event tree for loss of off-site power in a nuclear plant

We can assume that this procedure is being developed as part of risk assessment in a nuclear plant. In the left-hand column, the event tree shows the initiating LOOP event and associated frequency of its occurrence. The other headers show the safety function. This event tree includes the reactor protection system, the emergency power supply system, and the shutdown cooling system. The right-most column shows the consequences and of the LOOP event and its frequency. Normally, the last column also shows the accident sequence and its number; however, to keep the event tree simple this information has not been shown.

The "failure" and "success" of the safety function are shown on the event tree. The successful functioning of the safety function is depicted by an upward movement from the node, whereas the failure of the function is shown by a downward movement from the node. Given the failure probability $p_f$, the probability of success is shown as $(1 - p_f)$. Two sequences lead to the "safe" state as the safety functions; that is, the reactor shutdown and shutdown core cooling are achieved successfully. However, there are three accident sequences which lead to "unsafe" states. Accordingly, the three accident sequences (AS) are as follows:

AS1 = LOOP * SDC(Following the LOOP event, the SDC failson demand)
AS2 = LOOP * EPS * SDC(Following LOOP, EPS as well as SDC failson demand)
AS3 = LOOP * RPS(Following LOOP, the reactor protection/shutdown system fails on demand);

The next step is quantification of the event tree events. The failure frequency of the initiating events is estimated from the plant-specific data or generic data obtained from other plants. The estimation procedure for initiating event frequency is not discussed here. Further, estimation of failure probability of safety system is carried out as discussed in the fault tree section. There are many techniques to estimate the failure probability of the safety system; however, the fault tree analysis is a commonly used approach. Once the failure frequency of the initiating event and failure probability of the safety system is obtained, these values are included in the event tree diagram as shown in Fig. 4.9. The qualitative equation of accident sequences 1, 2, and 3 obtained earlier is used for obtaining the estimates for AS1, AS2, and AS3, by substituting the values of LOOP frequency, and RPS, EPS probabilities shown in the figure. Accordingly, the estimation is as follows:

$$AS1 = (1.1/\text{year}) * (4.0 \times 10^{-5}) = 4.4 \times 10^{-5}/\text{year}$$

$$AS2 = (1.1/\text{year}) * (4.0 \times 10^{-3}) * (4.0 \times 10^{-5}) = 1.8 \times 10^{-7}/\text{year}$$

$$AS2 = (1.1/\text{year}) * (3.1 \times 10^{-5}) = 3.4 \times 10^{-5}/\text{year}$$

As can be seen, accident sequences 1 and 3 are approximately showing the same contribution for the LOOP scenario. If we assume, considering the safety case of a nuclear plant, that these accident sequences result in reactor core damage (CD), then the core damage contribution from the LOOP can be given as:

Core Damage Frequency (LOOP) = AS1 + AS2 + AS3

$$= 4.4 \times 10^{-5} + 1.8 \times 10^{-7} \times 3.4 \times 10^{-5}$$

$$= 7.81 \times 10^{-5}/\text{year}$$

The contribution from LOOP for the total core damage frequency is $7.81 \times 10^{-5}$/year. It may be noted that this estimation has been performed to illustrate the event tree methodology, where the input data have been assumed. Further, details and related aspects of event tree and fault tree analysis (e.g., uncertainty analysis, sensitivity analysis) will be discussed in Chap. 6 on PRA and subsequent chapters on the subject.

It is important to note that in PRA modeling for nuclear plants, the fault tree and event trees are more elaborate and complex. The analysis for these complex scenarios is normally performed using computerized code because it provides the required automated environment not only for cut-set analysis but also performs many operations to estimate other parameters such as importance, uncertainty and sensitivity analyses.

## 4.6 Markov Model

Markov models are used to model the reliability of complex scenarios, for instance, where even single components can have multiple states. This approach is also called a state-space approach. The major features of the Markov approach are:

- It deals with two random variables, the state of the system ($P$) and the time of transition from one state to another, $t$. The reliability models that use the Markov process deal with discrete states and continuous transition times.
- The Markov model assumes that the probability of transition from one state to another depends on the current state of the system and not on the previous states. This means the transition process from one state to another state is based on memory-less function.
- The transition parameters (e.g., the repair or failure rates) are assumed to be constant. These earlier features make the exponential distribution an applicable distribution of the Markov model.
- The end state of the system is known as the absorbing state. Generally, the failed state of the system is referred to as the absorbing state of the system. The probability of reaching the absorbing state for $t = \infty$ is "1."

## 4.6.1   Markov Model for a Single-Component Non-repairable System

Figure 4.10 shows the general features of a Markov model for a single non-repairable component system having two states as follows:

State "1": Component is operating.
State "0": Component is failed.

Let $P_1(t)$ be the probability of the component in state "1" and $P_0(t)$ is the probability of the component in system state "0." The probability of the system being in state "1" at time $t + \Delta t$ is given by

$$P_1(t + \Delta t) = (1 - \lambda \Delta t) P_1(t) \tag{4.12}$$

Similarly, there are two scenarios for the component being in state "0." Either the component is in transit at time $\Delta t$ from state $P_0$ to $P_1$ or the component remains in state $P_0$ at time $t$. Hence, the probability of the component being in state "0" at time $t + \Delta t$ is given by

$$P_0(t + \Delta t) = \lambda \Delta t P_1(t) + P_0(t) \tag{4.13}$$

From Eq. (4.12), we have

$$
\begin{aligned}
P_1(t + \Delta t) - P_1(t) &= -\lambda \Delta t P_1(t) \\
\frac{P_1(t + \Delta t) - P_1(t)}{\Delta t} &= -\lambda P_1(t) \\
\frac{dP_1(t)}{dt} &= -\lambda P_1(t)
\end{aligned}
\tag{4.14}
$$

The above differential equation with dummy variable $x$ has the following solution:

$$
\begin{aligned}
\ln P_1(t) &= -\int_0^t \lambda dx + C \\
P_1(t) &= \exp\left[ -\int_0^t \lambda dx + C \right] \\
P_1(t) &= B \exp\left[ -\int_0^t \lambda dx \right]
\end{aligned}
\tag{4.15}
$$

**Fig. 4.10** Markov model for single-component, two-state system

It may be noted that the term $P_1(t)$ gives probability of being in the operating state for time $t$. Thus, this term defines the reliability of the single-component non-repairable system. Further, at time $t = 0$, the reliability of the system = 1, i.e., $P_1(t) = 1$. Hence, the constant $B = 1$.

Therefore,

$$R(t) = P_1(t) = \exp\left[-\int_0^t \lambda \, dx\right] \tag{4.16}$$

which is the general statement of reliability for a non-repairable component.

### 4.6.2   Markov Model for a Repairable System

Consider a single-component system with a constant failure rate $\lambda$ per unit of time and repair rate $\mu$ per unit of time. It is assumed that after repair the component is returned to an operating state. The component has two states: the operating state is denoted as state "1" while the failed state is denoted as state "0." The Markov graph representation for this system is shown in Fig. 4.11.

If the system is in state "1," then the transition probability that the component will fail between time $t + \Delta t$ is given as $\lambda \Delta t$. Similarly, the transitional probability from state "0" to state "1" between time $t + \Delta t$ is given as $\mu \Delta t$.

From the above formulation, the probability of the component in state "1" can be given as (a) either the component remains in state "1" at any time between $t + \Delta t$, given as $P_1(t)(1 - \lambda \Delta T)$ or (b) the component is repaired and the component is in transit from state "0" to state "1" with a probability $\mu \Delta t$. Accordingly, the equation for the component being in state 1 is given as:

$$P_1(t + \Delta t) = (1 - \lambda \Delta t)P_1(t) + \mu \Delta t P_0(t) \tag{4.17}$$

Similarly, the equation for the component in state "0" can be written by considering two components of failure and repair as follows: either the component remains in state "0" between time $t + \Delta t$ with a probability $(1 - \mu \Delta t)P_0(t)$ or the

**Fig. 4.11** Markov model for one-component, two-state system

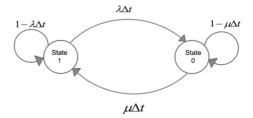

component is in transit from state "1" to state "0" with a probability $\mu\Delta t$. Hence, the equation for the component being in state "0" between time $t + \Delta t$ is given as:

$$P_0(t + \Delta t) = (1 - \mu\Delta t)P_0(t) + \lambda\Delta t P_1(t) \tag{4.18}$$

Rearranging Eq. (4.17) gives:

$$P_1(t + \Delta t) - P1(t) = -\lambda\Delta t P_1(t) + \mu\Delta t P_0(t)$$

or

$$\frac{P_1(t + \Delta t) - P_1(t)}{\Delta t} = -\lambda P_1(t) + \mu P_0(t)$$

Similarly, rearranging Eq. (4.18) as follows:

$$\frac{P_0(t + \Delta t) - P_0(t)}{\Delta t} = -\mu P_0(t) + \lambda P_0(t)$$

The above equation can be written in the form of a differential equation as follows:

$$\frac{dP_1(t)}{dt} = -\lambda P_1(t) + \mu P_0(t) \tag{4.19}$$

$$\frac{dP_0(t)}{dt} = -\mu P_0(t) + \lambda P_1(t) \tag{4.20}$$

The above equation can be solved using the Laplace transform method. The initial condition of the problem is: at time $t = 0$, the probability that the component will be in the operating state is "1." Hence, $P_1(0) = 1$ and $P_0(0) = 0$, i.e., the probability that the component will be in state 0 at time $t = 0$ is 0.

The Laplace transform of the first-order differential equation is given as:

$$\mathcal{L}\left\{\frac{dg}{dt}\right\} = sG(s) - g(0) \tag{4.21}$$

The term $g(0)$ is the value of the function at time $t = 0$, derived from the initial condition.

Taking the Laplace transform of Eqs. (4.19) and (4.20), we have

$$sP_1(s) - P_1(0) = -\lambda P_1(s) + \mu P_0(s)$$

Since at time $t = 0$, $P_1(0) = 0$, we have

$$sP_1(s) - 1 = -\lambda P_1(s) + \mu P_0(s)$$

or

$$sP_1(s) + \lambda P_1(s) - \mu P_0(s) = 1$$

$$(s + \lambda)P_1(s) - \mu P_0(s) = 1 \tag{4.22}$$

Similarly, considering equation

$$\frac{dP_0(t)}{dt} = -\mu P_0(t) + \lambda P_1(t) \tag{4.23}$$

Taking Laplace transform as follows:

$$sP_0(s) - P_0(0) = -\mu P_0(s) + \lambda P_1(s)$$

$$sP_0(s) = -\mu P_0(s) + \lambda P_1(s)$$

or

$$-\lambda P_1(s) + (s + \mu)P_0(s) = 0 \tag{4.24}$$

Solving Eqs. (4.22) and (4.24) for two-variable $P_1(s)$ and $P_0(s)$ as:

$$P_1(s) = \frac{1 + \mu P_0(s)}{(s + \lambda)}$$

Putting the value of $P_1(s)$ into Eq. (4.8)

$$-\lambda \left\{ \frac{1 + \mu P_0(s)}{(s + \lambda)} \right\} + (s + \mu)P_0(s) = 0$$

Solving for $P_0(s)$, we get

$$P_0(s) = \frac{\lambda}{s(s + \lambda + \mu)}$$

and

$$P_1(s) = \frac{s + \mu}{s(s + \lambda + \mu)}$$

Take the inverse Laplace transform as follows:
Denote $\lambda = x$, $\mu = y$, and $\lambda + \mu = z$.
Consider the following equation for $P_1(s)$.

$$\frac{s+\mu}{s(s+\lambda+\mu)} = \frac{s+y}{s(s+z)}$$

Finding the partial fraction as

$$\frac{s+y}{s(s+z)} = \frac{A}{s} + \frac{B}{s+z} = \frac{A(s+z)+Bs}{s(s+z)}$$

$$s+y = A(s+z) + Bs$$

Let $s = 0$

$$A = \frac{y}{z}$$

Again, putting $s = -y$

$$B = \frac{A(z-y)}{y}$$

But

$$A = \frac{y}{z}$$

We have

$$B = \frac{(z-y)}{z}$$

$$P_1(s) = \frac{s+\mu}{s(s+\lambda+\mu)} = \frac{y}{zs} + \frac{z-y}{z}\frac{1}{s+z}$$

Taking the inverse Laplace transform as follows:

$$\mathcal{L}^{-1}\left[\frac{s+\mu}{s(s+\lambda+\mu)}\right] = \frac{y}{z}\mathcal{L}^{-1}\left[\frac{1}{s}\right] + \frac{z-y}{z}\mathcal{L}^{-1}\left[\frac{1}{s+z}\right]$$

Substituting back the value of $y = \mu$ and $z = \lambda + \mu$, we have

$$\mathcal{L}^{-1}\left[\frac{s+\mu}{s(s+\lambda+\mu)}\right] = \frac{\mu}{\lambda+\mu}\mathcal{L}^{-1}\left[\frac{1}{s}\right] + \frac{\lambda}{\lambda+\mu}\mathcal{L}^{-1}\left[\frac{1}{s+\lambda+\mu}\right]$$

We know that

$$\mathcal{L}^{-1}\left[\frac{1}{s}\right] = 1 \quad \text{and} \quad \mathcal{L}^{-1}\left[\frac{1}{s-a}\right] = \exp(at)$$

$$P_1(s) = \left[\frac{s+\mu}{s(s+\lambda+\mu)}\right] = \frac{\mu}{\lambda+\mu} + \frac{\lambda}{\lambda+\mu}\exp[-(\lambda+\mu)t]$$

If $A(t)$ denotes the availability of the component, then $P_1(t) = A(t)$, which implies

$$P_1(t) = A(t) = \frac{\mu}{\lambda+\mu} + \frac{\lambda}{\lambda+\mu}\exp[-(\lambda+\mu)t] \qquad (4.25)$$

Figure 4.12 shows the availability graph of a repairable system. The figure also shows steady-state availability when $t$ becomes larger. This equation gives dynamic availability, that is, the value of instantaneous availability for various times $t$. When $t$ approaches infinity then the second part in the above equation becomes 0. Hence, steady-state availability is

$$A(\infty) = \frac{\mu}{\lambda+\mu} \qquad (4.26)$$

This can also be written as

$$A(\infty) = \frac{1/\text{MTTR}}{1/\text{MTTF}+1/\text{MTTR}} \qquad (4.27)$$

$$A(\infty) = \frac{\text{MTTF}}{\text{MTTF}+\text{MTTR}} \qquad (4.28)$$

MTTF is the uptime, and MTTR is the downtime of the component in a repairable system; therefore, the steady-state availability is

**Fig. 4.12**  Availability graph of repairable system showing steady-state availability when $t$ becomes larger

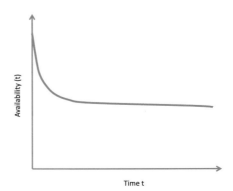

$$A(\infty) = \frac{\text{Uptime}}{\text{Downtime} + \text{Uptime}} \qquad (4.29)$$

The probability of state $P_0(t)$ can be derived from the existing Laplace equation for $P_0(s)$ as follows:

$$P_0(s) = \frac{\lambda}{s(s + \lambda + \mu)}$$

Denote $x = \lambda$, $y = \mu$, and $z = \lambda + \mu$.
Use the partial fraction to simplify the equation

$$\frac{x}{s(s+z)} = \frac{A}{s} + \frac{B}{(s+z)} = \frac{A(s+z) + Bs}{s(s+z)}$$
$$x = A(s+z) + Bs$$

Put $s = 0$, we get

$$A = \frac{x}{z}$$

Again, put $s = -z$, we get

$$B = -\frac{x}{z}$$

$$\frac{x}{s(s+z)} = \frac{A}{s} + \frac{B}{(s+z)} = \frac{x}{z}\frac{1}{s} + \left(-\frac{x}{z}\right)\frac{1}{s+z}$$

Taking the inverse Laplace transform to arrive at the solution of the final equation for "0" state

$$\mathcal{L}^{-1}\left[\frac{x}{s(s+z)}\right] = \frac{x}{z}\mathcal{L}^{-1}\left[\frac{1}{s}\right] - \frac{x}{z}\mathcal{L}^{-1}\left[\frac{1}{(s+z)}\right]$$
$$\mathcal{L}^{-1}\left[\frac{x}{s(s+z)}\right] = \frac{x}{z}1 - \frac{x}{z}\exp[-zt]$$

Substituting for value of $x$ and $z$ in terms of $\mu$ and $\lambda$, we have

$$P_0(t) = \frac{\lambda}{\lambda + \mu} - \frac{\lambda}{\lambda + \mu}\exp[-(\lambda + \mu)t] \qquad (4.30)$$

This equation for $P_0(t)$ defines the system unavailability, i.e., unavailability as follows:

$$U(t) = 1 - A(t) = P_0(t). \tag{4.31}$$

## 4.7 Advanced Approaches in System Analysis: An Overview

Even though the fault tree analysis and event tree analysis approaches are central to risk modeling, there is a growing feeling that these two approaches are static in nature and hence need to consider dynamic modeling. To this effect, considerable work has already been published on the dynamic aspects of system modeling, mainly employing the dynamic fault tree and dynamic event tree [7–12]. However, the dynamic fault tree and dynamic event tree approaches are basically computationally resource intensive, and the present state of the art is such that these dynamic approaches have been used at the developmental level to investigate some specific engineering problems that needed dynamic treatment. While the results have been encouraging, the static fault tree and event tree approaches form the mainstay in PRA modeling.

Similarly, developmental work has also been reported on an application of the binary decision diagram approach to fault tree analysis. Even though the binary decision diagram provides an improved framework for fault tree analysis, it is computationally intensive and requires the development of heuristics to address the issue of ordering and sequencing.

Keeping the above in view, the following section introduces the dynamic fault tree, dynamic event tree, and binary decision diagram approaches for the sake of completeness of the system modeling.

### 4.7.1 Dynamic Fault Tree

The dynamic fault tree is an extension of the static fault tree methodology with time requirements. Cepin and Mavko [13] have shown that dynamic fault tree evaluation contributes to improved nuclear plant safety by focusing on improvement tests and maintenance activities. Consider that analysts are required to evaluate the system unavailability over a period of time when the intermittent epoch imposes certain plant conditions, such as recovering a failed component, human error to recover a failed system, or human intervention that leads to failure of an operating system. Apart from this, the modeling of component unavailability as a function of time also requires system modeling in the temporal domain. Figure 4.13 shows one of the characteristics of dynamic modeling requirements. Consider that, as part of dynamic PRA, it was required that a safety system unavailability (i.e., the ECCS) is to be evaluated for certain component status changes, human intervention, or actuation of a recovery mechanism. It is given that after the initial epoch, i.e., at time $t = 0$ $(t_0)$, to any time $t_n$ $(n = 1, 2, 3, \ldots)$ these changed configurations over a period of time need to be evaluated. A problem of this nature cannot be addressed by the static fault tree approach and requires dynamic fault tree modeling because the system unavailability needs to be evaluated at a given epoch of time as shown in Fig. 4.13. For better appreciation, consider an example scenario that after LOCA initiation at time $t = 0$, the ECCS comes on demand as per the initial configuration and injects coolant from accumulator-1; however, the accumulator-1 inventory is exhausted at the epoch of time $t_1$, and accumulator-2 is pressed into operation at $t_1$; the change in system unavailability is reflected at $t_1$ and remains constant up to $t_2$. As per system design, accumulator-2 lasts up to time $t_3$; when the suppression pool water injection is initiated (either manually or automatically), the change in unavailability is reflected in the figure. This example illustrates the characteristics of dynamic modeling in the temporal domain.

The dynamic fault tree employs all the symbols of the static fault tree methodology. In fact, two symbols—house event and PAND gate—are used in the static as well as the dynamic fault tree. For example, the house event in the static fault tree is used to model switching in/out of certain configurations or success criteria to avoid the necessity of making more than one fault tree for the changed configurations or assumption. It is like performing a sensitivity analysis by considering an assumption. For example, the house event will allow the evaluation of 2-out-of-3 and 3-out-of-3 failure criteria in the same fault tree. While in the static procedure the switching is done manually to evaluate these criteria, in the dynamic fault tree it is achieved computationally and built into the simulation model.

There are a few additional operators/entities/gates we will refer to, hereafter, as dynamic gates, used in the dynamic fault tree, as shown in Table 4.6. The input characteristics as well as operations performed by each dynamic gate or entity also have been explained in the table. These dynamic gates are central to the dynamic fault tree modeling. However, the rather bigger challenge is how to evaluate the fault tree as the process is computationally very intensive. In fact, we are able to attempt the dynamic fault tree evaluation due to available computational tools and methods. However, the full potential of the dynamic fault tree is not realized, on the other hand, because further capacity build-up is required in terms of computational power.

The dynamic gates, viz. spare gate (SPAR), sequence enforcing (SEQ), functional dependency (FDEP), and priority AND (PAND), house event (H) as switching elements provide the necessary fundamental building blocks along with static OR, AND gates [14, 15].

**Fig. 4.13** Illustration of output of a dynamic fault tree

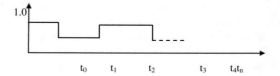

**Table 4.6** Dynamic gates

| Dynamic entity/ gate | Graphical representation of entity | Description of gate operation on input for generating an output and *failure criteria* | Significance/remarks |
|---|---|---|---|
| Priority AND gate (PAND gate) | Output  A  C  B | The output occurs if the input occurs in sequence as set by the analyst. For example, output $A.B.C$ occurs, if the events $A$, $B$, and $C$ follow the chronological sequence $A$ then $B$ and then $C$. Failure occurs at the gate output when the component at the input fails in this sequence, $A$, $B$, and $C$ | This gate has already been discussed in Sect. 4.5 on static fault tree analysis. This gate is normally used in dynamic fault trees |
| HOUSE or EXTERNAL event (H event) | Output (0,1)  H  A (0,1) | House events are extensively used in dynamic fault tree modeling for configuration management as a switch to make an input set to 0 or 1. These events are set to 0 or 1 manually in the static fault tree and dynamically in DFT | House events in static fault tree are used to avoid duplication of the fault tree when selected assumptions are to be reflected in the analysis as per the requirements |
| Spare gate (SPAR) | Output  SPAR  A  B C  Maininput  spareinput | This gate can have one or more main input(s) and one or more spare input(s). The gate has provision for defining the minimum number of components required to deliver an output as gate failure. When the main component fails, then the spare is switched on and the system fails when the failure criteria are met, viz. in this example if failure criteria are 2 out of 3 (F), then when the first spare fails the gate output occurs as gate failure | This gate represents a plant scenario where redundancy in the safety function is common and represents a dynamic condition. Failure of the main component and later the spare component leads to the failure criteria |
| Sequence enforcing gate (SEQ) | Output  SEQ  A  B  C | This gate is used when modeling requires failure to occur in a given predetermined sequence, to register a failure at the output of the gate | It is different from the PAND gate in that a sequence is predetermined unlike PAND gate where it follows a sequence from left to right |
| Functional dependency gate (FDEP) | Output  FDEP  T  A  B C  Dependent event | $T$ is a trigger event. If there is a trigger, then the dependent event occurs and causes failure registration of the FDEP gate output | The trigger event could be a basic or output from a gate that causes other dependent events (here, $A$, $B$, or $C$) to occur. Output from a Markov model can be used to cause the dependent event |

The available approaches for solution of the dynamic fault can broadly be divided into analytical and simulation approaches. The major analytical approaches are state space (normally employing Markov analysis), numerical integration, and Bayesian network. Each of these approaches has their pro and cons and is employed keeping in view the modeling and resource requirements. Hence, researchers have used one of the above approaches considering the specific characteristics of the problem to be solved.

In their 2002 study, Marko and Mavo [13] demonstrated that the application of dynamic fault tree reduces the system unavailability by arrangement of outages of safety equipment. In this study, house events are represented in a matrix where the rows represent the epoch of time chosen to represent the dynamic states and the columns represent the number of house events.

The modeling of a dynamic fault tree deals with (a) time-dependent component unavailability $Q(t)$ and (b) application of house events to the model, where case A deals with outage of equipment using the OR gate and case B deals with house event under AND gate to model more operational modes as shown in Fig. 4.14.

As can be seen, the time-dependent unavailability $Q(t)$ and the house event enable modeling of equipment outage and analysis of various operational modes of equipment, as shown in cases A and B, respectively. The available literature shows that the Boolean operations performed in a dynamic fault tree for complex systems, particularly as part of a PRA study, often become computationally prohibitive. Hence, the simulation approach provides an alternative approach to dynamic fault tree analysis.

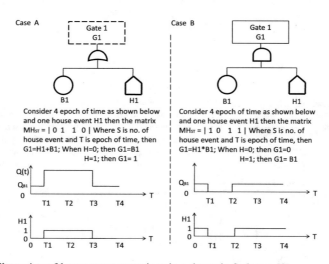

**Fig. 4.14** Illustration of house event operations in a dynamic fault tree [8]

## 4.7.2  Dynamic Event Tree Analysis

As discussed earlier, the static event trees are used in PRA for accident sequence modeling. In Level 1 PSA, the required safety functions (e.g., safety system, human action, and recovery) and their chronological order in which the safety function is demanded are prescribed by the analysts. Further, the branching of the nodal points in the static event tree is binary and is governed by the success and failure of the safety function.

However, the accident scenario evaluation for real-time situations can be modeled more realistically with treatment of the scenario in the temporal domain and provision of more than two branching points. The branching in the dynamic event tree follows the demands of the specific scenario and happens arbitrarily, but at a discrete point in the temporal (time) domain. Probabilistic dynamics enable us to fully account for the temporal evolution of the interaction of dynamics and stochastic in the evaluation of the accident consequences and their conditional (the condition is the initiating event) probabilities. Due to the nature of dynamics, the number of event sequences can be complex and very large, and to the extent they might pose computational constraints. Probabilistic dynamics operate on actual time/state space; however, discretization is required for the evaluation to be numerically manageable [9]. The computational resources could pose a constraint for this dynamic evaluation. The dynamic event tree analysis, as part of an accident sequence evaluation, requires data inputs such as process variables, information to assess changes in component status, and human actions and their effects on the system toward modeling the transient/change in plant status. Therefore, quantitative criteria/heuristics and plant variables are required for gainful utilization of resources.

We will consider the simplified methodology of dynamic event tree modeling formulated by Costa and Sui [7, 12] to illustrate modeling and analysis of dynamic event tree in this section. The dynamic event tree analysis method has been discussed to present the essential features and characteristics of dynamic event trees. The salient features of the dynamic event tree method (DETAM) are illustrated using an example of two systems A and B.

As can be seen in Fig. 4.15, only two systems, A and B, are considered for simplicity. However, in real time, there are more than two systems. At this point, it is important to understand that modeling and analysis of dynamic event trees require computational techniques, which in turn require formalization of computational functions and the associated set of variables or parameters and decision heuristics/rules controlling the course of computations (Table 4.7).

As we have seen, applications of the dynamic fault tree and the event tree approach are increasing as independent studies as well as part of risk assessment. We have also discussed the challenges that these approaches are facing when it comes to dynamic modeling, such as the requirements of computational resources, modeling of large fault trees, heuristics for fault tree reduction and its analysis, including identification of root and terminal events, selection of required sets of deterministic and probabilistic parameters in dynamic event tree analysis, and selection of time

**Fig. 4.15** Dynamic event tree illustration with two systems

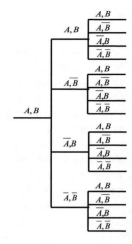

steps for modeling. However, extensive works have been reported in improving the dynamic approach either through the development of new structures [16] or through simplification or use of assumptions in the present approach [17].

**Table 4.7**  Characteristic set of variables defined by the DETAM application [7, 12]

| Characteristic set of variables/ rules/heuristics | Description | Implementation in DETAM and further interpretation |
|---|---|---|
| Branching set | Plant parameters/conditions that determine the status of a plant and safety function, operators identification of system status, and the proposed action in response to the plant diagnosis | $\{XA, XB\}$ is the branching set. Since there are two binary states, $XA$ can have value of 0 or 1 for successes and failures, respectively. The same argument applies to $XB$. When the number of safety functions increases, say to a number $n$, the branching set will be $\{XA, XB, XC,\dots, XN,\}$ |
| Plant status | Defined by branching variables and plant process variables | Each branch in the dynamic event tree has an end state, and the frequency value of $XA$ and $XB$ determines the probability of the respective end state, i.e., the plant state determined by that branch under consideration. For example, for the two systems there are 16 end branches that lead to four plant states<br>The plant status grows geometrically with the number of safety functions |

(continued)

**Table 4.7** (continued)

| Characteristic set of variables/ rules/heuristics | Description | Implementation in DETAM and further interpretation |
|---|---|---|
| Branching rules | Branching occurs at fixed points in time | In DETAM, the branching rule is a fixed interval $(t_1, t_2)$ for $XA$ and $XB$ However, keeping in view the requirements of specific problems, the rules for branching could be, for example, change in status of the safety function or an event enforced externally, such as an operator action. Generalization of DET requires algorithms to interpret this information from plant variables. The advantage of this generalized model is that it avoids computation resources because, for example, the use of a fixed interval will require a small $\Delta t$ to capture the instant of change in the plant/safety function status |
| Expansion rules | Cut-off criteria are used to terminate the accident sequence, below the predefined lower value of a frequency. This is to conserve computational resources | The cut-off criterion is one of the most used approaches to arrest the uncontrolled growth of accident sequences. Other criteria could be termination of plant into a safe state as defined by either termination of the postulated initiating event at the root or ensuring availability/implementation of safety function or based on a crew decision for a given plant state |
| Tools | Problem-specific models for plant behavior, stress, conditional frequency of operator state changes | There are two major categories of tools used in dynamic scenario evaluation— deterministic and probabilistic. The physical phenomenon for computation of a plant state for given plant transient or end state is performed by deterministic analysis, such as the value of some process variables. The probabilistic tools (e.g., PRA software) provide the estimate of frequency for a spectrum of plant states |

Dynamic fault tree and event tree analyses can be carried out either employing analytical approaches or using simulation methods. The existing four major methods are (a) state-space based methods such as Markov modeling, (b) stochastic methods, (c) combinatorial methods [18], and (d) simulation methods generally based on the Monte Carlo approach. For example, [17] used the Markov approach to capture dependency as part of dynamic modeling. Rao et al. [11] applied the simulation-based approach to address the complexity involved in dynamic modeling in PSA. For new approaches to the analysis of dynamic modeling, readers are directed to the related literature.

### 4.7.3  Binary Decision Diagram

The binary decision diagram (BDD) was introduced by Bryant in 1987 [19], and in 1993 Rauzy proposed an application of this approach for fault tree analysis [20]. In this sense, the BDD approach to fault tree and event tree analyses can be considered a recent development. The BDD approach provides an accurate solution because it does not require approximations inherent to traditional solution methods in classical fault tree analysis [21]. However, due to higher computational requirements, the present form of BDD has limitations when it comes to large-size fault trees typical to PRA studies due to combinatorial explosion.

It may be noted that the BDD approach is built on the fault tree methodology, and the only difference is that BDD provides an efficient solution particularly for small or medium size fault trees. So the first step in BDD analysis is reduction of the given fault tree by employing a reduction algorithm. The BDD analysis is largely governed by the efficiency of the reduction algorithms, and this is one of the major limitations of the BDD analysis because there is a single heuristic or algorithm that provides a uniform solution for different BDD models.

BDD is a bottom-up approach that starts with the root basic event and is based on Shannon's decomposition theorem for Boolean logic [18] as follows:

$$f = x_i \cdot f_{x_i=1} + \overline{x_i} \cdot f_{x_i=0} \quad 1 \le i \le n \tag{4.32}$$

where $f$ is Boolean formulation over a set of Boolean variables $x_1, x_2, x_3, \ldots, x_n$; further, it can be expressed as if-then-else (ite) rule or heuristic as follows:

$$f = \mathrm{ite}(x_i, f_{x_i=1}, f_{x_i=0}) \tag{4.33}$$

where $f_{xi=b}$ implies $f$ evaluated at $x_i = b$; and $b$ is a Boolean constant having value either as 0 or 1. Shannon's decomposition has "then-edge as 1-edge" or "else-edge as 0-edge" as shown in Fig. 4.16.

**Fig. 4.16** Shannon's decomposition of a Boolean node [22]

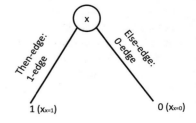

**Fig. 4.17** A fault tree basic
event or node representation
in a binary decision diagram

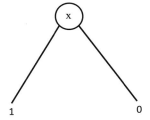

Let us consider the above formulation for the fault tree analysis. In the static fault tree, we have OR and AND gates as the basic and commonly used gates for system representation. The BDD approach used for fault tree analysis is referred to as reduced ordered BDD, i.e., BDD is in a reduced form. The NASA Fault Tree Handbook [23] provides an elegant and simple illustration of the BDD approach to fault tree analysis.

The process of making a BDD is a bottom-up approach. For each basic event, the BDD is drawn. From Shannon's decomposition, a basic event $X$ can be represented as shown in Fig. 4.17.

### 4.7.3.1 BDD for OR and AND Gates

Given that $X$ is the root basic event, the BDD representation of the addition of two basic events $X$ and $Y$, $X + Y$, or $X$ OR $Y$ is shown in Fig. 4.18.

Similarly, for the AND gate or product operation, the BDD will be as shown in Fig. 4.19.

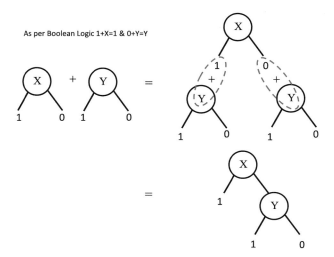

**Fig. 4.18** BDD for summation or OR gate

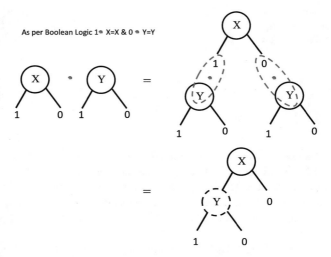

As per Boolean Logic 1● X=X & 0 ● Y=Y

Fig. 4.19  BDD for AND operation or product of two basic events

### 4.7.3.2  Simple Fault Tree BDD Construction

Consider the fault tree in Fig. 4.20. For constructing the BDD, the first step is to reduce or simplify the fault tree such that it is in its minimal form and no further reduction can be carried out [22].

**Fig. 4.20** Fault tree reduction and deduction of the BDD for $X$ $(Y + Z)$ configuration

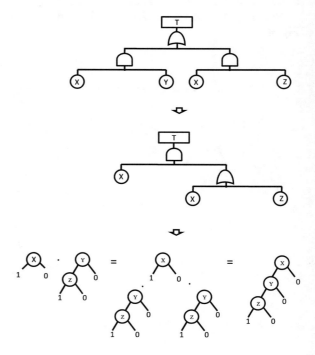

**Fig. 4.21** Fault tree
reduction and configuring the
BDD $X + YZ$ logic

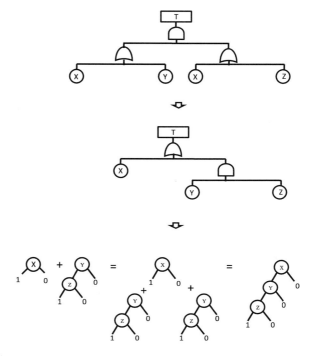

One of the simplest approaches is Boolean reduction. Figure 4.21 shows the reduced fault tree after applying the Boolean reduction logic. Consider for the purpose of illustration a simple fault tree having AND and OR gates and its conversion to BDD. Let us assume three components are connected in two configurations as shown in the two fault trees shown in Figs. 4.20 and 4.21.

The first step deals with reduction of the fault tree to its minimal form. We have the top event $T$, which can be represented as follows:

$$T = XY + XZ \tag{4.34}$$

$$T = X(Y + Z) \tag{4.35}$$

The reduced fault tree is derived based on Eq. (4.34).

Further, BDD is derived by identifying $X$ to be the root node with ordering scheme as $X \leq Y \leq Z$.

A similar procedure is employed for drawing BDD in Fig. 4.21 as follows:

The above two examples illustrate the broad procedure involved in deriving a BDD from a reduced fault tree. The key element of this procedure is ordering of the event, including identification of the root node and the terminal node. The root node is the one with which the BDD is initiated, e.g., the node $X$ is the root node and the last tail binary nodes (0, 1) are called the terminal node. The above example deals with a simple fault tree; hence, the procedure for fault tree structuring/reduction is relatively simple. For a large fault tree, the procedure is complex. Even though there are some applications of BDD for event tree analysis, this approach has not been popular as part of PRA. Based on this, it can be observed that further R&D is required to make the BDD approach an integral part of PRA.

The scope of this section was to introduce the BDD approach and discuss its potential for PRA applications. For detailed aspects of the BDD, such as quantification procedures for BDD [22, 24] and dynamic event tree analysis using the BDD approach as part of PRA [25] and improved computational procedure [26], the reader is advised to refer to the related literature on BDD and its application.

# References

1. Procedures for Performing a Failure Modes, Effects, and Criticality Analysis, MIL-STD-1629
2. Potential Failure Mode and Effects Analysis in Design (Design FMEA), Potential Failure Mode and Effects Analysis in Manufacturing, SAE J1739
3. Failure Modes and Effects Analysis and Critical Items List Requirements for Space Station, SSP 30234
4. International Standard, Analysis Technique for System Reliability—Procedure for Failure Mode and Effect Analysis (FMEA), IEC-60812 (IEC, 2006)
5. M. Modarres, M. Kaminskiy, V. Krivtsov, *Reliability Engineering and Risk Analysis—Practical Guide*, 2nd edn. (CRC Press Taylor & Francis Group, 2010)
6. International Atomic Energy Agency, IAEA Safety Standards—Development and Application of Level 1 Probabilistic Safety Assessment for Nuclear Power Plants, IAEA Safety Guide No SSG-3 (IAEA Headquarters, Vienna, 2010)
7. C. Acosta, N. Siu, Dynamic event trees in accident sequence analysis: application to steam generator tube ruptures. Reliab. Eng. Syst. Saf. **41**, 135–154 (1993)
8. T. Aldemir, A survey of dynamic methodologies for probabilistic safety assessment of nuclear power plants. Ann. Nucl. Energy **52**, 113–124 (2013)
9. E. Hofer, M. Kloss, B. Krzykacz-Hausmann, J. Peschke, M. Sonnenkalb, *Dynamic Event Trees for Probabilistic Safety Assessment*, http://www.eurosafe-forum-org/sites/default/files/euro-2-2-10-trees-probabilistic.pdf
10. L. Andreas, T. Norberg, L. Rosan, Approximate dynamic fault tree calculations for modeling water supply risks. Reliab. Eng. Syst. Saf. **106**, 61–71 (2002)
11. K.D. Rao, V. Gopika, S.V. Rao, H. Kushwaha, A. Verma, A. Srividya, Dynamic fault tree analysis using Monte Carlo simulation in probabilistic safety assessment. Reliab. Eng. Syst. Saf. **94**, 822–883 (2009)
12. C. Acosta, N. Sui, *Dynamic Event Tree Analysis Method (DETAM) for Accident Sequence Analysis, Final Report NRC-04-88-143* (USNRC, Washington DC, Oct 1991)
13. M. Cepin, B. Mavko, A dynamic fault tree. Reliab. Eng. Syst. Saf. **75**, 83–91 (2002)
14. D. Marko, S. Ilic, J. Glisovic, D. Catic, Dynamic fault tree analysis of Lawnmower, in *9th International Quality Conference, Center for Quality, Faculty of Engineering, University of Kragujevac*, June 2015
15. H. Boudali, P. Crouzen, M. Steolinga, *Dynamic Fault Tree Analysis Using Input/Output Interactive Markov Chains*
16. M. Guillaume, R. Jean-Marc, L. Jean-Jacques, Dynamic fault tree analysis, based on the structure function, in *Annual Reliability and Maintainability Symposium 2011, (RAMS 2011)*, Jan 2011

17. P. Bucci, J. Kirschenbaum, A. Mangan, T. Aldemir, Construction of event tree/fault tree models from a Markov approach to dynamic system modelling. Reliab. Eng. Syst. Saf. **93**, 1616–1627 (2008)
18. C. Shannon, A symbolic analysis of relays and switching circuits. Trans. AIEE **57**(12), 713–723 (1938)
19. R. Bryant, Graph based algorithms for boolean function manipulation. IEEE Trans. Comput. **35**(8), 677–691 (1987)
20. A. Rauzy, New algorithms for fault trees analysis. Reliab. Eng. Syst. Saf. **40**, 203–211 (1993)
21. A.K. Reay, J.D. Andrews, A fault tree analysis strategy using binary decision diagrams. Reliab. Eng. Syst. Saf. **78**, 45–56 (2002)
22. C. Ibanez Llano, A. Rauzy, E. Melendez, F. Nieto, A reduction approach to improve the quantification of linked fault trees through binary decision diagram. Reliab. Eng. Syst. Saf. **95**, 1314–1323 (2010)
23. NASA, *Fault Tree Handbook with Aerospace Applications Version 1.1* (NASA Office of Safety and Mission Assurance, NASA Headquarters, Washington, DC 20546, Aug 2002)
24. G. Daochuan, L. Meng, Y. Yanhua, Z. Ruoxing, C. Qiang, Quantitative analysis of dynamic fault trees using improved sequential binary decision diagrams. Reliab. Eng. Syst. Saf. **142**, 289–299 (2015)
25. J. Andrews, S. Dunnett, *Event Tree Analysis Using Binary Decision Diagram*, www.lboro.ac.uk/microsites/maths/research/preprints/.../99-25.pdf
26. R. Sinnamon, J.D. Andrews, Fault tree analysis and binary decision diagram, in *Reliability and Maintainability Symposium, 1996 Proceedings, 1996 IEEE. 07803-3112-5/96/*

# Chapter 5
# Life Prediction

> *The best way to predict the future is to create it.*
>
> Peter Drucker, www.sl-designs.com

## 5.1 Introduction

Many nuclear plants in the world have been in operation for over 30 years. In fact, in the USA, life extension certification for many NPPs has been obtained from 40 to 60 years and further R&D work is being performed to explore the possibility of extending life from to 80 years [1]. Of particular concern is to identify the non-replaceable structural and mechanical components and their degradation mechanisms, and then to develop methods to support estimation of remaining useful life.

Life assessment or "fitness for service" prediction for nuclear components is vital to nuclear system reliability and safety. Here, the efforts are required to predict the remaining useful life based on applicable and traceable degradation mechanisms. For the purposes of life prediction/assessment, engineering components can broadly be categorized into: (a) structural components, which include civil as well as steel structures, (b) mechanical components, (c) electrical, electronic, and process instrument components, (d) nuclear components, (e) software components as a part of digital systems. Software life cycle evaluation is not in the scope of this chapter as it requires a different treatment.

Even though rigorous qualification procedures are employed before inducting these components into the system, the life prediction/assessment approaches differ based on the category of these components as well as their safety classifications. For example, for structural and mechanical components a qualitative assessment is made based on the degradation mechanisms and life cycle loading conditions (e.g., thermal cycles, vibrations, radiation, shocks) for which it is designed. The evaluation of safety margin during the design stage followed by fitness for service evaluation during operational stages provides a judgment about remaining life and expected performance of these components. For electrical or electronic components, physics-of-failure models are needed and they may be supported by

© Springer Nature Singapore Pte Ltd. 2018
P. V. Varde and M. G. Pecht, *Risk-Based Engineering*, Springer Series in Reliability Engineering, https://doi.org/10.1007/978-981-13-0090-5_5

accelerated life test methods to provide estimates of remaining useful life of the components as part of qualification program. The basic idea behind these qualification procedures is the expectation of significant reliability and life assurance such that the risk as well as availability targets are achieved.

In an operational environment, the surveillance programs become the means of obtaining assurance for expected availability and safety of the components and systems be it structural, mechanical, electrical, electronic, or nuclear. The major approach to surveillance includes: (a) monitoring for any deviation from normal operations, (b) routine maintenance, which includes calibration checks, repairs, and replacements for electronics and instrumentation, (c) periodic performance testing, (d) in-service inspection for structural and mechanical components, and (e) condition monitoring.

The objective of quality assurance methods in design and operations meant to check fitness for service of systems structures and components on periodic intervals. Normally, the design lives of the SSCs are predicted qualitatively by employing the tools and methods that consider the rate of degradation and postulated life cycle service conditions, which might include various severe conditions. Similarly, although the surveillance program provides a qualitative assessment of system preparedness for acting on the postulated demand, there is no quantitative indicator that enables prediction of reliability or remaining useful life of the component [2].

It may be emphasized that the lifetime of the engineering SSCs is determined by the rate of underlying degradation mechanisms under the actual operational conditions. The degradation process reaches a critical level when key parameters that determine end of life are reached.

## 5.2   Literature Review

This section reviews the available literature keeping in the input required in implementing the life/failure prediction requirements in support of IRBE. From this point of view, this review can be divided into two categories, viz., (a) the reliability and life prediction for electrical and electronics components and (b) qualification of components considering projected material degradation under various operational and environmental stresses for structural and mechanical systems. In fact, the reliability and life testing as a subject are more commonly employed for electrical and electronic components [3]. The application of life-testing families, like accelerated life testing (ALT), highly accelerated life testing (HALT), highly accelerated stress screening (HASS) have found applications in not only safety- and mission-critical systems, like nuclear plants, process and chemical plants, space and aviation, transport systems but in many commercial and domestic products manufacturing systems in support of product performance tracking in general and warranty evaluation, in particular.

In electronic systems, the solder joint fatigue failure for microchips has been identified to be one of the major challenges, and Lee et al. have reviewed this subject in-depth and proposed a methodology to choose an applicable model and approach to address life prediction requirements against the fatigue-induced degradation of solder joints [4]. Yang provides a comprehensive treatment to life cycle reliability and associated testing techniques along with a commentary of degradation mechanisms and warranty aspects and finally validation techniques [5]. Sundaram and Kalivoda employ Weibull model to assess "plug reversal" phenomenon in a three-phase motor of a refrigerating machine compressor [6]. The accelerated life test requirements for consumer products, keeping in view the data collection and analysis, are unique in the sense that an interpretation is required to comprehend the data such that uncertainty in predictions can be reduced. In this context, Pham in his publication provides a collection of topics to address the requirements of accelerated life test for commercial products [7]. The goal of many accelerated life test is to obtain failure time distribution. The thesis of Ma proposes a new approach computing variance of maximum likelihood estimators for large sample with censoring and time varying stress-based cumulative model as also development of a useful test plans using small number of test units [8].Yin and Weng proposed a support vector machine-based approach to analyze life test data where parametric models are not known and demonstrate the applicability of their approach through simulation experiments [9]. Switch mode power supply is a vital component of digital systems in general and computer systems in particular and that is why reliability of SMPS is critical to system safety. Goodman et al. proposed a prognostic-based approach to predict the incipient failure in advance and therefore enable a health management strategy for SMPS, and it was demonstrated that ePHM (electronic prognostics and health management) can be an effective strategy for impending failures [10]. Remaining useful life (RUL) estimation is one of the goals of accelerated testing as part of aging studies. There are many approaches to RUL estimation and most of them are data driven approaches, i.e., use of available data to predict the RUL, however, there is no consensus on the one approach that can be employed for RUL estimation. Si et al. provide a review of data-driven approaches for RUL estimation and identify challenges and road map for future research [11]. Douglas et al. developed prognostic and health management approach for radio frequency system and demonstrated its successful implementation by performing accelerated test and shown that it was possible to predict the RUL within $\pm5$ thermal cycles. During an accelerated life testing, the product may fail due to one of the possible mechanisms or causes which need not be independent of other competing causes. Diwan and Kulathinal investigate the phenomenon of dependent competing causes of failure during accelerated life testing and validate this approach on real-time data [12]. Competing risk problems involving degradation failure has been one of the major area of research due to complexity of more than one mechanism. Wenbiao and Elsayed have modeled both degradation and catastrophic failures under accelerated conditions and the methodology is validated through a case study [13].

Mechanical and chemical properties of material, ductility, stress, strain, coefficient of thermal expansion, fatigue, creep wear provided the basis to demonstrate capability of the components against operational and environmental stresses be it structural, mechanical, electronics and electrical or nuclear components. Vable's publication on mechanics of material in general and the chapter on mechanical properties on materials in particular is relevant in this context. Be it mechanical, electrical, electronic, or nuclear components [14]. Fatigue-induced damage has been one of the critical areas of research as in many structural failure fatigue has been found to be one of the root causes. In this respect, Fatemi and Yang provide a survey of the available models and methods on cumulative fatigue damage and life prediction theories and concludes that even though many damage models have been developed, none of them enjoys universal acceptance while each model can only account for specific conditions and loads [15]. Upadhyay and Sridhara provide the model for fatigue life prediction by applying continuum damage mechanic and fracture mechanic approach for fatigue crack initiation and propagation, respectively, for EN-16 steel and 6082 Aluminum [16]. Extensive work is being performed in respect of application of damage modeling for life prediction particularly for high-temperature applications. Fujiyama and Fuziwara provide a review of damage models for steam turbine and gas turbine develop an integrated approach employing trend analysis, cumulative damage rule, damage parameter, and simulation analysis toward enabling application of risk-based approach [17]. The nuclear community is focusing on life extension program from the designed life of 40 years to 60 years and beyond, and in this context the white paper from ORNL on "Materials for Light Water Reactor" presents the state of the art in material degradation aspects particularly for non-replaceable core components and identifies, mechanics of degradation, mitigation strategies, modeling and simulation, monitoring and feedback (prognostics) and management as the areas for further research and development [18]. In this context, the USNRC document entitled "Expert panel report on proactive material degradation assessment" deals with (a) identification of reactor components that could reasonably be expected to material degradation beyond design life, (b) identify the degradation modes and mechanisms for these components, and (c) assess the state of the art on degradation toward developing mitigation strategies [1]. The deterministic models provide the needed framework for fatigue crack growth modeling; however, it is required that randomness associated with each parameter and prediction strategies need to be addressed to reduce uncertainty in predictions. This extensive work is being done on application of probabilistic approach to crack initiation and propagation. Raghu Prakash and Hariharan have worked on a multi-segment probabilistic fatigue crack growth model toward developing reliability factors that captures scatter in specific crack growth region [19]. NASA has incorporated stochastic fracture mechanics approach for evaluation of design fatigue life for metallic and ceramic components subjected to fluctuating stress [20]. It has been observed that while that stochastic analysis extends the deterministic models; however, the deterministic models are based on conservative factor of safety to account for the scatter in data while the probabilistic or stochastic approach provides improved insights on uncertainty in data and thereby in the results of the analysis. It was also observed that data from nondestructive testing (NDT) also forms part of fracture probability evaluation and the major concerns are related to "probability of detection" in respect of in-service inspection of data. IAEA document on "Nondestructive testing for plant life assessment" highlights the role of NDT technique in support of structural health assessment and thereby feasibility of evaluation of remaining life of the nuclear systems [21]. However, there are some issues, as pointed out in NASA publication that probability of non-coverage or non-detection remains a problematic area that need further research. As part of engine rotor life extension initiative, John et al. argue and evaluate role of residual stresses on the rotor life prediction and identify critical issues to be addressed during implementation of life prediction methods [22]. This paper further argues that significant increase in damage tolerance can be obtained if residual stresses are included in the life prediction methodology. Wu examines the complexity involved in life prediction of gas turbine material, particularly the ones subjected to high temperature and complex loading environment, and argues the application of physics-based life prediction methodology probability to support the prognostics and health management program of gas turbines [23].

Even though the review presented here is not exhaustive, it meets the purpose of providing the state of the art in life prediction of electronics and electrical systems through accelerated life test methods on one hand and the life prediction methodology for structural and mechanical systems considering various modeling approaches targeted to associated damage/degradation mechanisms. The following sections in this chapter focus on accelerated life-testing method and a introduction to selected damage modeling approaches.

## 5.3   Major Steps in Life Prediction

Life prediction for any engineering component involves following broad steps:

1. Identification of a life prediction method for a given SSC;
2. Identification of the critical failure modes vis-á-vis the precursor parameters that provide information on the level of degradation;
3. Identification of the critical value of the parameters to define the failure criteria, with acceptable levels of uncertainty that defines end of life;
4. Characterization of the degradation process for various SSCs;
5. Development of required fixture and data collection set up for the experiment;
6. Life test experiment or simulation considering a population size, test parameter, and decision criteria;
7. Effective monitoring of the degradation trend;
8. Evaluation of the influence of operational as well as environmental stresses, including predefined extreme shock conditions such as extreme temperature, mechanical shocks, or seismic loads for which the component has been designed;
9. Analysis of data and compilation of results.

The life prediction methods employed depend on the category of the component, viz., mechanical, electrical, or electronic. For example, for electronic and process instruments, the accelerated life test methods are the preferred approach. For structural and mechanical components, the qualification testing and degradation modeling due to life cycles loads like mechanical and thermal fatigue, creep, wear, and corrosion form the major approach for assessing the remaining useful life. An FMECA is performed to understand the applicable degradation mechanism and the precursors that need to be monitored. For a given experiment, it is necessary to develop the failure criteria. Development of failure criteria is a specific to a given application. For example, for a capacitor the increase in equivalent series resistance (ESR) or drop in conductance to a value that makes a given circuit fail form failure criteria. Often it is required to monitor the precursor parameter online during an experiment. Hence, a fixture is required to monitor the parameter and alarm when the parameter reaches the set limit. The acceleration factor for an experiment is determined considering the time to run the experiment for higher stress and the operational or environmental stress that will be seen by the component during its routine use. Normally experiments are performed at more than three stress levels to get the stress–life relationship. Finally, identifying applicable distributions for given tests, data analysis for life prediction forms the final step in life prediction approach.

## 5.4  Material Properties and Component Characterization for Life Testing

Material properties determine the application of a material for a given specific application. The life cycle of a component is determined by selection of material and projected degradation mechanisms against postulated environment and operational stress right during design stage; assessing the performance requirements during the operational phase and provision to be kept for its surveillance and maintainability provisions. The electrical properties relevant to electrical and electronics material are electrical insulation, permeability, resistance, capacitance, and inductance. If these properties change over a period of time or under the influence of environmental stresses (e.g., temperature and humidity) or operational stresses (e.g., voltage, current) then the life of the components is affected. The mechanical properties that affect the life of the components are strength, ductility, brittleness toughness, creep, fatigue, and mechanical wear. Apart from these properties for nuclear systems, different types of corrosion play a significant role in determining component life. For more details on material properties, particularly those that directly affect the life of the SSCs, readers may refer to the available literature on the subject. Detailed discussion on material properties is not within the scope of this book.

Life testing is a vital part of risk-based engineering. Even though the Chap. 13 on prognostics and health management deals with the categorization in a more explicit sense, this section deals with this categorization from a life-testing perspective. In fact, the results of life testing form the input to the prognostic framework for life/failure prediction. For the purpose of life prediction of the components in a nuclear plant set up, the SSCs can be categorized as follows:

## 5.4.1   Core SSCs

Here, the phrase "core SSCs" has been used to indicate those categories of SSCs that form an integral part of the systems that cannot be replaced and that essentially determine the life of the plant or systems. For example, the design life of nuclear plants can be considered to be primarily governed by reactor core components and structures that cannot be repaired or replaced. Reactor structural components include the reactor vessel and core structural components such as the end shield, primary coolant piping, and support structures. Apart from these components, the concrete structure, particularly the reactor pile block, and containment and spent fuel storage bays, also forms part of the component group that determines the design life of the plant. Periodic surveillance and inspection methods are used to assess the structural integrity of these components. These categories of SSCs require a structural engineering approach to testing and evaluation for determining the life or, more preciously, the remaining useful life of the SSCs.

## 5.4.2   Semi-Integral SSCs

These are structures that are expected to be replaced during the lifetime of the plant. For example, in nuclear plants this category includes steam generators, coolant channels, de-aerators, heat exchanges, primary coolant pumps, vehicle airlocks, and containment dampers. However, replacement/repair work of these SSCs is time and resource consuming and requires extended plant/system shutdown. Generally, it is recognized during design/construction stages of the plant these SSCs will be replaced/repaired at least once during the life time of the plant.

## 5.4.3   Repairable and Replaceable Systems

The design of the plant ensures that most of the components, barring the SSCs covered in previous Sects. 5.4.1 and 5.4.2, will need to be repaired on routine basis employing maintenance management program. In case the degradation level drops to such an extent that failure frequency of the component increases substantially and even with frequent and extensive repair it is not possible to restore the component, then it is concluded that replacement is necessary and the existing component should be replaced by a new one. Maintenance of plant availability and safety are the governing considerations, implemented through routine maintenance management program. Most of the mechanical, electrical and electronic components fall in the category, such as pumps, valves, diesel generators, switch gears (e.g., transformers, bus bars, breakers batteries, convertors and invertors). Control and protection system electronics along with shutdown devices might also fall in the category of replaceable components.

### 5.4.4  Electrical Systems

Most of the electrical components such as diesel generators, transformers, electrical buses, breakers and isolators, uninterrupted power supplies, and convertors might require replacements during one life cycle of the plant, for example, 40 years. The performance of these components is assessed over the life cycle and a decision is made to replace these components when reliability or safety issues demand these replacements. Even though the electronic systems in nuclear plants are considered to be monitored systems with diagnostic features, these systems also wear out and the signs of aging show up in the range of 10–25 years depending on the service conditions. More often these systems are replaced for reasons such as obsolescence. Since these systems form the core of the control and protection systems, their performance is monitored closely, particularly for common cause failures.

Traditionally, the available literature shows that a plant's life extension program is often based on qualitative criteria of safety and reliability. The conservative criteria governed by a defense-in-depth philosophy form the basis of evaluating the decisions related to repair or replacements. This chapter examines the role of life prediction in support of a risk-based approach.

## 5.5  Definition of Failure

In life testing, it is important to have a comprehensive definition of failure. It is required that before the SSC is subjected to testing or modeling, the competing failure mechanisms and thereby the precursor parameters that can provide information about the degradation should be monitored. The literature often conveniently shows this degradation by a line on the scale of time vs. the parameter being monitored. However, in actual practice, particularly for structural components this is not the case for two reasons: (a) most of the systems are not monitored online, and (b) there are many instances where the intermittent phenomenon/degradation is not tracked due to its nature. Hence, care should be taken while interpreting the data by postulating the various hypothesis and possibilities. The definition of failure should also take into account the practical application. For example, it may be acceptable that a pump with a normal flow of 10,000 lpm can deliver a flow of

4,000 lpm in a degraded state for one application because the process safety requirements are met even with this smaller flow. However, the same pump with similar degradation may not be acceptable if the failure criterion is a minimum low of 5,000 lpm. Similarly, a batter manufacturer's product may have a battery life of 20,000 h with a failure criterion of only 2 h service life at a given voltage after fully charging the battery. Another company might use the failure criterion of 1.3 h and might show a higher mean time to failure (MTTF) of 23,000 h. Obviously, the definition of failure criteria a must be defined very clearly as part of life prediction.

In life testing, MTTF or mean time between failures (MTBF) is extensively used to defining the service life of a component. First, the life of a battery as claimed by many suppliers is based on estimation of MTBF. However, because the batteries are, in the conventional sense, non-repairable components, this use of the term MTBF is not correct. MTFB is used only for repairable components. Second, MTTF statements are often based on assumptions that the component failure data follows exponential distribution, which may not be correct. Hence, it is important that the agencies using MTTF for determining the life of their product should provide a claim or documentary/test proof that the subject component follows exponential distribution. If this is not the case then applicable distribution should be used. Hence, it must be recognized that life prediction based on the MTBF is a subjective rather than objective declaration. Last, and most important, non-practitioner or decision makers often intuitively construe that a subject component will have the specified life. This is not so. For example, if a life test estimates that MTTF is, say 300,000 h, then only 63.2% of the components will have a life of 300,000 h and the rest may fail even before this specified period. As the evaluation of failure criteria is central to life prediction, a detailed failure mode and effects and criticality analysis (FMECA) is required. Hence, it is important to have a mechanics-of-failure or physics-of-failure approach to understand material degradation and limiting parameter values that defines the failure. For example, even if we are able to trend the corrosion induced degradation that causes loss in thickness of a piping, it is required that what is the minimum thickness that can hold the pressure in the piping system. Here, the challenge is to model the degradation and definition of failure criteria with acceptable level of uncertainty.

## 5.6   Material Degradation and Its Characterization

The basic material properties that determine the life of a given component for a given operational and environmental stresses can be categorized into physical, chemical, electrical, mechanical, and nuclear properties. The major properties that form part of life testing or degradation are given in Table 5.1.

The rate of material degradation over a period of time determines the life of the components. However, the failure criteria are determined using a factor of safety such that the SSC is taken out of service well before the actual failure of the components and the penalty or consequences of failure can be avoided or reduced.

**Table 5.1** Life limiting properties and associated parameters

| Category | Property | Applicability |
|---|---|---|
| *1. Atomic/Nuclear* | | |
| | Atomic mass Atomic number Atomic weight Density Resistance to radiation | These are fundamental properties that are considered for given nuclear/non-nuclear functional requirements. For most of the core and associated structural core components, the selection of material is based on resistance to irradiation damages, such as hydrogen embrittlement |
| *2. Chemical* | | |
| | Corrosion resistance Surface tension Reactivity (activation energy) pH Conductivity Hygroscope | Resistance to corrosion is a desirable attribute; however, corrosion is one of the major degradation materials and corrosion modeling is carried out to assess most of the structural components as also high-temperature electrical components/electrical connectors and joints. The activation energy of the process is based on the reaction rates for various environmental and operational stresses. Controlling pH and conductivity provides an effective mechanism for ensuring the life of the coolant systems |
| *3. Mechanical* | | |
| | Tensile or yield strength Comprehensive strength Fatigue limit Ductility Hardness Fracture toughness Shear strength Elasticity Plasticity Coefficient of friction Creep | Mechanical properties determine the life of structural components. Qualification testing and periodic tests and surveillance are carried out for adequate assurance of the life of mechanical components. Proof tests as per ASME code guidelines are followed by pre-service inspection and are carried out as part of SSC qualification. Cyclic tests to ensure adequate resistance to fatigue-induced failure forms an inherent part of component qualification. High-temperature components under mechanical stresses are also subjected to creep or creep-fatigue investigation |
| *4. Electrical* | | |
| | Electrical conductivity Dielectric constant Dielectric strength | Increase in electrical resistance for contacts and electrical circuits; delamination or decrease in dielectric strength for capacitors and breakers; decrease in insulation resistance for motor/alternator winding are some of the signs of degradation in electrical systems. High current in semiconductor devices induce electro-migration that results in metallization failure |
| *5. Thermal* | | |
| | Coefficient of thermal expansion Eutectic point Flammability Melting point | Temperature-induced degradation has been one of the major causes of micro-and power electronics and components. Failure of the mating part due to differential coefficient of thermal expansion also contributes to the failure of the component due to shear stress phenomenon |
| *6. Magnetic* | | |
| | Hysteresis Permeability | Induced magnetic field due to high voltage causes degradation or breakdown of dielectrics in semiconductor devices |

The phenomenon of degradation can be tracked/measured by identifying the precursor for a given applicable mechanism that defines the life of the component. Before we go to the discussion on various degradation mechanisms let us understand the four major factors that govern/affect the properties of materials are microstructure, chemical composition, crystal structure and texture, and lattice defects.

### 5.6.1 Microstructure

The microstructure of a prepared material surface or a film observed at higher magnification (>25×) affects material properties such as strength, hardness, ductility, creep properties, corrosion resistance, wear resistance, and temperature-dependent performance of material. Microstructure characterization forms the fundamental activity for a designer in material selection. For the higher magnification, a special optical microscope, a scanning electron microscope, or a polarized microscope is required to facilitate observing or digital recording of the surface features. The information on grain size and grain boundary along with other parameters is used for characterization of the material.

### 5.6.2 Chemical Composition

Chemical composition of a material determines the material properties. For example, the low carbon and presence of nickel and chromium as alloying elements makes the stainless steel SS304 that renders steel corrosion-resistant and determines its application in nuclear systems. Similarly, 0.5% niobium makes the alloy Zr a material for pressure tubes in the heavy water reactor. Even though various grades of carbon steel having different chemical composition are used for pipes and various structural materials, its degradation by corrosion is controlled chemically by the coolant for the pipe casing, whereas a protective coating is added to reduce outside corrosion due to environmental attacks. Doping of pure Si and Ge makes the material for semiconductor components such as diodes and transistors, whereas $SiO_2$ is a substrate material in electronic components.

### 5.6.3 Role of Crystal Structure in Material Failure

The properties of materials, specifically the mechanical properties, are governed by the arrangement of the atoms. This arrangement is called the material's crystal structure. A crystal's structure and symmetry play a role in determining many of its physical properties, such as cleavage, electronic band structure, and optical transparency. For understanding failure initiation or propagation at microscopic or nano-level requires understanding of material behavior at crystal level. There are mechanisms, like slip, lattice defects, and dislocations while at macro-level, the bulk defects like voids, preexisting cracks, blowholes provide the sites for failure initiation under stress/loading conditions [24]. The basic assumption in failure analysis is that the materials strength and fracture strength is governed by micro-structural and bulk. For example, plastic deformation of crystalline material occurs by slip which is defined as sliding of planes of one atom over another, while the edge dislocation which can be visualized as having one partway in a perfect crystal accommodating a part plan of atoms. Similarly, the formation of grain boundary during the liquid crystallization phase also determines the mechanical properties. For example, smaller grain size favors higher strength as the larger grain boundary provides favorable condition for crack initiation and propagation.

## 5.7  Life Prediction/Assessment Approaches

This section deals with life prediction as well as life assessment approaches. Traditional approaches such as qualification screening, surveillance testing, and conditioning monitoring, give decision makers a qualitative feel of the life expectancy and reliability of the components. Advanced methods such as PoF and PHM model uncertainty in life prediction, and quantified estimates generally form part of the prediction.

Presently, there is extensive R&D on the life prediction of SSCs. Figure 5.1 shows the major methods for predicting the remaining useful life of the components. The following subsections present the salient features of these approaches. The aim here is not to provide the exhaustive tools, methods, and models available for each technique. These details can be obtained from the available literature. Here, the aim is to discuss the associated fundamentals, advantages, and imitations of these techniques in respect of their application to risk-based engineering. Hence, it is important to understand the roles and limitations of evolving life prediction techniques.

There is no single approach that can be recommended for life prediction of engineering components. Every approach has some advantages and limitations. In this context, the analysts need to consider the nature of the problem and the resources that are available along with other considerations to choose one or a combination of more than one approach to solve the problem in hand.

The following section discusses the life prediction approaches presented in Fig. 5.1. It may be noted that strictly speaking, the NDT approaches are only in-service inspection approaches, like screening, in-service inspection, and condition monitoring; however, the data and insights available used from ISI and condition monitoring can be used as input for life prediction. The other categories like accelerated life testing that includes rate and stress acceleration are employed for life and reliability prediction, while the highly accelerated test is used to understand

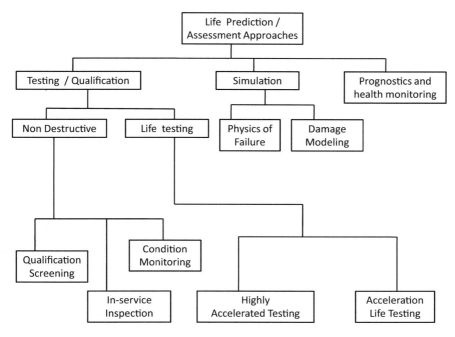

**Fig. 5.1** Major approaches to life prediction/assessment

dominant failure mechanisms. This helps is identifying the applicable degradation and corresponding precursor. Following sections discuss the role of individual technique in life prediction.

As shown in Fig. 5.1 under the category of testing and qualification, there are two subcategories namely nondestructive testing and life testing. The second category and third category of life tests are simulation approach and prognostics and health management. The following sections follow this hierarchy.

### 5.7.1   Nondestructive Testing (NDT)

The terms "nondestructive testing" or "nondestructive evaluation" refer to a group of scientific testing methods where the object or material is subjected to test to evaluate the presence of a flaw or defect either on the surface or in the bulk. This approach to testing/monitoring has found wider acceptance and application in many engineering systems as the integrity of the object to a large extent remains intact. Even though this technique is generally viewed as a quality control method, one of the significant benefits of NDT is that it increases the safety and reliability of the product during operation [25]. There are two major applications of NDT particularly for operating plants—in-service inspection and condition monitoring.

### 5.7.1.1  Qualification Testing and Screening

This is a traditional approach to ensure component service life and reliability. At the design stage, this approach requires ensuring that component meets requirements of establish codes and validation and verifications where proof tests at stress level higher level than the design loads is performed to get the confidence that the structure/component will be able to meet the operational requirements. Conservative safety margins and compliance with national and international codes provides required assurance for operational reliability and safety.

### 5.7.1.2  In-Service Inspection

During the commissioning and operational phase of the plant, the in-service inspection (ISI) technique is generally used for structural components such as concrete, steel structures, and piping networks in the coolant systems. The major techniques for ISI that are commonly used in nuclear and non-nuclear industries include visual inspection, liquid penetration test, magnetic particle test, and eddy current test. These techniques differ in their respective capability in terms of applicability to material type, defect type, defect size, and defect location. The visual and liquid penetration tests are versatile and are generally applied to surface defects, whereas the ultrasonic technique is the most widely used method for detection of defects in the bulk sitting deep inside the material. In fact, ultrasonic probes are widely used for measuring material thickness to assess degradation of SSCs due to local or uniform corrosion and also to track the growth of the existing flaws in the material. Similarly, the radiography technique is used when wide coverage is required. The magnetic particle test is generally used for magnetic materials, but it can only detect subsurface defects and is not capable of detecting buried defects in the bulk.

It is also recognized that although ISI and condition monitoring are resource-intensive techniques, safety cannot be compromised. This calls for prioritization of the SSCs based on safety importance. Even though traditional methods employ the qualitative notion of likelihood and consequences, the advent and maturity of risk assessment techniques such as probabilistic risk assessment has enabled development of risk-informed ISI programs for nuclear plants [26]. These ISI program can be optimized to address plant safety as well as availability aspects. Chapter 15 provides a case study of a risk-informed approach to nuclear plants.

### 5.7.1.3  Condition Monitoring

The term "condition monitoring" is associated with monitoring the performance of rotating machines by employing vibration, temperature, shaft run out, oil quality of bearing, and acoustics. Condition monitoring is carried out in two modes, off-line or online. The factors that go into deciding whether a given machine should be monitored online or off-line are: safety importance, maintainability aspects, cost, site-specific constraints, and net benefits. The limitation of condition monitoring is that it tracks the precursor parameter, such as vibration or temperature, but fails to predict the level of damage and in turn the information about the remaining useful life. The experience suggests that often the decisions are based on conservative criteria and the potential useful service life of the component is lost. On the other hand, the failure often occurs well before a prediction.

To overcome these difficulties, extensive work is in progress on development of PHM/monitoring. It may be noted that monitoring of changes/modification and generation of feedback is an integral part of the implementation of risk-informed applications [27]. This aspect will be discussed in the following section on PHM.

## 5.7.2 Life Testing

There are two types of life testing—the time terminated life test called Type I test and failure-terminated test called Type II test. These tests are carried out on a sample drawn from a population with an assumption that these samples represent the population. These samples comprising adequate numbers are subjected to test under use conditions. For example, 1,000 electronic capacitors are put to test for actual use condition, for instance maintaining the usage environment (e.g., temperature, humidity) and operational stresses (e.g., voltage, current). A record of failed components is maintained. Once the test is terminated either using Type I or Type II criteria, the MTTF estimated using the model discussed in Chap. 3. The failure data is plotted employing a probability distribution, such as exponential or Weibull distribution. Selection of the proper distribution is crucial to the quality and accuracy of the results. It may be noted that in these types of testing more data is collected by increasing the number of components in the sample to have good cumulative time as well as adequate failure data. However, in most of the practical situations the number of components available for testing is countably low. Hence, this life test may not be adequate under usage conditions. In such cases, the accelerated life test technique is used.

### 5.7.2.1 Burn-in Testing

Highly accelerated testing (HAT) is also referred to as "burn-in test," "elephant test," or "torture test." In much of the literature, this test is generally referred to as a highly accelerated life test (HALT); however, this test is not designed to provide any information on life expectancy of the product. In fact, these are qualitative tests and are designed to collect information on component failure mechanisms. The components are made to fail by applying very high levels of stress. Further, the failed components are subjected to root cause analysis to understand the material degradation and interfacing faults. The aim of these tests is not to collect time-to-failure data as applicable to previous or following cases; rather it is to collect information to improve the product design.

## 5.7.3   Highly Accelerated Stress Screening

Highly accelerated stress screening (HASS) or qualification tests are a well-established tool/procedure that involves subjecting the component to prede-termined higher stresses for specified time. HASS is a quality control procedure. Even though there is no information on the useful remaining life of the component after the screening qualification, there is general acceptance that this approach is effective in screening the components having manufacturing- and material-related defects. The components that pass through the screening tests are not expected to fail in the field or, to be precise, during the operational phase, and the only failure that can be expected is random failure.

However, it is also recognized that screening also consumes a part of the useful remaining life of the components because the screening also has potential to induce a new fault that may manifest during operational phase a little earlier than expected.

A risk-informed framework as an additional input has the potential to improve the life of the component in the operational phase. Application of the design of experiment (DOE) approach along with risk input can provide an effective framework to this end. Design of experiment will enable optimization of resources while maintaining the objective function for given constraints. Here, risk can be one of the additional inputs for designing the screening test parameters. However, the detail on how to design the test and its application is beyond the scope of this book.

From the risk perspective, even though the ISI and condition monitoring tech-niques provide valuable information on component health that enhances safety of the component and the plant, the challenges related to probability of detection of flaws or deterioration and coverage probability of a given campaign are the two aspects that need attention.

### 5.7.3.1   Accelerated Life Testing

The accelerated life-testing (ALT) technique is extensively employed while developing a new system or component for predicting the life and reliability of the components for given failure modes and mechanisms. Accelerated life testing is also used to predict the remaining useful life for existing or old components in support of decisions for their renewal/repair or replacement.

Why has accelerated testing become a common industry practice? The first and foremost reason is that it is not possible or practical to test the components for their useful period, which is generally very long, say, a couple of years and sometimes tens of years. At the same time, an added assurance of the life expectancy of these components is required. The fundamental principle involved in the ALT technique is that by increasing the stresses (operational and/or environmental stresses), the time for test duration can be compressed/reduced. The idea is to collect information on not only life or reliability but also competing failure mechanisms. Figure 5.2 shows the fundamental principle of accelerated testing.

The subject component is tested at higher stresses because more failure data can be collected in a relatively small time interval. Testing is performed at three stress levels as shown in Fig. 5.2, and these levels are chosen as the minimum three points required to obtain the approximation of the stress–life relationship. Curves A, B, and C underline the importance of the minimum three levels. Curve A is applicable when the stress–life relationship is linear; however, this may not be the case of the relationship could be around curves A or C. The evaluation of the stress–life relationship is crucial to life prediction at the use stress level as shown by estimates L1, L2, and L3. One can appreciate the importance of having an adequate number of data points in accelerated life testing; otherwise, the prediction may turn out misleading. The estimated acceleration factor is another parameter that affects the results of the ALT. It may be noted that the stress levels should be such that the applied stress accelerates only those failure modes that are applicable to the use conditions and should not induce new failure modes that would not be encountered when the component is operating underuse conditions. Figure 5.3 shows the relationship for various stress levels.

Figure 5.3 shows the use stress ranges, which are governed by day-to-day environmental limits, such as normal temperature and humidity, and operational

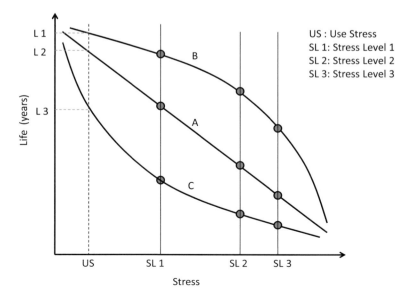

**Fig. 5.2** Stress–life relationship

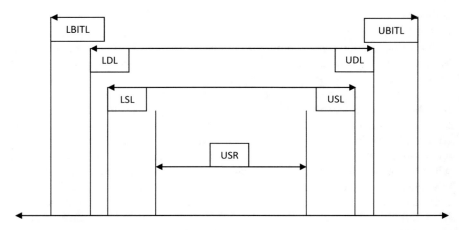

Legends: USR: Useful Life Stress Range; LSL: Lower Stress Level;
USL: Upper Stress Level; LDL: Lower Design Limit;
UDL: Upper Design Limit; LBITL: Lower Built-in Test Limit; and UBITL: Upper Built in Test Limit;

**Fig. 5.3** Stress levels and their inter-relationships

parameters, such as voltage and current levels for electrical/electronic components. The lower specification limits (LSLs) and the upper specification limits (USLs) as applicable are provided by the vendor or determined by the plant or product owner and are set by considering safety factors such that the design limits are not reached. Further, even though the product has been demonstrated to be safe within the design limits, the defense-in-depth principle requires that these limits are not reached.

Hence, to determine the test limits for stresses, the design of experiment should consider the input from physics of the material, geometry, and design parameters. These limits are all the more required when the number of test samples is limited.

**Acceleration Factor**

The acceleration factor (AF) relates the life or failure at high stress level at which the collection of component or a sample is subjected to testing and the life or failure rate at use stress level. The expression for AF is as follows:

$$AF = \frac{Life_{US}}{Life_{AS}} \tag{5.1}$$

$Life_{US}$   life predicted at use stress environment
$Life_{AS}$   life predicted at accelerated stress environment

AF can also be expressed in terms of failure rate ($\lambda$/unit time) estimates

$$AF = \frac{\lambda_{AS}}{\lambda_{US}} \tag{5.2}$$

### 5.7.3.2   Accelerated Life Stress Models

Accelerated life test modeling deals with developing a relationship between stress and life for engineering systems. Arrhenius model relates rate of reaction (or degradation) as a function of temperature as follows [28]:

$$r = A \exp[-B/T] \tag{5.3}$$

Where r is the reaction rate of reaction, A and B are constants, and T is temperature in Kelvins. The acceleration factor may be determined as follows:

$$AF = \frac{Ae^{-B/T_2}}{Ae^{-B/T_1}} = \exp\left[B\left(\frac{1}{T_1} - \frac{1}{T_2}\right)\right] \tag{5.4}$$

This empirical model was proposed by Svante Arrhenius in 1889 [1, 2], and he was awarded Nobel Prize in Chemistry for his work in 1903. The above equation is generic and enables evaluation of acceleration factor when the experiment is performed at two temperatures. This form of the equation provides effective solution for a specific experiment and is independent of any specific physical parameter.

However, the above equation is often used by relating reaction rate as function of the specific activation energy $E_a$ for a given process or configuration. The unit of rate constant depends on the global order of the reaction [30], the unit rate of the coefficient is "per second." At temperature $T$, the molecules have energy according to the Boltzmann distribution. Also, the fraction of collisions for energy greater than $E_a$ vary with $\exp(-E_a/kT)$.$K$ reaction rate/s.

The Arrhenius equation is also expressed as follows:

$$k = A \exp\left(-\frac{E_a}{kT}\right) \tag{5.5}$$

where $k$ = the rate constant (/s); $A$ = constant (a pre-exponential factor or frequency factor) that depend on material properties, surface parameters, geometry, and environmental and operational stress conditions; $E_a$ = activation energy (eV) per molecule (in the above equation if R—universal gas constant is used in place of the Boltzmann constant $k$ then $E_a$ is expressed in energy per mole); $k$ = Boltzmann constant $8.6171 \times 10^{-5}$ eV/°K; and $T$ = temperature in Kelvin.

Consider Eq. 5.5 and assume that failure occurs when a given amount of critical moles or chemical materials have participated in the reaction. Given that the failure criteria determine the life of the reacting material, the time to failure ($t$) can be given as follows:

$$t = \frac{1}{A \exp\left(-\frac{E_a}{kT}\right)}$$
$$t = A \exp\left(\frac{E_a}{kT}\right) \tag{5.6}$$

The Arrhenius model is used for the cases involving only temperature stress, i.e., only single stress. Since the scope of this book is not to discuss all of the life–stress models, Table 5.2 provides a summary of some selected models and associated details, which are commonly employed in risk analysis.

Consider the equation for acceleration factor (AF) for the Arrhenius model, which can be given as

$$AF = \frac{t_{US}}{t_{AS}} \tag{5.7}$$

where $t_{US}$ = time to failure at use stress condition corresponding to temperature $T_{US}$ and $t_{AS}$ = time to failure at accelerated stress condition corresponding to temperature $T_{AS}$. Consider Eqs. 5.6 and 5.7 for expressing AF for the Arrhenius model as follows:

$$AF = \frac{A \exp\left(\frac{E_a}{kT_{US}}\right)}{A \exp\left(\frac{E_a}{kT_{AS}}\right)}$$
$$AF = \exp\left[\frac{E_a}{k}\left(\frac{1}{T_{US}} - \frac{1}{T_{AS}}\right)\right] \tag{5.8}$$

As can be seen above, activation energy is one parameter which is critical for the accuracy of the results. Often, the activation energy values are difficult to obtain or not available for a given experiment. Even though the empirical models provide a way for acceleration of acceleration factor (AF), the evaluation of acceleration factor using generic model (which is independent of any specific constants, like activation energy) experimental life test methods should be the first choice for reducing the uncertainty in results. Table 5.2 provides more models for evaluation of acceleration factor. As can be seen that derived Eyring model evaluation acceleration factor when there is synergy of two stresses like combined effect of temperature and humidity. Similarly power law model is used generally non-thermal stresses, like mechanical fatigue.

**Table 5.2** Summary of commonly used models for acceleration factor (AF)

| Model | Original model, developer, year | Physics basis | Expression | Reference; model parameter |
|---|---|---|---|---|
| Arrhenius model | Svante Arrhenius, 1889 | Statistical thermodynamics | $AF = \exp\left[\frac{E_a}{k}\left(\frac{1}{T_{US}} - \frac{1}{T_{AS}}\right)\right]$ | [29]; $E_a$ = activation energy; $k$ = Boltzman constant; $T_{US}$ = use stress temp.; $T_{AS}$ = acceleration stress temp |
| Eyring model | Eyring, H. (1935) | Statistical thermodynamics | $AF = \exp\left[\frac{T_{AS}}{T_{US}}\frac{E_a}{k}\left(\frac{1}{T_{US}} - \frac{1}{T_{AS}}\right)\right]$ | [31]; Parameter as above. This model is modification to Arrhenius model; This model can, unlike Arrhenius model, take more than one parameter |
| Temperature-humidity model | Derived from Eyring, model, (1998) | Statistical thermodynamics | $AF = \exp\left[\frac{E_a}{k}\left(\frac{1}{T_{US}} - \frac{1}{T_{AS}}\right) + B\left(\frac{1}{H_{US}} - \frac{1}{H_{AS}}\right)\right]$ | [32]; All the terms common to previous model have the usual meaning; $H_{US}$ and $H_{AS}$; humidity at use stress and acc. stress level |
| Inverse power law model | | | $AF = \left(\frac{V_{US}}{V_{AS}}\right)^{n}$ | This model is generally used for non-thermal stresses |
| Temperature–non-thermal (T-N-T) model | Arrhenius and inverse power law | | $AF = \left(\frac{U_{AS}}{U_{US}}\right)^{n}\exp\left\{B\left(\frac{1}{T_{US}} - \frac{1}{T_{AS}}\right)\right\}$ | All the terms common to previous model have the usual meaning |

### 5.7.3.3    Evaluation of Probability Density Function Considering Arrhenius (Life–Stress) and Weibull (Life–Pdf) Model

The Weibull distribution is used extensively in predicting life or reliability of the components in life testing. Further, the Arrhenius model is also used extensively where temperature is the only stress against which the life estimation is to be carried out.

The pdf for two-parameter Weibull distribution is given by the following equation:

$$f(t) = \frac{\beta t^{\beta-1}}{\alpha^\beta} \exp\left(\frac{t}{\alpha}\right)^\beta \tag{5.9}$$

From equation

$$\alpha = A \exp\left(\frac{E_a}{kT}\right) \tag{5.10}$$

The Weibull pdf for temperature $T$ stress can be expressed as follows:

$$f(t, T) = \frac{\beta t^{\beta-1}}{\left(A \exp\left[\frac{E_a a}{kT}\right]\right)^\beta} \exp\left(\frac{t}{\left[A \exp\left|\frac{E_a}{kT}\right|\right]}\right)^\beta \tag{5.11}$$

This equation can be used to derive the various parameters of the Weibull–Arrhenius function, such as reliability and MTTF.

## 5.7.4    Simulation-Based Approaches

There are two broad approaches for life prediction in simulation categories viz., the physics-of-failure and damage modeling approach. In fact, both the approaches are at present remaining limited to application in laboratory environment; however, the models and methods are being used in the field for specific applications.

### 5.7.4.1    Physics-of-Failure

The physics-of-failure (PoF) approach provides tools, models, and methods for simulating the component and system behavior well before development of actual system. PoF is a science-based approach which employs deterministic or science-based models which can take random variable as input to model uncertainty in life prediction. In fact, the term PoF is mostly associated with electrical and electronics systems; however, there are application of this approach to structural and mechanical systems also. This subject has been discussed in detailed in Chap. 12. The damage modeling forms part of the PoF approach.

There are three major approaches for damage-based life prediction, viz., (a) damage trend analysis, (b) cumulative damage rule, and (c) damage parameter and simulation analysis. The objective of damage trend analysis is to correlate the damage data with operational history to obtain the damage trend curve under operational loading conditions. The cumulative damage rule has extensively been applied to fatigue-induced damage. The basic assumption is that fatigue damage increases with fluctuating cyclic stresses which may lead to failure of the structure due to fracture. This phenomenology was first expressed in 1945 by Minor's mathematical formulation, also referred as minor's rule as follows:

$$D = \sum_i \left( \frac{n_i}{N_{if}} \right) \tag{5.12}$$

where

D    Damage
$n_i$    number of cycle, and
$N_{if}$    number of cycles to failure

The available state of the art or on damage theories can be grouped into six categories, viz, linear damage rules, nonlinear damage curve, life curve modification methods, approaches based on crack growth concepts, continuum damage mechanics model, and energy-based theories [15]. The damage simulation is used for validating the damage model toward predicting the future damage trends [17]. The damage-based life prediction is directly relevant to implementation of risk-based strategies in maintenance management or in-service inspection or overall surveillance program. Further details of damage based simulation are not in the scope of this chapter.

### 5.7.5 Prognostics and Health Management

The technique for life prediction will not be discussed here as Chap. 13 provides detailed on the prognostics and health management for not only electronics but structural and mechanical components in nuclear plant

## 5.8 Conclusions

This chapter presents the life prediction approaches which are divided into two major categories, viz., life-testing approaches mainly comprised of family of accelerated life test methods and degradation/damage modeling methods particularly, for structural and mechanical failures. The accelerated life test approaches are being employed on routine basis for predicting remaining useful life of the electronics and electrical components and systems as part of aging management or qualification program. For mechanical and structural systems, even though condition-based maintenance or in-service inspection program still forms the major approach for life prediction. This chapter provides input for PoF modeling and prognostics and health management discussed in Chaps. 12 and 13. As life prediction is a past of simulation approach in these two areas.

## References

1. US Nuclear Regulatory Commission, *Expert Panel Report on Proactive Materials Degradation Assessment* (USNRC, Washington, D.C., 2006)
2. M.G. Pecht, *Prognostics and Health Management of Electronics* (Wiley, New Jersey, 2008)
3. V. Naikan, *Reliability Engineering and Life Testing* (PHI Learning, New Delhi, 2009)
4. W. Lee, L. Nguyen, G. Selvaduray, Solder joint fatigue models: review and applicability to chip scale packages. Microelectron. Reliab. **40**, 231–244 (2000)
5. G. Yand, *Life Cycle Reliability Engineering* (Wiley, New Jersey, 2007)
6. S. Sundaram and F. Kalivoda, Reliability life testing and evaluation of 3-phase motors, in International Compressor Engineer Conference, 1982
7. P. Hoang, *Springer Handbook of Engineering Statistics: Accelerated Life Test Models and Data Analysis* (Springer, Berlin, 2006)
8. H. Ma, New developments in planning accelerated life tests (2009)
9. P. Yin, C. Wang, Life prediction of accelerated life testing based on support vector machine (2011)
10. D. Goodman, J. Hofmeister, J. Judkins, Electronics prognostics for switched mode power supplies. Microelectron. Reliab. **47**, 1902–1906 (2006)
11. X.-S. Si, W. Wang, C.-H. Hu, D.-H. Zhou, Remaining useful life estimation—a review on the statistical data driven approach. Eur. J. Oper. Res. **213**, 1–14 (2011)
12. I. Dewan, S. Kulathinal, *Accelerated Life Testing in the Presence of Dependent Competing Cases of Failure* (Indian Statistical Institute, Delhi, 2005)
13. Z. Wenbiao, E. Elsayed, An accelerated testing model involving performance degradation, in *RAMS 2004*, 2004
14. M. Vable, Mechanical properties of materials (2014), http://www.me.edu/
15. A. Fatemi, L. Yang, Cumulative fatigue damage and life prediction theories: a survey of the state of the art for homogeneous materials. Int. J. Fatigue **20**(1), 9–34 (1998)
16. Y. Upadhyaya, B. Sridhara, Fatigue crack initiation and propogation life prediction of material, in *International Conference on Mechanical, Electronics and Mechatronics Engineering (ICMEME'2012)*, Bangkok, 2012
17. K. Fujiyama, T. Fujiwara, Damage in high temperature components and the life assessment technologies, in *ORAL Reference:ICF100868OR*

18. J. Busby, R. Nanstad, R. Stoller, Z. Feng, D. Naus, Material degradation in light water reactors: Life after 60, ORNL
19. R.V. Prakash, K. Hariharan, A multi-segment probabilistic fatigue crack growth model to account for reliability in design of components, in *IEEE-2010*, 2010
20. National Aeronautical and Space Administration (NASA), Preferred Reliability Practices: Fracture Mechanics Reliability, 1995
21. International Atomic Energy Agency, *Non-Destructive Testing for Plant Life Assessment* (IAEA, Vienna, 2005)
22. R. John, J.M. Larsen, D.J. Buchanan, N.E. Ashbaugh, Incorporating residual stresses in life prediction of turbine engine disks, in *RTO AVT Symposium on Ageing Mechanisms and Control: Part B—Monitoring and Management of Gas Turbine Fleets for Extended Life and Reduced Costs*, Menchester, 2001
23. X. Wu, Institute for Aerospace Research, National Research Council, www.intechopen.com. Accessed 13 Aug 2015
24. W. F. Hosford, *Mechanical Behavior of Materials* (Cambridge University Press, Cambridge, 2005)
25. International Atomic Energy Agency, Non-destructive testing for plant life assessment, IAEA Training course series 26, Vienna, 2006
26. United State Nuclear Regulatory Commission, *An Approach for Plant-Specific Risk-Informed Decision Making for In-Service Inspection of Piping* (USNRC, Washington, D.C., 2003)
27. United State Nuclear Regulatory Commission, *Risk-Informed Approach to Plant Specific Changes* (Washington, D.C., USNRC, 2000)
28. C.E. Ebeling, *An Introduction to Reliability and Maintainability Engineering* (Tata McGraw-Hill Education Pvt. Ltd., New Delhi, 2010)
29. S.A. Arrhenius, Über die Dissociationswärme und den Einfluß der Temperatur auf den Dissociationsgrad der Elektrolyte. Z. Physik. Chem **4**, 96–116 (1889)
30. S.A. Arrhenius, Über die Reaktionsgeschwindigkeit bei der Inversion von Rohrzucker durch Säuren, *ibid*, vol. 4, pp. 226–248 (1889)
31. H. Eyring, The activated complex in chemical reactions. J. Chem. Phys **3**(2), 107–115 (1935)
32. W.Q. Meeker, L.A. Escobar, *Statistical Methods for Reliability Data* (Wiley, New York, 1998)

# Chapter 6
# Probabilistic Risk Assessment

> *Every once in a while, a new technology, an old problem and big idea turn into an innovation.*
>
> Dean Kamen

## 6.1 Introduction

Probabilistic methods in the general probabilistic risk assessment (PRA) framework are at the core of integrated risk-based engineering and risk-based decisions. The probabilistic tools and methods are discussed in Chaps. 3 and 4 in this book. This chapter is dedicated to PRA, which is the analytical technique used for quantitative evaluation of risk for a complex scenario. This technique is also commonly referred to as probabilistic safety assessment (PRA). These two terms are used interchangeably among safety practitioners, but there is no distinction between them. At a comprehensive level, this technique is used in the nuclear [1], space and aviation [2], and chemical and process industries [3] and for applications in environmental protection [4]. In a limited sense, this technique is also employed in other systems such as railways for structural systems, dams [5, 6], and shipping [7] in support of design evaluation. The PRA/PRA methodology is also referred to as quantitative risk assessment (QRA). Here also the major framework remains the same as PRA/PRA, but the term QRA is used in some domains [8].

For the purposes of this book, the term PRA will be used because it directly connects to the title, namely integrated risk-based engineering. The IAEA Safety Glossary [9] defines probabilistic risk assessment as "a comprehensive, structured approach to identifying failure scenarios, constituting a conceptual and mathematical tool for deriving numerical estimates of risk." The various aspects associated with PRA are explained through references to nuclear power plants in this chapter.

© Springer Nature Singapore Pte Ltd. 2018
P. V. Varde and M. G. Pecht, *Risk-Based Engineering*, Springer Series in Reliability Engineering, https://doi.org/10.1007/978-981-13-0090-5_6

## 6.2    Basic Elements of Risk

Figure 6.1 depicts the fundamental concept of risk analysis, which involves asking the following questions: (a) What can go wrong? (b) How frequently can go wrong? (c) What is the hazard potential for a set of failures in the plant? These questions lead to the fundamental equation of risk as follows:

$$\text{Risk} = \sum_{i=1}^{n} \text{Likelihood}_i \times \text{Consequences}_i \tag{6.1}$$

As can be seen from this equation, the summation of frequencies of various accident sequences and associated consequences for various scenarios ranging from $i = 1$ to $n$ provides the statement of risk. The same philosophy is depicted graphically in Fig. 6.1. The consequences can be characterized as either spontaneous, such as the number of fatalities or injuries, or long term, such as the risk of cancer or any other ailments referred to as the stochastic effect. The consequences can also be defined as either loss of property, societal effects, or any other economic impact, such as loss of working hours.

The information on plant design, engineering, and operational aspects forms the input to reliability models such as fault trees and event trees to create the model of the plant. PRA is arguably one of the most significant and successful applications of reliability engineering. Further, among many risk-based frameworks, PRA provides the most robust technique for implementation of risk-based applications [10]. The major attributes that make PRA the backbone of risk-based engineering include:

- *Elegant framework*: PRA provides a holistic, systematic, and rational approach in support of plant design evaluation, operational parameter identification, evaluation of technical specifications, which otherwise are based on qualitative considerations such as engineering judgment and vendor recommendations like on allowable outage time, test and maintenance interval evaluation, etc.
- *Quantification*: PRA enables quantification of the performance level and, more importantly, "safety." Otherwise, these terms are treated as qualitative attributes.

**Fig. 6.1**  Basic elements of risk analysis

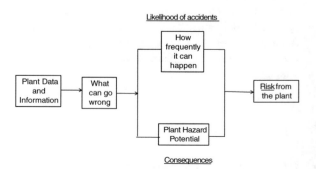

- *Power of integration*: The PRA approach provides a framework for representing an integrated model of the plant.
- *Considerations of human factors*: PRA's integration of human factors provides an opportunity for accounting for human performance along with hardware and software performance for assessing plant safety.
- *Uncertainty characterization*: PRA provides an effective mechanism for characterization of uncertainty by providing quantification of uncertainties in the estimates and results.
- *Improved regulatory review*: The risk-informed approach provides an effective and complementary framework for regulatory review.
- *Real-time applications*: PRA provides an efficient mechanism to develop real-time applications, such as risk monitoring, risk-based in-service inspection (RB-ISI), and human factor evaluation.
- *End application*: PRA forms the backbone of risk-informed/risk-based applications.

The real-time applications such as risk monitoring and risk-based ISI and maintenance are the testimony to the growing popularity of PRA for handling real-time issues. PRA deals with evaluation of risks from a plant (nuclear, chemical, or process) by postulating potential accident scenarios.

## 6.3   Three Levels of PRA

PRA studies are performed at three levels [11]. The available literature shows that Level 1 PRA has been performed for most of the nuclear plants globally, while there are good numbers of plants having Level 2 PRA while Level 3 PRA has been performed for only a few plants. It can be argued that in terms of benefits for routine applications, such as living PRA, risk monitor, risk-based ISI, maintenance optimization Level 1 PRA has been found very useful in the nuclear industry. Level 2 and Level 3 studies are used for accident planning and emergency preparedness including severe accident management.

For nuclear plants, PRA is performed at three levels as follows:

*Level 1 PRA*: The scope of PRA is to perform system analysis. This involves assessment of the plant's response for the selected initiating events and generating a list of accident sequences and its quantification. The result of Level 1 PRA is the integrated statement of core damage frequency (CDF) obtained by summation of the frequency of all the qualified accident sequences. Identification of the plant's strengths and weaknesses is one of the outcomes of Level 1 PRA. Results of Level 1 PRA form the input for Level 2 PRA.

*Level 2 PRA*: Level 2 PRA utilizes the results of Level 1 PRA to determine the plant damage states by identifying and grouping accident sequences. Containment scenario modeling for various plant damage states and source term evaluation form the mainstay of Level 2 PRA. Assessment of release frequency considering various source terms and containment response is performed as part of Level 2 PRA. The result of Level 2 PRA is estimates of release frequency for various source terms. The results of Level 2 PRA form the input for Level 3 PRA.

*Level 3 PRA*: Level 3 PRA deals with the consequence assessment. This involves assessment of radiation doses to the members of the public; for every release, once the activity is released in the atmosphere.

Figure 6.2 provides an overview of all three levels of PRA. The left-hand side of the figure provides a list of initiating events (IEs) that have been selected for analysis. Event trees are constructed for the IEs considered for detailed analysis. Typically, around 50 ETs are constructed as part of the accident sequence analysis for a nuclear power plant. Major categories of IEs are loss of coolant accident (LOCA), reactor transients (loss of off-site power (LOOP), and loss of regulation accident (LORA). The event trees are propagated by inducing the plant safety system response.

Hence, plant safety systems, human actions, and recovery probabilities form the headers in the event trees. The accident sequences are categorized considering the definition of the core damage states. Typically, there are 20 core damage categories. The CDF is evaluated for the plant, and the CDF statement forms the major result of Level 1 PRA. Uncertainty analysis and sensitivity analysis also form part of Level 1 PRA.

Level 2 PRA starts by suitably formulating plant damage states. Here, a review of not only accident sequences but also the complete Level 1 PRA model is required to further consolidate the plant damage states. The deterministic thermal-hydraulic and structural safety aspects are evaluated for considerations in

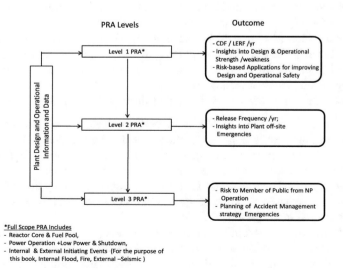

**Fig. 6.2** Interrelation of three levels of PRA and major results

Level 2 PRA. Often, additional analysis must be performed as part of Level 2 PRA. The major objective is to model the physical phenomena that define the core and coolant system for a given accident sequence. Containment event trees are developed considering the containment response, viz. continued containment integrity maintenance, containment isolation, emergency ventilation, and filtering system. The output of the containment event tree analysis is a set of release sequences. These sequences are binned (grouping of accident sequences characterized based on the nature and quantum of the releases) to further categorize the source terms for various releases. The statement of quantum and frequencies of release forms the main result of Level 2 PRA.

The result of Level 2 PRA forms the input for Level 3 PRA. Once the "containment source terms" are obtained, further modelings for assessment of risk to members of the public for various weather conditions are obtained. The major emphasis is the characterization of the risk levels at the exclusion zone (typically $\sim 1.6$ km radial distance around the plant for various accident conditions).

Level 1 PRA, which deals with plant systems modeling, provides the basis and framework for the development of risk-based applications, which include an integrated tool for risk-based design optimization, risk monitoring for plant configurational management, technical specifications, and prioritizing systems, structures, and components (SSCs) for asset management (maintenance and test schedules, aging management programs, etc.). This book focuses on aspects of Level 1 PRA but also introduces Level 2 and Level 3 PRA.

When regulatory decisions are based exclusively on the results and insights from PRA, the approach is called "risk-based," whereas when PRA is just one of the inputs along with a deterministic approach, the approach is called "risk-informed." However, the integrated risk-based approach proposed in this book considers that for any purpose the deterministic and probabilistic phenomena can provide a holistic and integrated picture. Such is the case with risk analysis, which, along with uncertainty, is inherent to the deterministic approach—the only difference being that it is qualitative and conservative. The PRA contribution provides an integrated, systematic, and quantitative risk model that also considers the human factor and common cause failure. Otherwise, all the scientific models and data that form the major inputs in PRA are solely based on deterministic analysis. Therefore, the proposed approach considers deterministic and probabilistic science together to arrive at risk estimates. The following section deals in brief with the role of PRA in IRBE, and the rest of the chapter is devoted to PRA methodology.

## 6.4 Role of PRA for Risk-Based Applications

Traditionally, complex engineering systems such as nuclear plants, process systems, aviation systems, and structural systems such as civil and mechanical structures are designed, maintained, and regulated based on conservative deterministic design procedures and criteria. Application of factor of safety along with multiple backup safety provisions and preconceived accident management plans are considered adequate to provide required safety assurance. Even though the deterministic approach is considered time tested and proven, the three major accidents, TMI, Chernobyl, and Fukushima, have provided vital insights and challenges that need to be addressed. There is a window of opportunities to strengthen safety given the backdrop of accumulated experience on design and operation of engineering systems that has logged in over the years. Apart from this, the new tools and methods that are available provide not only an improved perspective on safety but also have opened up new areas. This has led to realization of development of advanced concepts like design extension condition, severe accident management, and stress tests to evaluate safety levels beyond normal operating regime. Globalization introduces many challenges, the most important of which is to define "what is safe and how safe will be safe enough" while addressing the bottom line of the business (e.g., returns on electricity generation for nuclear plants). To this end, it is not possible to holistically assess safety based on the performance of nuclear plant components and systems with due considerations to external events. Compared to when these systems were developed, there is considerable accumulated operating experience on design and operation of components and systems of engineering systems. Advances in science and technology over the years have also influenced the way we look at systems. Evolutionary design approaches for system design, such as inherently safe design incorporating the reactor physics design that ensures, e.g., negative temperature and void coefficient, passive engineering design features like application of thermosyphoning for normal, and shutdown extended core cooling, have led to development of advanced nuclear plant where safety levels are higher compared to the conventional designs. It is quite obvious that there should be a review of traditional methods in light of the rational-based framework, such as PRA, that may allow optimization of both of the otherwise contradictory goals of safety and availability. Incorporation of digital technology has also enabled efficient designs for control systems.

This background and many other factors have directly or indirectly influenced the way we look at safety. The evolving scene has enabled us to have a rational-based optimum view of safety from the traditional "conservative" safety philosophy. At this point, it is very important that the role of PRA in design, operations, and regulation is noted to be increasing as part of risk-based or risk-informed decisions [10]. Probabilistic methods in general and PRA in particular provide the needed framework to address the above issues as follows:

- Quantified estimate of performance: PRA provides an effective strategy in support of risk-based decisions. The quantified estimate of the core damage frequency of the plant provides the level of safety of the plant. For safety systems, the statement of "system unavailability" provides the metrics for the safety level of the system.
- Failure probability: The concept of failure probability where the design models are analyzed considering inherent uncertainty in stress and strength parameters

instead of the factor of safety (in deterministic methods) provides an improved and effective metrics as part of risk-based design.

- Treatment of uncertainty: The confidence level (upper bound and lower bound) analyzed as part of the safety assessment provides added dimensions instead of point estimates, which do not provide variability in the estimates [12].
- Treatment of human performance: PRA provides a framework for integration of human performance into the plant model. Traditional methods do not treat this issue in a systematic manner.
- Common cause failure analysis: Apart from human factors, common cause failures can be considered to be the significant contributors to risk from the plant. The PRA framework provides a systematic evaluation and subsequent impact assessment of common cause phenomenon.
- Assessment based on component and system performance: The component and human reliability data form the basic building blocks of safety evaluation. The PRA framework is effective in providing the insights into safety by integrating SSCs and human performance, as reliability estimates. The deterministic approach lacks this integration.

The above features of probabilistic methods provide an effective framework for risk-based applications as follows:

(i) Integrated risk assessment of the plant: These models provide many applications, including regulatory reviews and comparison of available design options.
(ii) Plant configuration management: Here, operation and maintenance management of the plant using plant risk monitoring is one of the prime applications of risk-based technology. Other applications include prioritization of plant maintenance activities based on the safety significance of individual activity arranging from high importance terms in descending order [13]. Major operational applications include:

- Risk-based in-service inspection RB-ISI program. ISI program is developed considering the quantified notion of likelihood and consequences (core damage probability). The risk matrix of RB-ISI is developed using either EPRI/Westinghouse methodology.
- Development of risk-based operator support system for real-time scenarios in the control room of the plant.
- Incident or precursor event analysis.
- Plant technical specification optimization, which mainly includes surveillance test interval and allowable outage time of equipment and systems.
- Regulatory review of the safety cases.
- Integrated aging assessment studies.
- Public communication: It is very effective to compare different risk scenarios.

The available literature shows that there is growing interest in application of risk-based/risk-informed approach to address real-time issues. A framework is needed for PRA applications, which also include an administrative as well as regulatory setup that facilitates review based on national/international guidelines/ standards [14–16]. In this case, the application requires employment of the risk-based approach, such as risk-based ISI, and then, a systematic framework is required to logically and rationally address the decision process, for example, the one proposed in [17].

## 6.5  Quality in PRA and Its Applications

A PRA study must meet quality levels adequately such that it is commensurate with the given application. There are two major components to a high-quality PRA, viz. first, a quality assurance (QA) program that seeks management and organizational requirements to establish a QA program and, second, conformance to technical attributes to characterize the quality level of a PRA or its application.

The IAEA TECDOC-1151, while clarifying the difference between the terms "PRA quality" and "quality assurance," provides detailed attributes for various PRA activities and applications. "PRA quality" for a specific purpose refers to the technical adequacy of the methods, level of details, and data used to develop the PRA model [17]. To meet the PRA quality objectives, a quality assurance (QA) program is required that ensures application of state-of-the-art models, methods, and data, along with administrative provisions such as staffing, documentation, and verification and validation strategies in an adequate and controlled manner all through the stages of development of PRA and its applications. A QA program for a PRA covers all the activities necessary to achieve the appropriate PRA quality. The QA program sets forth the methods, resources, controls, and procedures and defines the responsibilities and line of communication affecting the quality of PRA [18].

A PRA project, whether it is developed to support a safety assessment or it is a PRA application, should envisage an explicit framework to cater to the QA requirement. To develop a quality framework, an in-house program of developing a quality document should be taken up during formulation of a PRA project. A policy decision should be made at the managerial level to work out the broad features of the QA program. The scope and objective of the PRA study along with national goals and policy will govern the choice and format of the program. Like any other project, it is a good practice that a PRA project report is prepared by the PRA team to bring out the broad features of the project, such as objectives, scope, target quality standards to which the PRA is required to conform, methodology, models and methods, and administrative setup, which includes responsibilities in management of the project, lines of communication and authority, project schedule, and provisions for change management.

Even though the IAEA-TECDOC-1101 report entitled "A Framework for a Quality Assurance Program for PRA" [18] deals exclusively with the QA aspects of PRA, the following list provides the major features of a QA program for a PRA study:

- *PRA working group*: At the beginning of the project, the working group should include a team leader and members having expertise in the area of PRA, design, operation and maintenance, and human reliability.
- *Definition of objective and scope of the study*: The objectives of the PRA study should be clearly defined. The project report should, as mentioned above, define the objectives and scope of the study, as well as the details or depth of the analyses related to issues such as human factors, common cause failure, and uncertainty and sensitivity analyses.
- *Quality conformance*: A decision should be made as to which standard/guide the PRA study should conform to. For example, IAEA TECDOC or ASME/ANS standard or any other national or international standard/guide can be identified for working out quality attributes for various stages or application of PRA.
- *Plans and schedules*: The project report should clearly define the manpower requirements and time required to complete the task based on the available resources. A detailed chart indicating various PRA activities and schedules should be included.
- *Establishment of criteria and technical requirements*: A system should be developed to cater to the technical requirement of PRA. This system mainly caters to the requirements of a QA program related to completeness of the model, consistency of the methods and procedures, and accuracy of the data and information. (In fact, this part is covered to a large extent by application of quality attributes as mentioned in item c.)
- *Multi-tier reviews*: The analyses carried out by the working group should be subjected to more than one level of review. This might include an internal review within the working group itself by members who are not directly involved in a subject task. The second-level review could be conducted by design, operational, and maintenance experts as the case may be or depending on the end product and stakeholders. The third-level review should involve peers comprised of reliability experts or PRA experts not involved in the development of the PRA model and preferably from other organization. As can be seen, before the PRA reaches regulators or a stakeholder, it has undergone many stages of review and this works toward enhancing the quality of the end product, i.e., the PRA study.
- *Input in support of PRA*: Provisions should be made in the QA program to ensure that the input from outside the PRA, for example, plant design and operational data, which includes thermal-hydraulic or core analysis, plant operating and emergency procedures, and component failure data, meets quality criteria.
- *Verification and validation of computer codes*: PRA modeling is a complex task and is carried out using a software code. The software should be capable of

offering the environment that is commensurate with the scope and objectives of the PRA study. Validation and verification of the software should be carried out using software for PRA modeling.

- *Uncertainty and sensitivity analyses*: The PRA results obtained are checked for erroneous inputs, suitability of initial assumptions of the analysis, and level of conservatism related to failure criteria. To achieve this, uncertainty and sensitivity analyses should be performed.
- *Documentation and information control*: The input data, the assumptions of the analysis, and the basis of different system failure criteria should form part of the project activity. Whenever possible, justification of the failure criteria and the subjectivity, if any, should be recorded. Other aspects of documentation and control are related to the traceability of the source of information/input data by providing the background information of raw data, such as component details, component boundaries, and sources of data and the rational for selecting a particular source. The PRA report should follow a standard format adopted by the available guides/standards.

The above features provide a general guideline and are not exhaustive. The PRA team will be required to develop a QA program that is commensurate with the objectives and scope of the study. The following section provides a brief outline for developing a quality attribute format for a given PRA/PRA application.

So for, we have discussed a system that will support development of a QA framework in PRA. However, we do not have a mechanism to check or rate the quality of a PRA. Ideally, a PRA study having adequate quality can only be considered for risk-informed/risk-based applications. In fact, the PRA document must state which standard/guide has been used to meet the "PRA quality" level. Hence, the PRA team should assess and determine which standard/guide, whether an international reference or a national standard/guide, the subject PRA study will conform to. The conformance to a standard/guide provides an improved understanding of the quality of a PRA study to either the regulator or the stakeholder, be it the NPP operator, designer, or user of PRA in public domain communication.

Well-recognized standards are available that can be used to develop quality checkpoints or attributes for a PRA project. These standards were published in 1995 by IAEA, and the latest was published by ASME/ANS in 2009. The selected standards/guide should meet the objectives, scope, and target application of the PRA project. Table 6.1 provides an overview of the available PRA standards in respect of their scope, QA approach, and broad framework.

The quality attributes are the conditions or requirements that need to be fulfilled to conform to the given quality criterion, for example, "initiating event list should be as complete as possible." This attribute is referred to as a general or higher-level attribute because it is related to the procedural element "initiating event analysis." Let us consider the quality framework developed in IAEA-TECDOC-1511. There are two types of attributes, the higher-level general attributes and the lower-level special attributes. The higher-level attributes are designed for procedural elements of PRA. In this document, the complete PRA procedure has been divided into nine

**Table 6.1** Overview of available guide/standard documents to support development of PRA quality attributes

| Document | Year of publication | Scope of the document |
|---|---|---|
| *Standards* | | |
| IAEA standard SSG-3 [19] | 2010 | Full-scope Level 1 PRA |
| IAEA standard SSG-4 [20] | 2010 | Level 2 PRA |
| Addendum A (ASME/ ANS RA-Sa-2009) [21] | 2009 | Level 1/LERF (large early release frequency) full-scope PRA (excluding low power and shutdown PRA) |
| Addendum B (ASME/ ANS RA-Sb-2009) [22] | 2011 | Level 1/LERF full-scope PRA (excluding LP and SD PRA) |
| ASME/ANS RA-S-1.2-2014 | 2014 | Level 2 PRA for light water reactors |
| ASME/ANSRA-S-1.3 (ANS/ASME-58.25) [23] | | Level 3 PRA |
| *Guides/technical documents* | | |
| USNRC regulatory guide RG 1.200 [24] | 2009 | Level 1 and Level 2 PRAs (internal + external hazards) |
| IAEA TECDOC-1511 [17] | 2006 | Limited-scope Level 1 PRA |
| NEI 05-04 [25] | 2006 | For operating reactors, limited-scope Level 1 (+ internal flood) |

elements in chronological order as initiating event analysis (IE), accident sequence analysis (AS), success criteria formulation (SC), system analysis (SA), human reliability analysis (HR), data analysis (DA), dependent failure analysis (DF), modeling and CDF quantification (MQ), and finally, results and analysis. As mentioned earlier, the attribute "the list of IE should be as complete as possible" is a higher-level or general attribute that has more than one special attributes. For example, one of the 13 special attributes listed in IAEA-TECDOC is "IE definition is clear and covers any plant disturbances that require mitigation to prevent the core damage." This special attribute requires the definition of IE as "an event which could directly lead to core damage or challenges normal plant operation and requires successful mitigation to prevent the core damage." What does it mean? In a given PRA study when the definition of the initiating event is clearly defined, then this quality attribute "is met or catered to." The complete subject of quality assurance of all the elements of a PRA is exhaustive, and details of the quality attributes are given in IAEA TECDO-1511.

Similarly, if the PRA team chooses to use the ASME/ANS framework to develop quality attributes, the ASME/ANS RA-Sa-2009 framework, as shown in Table 6.1, can be used for Level 1 PRA. The ASME/ANS standard evaluates the applicable technical requirements of PRA elements by determining the capability category. The standard defines three capability categories—Category 1 is the lowest, while Category 2 is higher and Category 3 is the highest capability. The standard provides an elegant mechanism to evaluate the capability category of each supporting requirement (SR) for a given technical element. Provision also exists in case a capability category could not be fulfilled by a PRA element. In such cases, supporting analysis can compensate for the lack of PRA capability of a PRA technical element. One of the approaches is by performing a deterministic analysis to support the case for a given PRA application. The ASME/ANS framework is targeted for implementation of the risk-informed approach in support of regulator review.

The scope and objective of the study should be commensurate with the scope of the standard/guide being used. From Table 6.1, it can be seen that for Level 1 PRA, particularly for limited scope and even full scope, there are a good number of options to use either IAEA or ASME standard. In fact, Level 2 PRA can also use IAEA [SSG-4] and ASME/ANS RA-S-1.2 standard, while Level 3 can use ASME/ANS RA-S-1.3 standard.

There can be a scenario where an organization or a national regulatory body, keeping in view safety requirements and resources needed, develops its own quality attributes. In such a scenario, apart from the ASME/ANS and IAEA-TECDOC-1511 quality attributes, other guides or technical documents like USNRC Guide 1.200, NEA PRA documents, or IAEA procedure guides, depending on the subject of application, can be used for developing PRA quality/capability attributes.

It may be noted that the treatment of the subject in this section is not exhaustive; hence, readers are advised to conduct thorough research before embarking on a quality policy for a given application.

## 6.6    General Elements of PRA

Before a PRA study is initiated, it is very important to define the objectives, scope, and purpose of the PRA. For example, for determining the scope of PRA the following objectives need to be considered.

### 6.6.1    Objectives of PRA

The general objectives of PRA are as follows:

- evaluate the safety level of the plant by estimating core damage frequency, release frequency, and overall risk to members of the public as part of Level 1, Level 2, and Level 3 PRA, respectively;
- identify important accident sequences;
- evaluate the plant's design strength and weaknesses;

- evaluate the plant operational safety;
- support risk-informed/risk-based applications;
- provide input, which mainly includes plant damage state scenarios;
- identify important plant damage states;
- provide input in terms of large radioactive release frequency, respective source terms from the containment;
- evaluate the overall risk to the members of the public. This input could be used for emergency preparedness;
- identify severe accident scenarios and support severe accident management programs.

As mentioned earlier, PRA can be initiated at any stage of the plant. However, PRA studies initial at early stages are more beneficial in terms of understanding and incorporating modification in design. Once the design is frozen and activities moves toward construction phase and later operational phase changes and modifications become more and more prohibitory. Accordingly, the broad objectives of PRA studies could be as follows:

- Evaluation of sitting criteria and considerations.
- Evaluation of various alternatives and designs for a comparative study can also be termed as evaluation of concepts and alternatives.
- Once a design is selected, a further objective is to create a PRA model of the plant based on given design information and data derived from the basic concept design.
- The general experience has been that the design further passes through two stages, primary and final. PRA at these stages can be a vital tool for configuration management and optimization of system configuration. In fact, the plant design changes take place through construction stages as many insights and constraints are known only when the plant is passing through the construction phase.
- To update PRA or to perform PRA at commissioning stages is a common experience because regulatory clearance for going into operation requires at least Level 1 PRA.
- PRA or more specifically a living PRA model is maintained all through the operational stages. The living Level 1 PRA model is periodically updated to reflect the current design and operating practices, particularly the plant's limiting conditions for operations (LCOs).
- PRA conducted to identify and prioritize SSCs in support of aging programs provides an effective mechanism for aging management and identification of areas for extended maintenance, retrofit, and safety upgrade to meet contemporary safety standards.

## 6.6.2  Scope of PRA

The scope of a PRA is determined by its objectives. The six major considerations/ factors that determine the scope of the PRA are as follows:

- *PRA levels*: It must be determined whether the scope includes only Level 1 PRA or Level 1 + Level 2 or all three levels. The scope of the PRA is usually determined by an application or regulatory requirements and by the resources ·and safety benefits that can be derived. The general observation has been that initially the utility starts with only Level 1 PRA, and later, if the need arises, then Level 2 and Level 3 are attempted. PRA is resource-intensive, and this aspect is one of the major considerations in determining the PRA level of for a given project.
- *Sources of hazard*: The potential sources of hazard must be identified. For example, for nuclear plants, the major hazard sources are the reactor core, the fueling operation, the fuel storage pool, and test and experimental facilities. Normally, the PRA is performed for the reactor core as the source of radioactivity; however, after the Fukushima accident, there is an increasing trend for performing PRA for the reactor pool also as a source of radioactivity. Other sources of radiation such as the fueling operations, and test and experimental facilities (for research reactors) also need to be considered when determining the scope of the PRA.
- *Scenarios covered*: The scope should define whether the PRA is being performed with consideration of internal and/or external initiating events. Most of the plants initiate PRA with only internal initiating events and later enhance the scope to include external events, such as seismic PRA, or external floods.
- *Plant operating state*: The operating states of the plants are full-power operation, transient-power maneuvering conditions, low-power operation, and shutdown conditions. Most of the PRA initiates with the full-power condition and later, depending on requirements, extends to low-power and shutdown conditions.
- *Stages of the plant*: To clearly understand the safety benefits for given options, the PRA should ideally start when the type of plant is being determined. This could be done by incorporating CDF levels as one of the specifications that determine the safety edge of various options. From an engineering point of view, the various stages are sitting, conceptual, preliminary design, design, construction, commissioning, operations, and aging. The advantage of conducting the PRA during the early stages, such as the sitting or conceptual stages, is that modifications can be easily made. As the plant goes into construction and operating stages, modifications become restrictive. Even though retrofitting is one of the options, the fundamental design constraints are challenging for retrofits.
- *Site PRA*: This PRA is performed to model risk estimates from a site and is also referred to as a multi-unit PRA. In a plant-specific PRA, risk estimate is given for a single reactor at the site. But the site can have more than one unit. Many

sites have up to 6–8 units or reactors. The objective of site unit PRA is to evaluate the risk to members of the public considering a scenario that has the potential to impact more than one site. For example, at Fukushima 3 units out of four located at the site suffered damages.

Even though the resource requirements are among the considerations for determining the objective and scope of PRA, every case requires specific treatment. Table 6.2 provides general guidelines for identifying the scope and level of details that may be considered. These considerations are not exhaustive, rather they are only indicative.

It is important to understand why the PRA study should be initiated. If a PRA is being performed to meet the regulatory requirements, then the objective is to demonstrate the safety of the plant against postulated initiating events. As far as possible, it is always desirable to use the updated plant-specific reliability data and information to reflect the as-built and as-operated conditions of the plant. However, if the PRA is being performed in support of risk-informed/risk-based applications (e.g., estimation of surveillance test interval, maintenance optimization, allowable outage time for equipment estimation, which directly impacts plant operating policy), then plant-specific data become a stringent requirement because specific situations require specific information to reduce uncertainty in decisions.

Hence, when a PRA study is conducted during the operational stages for implementing risk-based applications, then it is expected that: (a) plant-specific data are used, (b) the plant model represents the design and operational features adequately, (c) apart from the full-power operational state, the scope of the PRA includes the shutdown and low-power operational states, (d) the study is approved by the regulator for risk-based applications, (e) there are well-defined and accepted criteria for assessing the change cases, (f) the study provides estimates of uncertainties associated with the results, (g) the impact of the assumptions is evaluated and supported by sensitivity analysis, and (h) the application justifies the plant safety against identified common cause failures.

It should be demonstrated that the safety issues have been addressed with Level 1 PRA adequately and Level 2 and Level 3 PRAs may not be required. If it is felt that Level 1 PRA has not adequately resolved the issue, then the scope of risk-based application should cover performance of Level 2 and Level 3 PRAs.

## 6.6.3   Limited- and Full-Scope Level 1 PRA

Most PRAs can be categorized as limited-scope Level 1 PRA, whereas there is a noticeable trend to conduct full-scope Level 1 and Level 2 PRAs. Not many plants conduct Level 3 PRA. What is limited-scope Level 1 PRA? A Level 1 PRA study considers only the internal initiating event, with reactor core as the source of radioactivity and for full-power conditions. A limited-scope Level 1 PRA generally does not include the low-power and shutdown PRA, external events, or even

**Table 6.2** Major PRA requirements for various stages of a nuclear plant

| Requirements | Sitting/conceptual | Preliminary design/final design | Pre-commissioning | Operational |
|---|---|---|---|---|
| Objective | - Sitting considerations<br>- Regulatory approval for a site for locating a nuclear plant | - Design evaluation<br>- Balance of plant studies | - Regulatory consenting/approval<br>- Technical specification evaluation<br>- Assessment of emergency operating procedures<br>- General assessment of safety issues | - Performance and safety evaluation regulatory reviews<br>- Application of risk-based tools<br>- Risk-based O&M, risk monitor<br>- In support of change specifications.<br>- Risk-based ISI, precursor analysis |
| *Scope* | | | | |
| Levels | - At least limited-scope Level 1 PRA may be adequate | - Full-scope Level 1 and Level 2 desirable | - Level1/Level 2/Level 3 (desirable; however, Level 1 and Level 2 should be performed | - Level-1/Level 2/Level 3 (desirable; however, Level 1 and Level 2 should be performed |
| Events to be modeled | - Internal<br>- External (flood and seismic) | - Internal<br>- External and<br>- Combined | - Internal<br>- External<br>- Combined | - Internal<br>- External<br>- Combined |
| Source of radioactivity | - Reactor core | - Reactor core<br>- Fuel storage pool | - Reactor core<br>- Spent fuel pool | - Reactor core<br>- Spent fuel pool |
| Plant states | - Power operation | - Power operation<br>- Low-power and shutdown state | - Power operation<br>- Low-power and shutdown state | - Power operation<br>- Low-power and shutdown state |
| Purpose | - Sitting and design concept evaluation | - Design optimization<br>- Plant configuration and layout | - Integrated assessment<br>- Balance of plant studies<br>- Regulatory review | - Integrated assessment<br>- Living PRA<br>- Risk monitor application<br>- Risk-based application |

(continued)

**Table 6.2** (continued)

| Requirements | Sitting/conceptual | Preliminary design/final design | Pre-commissioning | Operational |
|---|---|---|---|---|
| Input required | - Project reports<br>- Information from similar systems | - Applicable standards<br>- Quality assurance program<br>- Design basis reports and plant layout; vendor information<br>- P&ID<br>- Control and protection logic drawings | - Technical specifications<br>- Safety report<br>- Design basis reports<br>- Control and protection logic drawings | - Commissioning reports<br>- Safety report<br>- Design basis reports<br>- Operating manual<br>- Emergency operating procedures |
| Reliability data source | - Generic data<br>- Similar plants | - Generic<br>- Similar plant | - Generic<br>- Similar plant<br>- Vendor input | - Plant-specific with Bayesian updating |

*Note* This table is indicative only and highlights the concept of requirements of PRA study and expectations for various plant stages. However, keeping in view the mandate or regulatory requirements for a given application, the PRA requirements at various stages also might change. For example, even during site evaluation all the levels of PRA might be required as the stakeholders might like to have an assessment on impact on public health during release from normal operation and as well as accident conditions. However, as expected, uncertainty in the final result is bound to be high because the study is based only on a preliminary design or on a concept and generic data internal fire and flood. A full-scope Level 1 PRA includes internal and external events, all operating states, and reactor core and fuel storage pool, as well as test and experimental facilities (for research reactors).

## 6.7  Methodology for Limited-Scope Level 1 PRA

The term "full scope" has a special meaning in PRA parlance. As part of the implementation of a PRA program, most of the PRA studies initiated worldwide are limited in scope because they only cover internal initiating events, full-power operational state of the plant, and reactor core as the source of radiation. These studies require an in-depth review to determine their adequacy for a given application, and they may not provide the holistic perspective that is required for an effective risk-based management approach. Hence, in addition to the above, the Level 1 PRA should include the shutdown state and low-power operational state of the plant, the reactor storage pool and other possible sources of radiation, and analysis of external events, such as flood, fire, seismic, aircraft crash, or impact assessment of an external object.

This section presents the procedural aspects of PRA that any organization or individual should consider before conducting a PRA project. The procedural steps covered in the following subsections are based on the USNRC Guide NUREG2815 [26] and IAEAPRA Guides for Level 1 PRA [19, 27]. There could be small variations in respect of different designs, like pressurized water reactor, boiling water reactor, Candu (Canadian design), or any advanced designs like advanced boiling water reactor, and advanced pressurized water reactor. However, the broad procedure remains the same.

### 6.7.1  Organizational and Management

Conducting a PRA study is resource-intensive and requires an organizational framework before the study itself can be initiated. The major organizational elements are as follows: (a) definition of the objectives and scope of the study, (b) identification and development of quality standard/procedures, (c) formation of a PRA team, (d) drafting of schedules, (e) organization of resources, (d) organizational framework with responsibilities and duties for smooth conduct of PRA, (e) review of policy and procedures, including the regulatory submission and review, and (f) documentation requirements.

The best approach to address the above aspects is to prepare a PRA project report that highlights and discusses the above organizational elements. This project report should clarify the purpose of initiating a PRA study. For example, the project could be fulfilling a regulatory requirement or the utility or facility could be proactively initiating PRA before developing applications such as risk monitoring, a risk-based ISI program, or safety evaluation. The objective and scope of the study might clarify, as mentioned above, the level of detail and the technical purpose of the study; for example, the requirement could be a limited-scope Level 1 PRA to begin with or a full-scope Level 1 PRA as a second phase.

The PRA team should decide which standard the proposed PRA will conform to. There are international standards for PRA such as ASME standard [20], IAEA standard [19], and national standards such as Canadian standard [14]. The regulatory body needs to know which standard a PRA is conforming to. In case the PRA study is not wholly conforming to a standard, it is beneficial to state which activities meeting the guidelines.

The PRA is an interdisciplinary activity and requires expertise on not only PRA, but design, operations, maintenance (viz. nuclear, mechanical, electrical, control, and instrumentation), and reactor physics and thermal hydraulics. Specific aspects require knowledge of structural engineering, particularly when seismic PRA also forms part of the PRA. A PRA team comprising experts from all these fields is vital for the success of the PRA study. The team should have one team leader, having expertise in PRA along with other experts, with a clear mandate and well-defined lines of communication and authority as part of the organization framework.

The team should develop a schedule that includes a time line for various activities. PRA projects are often outsourced to a consultant, but whether this is the case or the PRA is conducted in-house, certain aspects are worked out at the outset before initiating the project, such as the total man-hours required for PRA performance and review, facility or support infrastructure, and the computer software required to perform PRA. The benchmarked PRA computer software is prerequisite for implementing a PRA project.

The organizational activities of various procedural elements of the PRA should be depicted through a chart that defines information flow, lines of communication, and lines of authority (e.g., with whom the team leader interacts on a routine basis). The PRA study should be subjected to multi-tier reviews; for example, the team members among themselves can perform the first level of review. The second level could be the design or operations experts, while the third review can be performed by independent peers groups not directly associated with the PRA task.

After incorporating comments from independent reviews by the PRA project team, the final document can be produced for submission to regulators. There should be a well-defined template for preparing the documents. For example, a template is required for the system analysis and the initiating event analysis. These templates should provide guidelines in various subsections, such as the introduction, assumption of the analysis, and system analysis, (e.g., failure model and effect analysis, failure criteria, reliability data and quantification, fault tree analysis, common cause failure analysis, human reliability analysis, discussion, and presentation of results). The information in appendices should include, for example, a list of failure data, cut-set lists, and a description of any code used in the analysis.

The project report should bring out all the requirements, as mentioned above, clearly so that the PRA team as well as members of the organization has clarity on what to expect and what are the roles and responsibilities of the individuals.

## 6.7.2  Plant Familiarization

PRA, to a large extent, is about how well you know the plant. This includes detailed insights into the design, operation and maintenance, and regulation of the plant. Familiarization should not consist of merely participating in a lecture program on design and operation or even a plant walkdown if it is in operation. If the PRA is being conducted during the concept or design stages, then the only option is to learn from training/lecture courses; however, in such cases, involvement of designers in conducting the PRA will ensure that detailed design aspects are reflected in the PRA model. An updated status of the design and operations through tracking of modifications and changes ensures the incorporation of as-built and operated plant information in the PRA. Hence, the PRA team should have up-to-date knowledge of SSCs and various operational designs and configurations. The detailed information on SSCs is required to support selection of generic data from the available sources. The requirements and major activities that form part of plant familiarization are as follows:

- Plant management and organizational setup;
- Training and qualification of the PRA team in plant design and operational aspects;
- Plant walkdown to have knowledge of plant and equipment design features and layouts;
- Study of safety documents, like the safety analysis report, technical specifications and plant operating policy, and emergency operating procedures;
- Plant shutdown schedules;
- Surveillance, testing, and maintenance schedules and procedures;
- Individual system files to track changes in operating policies and modifications;
- Annual performance reports including safety performance indicators, significant events, and event report records;
- Correspondence with the regulatory authorities on modification proposals and changes in operating policies and their respective bases;
- Applicable quality assurance standards and procedures.

This list is not exhaustive and is mostly applicable to Level 1 PRAs. Level 2 and Level 3 PRAs might require additional documents such as containment leak test reports, core radioactive inventory for equilibrium core for given burn-up levels for Level 2 PRA, and feedback generated over the years on emergency drills and data on routine and emergency discharge for Level 3 PRA.

Keeping in view the above, plant familiarization tasks include (a) lectures and training programs on design and operational aspects, (b) plant walkdown for introduction to plant layout and equipment and system configurations, (c) library visits to study design and operations manuals, emergency operating procedures (EOPs), and safety reports, and (d) regular interactions with design and operations staff. Posting of PRA staff for in-plant training has been found to be an effective mechanism for imparting reasonable confidence to the staff in operational aspects of the plant.

## 6.7.3 Identification of Plant Hazards and Formulation of a List of Applicable Initiating Events

One of the parameters in formulating the scope of a PRA study is to identify hazard, which requires answers to the question, what types of hazards are we looking for? For example, for a process or chemical industry, the potential release of chemicals or gases, such as chlorine, hydrogen sulfide, ammonia, or potassium cyanide (the case in point is the 1984 Bhopal gas tragedy), either in the plant or outside the plant boundary is a hazard. For space missions, threats to the lives of the astronauts and the public during its mission could be a hazard. Unlike nuclear plants, where the source of radiation can be identified either as reactor core or as fuel storage, and a few other facilities, the hazard sources in a chemical plant are distributed and need special methodology, such as hazard and operability studies (HAZOP) to characterize not only the hazards in the plant but also efforts to mitigate the effects of hazard. However, to demonstrate this methodology, keeping in view the focus of the book, we will discuss the case of hazard identification and identification postulated initiating events.

In PRA, we model potential sources of ionizing radiation from a nuclear plant operation. The typical categories are the reactor core, the reactor spent fuel storage pool, and the test and experimental facilities (for the case of research reactors). Traditionally, the limited-scope Level 1 PRA considers the reactor core as the source of radiation. However, after the Fukushima incident in 2011, there is growing interest in including spent fuel storage along with reactor core as the source of radioactivity. The second aspect being investigated is site-level PRA or multi-unit PRA. In these PRAs, radiation sources originate from all the units (i.e., core and fuel storage pool can be considered for site-level PRA). Since site-level PRA is still in the R&D stage, this section will discuss only unit-level PRA.

Formulation of a list of initiating events requires a systematic approach that involves more than one strategy so it is as complete as possible. The major sources of information toward formulating a list of initiating events include:

- Master logic diagram;
- Safety analysis report and emergency operating procedures;
- Existing list of initiating events from a generic source;
- Existing list of initiating events from a similar plant
- Plant operating experience.

There are documents available in the literature that provides exhaustive information and guidance on selection of initiating event, viz. IAEA Technical Document [28]. A master logic diagram (MLD) enables the PRA team to logically analyze the plant scenario in respect of plant response systems for identifying the candidate initiating events. This deterministic study also postulates a list of initiating events, which is part of a safety analysis report. These events are plant-specific and can also be adopted suitably for PRA studies. There are generic lists of initiating events available in the open literature. These lists are formulated keeping in view the type of plant. The available list from an existing PRA for similar plant can often be a good source for formulating the list of initiating events. Before it is finalized, the list should be validated in light of available operating experience such that the list of postulated initiating events conforms to plant-specific data and information. The result of this step is a comprehensive list of initiating events.

## 6.7.4   Initiating Event Analysis

The available list of initiating event should be screened using quantified criteria, and the remaining events in the list are subjected to further analysis. The screening criteria could be a frequency of $<10^{-4}$ h for design basis events. Any event having a value more than these criteria qualifies for detailed qualitative event tree analysis. These criteria are either determined based on the experience of the analysts or a reference value used from the PRA procedure guide. Among these events, certain events may not require a detailed analysis and can be further screened out using qualitative arguments. The remaining list can be considered the final list that will be a candidate for detailed modeling using event tree analysis. The IAEA-TECDOC provides detailed guidance and screening analysis [28].

Initiating events can be combined in one group if:

- The events arise from the same/similar component failure in a system, such as LOCA, which is consequence of a breach in the coolant system boundary; hence, major medium and large LOCAs should be in the same group.
- The plant response, in terms of demands on the safety system, human action, or recovery mechanisms, is the same.
- The chronological order in which demand is placed on the safety system is the same.

For example, the broad category of LOCA will have subgroups, viz. small, medium, and large LOCA based on the size of the leak, the requirements of the safety systems, and the order in which these safety systems are demanded. Another category could be Class IV power failure, which will accommodate events leading to a station blackout scenario, and the safety systems required are reactor shutdown, decay heat removal, emergency power supply (Class III and/or Class II power supply).

A further analysis of the list of initiating events is conducted with the consideration of qualitative analysis. If the initiating event can be screened out by employing qualitative analysis and justification, then these events are taken out of the list of initiating events. The remaining initiating events are subjected to detailed analysis using event tree methodology.

## 6.7.5 Accident Sequence Analysis

Accident sequence analysis is performed in two stages, qualitative modeling and quantitative analysis. This section will discuss the qualitative aspects of event sequence modeling and introduce in a broad sense the quantification process to make this section more comprehensive. The detailed aspects related to quantification and analysis will be covered in the section on data analysis.

At this stage, it is required to identify and understand what forms a safety function and an initiating event in the plant. On a higher level, the plant can be categorized into two broad parts—the process system that supports regular plant operations (e.g., the main coolant system, power supply systems, regulation system) and the safety functions. The safety functions could be hardware functions that are normally referred to as a safety system or a safety support system, or they could be human actions involving recovery. The failure of the process systems leads to initiating events, whereas the response of the safety functions on demand determines the course of the accident sequence for a given initiating event and is analyzed as part of the process called accident sequence analysis.

The event tree analysis is the most commonly used technique for accident sequence modeling. Hence, this section discusses modeling aspects through the event tree analysis procedure. Event tree analysis is a graphical technique that uses deductive logic for propagating a response to an initiating event. In an event tree model, the safety functions form the header event arranged in their chronological order, i.e., in the order that the analyst perceives that the safety functions will be demanded. Even though event tree diagrams have been discussed in the Chap. 4 on system modeling, in this section, event tree analysis is discussed to highlight aspects related to accident sequence analysis.

Let us take the case of loss of off-site power (LOOP) failure analysis as part of Level 1 PRA analysis. Figure 6.3 shows the LOOP event tree. The LOOP event is introduced on the left side of the event tree diagram. It may be noted that off-site power failure can occur due to grid power failure or due to failure of on-site components, such as transformers and breakers. However, for the present event tree analysis, let us consider a loss of off-site power failure due to grid failure only.

Next, the plant has designed an automatic provision to cater to a power failure event without any immediate operator intervention. Hence, the sequence of events that call for actions are (a) on sensing the LOOP event, the reactor trips immediately, (b) since the main coolant system operates on off-site power, the main coolant system trips due to non-availability of the power supply, and the decay heat removal function is performed either by the feed water system or by the shutdown cooling system (for the sake of modeling, let us assume that the decay heat removal is being performed by shutdown cooling system), and (c) the plant logic automatically starts on the emergency power supply (let us assume three diesel generator sets) that caters to the emergency/essential loads in the plant.

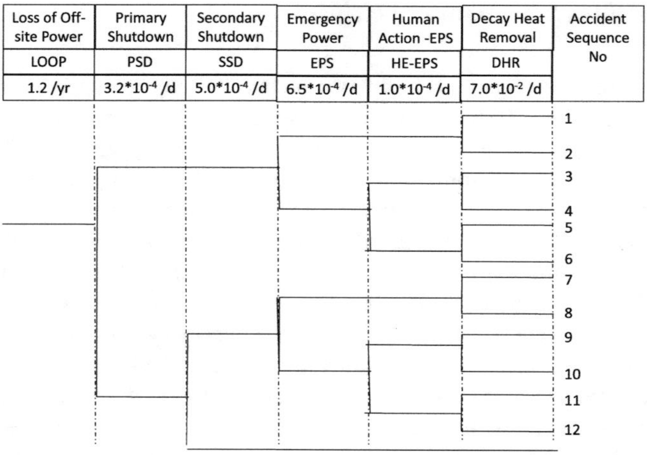

| Loss of Off-site Power | Primary Shutdown | Secondary Shutdown | Emergency Power | Human Action -EPS | Decay Heat Removal | Accident Sequence No |
|---|---|---|---|---|---|---|
| LOOP | PSD | SSD | EPS | HE-EPS | DHR | |
| 1.2 /yr | $3.2*10^{-4}$ /d | $5.0*10^{-4}$ /d | $6.5*10^{-4}$ /d | $1.0*10^{-4}$ /d | $7.0*10^{-2}$ /d | |

**Fig. 6.3** Event tree modeling for loss of off-site power accident sequence modeling

Further, let us assume a scenario in which there are two shutdown systems in the plant—the primary and secondary shutdown systems—and both the shutdown systems have the capability to trip or shutdown the reactor independently. The plant protection logic is such that if the primary shutdown system fails, then the secondary shutdown system is automatically actuated for tripping the reactor. We will consider an operator action, i.e., recovery of failed diesel generators. The scenario is as follows: Out of the three diesel generators, no diesel generator was started on demand and operator action was initiated to recover at least one diesel generator such that plant safety criteria are met.

The event tree shown in Fig. 6.3 considers the above provisions and criteria.

### 6.7.6  System Modeling

In Chap. 4, we discussed in detail various aspects related to system modeling, including various techniques such as failure mode effect and criticality analysis (FMECA), fault tree, event tree, and Markov, and extended this discussion to introduce the advanced methods dynamic fault tree and dynamic event tree, BDD. In this section, we will discuss system modeling using static fault tree or simply referred to as fault tree analysis. The objective is to deal with the modeling aspects of the system as part of PRA in a conventional sense, i.e., static PRA in support of LOOP modeling.

As can be seen in the event tree, the emergency power supply system of the plant we consider here is a Class III system comprised of two diesel generator systems and associated electrical buses and breakers. Let us consider the following simple system configuration, as shown in Fig. 6.4, for a Class III power system.

**Fig. 6.4** A simplified schematic depicting typical configuration of Class IV and Class III power supply

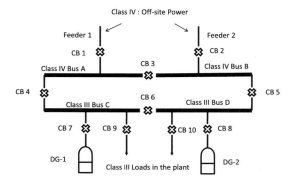

Failure criteria definition: The Class III power or emergency power supply is considered to have failed if the buses C and D experience under voltage for more than a predetermined time during a Class IV power scenario. Let us assume this time limit is 200 ms. This also means that even if one diesel generator (DG) starts on demand and feeds two buses, Class III power has met the success criteria.

The operational logic of the power supply system is as follows:

- The Class IV breakers, CB-1, CB-2, CB-4, and CB-5 trip and isolate Class IV bus from Class III bus. (Note that CB-3 and CB-4 are intertie breakers and have function to close when there is a power failure on one of the two buses and the failed bus requires power to be fed from the healthy bus.)
- DG-1 and DG-2 get a signal to start automatically and feed the Class III bus, by sensing under voltage on the Class IV bus.
- The DGs run to feed Class III loads for a given mission time.
- Once Class IV power resumes, the DGs are stopped and the Class IV power is normalized (by closing the Class IV breakers) manually.

Assumptions form an integral part of system analysis. Assumptions are made when there are complexities that need not form part of the problem and when certain information is not available. Assumptions also state the boundary conditions of the analysis being performed. It is the analyst's choice to formulate assumptions such that while meeting objective and scope of the problem, unnecessary complexities can be removed. Further, in case required, the impact of these assumptions can be studied when the system quantified model is ready.

In this case, to simplify the procedure let us make following assumptions:

- Class IV and control logic do not form the scope of the fault tree analysis.
- Once the DGs start, they operate successfully. Thus, the failure to run has not been considered in the fault tree analysis.
- Common cause failure (CCF) is included for the two DGs.

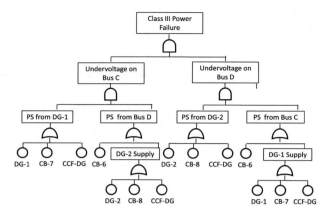

**Fig. 6.5** Simplified fault tree for Class III/emergency power supply system

Further, many more assumptions have been made for this illustration which is not explicitly mentioned and which will be discussed while discussing the fault tree for the Class III power supply system. The simplified fault tree for the Class III or emergency power supply power supply system is shown in Fig. 6.5.

The essence of fault tree analysis is that any one, or combination, of component failures that leads to top event, i.e., Class III or emergency power supply failure (i.e., under voltage on Bus C and Bus D), should be logically represented in the fault tree. The fault tree is a deductive approach to system modeling. The top-down approach continues until one reaches the basic components/human action. The definition of a basic component is governed by the aims and objective of the analysis. For example, in this example the DG is considered the basic component. However, applications requiring, let us say, analysis of the contributions of individual auxiliary systems to DG failure will require further analysis down to the individual auxiliary systems, such as the lubricating oil system, fuel oil system, jacket cooling water system, or control system. In fact, if the objective is to further assess which component of the auxiliary system contributes to DG failure (e.g., the fuel oil system itself, which is comprised of the fuel oil tank, pumps, filters, piping system, and instrumentation that includes level switches), then the definition of a basic component will include components of auxiliary systems. Hence, the analysts have to make a conscious decision about the depth of the analysis.

For the purpose of this illustration, it was considered appropriate to limit the analysis to the DG and circuit breakers (CBs) as the basic components. As can be seen, the power to the respective bus occurs by either DG or CB or common cause failure of DGs (CCF-DG). The DG and CBs are called an independent failure, while the CCF-DG is called a dependent failure. What it means is that one single cause can fail the redundant component, here the DGs. The CCF or dependent failure will be discussed in detail in Sect. 6.7.11, in this chapter.

The fault tree can be drawn either manually or using software, particularly a commercially available benchmarked software. The objective of the fault tree analysis is to obtain a list of cut sets at the qualitative level. The list of cut sets provides insight into component or combination of component failure that leads to top event, i.e., system failure. The fault tree analysis can be done using a Boolean logic reduction approach or any other method; however, the computerized approach provides an efficient solution.

Even though the quantification and data aspects have been considered separately, in this example, the failure probability for DGs, CBs, and CCF-DG has been included to demonstrate the quantification aspects, including the failure probability of cut sets, compared to Class III system failure probability. Accordingly, Table 6.3 shows the component failure probability assumed in this analysis.

It can be seen that CCF-DG has been assigned a probability of 10% of independent failure probability of DG-2, i.e., 5.0E−3. Here, a β-factor model has been assumed, which will be discussed in the following section on dependent/common cause failure.

Table 6.4 provides the result of this fault tree analysis, i.e., a list of cut sets or component combinations that lead to system failure associated probability and Class III system failure probability.

If we look at the cut set, we have CCF-DG as the only single-order cut set. The other cut sets are a combination of independent components in two parallel systems that can be referred to as two trains of the DG system, viz. train 1, which has DG-1 and CB-7, while train 2 has DG-2 and CB-8. Hence, the reduced fault tree based on the insights available from the cut sets given in Table 6.4 is shown in Fig. 6.6.

What does this analysis mean? Let us discuss both system-level insights followed by accident sequence aspects.

As indicated in the fault tree, the top event (i.e., the Class III power supply system failure probability) is 3.2E−3/demand. This value, when compared with a

**Table 6.3** List of component failure probabilities

| Component | Description | Code | Failure probability/demand |
|---|---|---|---|
| 1 | Diesel generator-1 | DG-1 | 2.0E−2 |
| 2 | Diesel generator-1 | DG-2 | 5.0E−2 |
| 3 | Circuit breaker-7 | CB-7 | 3.0E−3 |
| 4 | Circuit breaker-8 | CB-7 | 3.0E−3 |
| 5 | Common cause failure of DGs | CCF-DG | 2.0E−3 |

**Table 6.4** List of cut sets

| Cut set No. | Cut set | Unavailability |
|---|---|---|
| 1 | CCF-DG | 5.0E−3 |
| 2 | DG-1*DG-2 | 1.0E−3 |
| 3 | CB-7*DG-2 | 1.5E−4 |
| 4 | DG-1*CB-8 | 6.0E−5 |
| 5 | CB-7*CB-8 | 9.0E−6 |

**Fig. 6.6** Reduced fault tree
for Class III/emergency
power supply system

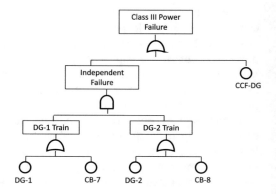

target value of 1.0E−3/demand, provides an assurance that the system performance can be considered satisfactory. However, efforts can be made to further achieve a target or lower figure by improving the component maintenance procedure such that the failure probability of an individual component can be reduced. This can also be done by enhancing maintenance efforts on the most contributing component; for example, DG-2 failure probability is more than DG-1. Further, the list of cut sets provides a combination of components and their failure modes; for instance, for CBs, the applicable failure mode is "fail-to-close on demand" or "fail-to-remain closed" on demand, whereas for DGs, the failure mode is "fail-to-start" on demand which leads to system failure. Investigation of CCF is the most important aspect that needs to be addressed, and design provision should exist to reduce the chances of CCF events, such as physical separation for protecting the redundant system from fire or an independent source of power supplies for auxiliaries.

As part of Level 1 PRA, the top event probability, i.e., Class III or emergency power supply system failure probability, per demand forms an input to the respective header event for the LOOP event tree as shown in Fig. 6.2, in quantitative aspects of accident sequence analysis.

There are two approaches for fault tree analysis—event-based and symptom-based. The event-based approach involves logical representation of component failure that leads to higher-level events, and this process continues until one reaches down to the basic components in the system. In the symptom-based approach, the systems govern the deduction process and not the component failure. Readers can refer to the related literature. The event-based approach is used when the analysts have a relatively improved understanding of the system operation logic, whereas the symptom-based approach is intuitive in nature.

Figures 6.4 and 6.5 present the event-based approach. For example, the symptom-based approach would proceed in asking questions as follows: It starts from top event "Class III power failure" and considers symptoms like "under voltage on bus C'" and looks for reasons. The reason is no input (voltage) from CB-7; the cause is this symptom or a CB-7 failure OR no input to CB-7. Why is there no input to CB-7? The terminal event is DG-1 failure. Even though the event-based and symptom-based fault trees might have different structures, the cut sets produced by both approaches are the same. Hence, the user has a choice to select any one of the approaches for system analysis.

## 6.7.7   Failure Criteria Evaluation

Definition of success or failure criteria is critical and a vital aspect of PRA. In this chapter, we will discuss in terms of failure criteria as it is in line with failure analysis or risk analysis character of the subject we are dealing with. At the outset, it should be emphasized that the characterization of failure criteria should, as far as possible, pass the following requirements:

- The definition of failure criteria should be crisp and clear.
- While PRA modeling requires that failure criteria should be defined in terms of minimal component or system configuration, e.g., 2-out-of-3 pump failures as the definition of failure criteria, this configuration should be linked to the requirement of deterministic thermal hydraulics in terms of process variables, such as the minimum flow and pressure required to keep the temperature within the predefined limits for the success of the system. For electrical systems, the minimum plant loads to address emergency requirements for, e.g., 3-out-of-3 DG failures for defining a Class III or emergency power supply system failure, while for control systems, the minimal configuration that will meet the protection requirements, e.g., 2-out-of-4 channel requirement to be called a system failure, is some of the examples of failure criteria.
- The definition of failure criteria starts at the component level and proceeds to the system and finally plant level. For example, the definition of DG failure might include either only "failed to start" or also include "failed to run." At the system level, let us say, for the Class III system, the definition of failure criteria might be worked out as the power supply to a given set of electrical buses in terms of under voltage levels determined based on successful system operation, or in terms of, say, 2-out-of-3 DG failure; at the plant level, the definition of core damage and associated level of degradations form the definition of failure criteria.
- The definition of failure criteria requires definition of component/system/plant boundary for which this definition of failure/success applies. For example, it should be explicitly made clear that the actuation or protection system deals with digital or analog control logic and whether it includes the protection devices such as shutoff rods/control rods or any other poison injection equipment such as solenoid valves or pumps.
- The failure criteria for human action as a safety function as part of an accident sequence evaluation after the initiating event require determination of the available time window and task analysis along with evaluation of various performance-shaping factors. Similarly, other human latent actions, such as test

and maintenance errors and calibration errors that remain latent and are revealed when the actual or subsequent testing is carried out, also need failure criteria defined considering related aspects such as the level or severity of fault system coping capacity.

- Mission time requirements of should form part of the definition of failure criteria. For example, mission time for long-term operation of emergency core cooling system in terms of hours or days should be specified.
- Apart from the above, the definition of failure criteria should also meet the specific requirements of the analysis and the subject being addressed.

Let us discuss in brief the aspects associated with the definition of core damage as the failure criteria at the plant level. Ideally speaking, the definition of core damages should be traceable to the risk to the individual members of the public or society at large. However, in practice, this information is available only when Level 3 PRA is conducted. But the process of PRA study starts with Level 1 PRA followed by Level 2 and then Level 3. Given the current scenario, for most of the NPP designs, such as PWRs, BWRs, and PHWRs, a general definition of core damage in terms of plant variables is available. For example, for PWRs the definition of core damage can be generally expressed as "Uncovery of Reactor Core" below a preset level. The common practice is to define core damages based on insights from the deterministic analysis in terms of parameters like coolability criteria, oxidation of cladding, limits on fuel temperature, and failure of structural systems.

For example, in time domain, it could be prompt to early core damage and delayed core damage, while in terms of severity of core damage, it could be from no core damage, degraded core, core damage, or severe core damage. When PRA studies were initiated in the early 1970s, it was common practice to define core damage by assuming conservatively that "any accident sequence that does not terminate in safe state can be categorized as core damage sequence." This definition provides a conservative basis to define the core damage and did not require elaborate knowledge of thermal hydraulics or structural analysis. However, the insights into severe accidents, the availability of computational tools, and advance modeling methods (including uncertainty characterization framework) have made it possible to derive a definition of core damage from physical core degradation parameters.

Further, the last but most important point in respect of the definition of failure criteria is that it should be benchmarked against the quality attributes or guidelines given in national or international standards [17, 22]. Hence, it can be inferred from the above discussions that the definition of failure criteria should be based on sound deterministic insights, severe accident phenomenology, requirements for the configuration and mission for SSCs, and validation and verification of quality attributes.

## 6.7.8  Data Collection and Analysis

The adequacy and accuracy of component failure rate data in a PRA study is one of the important factors that determine the quality of results and insights of PRA. The sources of data could be (a) generic, (b) plant-specific, (c) test data that includes accelerated test data particularly for new components where neither plant-specific nor generic sources are available, (d) handbook sources, like for electronic components the MIL-217 handbook [29], and (e) physics-of-failure models, particularly for electronics components.

### 6.7.8.1  Generic Data

Generic data available from various sources, including IAEA-TECDOC-478 [30] and NUREG-6268 [31], are used for quantification of the model when the PRA study is conducted at the design/construction or commissioning stages of the plant. Mostly, the data that can be obtained from generic sources come in the form of mean/median value of failure rate ($\lambda$ per unit time) or fixed probability, i.e., demand failure probability. Often, the data source specifies the confidence bounds, viz. lower and upper and lower bounds. Data on human error from handbook sources can also be categorized as generic.

Even though the use of generic data replaces the complex process of data collection and analysis, the following guidelines should be adhered to while using data from generic sources:

- The subject component detailed specification should be made such that component with the nearest specification can be selected from the database that to a large extent meets the component specification.
- It should be recognized that there are two broad categories of components, standby, and continuous operating. While selecting the failure rate from generic sources, this aspect should be considered.
- The root sources and accumulated experience and operating environment that form the basis of generic data should be evaluated to ensure adequacy of the data.
- It is always desired that data, apart from mean/median value, with confidence bounds and distribution are selected such that this information is effectively used for propagating uncertainty at the system and plant levels.
- Data having large uncertainty bands should not be used.
- The failure mode of the subject component and the generic data sources should be consistent. For example, if the failure mode for an air-operated valve is "fail-to-open on demand," then one should choose the component that has "fail-to-open" as the failure mode and not "fail-to-close." Often, the database provides a failure rate for "all modes"; i.e., the population has failed to open as well as failed to close, both the modes. When data for a given mode are not available, then conservatively "all the modes" are also selected.

### 6.7.8.2  Handbook Approach

The handbook approach is commonly used for electronic components. For example, MIL-HDBK-217F [29], RIAC-Plus [32], Electronic Parts Reliability Data—EPRD-2014 [33], TELCORDIA-SR-322, Parts 1–4 [34], BELCORE-TR-332 Version 1997 [35], IEEE-500 [36], and IEC-62380 [37] are some of the most commonly used handbook or data sources that are employed for reliability evaluation of electronics and control systems. These handbooks provide multiplicative models and reference failure data, which enable evaluation of failure rates for specific applications using some application-specific factors, also called $\pi$-factors or modifying factors, such as quality of components; environmental stresses, viz. temperature including junction temperature; and construction features such as die complexity, number of pins for semiconductor microchips, type of packaging. These handbooks provide the base failure rate for various types and ratings of devices in a component category that is modified using these factors to arrive at the failure rate of the subject component. These failure rate evaluation models are empirical in nature and derived from industry average failure data.

For example, the model for microchip—metal oxide semiconductor (MOS), gate logic array, in MIL-HDBK-217F [29]—is given as:

$$\lambda_P = (C_1\pi_T + C_2\pi_E)\pi_Q\pi_L \text{failure}/10^6 \text{ h} \tag{6.2}$$

where $\lambda_P$ is the part failure rate; $C_1$ is the die complexity factor; $C_2$ is the number of pins; $\pi_T$ is the temperature factor (ambience temperature); $\pi_E$ is the environment factor (ground benign, industrial, etc.); $\pi_Q$ is the equality factor (MIL-grade, MIL-screened, commercial/industry grade); and $\pi_L$ is the learning factor (experience). The MIL-217F provides tables and plots that facilitate selection of numerical value for these factors.

Handbooks have also been developed for mechanical components, such as the NSWC-98/LE1 handbook entitled "Handbook for Reliability Prediction Procedure for Mechanical Equipment," developed by the Carderock division of the US Naval Surface Warfare Center [38]. This handbook has been developed for reliability and maintainability prediction of design evaluation for naval warfare systems. However, the handbook approach has potential applications in other industry environments, including nuclear plants. Given the reliability prediction scenario in the nuclear industry, it can be observed that for mechanical and electrical components the generic component reliability data sources, such as IAEA-TECDOC-478 [30] and OREDA Hdbk [39] (Offshore and Onshore Reliability Data), are more commonly used for PRA modeling.

### 6.7.8.3  Physics of Failure

There are critiques of the handbook approach as it has some inherent limitations. It assumes a constant failure rate; hence, an applicable model is exponential distribution. The empirical models often produce results that may not correlate with field data because the effects of environmental conditions and actual operational stresses are not reflected in the model. Thus, these models are based on industry average performance and do not account for vendor, device, or event-related aspects [40]. The physics-of-failure (PoF) approach has been argued to be superior to the handbook approach because it is science-based [41]. The PoF approach relates the failure rates or time to failure with parameters related to environmental and operational stresses, such as temperature, humidity, duty cycles, current density, and voltage. The PoF framework enables the use of deterministic damage/degradation models to predict the failures. The computational capability and advances in intelligent systems enable complex modeling of corrosion, wear, fracture, fatigue, and delamination. Even when extensive efforts are underway to develop advanced PoF models, the PoF approach still finds wider application outside of the laboratory environment.

Given that that the PoF-based approach is still evolving and remains in the realm of elite laboratories, while the handbook approaches are modified from time to time not only in terms of including new components such as photonics and new IC technology, but also in terms of employing improved models, the PRA practitioners still favor the handbook, for example, RIAC+ handbook in its modified form [32], due to practical constraints/conditions. The handbook approach provides the reference failure rates that are modified using various factors that account for use of operational and environmental stresses and for quality level, construction, and complexity of design.

Electronic systems have evolved from classical valve-based to relay-based and further to solid-state and now digital systems. Digital systems enabled design of controls having embedded systems that employ complex chips such as field-programmable gate array (FPGA) where software is a new entity. The presence of software and the design and construction features (e.g., millions of transistors connected through a complex network of metallization, with multi-level designs enabled by vias) requires a new approach that should address new failures introduced by software as well as hardware. Hence, it is only a matter of time before PoF-based approach will replace the handbook approach. This subject has been discussed in detail in Chap. 12.

### 6.7.8.4 Plant-Specific Data

The PRA study should always use as much plant data as possible. There is no consensus as to how much operating experience is considered as adequate for having reasonable confidence in the data. It can be argued that certain components like air- and motor-operated valves, centrifugal pumps, electromagnetic relays, switches, connectors, piping system, and electrical circuit breakers have a population in the plant such that plant operating experience of more than 10 years will provide a reasonable accumulated operating experience adequate for PRA system modeling. However, more than 10 years is always desirable. The integration of experience obtained from a similar plant will further reduce the uncertainty in the final estimate for each component. This plant-specific performance of the equipment is used for generating reliability data, such as the failure rate or demand failure probability.

The data obtained either from a generic source or from a plant-specific source are used as input for either working out the initiating frequencies or evaluating component unavailability.

## 6.7.9  Initiating Event Frequency Quantification

The initiating event frequencies are also evaluated by some special techniques, such as probabilistic fracture mechanics approaches for evaluating piping system failure frequency as part of LOCA analysis. Initiating event frequencies can also be obtained from generic sources. For example, IAEA-TECDOC-719 [28] and NUREG CR-6268 [31] provide frequency for LOCA initiating events. However, it is always preferable that the initiating event frequencies are evaluated using plant-specific data for the operating plants. For example, Class IV power failure frequency evaluated from plant-specific data reflects the reliability of the grid supply, the plant configuration, and the quality of the maintenance program at the plant.

A clear and concise definition of failure criteria is a first step in quantification of initiating event frequency. IAEA-TECDOC-719 [28] can be referred to for details on evaluating the frequency of initiating events.

## 6.7.10  Major Component Categories and Model
       for Estimating Unavailability

From the perspective of PRA modeling, the plant system and equipment can be categorized into three types of systems: continuously operating equipment/systems, safety systems, and emergency systems. Continuously operating equipment/ systems are also referred to as process systems and are required for operation of the plant. Examples are the primary cooling water systems, regulation systems, and normal or Class IV power supply systems. Failure of these systems takes the plant to a shutdown state. Safety systems remain in standby condition and actuate to either shutdown the plant or remove decay heat. Even for the purpose of PRA modeling, the safety support systems, such as Class III power supply, and compressed air system can also be put in this category. The emergency systems include engineered safety features, such as the emergency core cooling system (ECCS), containment and emergency exhaust, and hydrogen re-combiners, which are tested periodically and automatically actuated during accident conditions.

The reliability parameters required to estimate the failure probability component/system unavailability are given in Table 6.5.

The unavailability estimates obtained from the models in Table 6.6 are used as point estimates in fault tree analysis. However, uncertainty evaluation (i.e., evaluation of upper and lower bounds) forms part of the fault tree analysis. Hence, it is required to include information on applicable probability distribution along with the error factor, if the distribution is log-normal, mean and standard distribution if the distribution is normal, and failure rate along with time (mission time, test interval, repair time) if the distribution is exponential distribution. For the case of exponential distribution, $\chi^2$—distribution is assumed to evaluate the uncertainty bounds.

## 6.7.11 Dependent Failure

For two $A$ and $B$ events, having their failure probabilities $P_A$ and $P_B$, respectively, to be independent, the following formulation should satisfy:

$$P_A \cap P_B = P_A \bullet P_B \tag{6.3}$$

This formation enables use of the "AND" gate with a consideration that events $A$ and $B$ are independent. However, for dependent events, this formulation is not applicable, i.e.,

$$P_A \cap P_B \neq P_A \bullet P_B \tag{6.4}$$

When events $A$ and $B$ are dependent, then this expression can be expressed by conditional probability as follows:

$$P_A \cap P_B = P_A \bullet P_{B|A} \tag{6.5}$$

$$P_A \cap P_B = P_A \bullet P_{A|B} \tag{6.6}$$

Hence, dependent event modeling requires special considerations. In the context of PRA modeling, the dependent phenomena require treatment in two broad categories, viz. intrinsic dependency and extrinsic dependency.

Intrinsic dependency or functional dependency is explicitly modeled in PRA. For example, dependency of a system comprising three redundant operations on its common power supply is an example of intrinsic dependency because all three pumps will fail if the power supply is not available. The fault tree modeling explicitly models this type of dependency. Modeling of this type of dependency is easily captured by functional analysis.

One of the special categories of dependent events referred as a common cause failure (CCF) event occupies a central impetus in PRA modeling. CCFs are single events or a combination of events, malfunctions, or failures that fail more than one

**Table 6.5** Parameters required for estimation of failure rate/unavailability in PRA

| S. No. | Estimation | Unit | Symbol | Remarks/description |
|---|---|---|---|---|
| 1 | Initiating event frequency | Per unit time (e.g., per year) | – | Derived from data on process system performance |
| 2 | Unavailability | Per demand | $U$ or $Q$ | Unavailability statement is used to indicate summation of all applicable failure probability or probability that system will not actuate when demanded |
| 3 | Operating failure rate | Per unit time (e.g., per hour) | $\lambda_o$ | For continuously operating/process system |
| 4 | Standby failure rate | Per unit time (e.g., per hour) | $\lambda_s$ | For systems remaining in standby condition and actuates on demand. What constitutes standby failure can be defined as failure during standby condition of the plant, but how to get these failures is a question. This parameter assumes that it is possible to monitor the status of the component, and provision exists to monitor the component in standby condition such that failure that will not allow the component to actuate when the demand is placed exists in the system. A common practice is to extract this information from demand failure probability estimate |
| 5 | Demand failure probability | Per demand | $q$ | This is a time-independent parameter and provides probability that the component will not activate or start on demand. PRA application development requires this information to be used to get standby failure rate as time-independent parameters do not form an effective input when it comes to development of PRA application, e.g., surveillance test interval optimization, allowable outage time estimation |
| 6 | Mission time | Time (h) | $t_m$ | Derived from safety analysis. How long equipment is required to run to cater to safety requirements |
| 7 | Repair time | Time (h) | $t_r$ | Also referred to as mean time to repair (MTTR) is used to estimate repair-related unavailability |
| 8 | Surveillance test interval | Time (h) | $t_T$ | Periodicity of testing, e.g., weekly, monthly. Derived from plant technical specification |
| 9 | Maintenance interval | Time (h) | $t_{ma}$ | Periodicity of maintenance, e.g., weekly, monthly. Derived from plant maintenance schedules |

The above parameters are used to model average unavailability as given in Table 6.5 taken from NUREG-2815 [26]

**Table 6.6** Average unavailability model for component unavailability [26]

| S. No. | Component type | Time averaged unavailability model | Applicable parameters | Data requirements |
|---|---|---|---|---|
| *Online/process/continuously operating types* | | | | |
| 1 | Non-repairable or mission mode | $1 - \exp(-\lambda_o t_m)$ | $\lambda_o$: Operating or mission failure rate; mission time | For $\lambda_o$: Accumulated time to failure; number of failures in accumulated time. For $t_m$: Mission time |
| 2 | Repairable components | $\dfrac{\lambda_o t_r}{1 + \lambda_o t_r}$ | $t_r$: Mean time to repair | For $t_r$: Average of observed time to repair and number of repairs performed |
| *Standby components* | | | | |
| 3 | Tested standby | | | |
| 3.1 | Random failure during standby condition | $1 - \dfrac{1 - \exp(\lambda_s t_i)}{\lambda_s t_i}$ | $\lambda_s$: Standby failure rate; $t_i$: Test interval | For $\lambda_s$: Observed failures during standby condition can be derived from demand failure data for a given mission time. For $t_i$: Obtained from the technical specification or time interval followed for testing |
| 3.2 | Contribution from test duration | $\dfrac{\tau}{t_i} q_o$ | $\tau$: Average test duration; $q_o$: Override unavailability | For $\tau$: Data on time taken for test and number of tests during a given period. For $q_o$: Operations and engineering insights |
| 3.3 | Contribution of repair during outage | $\lambda_s t_r$ | $t_r$ = Mean time to repair | For $t_r$: Obtained from accumulated time of repair divided by number of repairs |
| 3.4 | Contribution from scheduled maintenance | $f_s t_m$ | $f_m$ = Frequency of schedule maintenance; $t$ | |

redundant component instantaneously or in such a short interval that leads to failure of the safety function or mission. The IAEA Safety Glossary defines CCF as "failure of two or more SSCs due to a single specific event or cause" [9]. Even though in recent times there is no specific interest in distinguishing between CCF and common mode failure, for the sake of clarity, the common mode failure is defined in the IAEA Glossary as "failure of two or more SSCs in the same manner or mode due to a single event or cause." The glossary further elaborates that common mode failure is a type of common cause failure in which the SSCs fail in the same way. However, what is important is the concern that is associated with CCFs, and this is failure of redundant system instantaneously or in a short time such that the safety function is not met successfully or leads the system to a degraded condition. Hence, there is need to discuss this aspect in respect of its major character or elements. There are four characteristic features that characterize the CCF as follows:

- *Coupling mechanism*: A coupling mechanism does need not to be a failure, but a phenomenon that leads to failure of all the redundant components in a CCF group (CCFG). For example, increase in temperature and humidity in an area can lead to failure of electronics due to shorting.
- *Root cause of failure*: The root cause is the basic reason that components fail, i.e., a root cause of redundant component failure in a CCFG. There can be one or more root causes that lead to failure of more than one unit in redundant components in the group.
- *Instant of failures*: The instant of failure is an important attribute that requires consideration of the consequences of failure in terms of adverse effects of failure of components in a CCFG. For example, in the case of shutdown devices in a reactor (i.e., the shutoff rods), instantaneous failure of more than one shutdown device on a reactor trip demand qualifies the failure of more than one device as the definition of shutoff rod CCF. If the shutoff rods are failing during different demands, then this need not qualify as CCF, even if the root cause of failure might be same. For example, if one rod is failing in the $n$th demand and a second rod is following in $n + k$ demand even on the same reason, viz. damper failure, since the instant of failure is not same, these failures might be having same mode but do not qualify as CCF.
- The severity and frequency of CCF depends on many factors such as proximity, sharing of same environment, level of separation, level of dependency on a common source. This requires provision of defenses against CCFs, viz. physical separation or isolation from coupling mechanisms, functional independence, diverse designs.

The major root causes of component failure are as follows: (a) design, manufacturing, and layout; (b) operational; (c) external environment; and (d) quality related or internal to the component. One more category is created in all the CCF databases, i.e., unknown causes. This could be related not only to the operation, but also to the design, manufacturing, installation, and commissioning. The coupling factor can broadly be categorized into hardware quality, design based, operations and maintenance, and environment [42]. The experience is that it is not possible to identify all the CCF causes a priori. Often, the root cause of any failure or particularly CCF has been identified to be human error. Similarly, human error in design, operation, and maintenance can also provide a coupling mechanism or a factor responsible for CCFs. The point being made here is that often, human error is associated with operation and maintenance; however, root cause failure analysis shows that for many events human error in design, construction, and commissioning, or even regulatory processes, is responsible for CCFs.

A general observation is that CCFs are major contributors to the unavailability at the system level and CDF at the plant level. Therefore, an attempt should be made to demonstrate safety of the plant through defense mechanisms built into the plant. Many PRA studies recommend provision of physical separation, layout changes, changes in plant operations, and maintenance procedures to reduce dependency on common cause elements. Dependency can be of two types, viz. intrinsic and extrinsic.

However, even after thorough modeling for extrinsic dependency or CCF, assurance that all types of potential phenomena or situations have been modeled is not possible. New situations or phenomena could be a new root cause or failure that might affect redundant components. This situation has led to comprehensive treatment or modeling of common cause failures. The salient features of this CCF modeling procedure include the following:

- Identification of redundant or common cause component groups at the system and plant level.
- Identification of defenses built into the plant against CCF events, such as defenses against fire, flood, and human error while executing calibration or operational configurations.
- The potential phenomenology that could affect more than one component in a CCF group.
- For every scenario or what constitutes a CCF, a definition of CCF is worked out that includes definition of failure criteria.
- Selection of CCF models that quantify CCF group.
- Appropriate quantification that reflects the realistic requirements. For example, a low value of common cause factor, say $\beta$-factor, may not reflect the plant conditions realistically. Hence, it requires judgment and insights into generic experience as input for working out these factors.
- A sensitivity analysis for assumption of the models and associated values considered for CCF factors at the system as well as plant levels to assess the impact of these CCF events.
- CCF evaluation needs to be revisited when the PRA modeling moves from Level 1 internal event to internal and external hazard modeling because seismic, fire, and flood phenomena have been found to be the major sources of CCF events.

Incorporation of CCFs requires identification of the empirical model and letters its quantification. Table 6.7 shows the commonly used CCF models and associated parameters and remarks.

**Table 6.7** Common cause failure models

| S. No. | Name | Model | Parameters | Use and associated details |
|---|---|---|---|---|
| 1 | β-factor model | $\beta = \dfrac{U_{CCF}}{U_{CCF}+U_I} = \dfrac{U_{CCF}}{U_T}$ | • $U_{CCF}$: Unavailability due to CCF failures<br>• $U_I$: Unavailability due to independent failures<br>• $U_T$: Total unavailability | This is the most commonly used model and is actually meant for two-component standby systems. This model provides very conservative estimates of CCF. The available literature suggests that a β value ranges from very low from 0.01 for a system where good operating experience is available to a new design where β value can vary from 0.08 to around 0.1 |
| 2 | MGL (multiple Greek letter) model | $U_n = \dfrac{1}{\begin{bmatrix} m-1 \\ n-1 \end{bmatrix}} \left( \prod_{i=1}^{n} \eta_i \right) (1 - \eta_{n+1}) U_T$ | • $U_n$: Unavailability of nth order CCF group<br>• $U_T$: Total unavailability<br>• m: Group size<br>$\begin{bmatrix} m-1 \\ n-1 \end{bmatrix} = \dfrac{(m-1)!}{(m-n)!(n-1)!}$ | This model is a three-parameter ($\beta$ (beta), $\gamma$ (gamma), and $\delta$ (delta) model<br>$\eta$ MGL parameter and<br>$\eta_1 = 1; \eta_2 = \beta; \eta_3 = \gamma; \eta_4 = \delta; \ldots \eta_m + 1 = 0;$<br>The MGL model provides estimates of unavailability considering independent as well when the same component appears for different sizes of CCF groups. For example, the MGL model for four-component areas $P, Q, R,$ and $S$ follows:<br>$U_P + U_{PQ} + U_{PR} + U_{PS} + U_{PQR} + U_{PRS} + U_{PQS} + U_{PQRS}$ |
| 3 | α-factor model | $U_n = \dfrac{n}{\begin{bmatrix} m-1 \\ k-1 \end{bmatrix}} \dfrac{\alpha_n}{\alpha_T} U_T$ | • $U_n$: Unavailability of nth order CCF group<br>• $U_T$: Total unavailability<br>• m: Group size<br>$\begin{bmatrix} m-1 \\ n-1 \end{bmatrix} = \dfrac{(m-1)!}{(m-n)!(n-1)!}$ | This is a four-parameter model given as $\alpha_1, \alpha_2, \alpha_3,$ and $\alpha_4$. This model is also used in fault tree modeling when it is required to estimate unavailability from CCFG having order 2, 3, and 4 |

Depending on the need of the analysis, a model is chosen to estimate common cause contribution from various CCF groups of components.

### 6.7.12 Human Reliability Analysis

Human reliability analysis is performed in PRA to provide input and insights on human error probability. Human error modeling is performed as part of system modeling, e.g., basic event in fault tree analysis and as header event in event tree modeling as part of accident analysis. For example, human actions as recovery safety functions form part of accident sequence modeling. Human error probabilities also form basic events in fault tree, for example, manual action to start a failed pump from control room. Human error data can be obtained either using the handbook approach (e.g., THERP [43], HCR [44], or the most latest one ATHENA [45]) or if adequate plant-specific experience is available, then plant-specific data can also be used. Normally, plant-specific human reliability data are preferred because it reflects the plant's safety culture and represents specific hardware features and procedural aspects of the plant that work toward reducing uncertainty in the human error probability estimates. A common experience is that plant-specific data are not adequate and the handbook approach provides an effective option for human reliability modeling in PRA.

In human reliability parlance, human errors are categorized in five types: (a) latent human error, such as a calibration error or not reverting the valve status to normal or as desired after a surveillance test, which gets revealed when either there is an actual demand or during the next test; (b) action that causes an initiating event to occur, e.g., a loss of regulation incident (LORI) caused by an error during reactor start-up; (c) human error in implementing plant procedures such as a wrong action that leads to failure of procedure; (d) human error in not recovering failed equipment, and (e) human error performed during accident management further aggravated in plant condition.

Further details on human reliability analysis are discussed in Chap. 10.

### 6.7.13 Accident Sequence Quantification

We have already covered the following major aspects that are a prerequisite for accident sequence quantification:

- Event trees for various initiating events;
- Fault trees for safety systems;
- Techniques and models for assessing failure rate and failure probability for various events and components, which include standby and tested components

and components in process systems which include repairable and non-repairable components;

- Sources of generic and plant-specific data.

Even though the fundamental process of quantification remains the same, the PRA modeling process to a great extent is influenced by availability/ non-availability of the PRA software, which affects simulation capabilities. We can imagine three scenarios for creating the event tree and the fault tree. (a) The event tree and the fault tree are drawn manually (non-availability of PRA software environment). (b) APRA software environment facilitates limited capability for creating the fault tree and event tree model and offers logic model building only in text mode (i.e., graphical user interface (GUI) is either not available or is not working). (c) PRA modeling is completely done using a commercial environment that offers all the facility for creating fault and event tree analysis, basic event quantification, and performance of analysis (qualitative as well as quantitative) where features are available for uncertainty characterization.

For the purpose of this section, we will be focusing on the last scenario where commercial software for PRA modeling is used. The manual approach to PRA modeling becomes prohibitive because the Boolean analysis results in a combinatorial explosion of accident sequences that cannot be handled manually, and even if we attempt this, the correctness of the list of accident sequences and its quantification remains a challenge. The second option was offered by IAEA PRA software PSAPCK 4.8, which was extensively used by PRA practitioners during the late 1990s or early 2000. However, this software was not very user friendly and did not have many of the features that are offered by commercial software, and the analysts had to depend a lot on the documentation.

The general trend is to use commercial software or in-house developed software that provides the following: (a) an efficient GUI for fault tree, event tree, and Markov model and even at times a module for binary decision diagram modeling; (b) provision for linking the output of the Markov analysis as input to the fault tree analysis; (c) features for maintaining the component models and associated component reliability databases used in a PRA project; (d) automated analysis; (e) features for CCF and house event modeling; (f) features for importance and sensitivity analysis; and (g) presentation of results that include a list of accident sequences along with its quantification; a consolidated statement of core damage frequency, and a facility for presentation of major insights, such as the contribution of initiating events to the core damage frequency; the importance of various components and accident sequences; and the results of uncertainty and sensitivity analyses.

Hence, in this section, the discussion assumes that commercial software has been employed for PRA modeling. The accident sequence quantification is performed in two stages, viz. initial quantification and final quantification. During the initial quantification, the fault tree and event tree model simulation is carried out and the results are analyzed. In case some anomaly in modeling is observed, corrections that might include modification in the logic model, data, and assumptions are carried out as well as further simulations. Once the model is in an acceptable form, further simulations are conducted as part of the final quantification to study different plant configurations and cases as per the requirements of the project.

Uncertainty and sensitivity analyses are desired as part of a PRA study. Apart from this, the importance analysis enables prioritization of not only components but also accident sequences. Importance analysis is vital for development of PRA applications as part of the risk-based/risk-informed approach. The need for an uncertainty analysis arises due to inherent variability in the data and model as well as a lack of data and information, whereas a sensitivity analysis is required to evaluate the impact of various assumption and uncertainty in data, particularly for critical aspects (components/assumptions) of the modeling on the overall results of the analysis.

## 6.7.14  Uncertainty Analysis

Uncertainty characterizes the context of variability in the input data and model toward an evaluation of overall variability in the output or results of the study [12]. The typical process of characterizing uncertainty in input data is characterized by the probability distribution used to represent the data. For example, the failure rate of a motor-operated valve as input to the fault tree is defined to include: (a) distribution considered (e.g., log-normal), (b) the median value of failure rate $\lambda$, (c) distribution of time elements $t$ (e.g., mission time), and (d) error factor for $\lambda$ and $t$, along with other supporting parameters as per the requirements of the analysis. Monte Carlo and Latin Hypercube approaches are employed for simulation process. The number of iterations required to get a smooth distribution fit depends on the choice of the analyst. Normally, the heuristic followed for simulation is: If the failure probability is in the range of $10^{-3}$, then a minimum of 1,000 iterations is required.

Chapter 9 provides details on uncertainty analysis, including different types of uncertainty and various approaches for uncertainty modeling.

## 6.7.15  Sensitivity Analysis

The analysts define the final result of the analysis by his choice in respect of assumptions and boundary conditions of the analysis, initiating event frequencies, safety systems unavailability, based on SSCs independent and common cause failure probabilities, human error probabilities and associated uncertainty, models, etc., toward giving the results of the analysis. In this sense, there is a defined vector or pattern of these numerical values used for estimating the results of the analysis, e.g., core damage frequency. Subjectivity is associated with these results. In other words, if a sets of parameters that the analyst has assumed to be adequate, then this vector determines the result, say CDF result of the analysis. However, there is need to evaluate the impact of (a) the broad assumptions at the system and plant levels; (b) the relatively large uncertainty associated with some components, such as piping systems, human error, and software; (c) the effects of common cause failures; and (d) the effects of various plant configurations, such as consideration of safety shutdown system A and redundant safety shutdown system B together and one system alone.

Normally, formalization of the base or reference plant model is a prerequisite, which includes final plant configurations, data model, assumptions, and uncertainty, for the sensitivity analysis. Most importantly, the reference model should also meet, depending on the requirements of the analysis, the quality attributes to qualify as the accepted reference or base model. The objective of the sensitivity analysis is to demonstrate the safety of the plant against various assumptions and uncertainties or lack of data. In case any instances are observed when the results are higher or lower than the acceptable criteria, recommendations may be made for corrections either in respect of plant modifications, plant operating procedures, or repair/test policies such that, e.g., CDF value can be lowered to an acceptable level.

### 6.7.16  Importance Analysis

Apart from the sensitivity analysis, the importance analysis also forms part of the PRA study. The most widely used importance models include Fussel-Vesely, risk reduction worth, risk achievement worth, and Birnbaum importance measures [11]. The importance analysis is vital for prioritizing the components as well as the accident sequence. In fact, the importance analysis is one of the vital tools for developing risk-based or risk-informed applications. Importance analysis also supports the identification of accident sequences and components for sensitivity analysis.

### 6.7.17  Core Damage Frequency-Related Aspects and Formulation of Results and Their Interpretation

The accident evaluation program deals with binning of accident sequences to various categories depending upon the potential for severity of core damage. Often, guidance is available in standards for this categorization; for example, category A will have the highest severity, B will have the lowest severity compared to A, and category C will have a lower severity than B. Physical parameters and phenomena form the basis for this categorization, e.g., fuel/clad temperature, potential for violation of coolability criteria, early/delayed core damage, and reactivity phenomenon.

The above argument requires the definition of core damage for a given plant design. For PWR plants, a general definition of core damage is unrecovery of reactor core levels below a predefined criteria. However, arriving at a definition of core damage is not in the subject of this book. When detailed thermal hydraulics and structural degradation in the core and related components are not available, then these criteria, i.e., "Any accident sequence not leading to safe state" can be categorized conservatively as core damage. However, as soon as the safety analysis results are available, this definition can be further be fine-tuned to have best estimate of core damage. The main risk metric for Level 1 PRA is core damage frequency; however, the accident evaluation results are categorized into more than one category, as follows:

- The accident sequence having potential for leading to high severity of core damage, e.g., major LOCA and medium LOCA. These events further form subcategories such as A1, A2 ....
- Accident sequence having potential for leading to early core damage and delayed core damage based on the timing of core damage might fall into A or any other lower B categories.
- Enmasse fuel failure beyond the preset criteria that has potential for release beyond acceptable levels.
- Enmasse fuel oxidation beyond a set criterion.
- Level of damage to core and related structural components such that chances of plant returning to normal condition are reduced.

Analysts make a conscious decision to determine which accident sequence categories will be included to characterize the statement of core damage frequency. For example, the accident sequence leading to severe core damage where the core damage was predicted during the early phase after initiation of an accident normally forms part of the core damage, such as unsuccessful shutdown of the reactor after a LOCA event or Class IV power failure. Loss of flow accident into a couple of hours of cooling may form delayed core damage because the decay heat levels have significantly reduced (often $\sim 2$–$3\%$), and there is a possibility of human action to inject coolant either on the primary or on secondary sides.

As mentioned in Sect. 9.10, the sensitivity analysis forms part of the detailed evaluation. Sensitivity analysis is performed in the final phases of the analysis to evaluate the impact of various assumptions. For example, the analysts credit both shutdown systems (i.e., primary and secondary) for bringing the reactor to a safe state. However, there could be an argument that for certain postulated initiating events, the secondary shutdown system may not be adequate. In such cases, safety needs to be demonstrated even without crediting the secondary shutdown system. Hence, a sensitivity analysis should be carried out without crediting the secondary shutdown system. What practically is that, the PRA code is run by disabling the event tree header event for shutdown cooling system. The increase in core damage frequency so obtained will provide an estimate of core damage frequency with only one shutdown system. Similarly, sensitivity analysis can be performed by considering the upper bound (i.e., 95%) confidence value of a safety-significant parameter to account for parameter uncertainty for the given case.

The three important components, i.e., uncertainty analysis, sensitivity analysis, and important analysis, form part of the fine-tuning of the core damage analysis. It is desired in the PRA study that the sensitivity analysis should be conducted to support the arguments of conservative assumptions where the important analysis should be conducted to extend PRA applications.

### 6.7.18  Documentation

The following observations on documentation as part of PRA program:

- A system should be developed right in the initial stages of the PRA program for documentation, information flow, lines of communication, and lines of authority.
- The format of the system analysis should be drafted depending on the analysis requirements and circulated to all the members for consistency of reporting.
- The PRA report has two major parts—a main report that extends to many appendixes and an executive summary of $\sim 10$–$15$ pages providing the introduction, major features of the analysis including major assumptions, and results of the analysis.
- The IAEA Level 1 PRA procedure guide P-4 [11] provides the format of the main Level 1 PRA that can be adopted for full-scope Level 1 PRA.
- Quality attributed related to documentation as given in either IAEA-TECDOC or ASME document should be considered [17, 21].
- Documentation should also be maintained for comments obtained from the reviewers, at various stages, and how they were addressed.
- A control sheet should be part of the documents that keeps records of successive revisions.

## 6.8  Beyond Limited-Scope PRA—Other Major Modules for Full-Scope Level 1 PRA

When a PRA program is initiated in an organization or a country, a common practice is to perform a PRA with limited scope considering only internal initiating events, full-power operation, and reactor core as the source of radioactivity. Most of the countries have now enhanced the scope of PRA to full-scope Level 1 PRA, which extends the work to include external hazards and low-power and shutdown conditions. In terms of risk-based application, this book focuses on limited-scope Level 1 PRA; however, for the sake of completeness, the extended topics, such as low-power and shutdown PRA and external hazards, will also be discussed in brief. Further, Level 2 and Level 3 PRAs will be discussed in this chapter.

### 6.8.1  Low-Power and Shutdown PRA

USNRC evaluation of shutdown risk and the history of incidences during the shutdown mode of NPPs studied by the USNRC staff were perhaps one of the drivers for growing interest that followed on LPSD PRA [46]. In fact, the Chernobyl accident in 1986 highlighted the risk from LPSD where a decision to continue with low-power operation to facilitate conduct of a turbine-generator coast down experiment resulted in core melt [47].

The traditional approach to nuclear plant operation was built around the considerations that in case of a deviation from normal operation, however small or notional, the plant should be brought to the shutdown state. The basic assumption was that in a shutdown state the risk from the plant reduces to negligible or the plant poses no risk. However, the knowledge gathered over the years and recent risk assessment studies provided some other insights. For example, even in shutdown condition, the plant poses risk, and in some cases, it is even a significant fraction of the total risk. This realization formed the genesis of low-power and shutdown PRA (LPSD PRA).

A realization that is endorsed by LPSD PRA is that risk from plant conditions other than full-power operation, i.e., LPSD, is not small but significant. In some studies, the contribution to total risk from the shutdown condition has been found to be comparable to full-power operating conditions. Further, efforts are on two make LPSD PRA in support of risk-informed/risk-based decision [48]. One of the landmark LPSD PRAs performed for French 900 and 1300 MWe in the 1980s and many PRAs performed in recent times highlighted this aspect [49]. Perhaps this observation was the main motivation for the active interest world over to have a LPSD PRA. The available literature shows that over a couple of tens of LPSD PRAs have been conducted for nuclear power plants globally [50], and there is a particular interest in implementation of maintenance rules through risk monitors where LPSD forms the scope of the risk monitoring [51]. Like any other PRA model, LPSD PRA is a complex and resource-intensive option; therefore, there are instances where simplified LPSD PRA models have been developed to understand defense-in-depth provision during planned outages [52]. Further, Chu et al. [53] discussed the methodology and highlighted the need for internal flood and fire PRA through a study on the shutdown mode for mid-loop operations for Surry Unit-1. The state of the art matured for LPSD technology such that, apart from technical documents, efforts were visible to develop standards for LPSD PRA, one such movement was ANSI/ANS-58.22 [54].

The big question is: Why should the LPSD condition have such a high-risk component? The available literature provides technical insights; however, the potential major contributors to LPSD risk are as follows:

- Even though a plant is subcritical in terms of neutronic considerations, the possibility of reactivity-induced accidents particularly due to inadvertent action cannot be ruled out.
- The process of shutting down the plant involves introduction of a transient system and puts a demand on safety shutdown devices and shutdown systems.
- The shutdown state of a plant reduces safety margins. For example, the core cooling depends on decay heat removal system or any other system such as the steam generator system, while the main coolant loop is open for maintenance activities.
- The plant has moved from reference operating condition to a shutdown state where most of the maintenance/test activities are proceeding in parallel; hence, there is an enhanced cognitive load on the operating staff to comprehend the safety status. This situation has potential for error and mistakes.
- Human reliability is the cornerstone of safety of the plant in the shutdown state. Inadvertent actions, errors of communication, lapses, and institutional/ organizational failures pose safety implications.
- The possibility of fire and flood during most of maintenance activities can be potential initiating events.
- The test/maintenance activities or human error has the potential for common cause failure, e.g., physically blocking redundant safety trains and calibration errors, which remain undetected while the plant goes into operation.
- Some initiating events have their own risk component. Examples include Class IV power, particularly extended failures beyond a plant's coping capability, potential for leaks, and minor LOCA due to maintenance activities.

It was observed from the literature that LPSD PRA methodology requires harmonization and improvement. The NEA's Working Group Risk (WGRISK) developed a document that was based on surveys and feedback of member countries, which included LPSD methodology, data, and assumptions used in respective member countries. This document provides a summary of current approaches and good practices as well further research needed on LPSD PRA [55].

Based on the observation made in the preceding section, the major procedural elements are discussed as follows:

- Procedure to understand and identify the plant outage states;
- Identification of initiating events as applicable to LP and SD conditions;
- Requirements of quantification of system unavailability and initiating event frequency;
- Identification of plant operational states;
- Identification of human actions and associated analysis important to safety.

The major steps in LPSD PRA are similar to Level 1 on-power PRA, viz. plant familiarization, identification of initiating events, initiating event grouping and analysis, system modeling, qualitative accident sequence evaluation, data collection and analysis, human factor and common cause failure analysis, accident sequence quantification, uncertainty and sensitivity analyses, and documentation. However, there are some specific features of the analysis that require special attention from the perspective of LPSD aspects of the plant because each plant has some specific aspects of LPSD configurations.

As can be seen, the major steps involved in LPSD PRA are similar to the Level 1 PRA for full-power operation; hence, in this section, the specific features of LPSD-Level 1 PRA, as listed above, will be discussed.

### 6.8.1.1   Identification of Plant Outage Configurations

There are a host of activities planned in shutdown, and a framework is required to categorize these activities into a few manageable plant outages. If we look at the schedule of shutdown activities from LPSD PRA point of view, it can be seen that they are categorized into or carried out as part of a low-power state, a fueling outage, planned activities, and unplanned activities. The second aspect related to defining the outages is the boundary between one outage from the preceding or following outage category.

For example, low-power operation can begin when the control of the reactor is switched from normal power to low power, while earlier stages can go to full power as well. This analogy applies for dealing with the definition of the boundary between two plant operational states (POSs) in the following section. Each outage requires characterization. For example, a fueling outage might not allow any other maintenance or calibration activity related to reactor core, or any activity that has potential for reactivity changes or any neutronic/regulation instrumentation activities. Similarly, during fueling outages, there could be restriction by technical specification on performing any activity on the coolant (primary/shutdown system).

For those plants having on-power fueling, fueling outage forms part of full-power PRA. However, those failures in fueling that require plant shutdown should be identified, and accordingly, the fueling and on-power boundary are worked out.

Unplanned outages might result from certain equipment failures that may or may not require plant shutdown. The unplanned outage might result from some accident condition where plant shutdown is required. The equipment outage and associated maintenance will require determination of the operational and maintenance activities from the perspective of maintaining the basic safety functions, viz. reactivity, core cooling, radioactivity containment, and radiation protection considerations.

Figure 6.7 shows the stages of plant outages characterized through plant parameters against various plant operating states for a reference plant where fueling and ECCS testing are performed in reactor shutdown condition with coolant system depressurized, while the reactor building ventilation system is operating in normal mode. This outage chart provides requirements for the availability of safety functions such as shutdown cooling, shutoff devices in the safety bank of the reactor protection system (RPS).

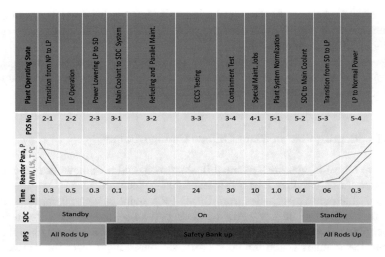

**Fig. 6.7** Simplified mapping of plant operational states in one of the *n*-outages

### 6.8.1.2  Identification of Plant Operational States

The determination of plant operational states (POSs) is a complex task, for there are a host of jobs scheduled in each outage. Formulating a manageable finite number of plant operating states requires a detailed analysis of each activity against identified parameters. From the perspective of LPSD PRA, it is required that plant operation and maintenance staff is involved in identifying various POSs. Defining each POS requires mainly the following considerations: in-core or out-of-core activity, core reactivity phenomena/status, requirements of shutdown reserves, decay heat levels, technical specification requirements, duration of the activity, containment and ventilation requirements, and system configuration/alignment. To reduce the complexity of POSs, often it is required to first create a pre-POS and finally combine these pre-POSs into a representative POS.

The above discussion highlights the point that each POS should be defined in terms of the entry and exit point to/from the given POS, the duration of the POS, and characterization in terms of reactor parameters. Even though each POS might appear in various outages, it will not be practicable from the point of view of the LPSD PRA to map all the POSs on all the outages; hence, a representative outage scheme that accounts for all the POSs may be considered to create an LPSD PRA model. For example, Table 6.8 shows 12 POSs formulated for a reference plant. In this plant, the fueling is performed in the shutdown state with the primary coolant loop in a depressurized condition. Other major surveillance activities include periodic testing of the ECCS and integrated containment leak test. At times, the scheduling is also performed for a major modification or planned repair or replacement activities. The shutdown activities follow the technical specification requirements.

**Table 6.8** Plant operational states during outage in the reference plant

| Plant state | POS | POS code | Description |
|---|---|---|---|
| Transition from full power to low power | Transition from power operation to low power (LP) | 2-1 | The reactor thermal power from full power is reduced as a consequence heat production decreases full power such that cooling requirement reduces and is insignificant at LP |
| | LP operation | 2-2 | Neutronic power <1% of FP; core $\Delta T \sim 0$; $K_{\text{eff}} = 1$; reactor is critical. Reactor main and secondary coolant system is in operation |
| | Power lowering from LP to shutdown | 2-3 | Reactor is made subcritical; all the shutoff rods are lowered to achieve complete shutdown state. Normal shutdown is achieved by raising the safety bank and energizing the trip circuit |
| Outage | Core cooling is switched from main coolant to shutdown cooling system (SDC) | 3-1 | Decay heat removal requirement reduces from initial 10% for 10 min to reduce to <5% of normal power. Coolant system pressure reduces from $\sim 10$ to <2.0 kg/cm$^2$; coolant temperature <32 °C; $\Delta T < \sim 0$ °C |
| | Refueling and parallel maintenance program (except activities that requires >12 h of reactor shutdown | 4-1 | Refueling activities using fueling machine A. Heavy water system in depressurized condition. The administrative procedure ensures no maintenance activities that affect or having potential to introduce transient into the heavy water coolant system and shutdown cooling; as also ventilation system |
| | Quarterly ECCS testing | 4-2 | Heavy water system in depressurized condition. ECCS on manual. RB ventilation system should be on. Shutdown cooling system is maintained |
| | Half yearly integrated containment leak test | 4-3 | No maintenance activity in reactor building. Containment is pressurized to required pressurized for integrated leak test. Shutdown cooling system is maintained |
| | Special maintenance activities | 5-1 | Depending on requirements of the task, e.g., maintenance on main or shutdown coolant system, ventilation system, fuel transfer buggy, fueling machine; the required plant configuration is maintained |
| | Plant system normalization | 5-2 | Aligning plant systems for reactor start-up |

(continued)

**Table 6.8** (continued)

| Plant state | POS | POS code | Description |
|---|---|---|---|
| Restart | Switchover of shutdown cooling to main coolant system | 6-1 | Coolant system pressure increases from 2 kg/cm² to >8 kg/cm² and flow from ~<10 klpm to ~63 klpm |
| | Transition from shutdown mode to lp operation | 6-2 | After completing pre-start-up checklist and ensuring other conditions, particularly in respect of maintenance activities, reactor start-up procedure for LP operation is taken up |
| | LP to power operation | 6-3 | Power raised in steps to full power; coolant system neutronic power = or <NP 100%; flow ~63 klpm, pressure: ~8 kg/cm², ΔT: ~20 °C |

### 6.8.1.3   Identification of a Set of Applicable Initiating Events

Let us first consider the following definition of an initiating event: "An initiating event is an event which leads to termination of normal plant operation, requiring protective action to prevent or limit undesired consequences." This definition caters well to the full-power or even low-power operation; however, the question is: What is the definition of normal plant operation? At the outset, it needs to be clarified that even the shutdown conditions of the plant, which could be a consequence of the termination of plant criticality, at times, could be termination of plant deviation; also form part of the normal operation. The only difference compared to full-power/LP operation is that the safety margins are reduced. For example, in shutdown the shutdown cooling is pressed into operation due to unavailability of the main coolant loop. Hence, the shutdown system, which acted as a safety reserve, is operating to cater to decay heat removal requirements. In case this system develops a problem, then there could be safety implications from the decay heat removal if no alternate system is available.

With the above background and clarifications, the broad procedure to identify the applicable list of initiating events for LPSD PRA is in line with full-power Level 1 PRA. One major difference compared to full-power PRA, where failure of process systems normally forms part of the initiating event, is that in LPSD PRA the maintenance action that led to unavailability of the safety function also forms a candidate initiating event in LPSD PRA, for example, failure of decay heat removal system.

The procedural steps to formulate a list of initiating events are in line with full-power PRA; however, to ensure that the initiating event list is as complete as possible, the following checks are required:

- Use general guidance from national and international guides, such as IAEA TECDOC-1144 [50]. These guides provide criteria that include events which threaten normal heat removal, primary circuit integrity, and/or loss inventory (potential for LOCA), reactivity control function.
- Review of full-power PRA initiating event list for its applicability to LPSD PRA.
- A review of LPSD PRA initiating event list from generic sources, particularly from similar plants.
- Review of operational and maintenance records to look for events or precursors that have potential to be candidate initiating events for LPSD PRA.
- Evaluation using master logic diagram.

Once an exhaustive list is created, the next step is screening to formulate a final list. This final list of 8 initiating events is used for detailed LPSD PRA analysis.

- Class IV power failure;
- Shutdown cooling system failure;
- Significant leak in main coolant system;
- LOCA;

- Fueling-induced accidents;
- Inadvertent reactivity addition;
- Single channel blockage;
- Emergency cooling line failure.

The next and vital step in LPSD PRA development of an IE-POS metrics: For the reference plant, eight POSs have been figured out after reviewing the shutdown schedules and considering the guidelines. Table 6.9 shows the dependence metrics for 8 LPSD initiating events and the 12 POSs.

### 6.8.1.4  Quantification of Initiating Event Frequency

Initiating event frequency estimation in LPSD follows a similar procedure as full-power PRA; however, there are some specific aspects that need consideration to avoid error in estimation:

- The initiating event frequency for LPSD PRA, even though estimated for a given POS, needs to be evaluated for per year frequency.
- The duration of POS should be used as one of the parameters in frequency estimation.
- The frequency estimation should be performed as far as possible from plant operational experience data.
- For plants in design stages, the frequency estimates are required to be derived either from generic data or from a similar operating facility.
- Analysts might like to use full-power PRA frequency as input to workout apportionment of shutdown or POS. For example, considering a typical availability factor, one can directly use an apportionment for shutdown PRA. However, this approach should be used only as the last option as chances of error cannot be ruled out unless great care is taken for evaluating the frequency for a given POS.

Three different conceptual models can be applied for the IE frequency estimation in LPSD PRA, in order to generate "per calendar year" frequencies [50] (Table 6.10).

### 6.8.1.5  Human Error Modeling

Shutdown is an human-intensive activity that mainly includes (a) operator action from the control room for maneuvering reactor control from normal power to low power and shutdown condition and after outage is over, then again back to normal power level; (b) plant has deviated from its reference configuration/condition due to a host of test/maintenance activities; (c) heavy cognitive loading of control room staff; (d) control and coordination require communication on a sustained basis, be it oral or written; (e) handling fueling operations often requiring manual operations

**Table 6.9** IE-POS metrics showing applicability of given plant operational states to initiating event(s)

| Initiating event | | Plant operational states | | | | | | | | | | | |
| --- | --- | --- | --- | --- | --- | --- | --- | --- | --- | --- | --- | --- | --- |
| | | Shutdown | | | Outage | | | | | | Restart | | |
| | | 2-1 | 2-2 | 2-3 | 3-1 | 4-2 | 4-2 | 4-3 | 5-1 | 5-2 | 6-1 | 6-2 | 6-3 |
| *Transients* | | | | | | | | | | | | | |
| T-1 | Class IV power failure | X | X | X | X | X | X | X | X | X | X | X | X |
| T-2 | SDC failure | | | | X | X | X | X | X | X | X | | |
| T-3 | Significant leak in MCS | X | X | X | X | | | | | | X | X | X |
| T-4 | LOCA | X | X | X | X | | | | | | X | X | X |
| T-5 | Fueling failure | | | | | X | | | | | | | |
| T-6 | Reactivity accident | X | | | | X | | | | | | X | X |
| T-7 | Single channel blockage | X | X | X | | X | | | | | | X | X |
| T-8 | Emergency line failure | | | X | X | X | X | X | X | X | | | |

**Table 6.10** Conceptual models for initiating event frequency estimation [50]

| S. No. | LPSD IE frequency evaluation model | Remarks/applicability |
|---|---|---|
| 1 | $f_{annual} = f_{hourly} \times t_{POS}$ | The data on hourly rate of occurrence of initiating event frequency in a given POS and duration of POS can be obtained from plant records. The frequency obtained is proportional to the duration of the POS |
| 2 | $f_{annual} = fp_{hourly} \times P(IE|p) \times t_{POS}$ | This model is used when data on the initiating event is not available; however, the record shows that data on a precursor are available, for example, human error leading to blocking of ECCS trains. In such estimation apart from precursor frequency, a conditional probability of IE given that a precursor has occurred is also required, which again can be estimated from plant records. This frequency is proportional to the duration of a given POS |
| 3 | $f_{annual} = np_{POS} \times fPOS_{yearly} \times P(IE|p) \times t_{POS}$ | This model again provides estimates based on precursor events. The initiating event arises due to error of commission or omission following an event which occurs a fixed number of times in a given POS. For example, an injection valve is opened four times during ECCS testing. Probability of error—a precursor will lead to initiating event latent ECCS blockage |

where

$f_{annual}$: "Per calendar year" frequency of occurrence of initiator in a given POS (/year)

$f_{hourly}$: Hourly rate of occurrence of initiator in a particular POS (/hour)

$t_{POS}$: Typical or representative estimate for a given POS in (hours per year)

$fp_{hourly}$: Precursor occurrence frequency or rate in a given POS (per/hour)

$P(IE|p)$: Probability of an initiating event given the occurrence of its precursor

$np_{POS}$: Typical or representative number of a given precursor expected to occur in a given POS (np/POS)

$fPOS_{yearly}$: Frequency or expected number of entry in a given POS yearly (/year)

through administrative/checklist procedures; and (f) equipment isolation and recovery activities requiring attention. Hence, the chances of human error increase during shutdown. Even though written procedures are issued in support of operations and maintenance/testing activities, the general experience is that utmost care is required to avoid human errors.

Human error modeling is one of the specific activities that require special attention for LPSD PRA, as the human error apart from the local effect can lead to common cause failure. Involvement of a human reliability expert in LPSD PRA along with an operations and maintenance expert is essential to reflect as-built and maintained status of the plant in LPSD PRA.

## 6.8.2 Fuel Storage Pool PRA

The fuel storage pool (FSP) is a facility attached to a nuclear plant where the irradiated or spent fuel removed from the reactor core is transferred and stored until its decay heat and radiation level are reduced to a level such that it can be handled in dry, i.e., without cooling. Traditionally, the design and operation of the SFP has been governed primarily by deterministic methods. The role of PRA as a complementary tool particularly for assessing the frequency of various initiating events was recognized by IAEA Safety Series 118 on the subject published in 2013 [56]. Prior to the Fukushima event, FSPs were generally not considered a high-risk component to overall risk from an NPP [57]. The two accidents in 2011, i.e., the terrorist attack on twin world trade center structures in USA and the Fukushima Dai-Ichi, changed the way the issue of safety was being considered traditionally. After the terrorist attack, the NRC issued orders to plant operators requiring several measures aimed at mitigating the effect of a large fire explosion or accident that damages FSP [58] toward ensuring the safety of SFPs in a deterministic sense. Even though the reactor core and SFP in a nuclear plant have been recognized as the major source of radioactivity in PRA, the PRA modeling for FSPs got a special focus only after the Fukushima accident. The events surrounding the Fukushima Dai-Ichi plant in Japan provided confirmation of the robustness of the FSPs but also heightened awareness that they could be subject to challenges that might ultimately lead to a radionuclide release [59].

The present scenario is that the nuclear industry is actively considering performance of a PRA for FSPs. To address current regulatory requirements for and studies about FSP risk evaluation, Westinghouse has developed an FSP PRA model [60]. This PRA model helps identify the important risk in regard to FSP equipment and human errors associated with fuel handling operations to identify potential vulnerabilities toward enabling safety improvements. EPRI has performed PRA for Mark 1 and Mark 2 BWRs [57]. It may be noted that the Fukushima Dai-Ichi plant was a Mark 1 design, and hence, this design formed a candidate design for this detailed study. EPRI study integrated the spent fuel pool risks into the plant model, i.e., PRA model of the plant.

The procedural elements of SFP are similar to the reactor core; hence, this section will not elaborate on procedural elements of SFP PRA. However, there are some features specific to SFP modeling that need to be considered as follows:

- An SFP cannot be modeled in isolation or as an independent facility, and the evaluation should take into account that the fuel pool has communication with the containment with transfer trenches (of course, with isolation barriers and gates) and an accident in the reactor containment does affect the pool and vice versa. An example is an interfacing LOCA or break outside the containment or energetic containment failure and high dose/release from containment that inhibits entry of operating staff into the reactor building.
- The major safety features of the pool include the pool make-up systems, the decay heat removal system, reactivity monitoring (if required based on reactor physics evaluation), and the LOCA monitoring system, mostly including pool level monitoring and leakage detection system.
- Spent fuels have robust and passive features. A pool is made of reinforced concrete several feet thick, is lined with steel lining, and has a water level that keeps the fuel submerged adequately to serve the purpose of passive decay heat removal/cooling and provides shielding against radiation.
- The dry storage bay or cask forms part of the spent fuel pool where fuel is transferred after ensuring that the decay heat level and the radiation level have come down to a level such that fuel can be handled dry with required shielding for further transfer/processing.
- The fuel handling system is an integral part of the SFP, and the initiating events formulated for the SFP should also account for handing aspects.
- The initiating event "heavy load drop" either in the pool or the dry bay should be analyzed as part of the SFP model.
- Even though the probability of structural failure in the pool is considered to be low, the possibility of drop in level due to a LOCA-type situation should form part of the accident sequence analysis.

The available literature shows that EPRI performed SFP PRA by integrating the assessment of SFP risk into the PRA model of the Mark 1 and II BWR reactor core. The study concludes that the risk component from SFP is very low, but not negligible. It also points to relatively large uncertainties creeping into PRA result from seismic events, heavy load drops, and interaction of severe accident progression in the containment and related adverse impact on the SFP [57].

Hence, the final remark on the decisions related to inclusion of SFP in the scope of PRA is that the Fukushima event along with the EPRI study suggests that SFP should form part of a full-scope level 1 PRA.

## 6.8.3   Internal Hazards

The available literature shows that there are two major internal hazards, internal fire, and internal flood that normally form part of a full-scope Level 1 PRA [53]. Studies are focused on evaluating the adequacy of protection against these two hazards [61]. The following two sections provide the salient features of analysis of these hazards as part of full-scope Level 1 PRA.

### 6.8.3.1   Internal Flood PRA

Flood PRA forms part of a full-scope Level 1 PRA. Even though the general experience and the available literature show that contribution from flooding PRA (FL-PRA) is significantly less compared with LPSD PRA, the requirements to ensure protection analysis to this effect are an industry norm [62]. Nevertheless, FL-PRA should be conducted, as the flooding phenomenon is intimately, like fire, related to plant-specific features and operation and maintenance practices, and hence, a systematic study can bring out weaknesses in the plant design, construction, and operations, which can be addressed to improving overall plant safety.

Internal FL-PRA deals with a systematic study where the major elements are mostly in line with limited-scope PRA, which is comprised of the following major steps:

1. Familiarization with plant design and operations aspects, particularly with layout.
2. Dividing the plant into different potential flooding zones, looking for the possibility of flooding source, quantum, and causes of flooding;
3. Initial screening of initiating events against set criteria with considerations of frequency and potential impact.
4. Event tree analysis for identified initiating events where the flood protection measures form the header event along with human mitigation/recovery actions.
5. Accident sequence analysis.
6. Fragility analysis toward assessment of failure probability of components, barriers, etc.
7. Quantification.
8. Documentation.

The PRA features that are specific to FL-PRA are as follows:

- *Plant walkdown*: The primary objective of a plant walkdown is to look for scenarios that have potential for safety functions, actions in general, and potential for common cause phenomenon in respect of redundant provision of the plant. A plant walkdown is a vital component of plant familiarization, toward assessing the actual design as well as prevailing realities that have potential for flooding as well as mitigation aspects, such as condition and upkeep of the drains, protection of cables and equipment from water jets, condition of barriers, and capability of the doors and sealing to affect the isolation. A walkdown should be a systematic process where the provision of checklist and handling of feedback should be validated during the final analysis.

- *Identification of flooding zones/compartments*: The flooding zones should be identified keeping in view the physical features, which include water collection capacity of the area (this should account for the available volume considering features that would add to or reduce the overall volume; e.g., structural features like pillars and foundations would reduce, while a pit or a sump space would add to available volume of the main area or compartment under consideration); its isolation from other areas, and mechanism for propagation of flooding. Areas that house or have potential to adversely affect to plant safety features get priority.
- *Characterization of potential flooding events*: This step deals with (i) identification of possible sources in a compartment (e.g., pipe, tanks, valves, water tables outside the compartments, as applicable to underground or sub-basement elevations); (ii) flooding mechanism (pipe or tank ruptures, underground structural leakages, spurious actuation of fire sprinklers, human error in maintenance); (iii) parameters related to severity of flooding (e.g., dumping capacity from the source in $M^3$, flow in $M^3/h$, flood expected rate of level rise, and total time required for critical levels); (iv) provision and information on automatic indication alarms, leak detectors like floor drains and beetles; and (v) most important critical flooding height for all the compartments that will have safety consequences [19].
- *Screening analysis*: Screening analysis is one of the major activities in FL-PRA. The available guides and standards on Level 1 PRA provide detailed guidance on screening of potential events for detailed analysis.
- *Plant flood procedures and human reliability analysis*: Analysis of plant flood handling/mitigation procedure should be checked against the available time for action, the availability of provisions, the availability of communication links, and the potential for adverse effect on the immovable or movable logistics (e.g., diesel generators).
- *Fault tree and event tree analysis*: In FL-PRA, the general approach is to use the full-power Level 1 fault tree for modifications as per the requirements of projected flood scenario for arriving at a flood fault tree.
- *Combined scenario*: The scope of FL-PRA procedures should include scenarios where the other internal events or hazards form part of what is normally referred to as "combined event or hazard," for example, flood and Class IV power failure due to heavy rains, seismic and flood event. Of course, analysis and modeling of combined events are one of the focuses after the Fukushima accident, where a tsunami-induced flood along with many other factors led to the catastrophe affecting not only one plant but three plants at a site. The message here is to keep the scope of FL-PRA such that other hazards or events are adequately addressed.
- *Dependent event/common cause failure*: The inherent process of FL-PRA provides an opportunity to analyze the functional relationship of SSCs though fault

tree analysis and the potential of flood events for common cause failure as part of design and operations. In fact, the analysis can provide vital insights on elevation for location of redundant equipment; for example, three redundant cooling pumps located in same location/elevation have potential for safety implications. Hence, it is required that these pumps are located if the design permits preferably at different elevations or at least physically separated so that flooding events do not affect the three trains instantaneously.

- *Accident sequence evaluation*: As part of the FL-PRA, it is required that the list of cut sets obtained from full-power PRA be scrutinized for flood, independent events in a cut set, having common cause failure potential. The experience has been that this exercise provides useful insights in terms of discovery of cut sets that form the input for FL-PRA as a common cause cut set in the FL-PRA fault tree.

- *Fragility analysis*: When the flood level rises above the component or system bottom elevation, the component is assumed to be submerged and to have failed. The fault tree of full-power Level 1 PRA is used, and unavailability for these failed components is set to 1. The electronics and electrical systems, including the communication network of the plant, require testing and analysis against a rise in humidity or splashing water in the area to characterize its fragility.

- *Initiating event frequency*: The initiating event frequency of catastrophic piping failures is obtained from generic sources; however, these data come with relatively large uncertainty. Hence, Bayesian updating of plant experience with generic priory data provides an effective mechanism to reflect plant-specific operating experience [61]. The analysis of failure probability of structural systems like concrete tanks and metal tanks should be based on plant-specific experience and probabilistic fracture mechanics approach.

- *Human actions and flood EOPs*: Flood EOPs are available in the plant. As part of FL-PRA, human reliability analysis should review these EOPs. In fact, these EOPs should be seen in the context of flood scenarios, in the sense of obstructions, unavailability of certain areas and passages, for personal and logistic movement in the plant. Feedback from flooding incident reports in the plant and from other plants particularly as well as training and refresher courses can provide valuable data and information for FL-PRA.

- *Shutdown condition*: The procedure for FL-PRA for full-power and LPSD PRAs is similar; however, some of the shutdown aspects may require special attention for FL-PRA in shutdown condition. These include maintenance and testing jobs that can contribute to initiating events and block safety functions; re-assessment of flood EOPs, as EOPs are generally written for full-power operations; data for shutdown PRA, including frequency of initiating events; and available time in shutdown due to depressurized systems and low heat content, which can be expected to be more for human actions.

- *Plant operation states*: The FL-PRA should be performed for all POSs and contributions to CDF that account for the POSs. For example, the 12 POSs identified as part of LPSD PRA should also be mapped for FL-PRA.

- *External flood PRA*: Even though external FL-PRA is conducted as part of an external hazard analysis, however, for consequence evaluation the fault tree and event tree models as part of internal FL-PRA are used, while additional initiating events from external sources along with containment as a barrier and penetration failure probability form part of the eternal FL-PRA. External flooding is particularly induced by torrential rains. Hence, it is required that plant EOP should be available for design basis flood as well as beyond design basis flood.

Figure 6.8 shows a typical event tree derived from a complex flood event tree for a reference plant. As can be seen, the rupture of a header/pipe forms an initiating event in the identified compartment or area of the plant. The first two headers highlight the importance of alarms or direct and indirect information available from the plant. At the outset, let us understand that the timing of the progression of the flood event directly depends on the size of the leak/rupture. If the rupture is catastrophic, then submergence will be relatively quick and the time available for action/diagnosis will be very restrictive and will determine the consequences. However, if the leak or rupture size is such that couple of hours are available for the action, which is an initial condition for this flood event analysis, then one can see the propagation and final consequences can be identified based on the plant response/human actions. Except for accident sequence 1, all other 6 accident sequences have potential for core damage because reactor tripping and core cooling are the two functions that need to work to save the core from degradation or damage.

The other aspect that we need to notice is that this event demands intense human actions/diagnosis. Hence, flood human reliability analysis should revisit the plant flood EOPs and reflect the insights and lessons learnt from the FL-PRA. The last and most important point is why one header failure should have safety implications. The lesson is related to the design and layout of plant. All the safety-related equipment should be located such that one common cause like a fire or flood event

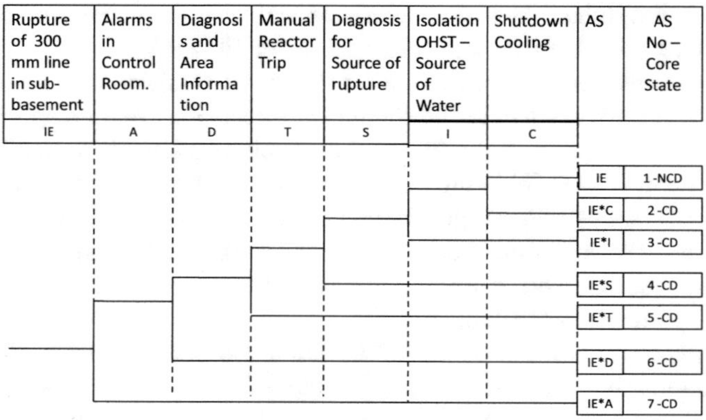

**Fig. 6.8** A simplified flood event tree

should not knockout all the redundant components/trains. Further, quantification of the accident sequence is in line with the general approach followed in Level 1 PRA for full-power operations.

However, the argument in favor of the present layout could be that catastrophic failure is a rare event and small rupture may have safety consequences because adequate time is available for gathering information, diagnosis, and human actions particularly related to reactor tripping and source isolation.

Even though obtaining the point estimates of unavailability of systems and safety functions and event tree analysis to identified flood accident sequences and final characterization of core damage frequency, the uncertainty characterization is a challenge in FL-PRA. The factors that make this aspect complex are human actions, the reliability of human action, the behavior of structural systems, and components in the flood scenario (there could be surprises in the sense that an SSC which was identified to be working has failed, while on the other side, an SSC which was considered healthy has failed). These aspects could lead to large uncertainty in the final results. Hence, apart from the median value, the uncertainty associated with FL-PRA results should also be checked while giving results of the full-scope PRA. One of the ways to address this aspect is to perform sensitivity analysis and use a conservative approach for giving final results of the FL-PRA.

### 6.8.3.2 Internal Fire PRA

NPP operating experience worldwide has shown that fire can be a safety-significant hazard, and therefore, regulators expect the licensee to demonstrate that NPPs meet fire safety goals and objectives [63]. Traditionally, NPPs are licensed to deterministic fire protection rules and regulation. In 2004, the USNRC amendment to allow/encourage licensees to voluntarily adopt National Fire Protection Agency's NFPA-805 standard [64], as a risk-informed performance-based alternative to the deterministic fire protection requirements, can be seen as one of the major shifts toward encouragement of the risk-based/risk-informed approach. This development is in line with the endorsement that fire PRA provides a systematic and effective framework to integrate deterministic and probability modeling toward establishing qualitative as well as quantitative goals, viz. contribution of fire core damage frequency (CDF) to net CDF of the plant.

In this context, Table 6.11, adopted from reference EPM presentation, provides rationales in favor of risk-based approach against certain criteria for fire risk modeling [65]. Figure 6.9, adopted from SSG-3 [19], provides an overview of the process of development of Level 1 fire PRA (L-1-FR-PRA or FR-PRA).

At the outset, a decision should be taken at the organizational level as to what procedure and guidelines to be followed while performing FR-PRA of the plant. In this context, the IAEA safety standard on Development and Application of Level 1 PRA for NPPs [19], NUREG/CR-6850 on Fire PRA Methodology for Nuclear Power Facilities (Vols. 1 and 2) [66], and NUREG-1934/EPRI 1023259 on Nuclear Power Plant Fire Modeling and Analysis (NPP-FIRE MAG) [64] are some of the

**Table 6.11** Rationales in favor of risk-based fire modeling and protection

| Criteria | Traditional engineering failure analysis | Risk-based approach for fire modeling |
|---|---|---|
| Initiating events | Qualitative aspects of fire events | Quantified estimates of initiating event |
| Fire protection system performance | Qualitative notion of expected performance | Quantitative estimates of protection system unavailability |
| Assumptions | There is an inherent assumption that consequence will result in loss of function | The risk-based approach evaluates the likelihood of consequences of the failure of all components |
| Component state | Component has binary, either fail/success states | Best estimate failure rate for each component |
| Common cause failure | No systematic, documentable, and comprehensive treatment of common cause failures | Systematic analysis and empirical characterization of common cause failure phenomenon and integration of these aspects in the fire modeling |
| Human error modeling | Limited and qualitative considerations of human actions | A detailed and comprehensive modeling of human errors with due considerations of performance-shaping factors toward characterizing quantified estimates of human error probability |

**Fig. 6.9** Major procedural steps in fire PRA

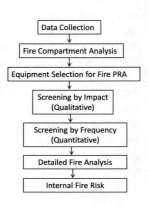

references that can be used to formulate FR-PRA procedure for specific applications. The insight obtained from many FR-PRAs suggests that the most safety-significant areas from FR-PRA analysis point of view include the control room, the cable spread room, Class III equipment, particularly DG building and diesel storage area, power switchgear yard or electrical substation areas, and turbine-generator building. However, a systematic analysis is required to identify various fire compartments to assess their fire potential and associated consequences.

The following section discusses the salient features of FR-PRA specific requirements, which are based on a broad framework of IAEA-SSG-3 guidelines on FR-PRA, in brief:

- *Data collection*: The scope of data collection involves preparation of a checklist in respect of what information to be collected, which includes plant layout, characteristics of plant areas in terms of fire potential, general design provisions of fire detection and protection, areas of fire loads, and control and power cable routings. Insights available from root cause analysis of fire incidents in terms of potential for overall safety of the plant is crucial for FR-PRA. The plant walkdown is a critical step in data collection. Of course, for a plant that is in the design stage, a walkdown is not envisaged; however, if the plant is in the construction, commissioning, or operational stages, then a walkdown is an integral part of data and information collection. The objective of a plant walkdown includes validation or confirmation of the data and information available from the design and operation manual, engineering/layout, or process and engineering drawings; collection of information on maintenance and housekeeping aspects, safety culture, and general organization management (e.g., unwanted material available which is potential fire loads, unauthorized storage of material); and degraded or unfinished condition of fire-retardant coating or degraded condition of fire mitigation system or a provision that was made on a contingency or ad hoc basis and has continued as a permanent feature.
- *Fire compartment analysis*: For the purpose of FR-PRA, the plant areas are partitioned into a suitable number of compartments. The compartments are created such that each compartment is independent in terms of fire barriers, which includes walls, sealed penetrations, capacity of isolation of entry/exits and ventilation systems, and floor drains and provision of fire detection and fire protection measures. Even though the detailed characterization of fire compartments is performed as part of a "detailed fire analysis," at this stage a preliminary characterization of a fire compartment is performed considering total fire loads, sources of initiation of fire, conditions that facilitate fire growth (e.g., ventilation or availability of highly flammable material like diesel oil, or wax bricks), types of fire or smoke detectors, routine testing schedules on these detectors, and testing of fire protection systems and layout of cable trays, which includes power and control cables.

Identification of fire-initiating events is a vital part of FR-PRA. However, the availability of data and information on fire-initiating events from plant records poses a challenge. On the other hand, it is desired that initiating events and their associated frequencies should be based on plant-specific characteristics. Hence, efforts should be made to extract the information from the list of initiating events for identified compartments keeping in view the potential for possibility of short circuits, overheating, loose contacts, maintenance, and human-induced events, while initiating event frequency can be obtained from the available data. This data can be updated with generic prior data obtained from generic sources or data from other similar plants by employing a Bayesian approach. This exercise will provide initiating event frequency for each compartment.

- *General approach to estimating FR-CDF from inputs from fire compartments*: The fire Level 1 PRA should be comprehensive enough in the sense that the scope of analysis should include, apart from full-power conditions, the LPSD scenario also. Hence, the fire compartment analysis should be evaluated in the context of applicable POSs toward giving estimates of fire core damage frequency. The following metrics, as shown in Table 6.12, for various fire compartments numbering, $n$, and POS numbering, $m$, and the formulation provide the overview of complete FR-PRA [63].

$$\text{FCDF} = \sum_{i=1}^{n} \sum_{j=2}^{m} f_{ij} \qquad (6.7)$$

FCDF: fire core damage frequency/r-y.

A threshold value of 1.0E−7/r-y has been used for FR-PRA for full-power modes [63].

- *Equipment selection for FR-PRA*: The customary practice is to perform Level 1 PRA for full-power operations. Hence, Level 1 PRA provides an effective source of input toward identifying SSCs for FR-PRA. The importance measure of the SSCs underlines the safety significance of the SSC, which is useful input for FR-PRA. However, additional efforts are required to understand the document component boundary. For example, information about the power and control cable connections and the types of connectors or sockets is normally not included in Level 1 PRA; however, this information is critical for FR-PRA. Also, the information on types of cables, whether it is a fire-resistant or fire-coated cable, should be available for fire modeling. Further, the data on different types of sealing, particularly on rubber or polymer sealing, should be recorded for equipment.

**Table 6.12** Metrics of FR-PRA compartments versus POSs

| Compartment $i = \{1, 2, ...n\}$ | | Plant Operational States (POSs) $J = 1, 2, ....m$ | | | | | | |
|---|---|---|---|---|---|---|---|---|
| | FP | $\xleftarrow{\hspace{2.5cm}}$ LPSD-POS $\xrightarrow{\hspace{2.5cm}}$ | | | | | | |
| 1 | 1 | 2 | .. | .. | .. | $m$ | $\sum j$ |
| 2 | $f_{11}$ | $f_{12}$ | .. | .. | .. | $f_{1m}$ | |
| .. | $f_{21}$ | -- | .. | .. | .. | $f_{2m}$ | |
| .. | .. | .. | .. | .. | .. | | |
| $n$ | $f_{n1}$ | | .. | .. | .. | $f_{nm}$ | |
| $\sum i$ | $f_{i1}$ | $f_{i2}$ | .. | .. | .. | $f_{im}$ | FCDF |
| | $f_{\text{FP}}$ | $f_{\text{LPSD}} = \sum_{j=2}^{m} f_{ij}$ | | | | | | |

Finally, a list of component is created such that the fire-induced failure of the subject SSC leads to an initiating event, failure of a safety/mitigation function, adversely affecting recovery actions, and spurious actuation of unintended and undesired functions that aggravate the situation further.

- *Qualitative screening of compartments*: From the area analysis, a list of compartments is made; however, from the FR-PRA point of view, a screening is required, so that compartments having higher safety significance can be selected for further analysis. This screening is based on qualitative arguments regarding safety significance in terms of fire load density below a threshold (a conservative criteria) OR all of these conditions: (a) no source of fire ignition in the compartment AND, (b) no safety system or safety-related system or their control or power cables are located in the compartment AND, (c) the potential for spread of fire to other compartment is very low. For a multi-unit site, the potential for spread of fire to other units should be considered.

- *Quantitative screening of compartments*: The screening done in the previous stage is based on conservative assumptions; however, all the compartments screened are not safety-significant; hence, there is a need to remove this conservatism to reduce further the list of candidate compartments for detailed fire analysis. This screening is called screening by contribution to the core damage frequency. Hence, the Level 1 PRA model is required for screening. Let us assume that the frequency of initiating events for compartments is known and conditional core damage probability (CCDP) can be evaluated for a given compartment by conservatively assuming the worst-case scenario, e.g., unavailability of all the safety equipment in the area. A CDF estimation using this case will give us a conditional probability of core damage considering unavailability of safety trains in compartment I due to a given initiating event in the respective compartment. The model for estimating contribution of fire on CDF is as follows [19]:

$$f_{cdf} = \sum_i f_{fire-i} \times CCDP_i \qquad (6.8)$$

where $f_{cdf}$ is the contribution from fire to the core damage frequency; $f_{fire-i}$ is the frequency of fire IE in compartment $i$; and $CCDP_i$ is the conditional core damage probability for compartment $i$.

This model provides a quantitative methodology for prioritizing a final list of compartments for detailed analysis. The threshold for screening the compartment should be sufficiently low to screen in all the safety-significant fire events and high enough to allow some screening out. Further, the cumulative core damage of the entire screen out event should not be more than a preset low threshold.

- *Detailed fire analysis*: For the major part, this detailed analysis deals with deterministic modeling of fire scenarios in various compartments [67]. The major scope and aims of detailed analysis are as follows: (a) remove the level of conservatism in the fire scenario identified during the screening process and thereby focus on the most safety-significant fire compartments, (b) use more realistic models for fire modeling, i.e., move from algebraic and empirical models to computational fluid dynamics CFD-based simulation approach for fire initiation and propagation, (c) fine-tune human reliability models for actions/recovery, (d) characterize the fire scenario in the time domain so that input is available to the operations and maintenance team for the available time window for recovery or any specific actions, (e) modify Level 1 PRA fault tree to reflect unavailability of components/safety system, and (f) modify the Level 1 PRA event tree to reflect the fire scenario toward arriving at a statement of contribution of fire to net CDF.

The major methodological steps, as a recommended methodology to arrive at the above objectives, can be summarized in the following six steps: (1) define fire modeling goals and objectives, (2) characterize the fire scenarios, (3) select fire models, (4) calculate fire-generated conditions, (5) conduct simulations to arrive at fire core damage frequency and sensitivity and uncertainty analyses, and (6) document the analysis. For details, refer to the related literature [19, 64].

Traditionally, the fire propagation and consequence analysis is carried out by simplified methods, flow models, and zone model-based codes [68]. However, advances in computational fluid dynamics have enabled development of state-of-the-art CFD codes, which facilitate advanced fire analysis and propagation modeling such that estimations are nearer to real-time conditions. Therefore, CFD codes are finding extensive application in fire modeling in nuclear plants along with conventional algebraic equations. A brief note on the evolution of fire modeling, for the sake of completeness of the scope of this chapter, is presented. This section is based on the information available from the presentation in fire modeling examples from the nuclear power industry [65]. A detailed fire modeling comprised of three-tier approaches is as follows:

- *First level*: Broadly based on closed form correlations. This approach is basically conservative in nature and generally used when resources or time is limited. The fire characterization is performed employing empirical or physical formulations, e.g., fire flow formulation, estimation of fire loads, plume temperature estimation. For example, the fire-induced vulnerability evaluation (FIVE) approach and fire dynamic tool (FDT) methodology are examples of first-tier models.
- *Second level*: This tier is characterized by being more realistic and less conservative than the first tier. The conservative assumptions are studied with additional and real-time inputs such that over-conservatism can be eliminated in modeling the fire scenario. The areas and components form the focus of detailed analysis. Human reliability aspects are treated more realistically. Fine-tuning of fire protection aims to avert severe damage to safety-significant components or targets. In short, fire initiation and propagations are studied in greater detail.
- *Third level*: This level is characterized by use of zone and field models. This level provides realistic estimates and scenario; however, the procedure is

time-consuming. The estimation procedures move from algebraic correlations to use of differential equations. Examples of this level are modeling in CFAST and FDS environment. For instance, the compartment fire modeling scenario, typically the case of FR-PRA analysis, involves analysis of potential for damage to targets in the compartment. Damage time calculation, multi-compartment analysis, and main control room fire modeling are some of the examples of third-tier analysis.

## 6.8.4  External Hazards

The state of the art of PRA for external events has evolved, and significant advances have been made over the last three decades [69]. The major factors for this growth are the insights from accidents and advances in computational tools and methods. A systematic approach on the subject by international and national bodies, for example, IAEA documents and workshops, has also contributed to the growing interest [70, 71].

A comprehensive list of external hazards should be formulated for the specific site based on the site characteristics. A reference to generic lists forms part of this review. The PRA modeling team is required to go through the applicability of these hazards to a facility under consideration. The screening process for external hazards should record the rationales, qualitative as well quantitative, for screening out the hazards from the scope of external event PRA. Performing external PRA for all the hazards is a resource-intensive proposition, while the experience has been that performing detailed analysis for selected events provides an effective strategy. A bounding analysis is performed with the aim of reducing the list of external hazards. The bounding analysis involves application of qualitative criteria, for example, potential likelihood and impact of the hazard on the core damage probability, plant-specific parameters (e.g., location of the plant if it is located near a river or sea), tectonic parameters of the site, distance and landing and takeoff routes if the airport is located in a certain region near the plant, and rain fall and cyclonic characteristics of the site.

For the screened-in external hazards, a detailed PRA is conducted. However, the general experience is that in many cases, the following external hazards cannot be excluded from the scope of analysis: seismic hazards, high winds, external floods, and human-induced hazards [19]. The general aspects that need to be considered in respect of performing Level 1 External Event PRA for these hazards can be briefly summarized as follows:

- *Data and information*: Plant-specific data should be used, as far as possible, for the Level 1 PRA for external hazards.
- *Parameters for characterization*: Each hazard is required to be characterized in terms of intensity of damage potential to the plant. Often, this characterization requires more than one parameter; for example, for seismic hazard, the ground

motion is characterized by acceleration, velocity, and displacement; damage potential can be characterized by water level, discharge rate, velocity, pressure differential, and duration of wave action.

- *Frequency of hazard*: The return period or frequency of hazards and period for these hazards, including hazard trends particularly in the recent past, should be site-specific. Even if the trend is showing signs it is decreasing and is below certain intensity, the screening of this hazard should be weighed against other applicable criteria.
- *Walkdown*: Plant walkdown forms an integral part of external event analysis. The objective is to gather plant-specific and real-time conditions of the SSCs and verification of design with as-built and maintained systems.
- *Fragility evaluation*: Even though the procedure for fragility evaluation might vary depending on the type of hazards, the fragility evaluation should be based on plant-specific data and should employ conservative criteria when there is uncertainty in respect of subject SSC meeting its performance given that the hazard has occurred.
- *Initiating events*: Apart from the Level 1 internal event initiating events, additional events associated with each hazard should be evaluated as part of accident sequence modeling.
- *Identification of SSCs*: The list of SSCs affected by the external hazards need not remain limited to Level 1 PRA for internal events and should extend to include those SSCs that have the potential to adversely affect the plant safety directly as well as SSCs that might cause failure of SSCs already part of Level 1 PRA.
- *Uncertainty analysis*: Uncertainty analysis including aleatory (due to randomness in parameters) and epistemic (due to lack of knowledge) should be included in the scope of external event PRA.
- *Sensitivity analysis*: The impact of major assumptions of the analysis should be evaluated by sensitivity analysis.
- *Interpretation of results*: Contribution of core damage from the considered hazards should be evaluated as well as the major contributing factors identified for further considerations/improvement.
- *Documentation*: A guidance document, be it national and/or international, should be followed to meet major quality and procedural aspects of the analysis.

The above bullets provide only an overview of salient procedural and requirement aspects and are not exhaustive. The treatment of all the hazards listed, as above, is not in the scope of this chapter; therefore, there is a conscious decision to dedicate the remaining section on seismic hazard to bring out the salient feature of external event PRA.

## Seismic Level 1 PRA

Nuclear power plants are designed and built to withstand environmental hazards including earthquakes [72]. The Seismic PRA (SPRA) forms part of the risk-informed approach for demonstration of earthquake resistance during the design phase of an NPP along with deterministic evaluation [73]. A rough estimate shows that SPRA has been conducted by many countries and over 40 SPRAs exist worldwide. A general experience in PRA studies of operating NPPs is that the seismic event contributes significantly to core damage frequency [74].

Further, the two rather recent incidents in Japan gave the world some important lessons. The earthquake of July 16, 2007, stuck the Kashiwazaki-Kariwa (KK) NPP, located in Niigata prefecture. Consequently, all the operating units were automatically shutdown safely. The earthquake intensity was significantly high—almost double of the seismic design basis; however, all the safety-critical systems, particularly the one classified as category 1, remained healthy, while there was some minor damage to other non-safety classes. This earthquake endorsed the safety classification scheme which was verified by this event. The second earthquake of magnitude 9.0 occurred on March 11, 2011, at Fukushima Dai-ichi. The control rods were inserted automatically in units 1, 2, and 3, which were operating at the time of earthquake. As a consequence of earthquake, the grid power supply failed, and, as intended, the emergency diesel generators started and emergency core cooling system began operating. However, after 1 h of earthquake, a tsunami wave of 14 m height struck the site and failed the emergency diesel generators, seawater pumps, and the decay heat removal from the reactor core. In spite of injecting water into the core, the fuel failure, followed by hydrogen explosion, occurred in the core. As a consequence, radioactive material was released from the plant [75].

The Fukushima incident, unlike the incident of 2007, attracted the attention of the safety community worldwide. The immediate action was a procedure called stress tests, which are performed on NPPs globally to ensure that there is adequate capacity in the plant to mitigate the consequences of potential extreme events. In Fukushima accident, the issue of combined event attracted attention. As Fukushima site experienced the earthquake, followed by tsunami and then station blackout scenario that led to extreme severity of the accident.

R&D is being conducted to improve understanding of a Fukushima-type scenario. One of the research areas includes the reliability characterization for station blackout scenarios; however, this incident introduced some fundamental changes in how we look at safety. There was a clear paradigm shift from the past engineering discipline that NPPs are to be designed to prevent any accident to the new discipline that they are to be designed to mitigate accidents, based on risk concepts. In this context and scope of this work, we better put it as risk-based concept where PRA plays the key role. This observation sets the ball in motion for our journey to understand the fundamental features of SPRA.

**Procedural Element of Seismic PRA**

The complete SPRA procedure can be broken down into five steps, as follows: (i) data and information collection, (ii) seismic hazard analysis or, more appropriately, probabilistic seismic hazard analysis, (iii) fragility evaluation for the plant SSCs, (iv) accident sequence evaluation, and (v) risk characterization [76, 77].

### 6.8.4.1   Data and Information Collection

Collection and analysis of plant information along with plant walkdown are central to this procedural step. Apart from this, information required for probabilistic hazard analysis, like development of database of earthquake histories, associated intensity, return period, for a considered period and distance, tectonics of the region, soil capacity, for source and ground motion modeling. A database is developed in respect of input required for the fragility analysis, such as characterization of concrete structures, foundation, and SSC capacity modeling. Level 1 PRA is a prerequisite for the SPRA. The list of candidate components derived from Level 1 PRA for fragility analysis should be developed. Apart from this, the additional SSCs that are identified for fragility analysis should be prepared. A generic database of component fragility needs to be developed. This database will be required when it is not possible to generate a fragility value for certain components, due to either limitation of resources or other logistic constraints (Fig. 6.10).

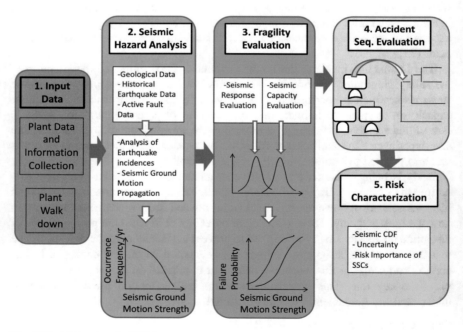

**Fig. 6.10** Major steps in SPRA

### 6.8.4.2  Probabilistic Seismic Hazard Analysis

The aim of the probabilistic seismic hazard analysis (PSHA) is to develop hazard curves which characterize the seismic exposure of a given site to the primary including the instrumentation recordings, as well as the geology of the region (e.g., soil property, fault lines) including physical evidence of the past seismicity [74].

The following procedure for PSHA evaluation is based on a publication by A. Primer explains the fundamentals and procedural aspects of PSHA [72]. The complete procedure of PSHA is implemented in three steps: (1) development of a seismic hazard source model, (2) characterization of a ground motion model, and (3) probabilistic calculation for hazard curve [78]. The evaluation aleatory and epistemic uncertainty for various parameters also form part of this procedure.

The setting of the earthquake source model is largely performed using probabilistic modeling where the major parameters of the model are focal location/geometry, magnitude and frequency of earthquake, and distance of the epicenter. Sources are normally divided into specific sources and regional sources. Each type of source is defined by parameters of source location, fault geometry, earthquake magnitude, and earthquake occurrence frequency based on active fault data ($10^5$ year) and historical earthquake data ($10^3$ year). The specific source model is based on source location and earthquake scale. Inland active faults and plate interface faults are also considered in the specific source approach. Regional source model operates on input on series of earthquakes that occur in a region with a certain area.

To put it in simply, the source model is a list of scenarios, each with an associated magnitude $m$; location $L$; and effective rate, $r$: For example, a source model can have $N$ scenarios ($E_n$) and can be expressed as follows:

$$E_n = E(m_n, L_n, r_n) \tag{6.9}$$

The second step of PSHA modeling is characterization of a ground motion model or attenuation relationship. Given the fact that a large number of earthquakes have been recorded, computing attenuation of seismic strength for a given distance from the source to the site can be achieved. In its basic form, the model provides the relationship of ground motion, e.g., peak ground acceleration (PGA) ($g$), with the distance for a given intensity, as shown in Fig. 6.11.

**Fig. 6.11** Attenuation relationship showing the median curve and confidence limits

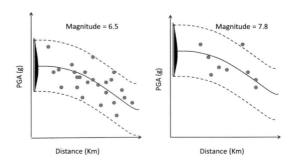

Let us perform PSHA estimation for a given site. With the availability of the source model and the attenuation model, it is possible to perform probabilistic calculation for arriving at a hazard curve. From the source data, we have $N$ data points for magnitude $m_n$; location $L_n$; and rate $r_n$. Let us assume that with the available data, the distance $D_n$ to the site from the source has been calculated. The attenuation relationship as obtained above enables formulation of ground motion using lognormal distribution as follows:

$$p_n(\ln PGA) = \frac{1}{\sigma_n\sqrt{2\pi}} \exp\left[-\frac{1}{2}\left(\frac{\ln(PGA) - g(m_n, D_n)}{\sigma}\right)^2\right] \qquad (6.10)$$

where the $g(m_n, D_n)$ and $\sigma_n$ are mean and standard deviation of $\ln(PGA)$ given by the attenuation relationship. The lognormal distribution for ln PGA and probability of exceeding each ln PGA are shown in Fig. 6.12a, b, respectively.

$$P_n(\ln PGA) = \frac{1}{\sigma_n\sqrt{2\pi}} \int\limits_{\ln PGA}^{\infty} \exp\left[-\frac{1}{2}\left(\frac{\ln(PGA) - g(m_n, D_n)}{\sigma}\right)^2\right] d\ln PGA$$

$$(6.11)$$

Figure 6.12b gives the probability of exceedance for each ln PGA, for a seismic event, hence, multiplying Eq. 6.11 with the rate of occurrence $r_n$, we get the annual rate at which each ln PGA is exceeded, say $R_n$, due to the scenario:

$$R_n(> \ln PGA) = r_n P_n(> \ln PGA) \qquad (6.12)$$

We have total $N$ scenarios, and hence, the cumulative rate of exceedance for a given PGA can be given by a summation up to the given value of PGA as follows:

$$R_s(> \ln PGA) = \sum_{n=1}^{N} R_s(> \ln PGA) = \sum_{n=1}^{N} r_n P_n(> \ln PGA) \qquad (6.13)$$

The hazard curve, giving an exceeding rate versus PGAs, as per Eq. 6.13, is shown in Fig. 6.13, the required product of PSHA.

**Fig. 6.12 a** Probability distribution of ln PGA and **b** cumulative distribution function or probability of exceeding each ln PGA or ground motion strength

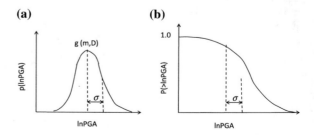

**Fig. 6.13** Seismic hazard
curve giving frequency of
exceedance for range of peak
ground acceleration or ground
motion strength

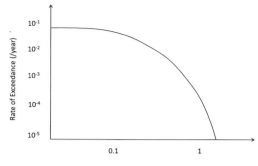

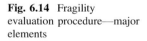

Peak Ground Acceleration (g), m/sec²

However, the treatment of uncertainty requires special attention in seismic modeling in general and modeling of PSHA in particular. The aleatory uncertainty (i.e., uncertainty due to randomness in the parameters), which is inherent to the estimates and cannot be reduced, is in practice captured by the probability distribution of the parameter, whereas the epistemic uncertainty (i.e., uncertainty due to lack of data and information) is captured by logic models and treated as a discrete input for a given parameter. For details on uncertainty, see Chap. 9.

### 6.8.4.3   Seismic Fragility Evaluation

Fragility is the conditional probability of failure of a structure or a component for a given PGA. Fragility is used for (a) conditional probability of initiating events, for example, LOCA or loss of off-site power, (b) re-evaluation of system unavailability by considering conditional failure probability of component failure that leads to system failure, and (c) estimation of accident sequence re-evaluation as part of SPRA. Plant-specific fragility estimates are always required in SPRA. The major procedural steps for fragility evaluation are shown in Fig. 6.14 [76, 79].

**Fig. 6.14** Fragility
evaluation procedure—major
elements

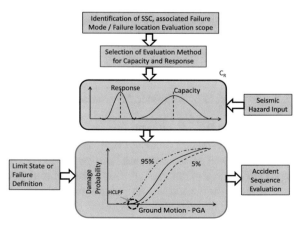

The objective of fragility analysis is to generate fragility curves as shown by the last item in Fig. 6.14. The result of fragility analysis forms the input for the accident evaluation program. The input from PSHA, i.e., ground motion, e.g., in the present case from PGA, is used along with other parameters for fragility modeling. The SSCs identified in Level 1 FP-PRA along with additional SSCs identified through walkdown and analysis are subjected to fragility analysis. As can be seen in Fig. 6.14, two elements comprise fragility, viz. seismic response and seismic capacity of the SSC. The mechanical parameters, strength, and ductility are used to characterize structural capacity, whereas the scaling of the design analysis results and the median factor of safety determine the response. Failure in respect of fragility analysis is defined as an event when an element reaches its limit state. For example, for a structure, elastic deformation constitutes failure; for piping, fracture or collapse of the pressure boundary, constitutes failure of support; for soil—liquefaction and slop instability constitute failure.

In the factor of safety approach, the fragility curve and associated uncertainty are expressed by three parameters, $A_m$, $\beta_R$, and $\beta_U$.

Hence,

$$A = A_m \varepsilon_R \varepsilon_U \tag{6.14}$$

where $A$ is the ground acceleration corresponding to any given frequency of failure; $A_m$ is the median ground acceleration capacity; $\varepsilon_R$ and $\varepsilon_U$ are the random variables with unit median and logarithmic standard deviations $\beta_R$ and $\beta_U$.

$$\beta_C = \sqrt{\beta_R^2 + \beta_U^2} \tag{6.15}$$

High-confidence, low-probability failure (HCLPF) capacity

$$\begin{aligned} \text{HCLPF capacity} &= A_m \exp[-1.65(\beta_R + \beta_U)] \\ &= A_m \exp[-2.33\beta_C] \end{aligned} \tag{6.16}$$

The fragility cure so obtained is used as input for accident sequence quantification. A database of fragility parameters should be maintained for reference and future applications.

### 6.8.4.4  Accident Sequence Evaluation

The Level 1 PRA model with internal events is adopted for SPRA to incorporate/ reflect the seismic aspects as well as additional aspects such as new initiating events, analysis of seismic isolation, protection and sensing, and actuation system in Level 1 PRA of internal events. Like fire and flood analysis, the seismic event can occur in the shutdown state of the plant. Hence, seismic initiating events should be mapped for all the applicable POSs, as fueling and maintenance activities can have additional requirements of modeling.

The major steps in accident sequence evaluation are [77]:

1. Data and information collection on plant accident sequences (from Level 1 PRA for internal initiating events);
2. Setting of initiating events;
3. Modeling of accident sequences;
4. Modeling of systems;
5. Analysis of containment system availability and reliability.

Some considerations that could be a good input toward formulating a list of initiating events for a seismic PRA are summarized as follows:

- Seismically induced failures of large components, e.g., reactor vessel, coolant common inlet and outlet header/plenum, heat exchanger or steam generator, pressurizers, fueling machines;
- LOCA (major/minor);
- Large header failure in secondary system in reactor building leading to large-scale flooding adversely affecting primary as well as shutdown cooling system;
- Structural damage in shutdown devices that blocks free insertion of these devices in the reactor core;
- Loss of off-site power;
- Transient leading failure of support systems, such as Class III PS system, Class II, and Class I systems;
- Structural failure leading to failure of ultimate heat sink.

Based on the fragility analysis, the safety systems and engineering features should be re-evaluated as part of seismic PRA with degraded capacity due to failure of some critical systems like pumps and power supplies. The accident sequence analysis should be carried out based on the new initiating events and degraded state of the plant. For quantification, the updated database based on fragility analysis should be used for evaluating the core damage frequency.

The final and quantified list of accident sequences which captures all seismic-induced initiating frequency and reduced safety system availability is used to arrive at core damage frequency. The Level 1 criteria are that for new plants the overall CDF of Level 1 study considering all initiating events (internal/external), full-power, as well as low-power shutdown state and reactor core as the source of radioactivity, should be less than $1 \times 10^{-5}$/r-y. Hence, the contribution of the SPRA to net core damage along with other states should be less than this target.

## 6.9   Level 2 PRA

### 6.9.1   *Background*

We have discussed the essential features of full-scope Level 1 PRA, which requires considerations of full-power operation, low-power, and shutdown; internal and external initiating events and reactor core as the source of radiation (for some reactors, the fuel storage bay has been included) are directly relevant and are required for implementation of integrated risk-based engineering in a complex and safety-critical systems. Some of the topics directly supporting the subject or topics requiring background areas for IRBE have been covered in Chaps. 1–5 of this book. PRA is conducted at three levels, and the scope of these three levels, viz. Level 1, Level 2, and Level 3, broadly covers system analysis for giving a statement of core damage frequency, containment phenomena modeling for estimation of release frequency, and consequence analysis for evaluating the risk to the public.

The risk-based applications, e.g., integrated design, or operational safety analysis; or evaluation/optimization of any specific aspects like surveillance test interval, allowable outage time, plant emergency operating procedures, precursor analysis; or development and implementation of risk monitor, in-service-inspection program, require only Level 1 PRA. Further, if we look at the status of PRA for NPPs, the available literature shows that Level 1 PRA has been performed for all NPPs, Level 2 study is being done for many plants, and most countries have at least one Level 2 PRA, while Level 3 PRA has been available for only a few NPPs [20]. Applications of Level 2 and Level 3 PRAs have been reported in for emergency scenarios and accident management programs [80].

WASH-1400 was the first comprehensive landmark study covering all three levels of PRA. The state of the art in Level 1 PRA has matured and finding many applications. At the same time, extensive work is being performed on application of levels 2 and 3 in support of severe accident management and emergency preparedness. This necessitated experiments for generation of data, models, and methods to reduce uncertainty in prediction of accident scenario. The NEA report documents these developments, including the computer code developed for Level 2 PRA [81]. These documents have gone a long way toward providing the general guidelines on a procedural framework that is crucial for performing Level 2 PRA; however, the comparison of the available Level 2 PRA studies suggests that a comprehensive document on common best practices and specific detailed criteria is required to harmonize the Level 2 PRA procedures. This document led to a work on development of best practices guidelines for Level 2 PRA development and application as part of Advanced Safety Assessment Methodologies referred to as ASAMPRA2 [80]. This document is compiled into 4 volumes and is expected to provide detailed guidance on performance of Level 2 PRA and will support the existing IAEA and ASME/ANS standards/guides.

With the above in view the authors made a decision to limit the scope of this book, by covering all the detailed aspects of Level 1 PRA, while Level 2 and Level

3 will be dealt with briefly for the sake of completeness. Hence, the following sections provide the salient features of Level 2 and Level 3 PRA.

## 6.9.2   Level 2 PRA Methodology

The Level 2 PRA deals with analysis of chronological progression of core damage sequences identified in Level 1 PRA toward providing the quantified statement of the likelihood of release consequent to severe damage to reactor fuel. The Level 2 PRA procedure deals with arriving at plant damage states based on the binning of accident sequences from Level 1 PRA; analysis of containment scenario considering the extent of damage to the fuel and subsequent release of radioactivity in the containment; modeling of source terms from the containment and analysis of release frequency for various source terms; and evaluation of large early release frequency—a major result of Level 1 PRA.

There is a minor difference in the approach to Level 2 PRA; viz., when Level 2 PRA is performed as an integrated exercise with Level 1 PRA, the Level 2 PRA requirements form part of the Level 1 analysis. For instance, analysis of scenarios involving their effect on containment systems forms part of Level 1 PRA. However, when Level 2 PRA is performed as an extension of Level 1 PRA (i.e. Level 1 PRA is already available while initiating a Level 2 study), then Level 1 PRA needs to be revisited such that Level PRA requirements are met (e.g., analysis of Level 1 accident sequences that might affect the isolation capability of the containment or implication of Level 1 scenario affecting performance of containment venting or emergency exhaust system fans/blowers).

Similarly, the scope of accident sequences from Level 1 PRA should also include full-scope Level 1 PRA (i.e. considerations of, apart from full-power, low-power and shutdown operational modes as also, along with internal event, considerations of external events). In summary, the accident sequences that form the input for analysis of plant damage states should be based on full-scope PRA. The reference to an applicable international or national standards/guide like IAEA safety standard SSG-4 entitled "Development and Application of Level 2 PRA for NPPs" [20], ASME/ANS standard "Severe Accident Progression and Radiological Release (Level 2) PRA standard for Nuclear Power Plant Applications for Light Water Reactors (LWRs) [81] should be referred to toward meeting the safety requirements of a Level 2 PRA.

The major objectives of Level 2 PRA are as follows:

- Analysis of accident progression toward obtaining a chronological understanding of physical chemical and radiological phenomenon in containment.
- Evaluation of containment-associated engineering features performance toward effectiveness in containment or controlled discharge of radioactivity.
- Evaluation of the source term and associated frequency for identified scenarios.
- Evaluation of the large early release frequency.

- Identification of containment system strengths and weaknesses and accordingly provide recommendation for improving accident mitigation strategies.
- Insights for emergency preparedness and severe accident management program.
- Input for Level 3 PRA modeling along with Level 3 PRA result on risk provides input for further fine-tuning criteria for defining core damage frequency.

The major procedural steps in Level 2 PRA are (1) interfacing Level 1 PRA for development of plant damage states, (2) containment performance and modeling for accident progression analysis, (3) analysis, (4) source term analysis, (5) estimation of likelihood for various source term and large early release, and (6) sensitivity and uncertainty analysis. The following subsections provide a brief overview of these steps.

### 6.9.2.1  Plant Damage States

The spectrum of accident sequences obtained from Level 1 PRA is very large and requires grouping to reduce the number of plant damage states (PDS) that are manageable from the point of view of detailed accident progression analysis. These PDS form the input for Level 2 PRA. The major heuristics or guidelines for grouping of accident sequences into plant damage states are as follows [82]:

- Initiating events of similar type/nature, e.g., LOCA events;
- The enthalpy/pressure of the reactor core and primary coolant system accident scenario initiated in full-power operation/shutdown state;
- Status of the safety systems, e.g., emergency power, reactor protection system, ECCS;
- Availability of the containment safety features and containment integrity;
- Scenario having potential for release/no release.

Generally, the experience in Level 2 PRA is that the spectrum of accident sequences more than say 10,000 or so converges into plant damage states ranging from 10 to 50 or little more. These PDSs form the starting point for accident progression modeling employing containment event tree analysis approach.

### 6.9.2.2  Accident Progression Analysis

A comprehensive list, as complete as possible in terms of covering the potential accident spectrum, is a prerequisite for accident progression modeling using the containment event tree approach. Apart from this, detailed modeling of thermal-hydraulic, physical, and radiological phenomenon forms the major input to address the nodal bifurcation in containment event tree (CET) modeling. Event tree is developed for each damage stage to model radiological behavior in containment.

End states of the containment event tree are characterized using attributes such as time elapsed since the initiation of the accident, elevation or height of discharge, energy component of the release, and release rate.

Accordingly, the containment end states for three phases, viz. (a) the early phase of in-vessel core damage generally varies from an initial period up to 2 h into the initiation of the accident, (b) the late period of damage progression range from 2 to 10 h, this includes breach of reactor vessel and (c) the long-term response of the plant is more than 10 h. The early-phase questions which form header events in the containment event tree are related to response or performance of the containment (e.g., containment confinement and isolation, containment leakage or bypass etc.) and operation of engineering safety features, which includes containment venting or emergency exhaust system iodine filtering system. In the late period, there are questions related to assessment of cladding failure and whether it was extended to fuel damage and consequent release from the reactor, core submergence and cooling requirements, any leakage of core material and questions related to metal water reaction. For extended accident scenario analysis modeling aspects might include availability of AC power for extended period, operation of ECCS in extended mission time, capability of containment to confine the radioactivity or question related to loss of containment integrated for sustaining the confinement function. Apart from this there are questions related to human reliability in or implementation of plant accident management procedures and recovery of failed systems.

The elevation factor is characterized by discharge and ground level or through the stack, as also the energy of release is characterized by attribute high or low. The release rate or mode could be characterized by attributes such as rapid puff release, slow and steady discharge, or multiple plumes.

The event tree modeling requires extensive input from severe accident management analysis in support of branch point decisions while CET modeling. Major categories of analysis support are thermal-hydraulic analysis to estimate thermal stresses on cladding and structural material, core and debris behavior potential in terms of vessel penetration and interaction with concrete, prediction of containment performance due to severe accident loads/environment, and radiological analysis toward predicting radioactivity behavior in the containment and formulation of net containment source terms for various accident progression scenarios. These complex computations are achieved through the use of computer codes.

The outcome of this step is a list of containment event sequences giving event progression sequences, and these sequences form an input for source term analysis.

### 6.9.2.3  Source Term Analysis

The large number of accident sequences or the end point sequences of the generated containment event tree analysis needs to be grouped toward making the source term characterization for consequence evaluation in Level 3 PRA more manageable. This process is called binning of containment end points of sequences. The binning is carried out in two steps. First, the source term categories are developed based on the similarity of the source term phenomena. The outcome of this step is binned severe accident sequences that lead to similar release in terms of radioactivity, timing, quantum and similar release to the environment. In the next step, as per the requirements of Level 3 PRA, the release categories are further binned to provide release categories for off-site consequence. The binning of end points of CET is performed based on certain parameters/conditions or heuristics. The major considerations are the mode of ex-vessel release, the containment leakage rate, the effectiveness or reliability of emergency exhaust system toward filtering out the major isotope of iodine, and the timing or duration of the release (Table 6.13).

One of the major results of Level 2 PRA is a statement of large early release frequency or release frequency per year. This statement is a result of the containment event tree end states, which provides the frequency for each accident progression sequence. The source term analysis provides information that defines the release in terms of the rate of release of radionuclide for the given containment sequences, referred to the as source term. This source term is evaluated by considering the time of release, the coolant system pressure, the mode of containment leakage, and the elevation of release. Hence, the radiological source term can be characterized for release category, (a) frequency/year, (b) bin attributes (time of release, coolant system pressure, mode of containment failure, etc.), and (c) radioactivity release in terms of percentage of reactor core net inventory (for a given fuel burn-up level), viz. 100% of noble gases like xenon, 1% of iodine and 1% of particulate, and other fission products. From this exercise, we will have a statement of frequency for give source terms. When we add the required source term frequency, we get the statement of release frequency. This procedure is similar to what we followed in estimating the core damage frequency estimation.

However, one of the major objectives of Level 2 PRA is to provide input for Level 3 PRA. For Level 3 analysis, all the release categories will be transported as input for Level 3 PRA analysis.

**Table 6.13** Radionuclide groups [27]

| Group | Elements in group | Representative element in group |
|---|---|---|
| Noble gases | Xe, Kr | Xe |
| Halogens | I, BrI | |
| Alkali metals | Cs, RbCs | |
| Alkaline earths | Ba, SrBa | |
| Tellurium group | Te, Sb, SeTe | |
| Refractory/noble metals | Ru, Mo, Pd, Tc, Rh | Ru |
| Lanthanides | La, Y, Nd, Eu, Pm, Pr, Sm Zr, Nd, Nb, Cm, Am, La | |
| Actinides/cerium group | Ce, Pu, NpCe | |

#### 6.9.2.4  Documentation

The IAEA Guide SSG-4 [20] provides adequate guidance on requirement of documentation. The major aspects that need to be covered in documentation are as follows:

- Complete clear linkage, from Level 1 PRA in terms of plant damage states to Level 3 PRA in terms of data, major assumptions and sensitivity analysis, uncertainty associated with source terms, and associated frequency.
- The results of the analysis and its interpretation should be elaborated such that the available insight helps in the development of Level 3 PRA.
- Recommendation of the analysis should be part of documentation.
- Use of Level 2 PRA in support of evaluating mitigation provision and severe accident management.

## 6.10  Level 3 PRA

### 6.10.1  Background

The release of radioactivity from the stack at higher elevation or ground level from the reactor containment and its propagation following an accident depends on many factors, mainly the source term category, the dominant propagation routes, and the exposure pathways. The Level 3 PRA deals with evaluation of consequences in terms of health risk or fatalities to individual members of the public; a group of population or society; adverse effects on the environment; and economic loss. Other intangible consequences might be notional but affect the public conscience or opinion as a consequence of an accident. For example, the Fukushima accident raised public opinion that led to shutdown of all the NPPs in Japan and raised public concerns in other countries.

The current status is that unlike Level 1 and Level 2 PRAs, not many Level 3 PRAs for NPPs have been performed. One of the major reasons could be that performing PRA studies is a resource-intensive proposition and requires sequential processing. The nuclear industry first concentrated on completing Level 1 PRA and then moved to Level 2 PRA. The attention has been recently drawn to Level 3 PRA. The second reason is that after the Chernobyl (1986) and Fukushima (2011) accidents, the need to critically model the Level 3 PRA was felt all the more. Good data and insights were available from these two accidents, and the results of Level 3 PRA provide an effective and systematic approach to understanding severe accident consequences and off-site emergency management. The general description of the phenomena associated with the release is formation of plume, depending on the weather conditions, dispersion into the atmosphere though transport and diffusion, and radiation consequences to members of the public through various external and internal pathways.

## 6.10.2    Overview of Methodology

Figure 6.15 provides an overview of the methodology of Level 3 PRA. The basic information on source term is available from Level 2 PRA. The following step deals with evaluation of propagation radioactivity to the environment. The dispersion into the environment depends on meteorological conditions for which weather data sampled periodically, preferably on hourly basis, forms a major input. A dosimetry model is used for dose evaluation to the members of the public. The countermeasures considered, such as intervention in the food supply chain, shelter, and iodine tablet distribution, are vital inputs for dose evaluation. Depending on the site location, the information/data on population distribution, agricultural practices are used to determine the impact on the health of members of the public. The data on agriculture distribution are utilized to evaluate the economic impact of the accident. The last but an important aspect of an accident particularly from long-term considerations is the impact on the psyche of the people in any country or at an international level, which determines the future course of policy decisions at a national as well as at an international level.

It may be noted that due to complexity in consequence analysis the modeling and analysis for Level 3 PRA is carried out using consequence analysis codes. Many of the activities in Level 3 PRA are carried out simultaneously. For example, data collection, assessment of dose evaluation, economic impact assessment can be performed either simultaneously or can overlap and need not be sequential activities.

The following section discusses the major salient features of Level 3 PRA.

**Fig. 6.15**  Overview of Level 3 PRA methodology

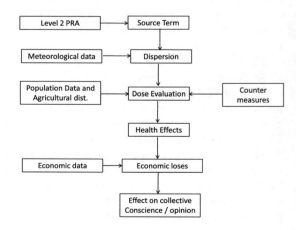

## 6.10.3   Source Term

The source term characterization in Level 3 PRA requires input on: (a) the magnitude of release in terms of fraction or percentage of the net reactor core inventory, (b) the timing of release, (c) the frequency of release, (d) the elevation of release, (e) composition of the release group in terms of release of noble gases (e.g., xenon), volatile iodine and particulates (carrying isotopes like cesium), (f) physical and chemical form of the radioisotope, volatile gases, or aerosol, (g) energy of the release, and (h) the thermal components of the release. Collection of meteorological data requires sampling on an hourly basis at least for a year to have effective modeling that forms an input for dispersion modeling. The source terms obtained from Level 2 PRA are aggregated into groups of similar source terms. The major parameters for this grouping are quantum of release, time of release, physical and chemical nature of the isotope, and the severity of the consequences. This grouping is a first step in the probabilistic consequence analysis or Level 3 PRA.

## 6.10.4   Meteorological Data and Sampling

This includes collection of data and information on weather stability characterization, wind, and rainfall data. There are six categories of weather stability classes, viz. *A*, *B*, *C*, *D*, *E*, and *F*, where *A* is the most unstable and *F* is the most stable. The atmospheric stability along with associated probability with each weather class is an important consideration in evaluation of atmospheric dispersion and propagation of the radionuclide. The weather data are characterized by a wind rose diagram which shows the distribution of wind velocities, in angular or cardinal directions that form the input for dispersion modeling. Similarly, the effect of rainfall near the source and away from the source is required to be evaluated because rainfall impacts the dispersion results. Generally, the sampling of meteorological details is to be carried out on an hourly basis for at least for a calendar year for good results.

## 6.10.5   Agricultural and Population Data

The agricultural and population data around the source at various radius can be obtained from the census documents and represented graphically on a polar or cardinal format, as shown in Fig. 6.16, such that resolution is higher, say ∼00 m, near the site and reduces as we move away from the site. The information contained in each sector or zone is location of the houses, use of forming land/crops, and routes for evacuation.

**Fig. 6.16** Polar grid system
to graphically show
population/agricultural
demography around the
source or plant

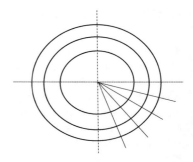

## 6.10.6   Atmospheric Dispersion and Propagation

Even though considerable advances have been made in modeling of atmospheric
dispersion and propagation right from plume modeling to transport of radionuclide
and further propagation finally leading to assessment of dispersion and ground
contamination. However, the Gaussian plume diffusion model provides an effective
approach, with modest computational requirements and input data, and enables a
large number of simulations to be performed. The Gaussian formulation for concentration of radionuclide C is a function of Euclidian coordinates: $x$, $y$, $z$ and the
effective stack height $H$ [83, 84] and given as follows:

$$c(x, y, z, H) = \frac{Q}{2\pi\sigma_y\sigma_z u} \left[\frac{-y^2}{2\sigma_y^2}\right] \cdot \left(\exp\left[\frac{-(z-h)^2}{2\sigma_z^2}\right] + \exp\left[\frac{-(z+h)^2}{2\sigma_z^2}\right]\right) \quad (6.17)$$

The parameters of the Gaussian plume model and its interpretation are shown in
Fig. 6.17. The plume that rises above stack height $h$ is shown by parameter $H$. The
plume height has significant effect on near source concentration and has less impact
on the far source region. The direction of propagation is in line with the $x$-axis,
while the $z$-coordinate shows the dispersion elevation (shown by the imaginary
centerline); $Q$ is the mass of emission and the standard deviation of the plume in
directions $y$ and $z$; i.e., crosswind is given by $\sigma_y$ and $\sigma_z$, respectively. Fig. 6.18

**Fig. 6.17** Illustration of
Gaussian plume model [85]

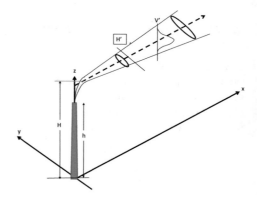

**Fig. 6.18** Simplified illustration of atmospheric dispersion and propagation in polar grid format; red indicates highest density of radionuclide which reduces successively as shown in blue, dark yellow, and light yellow during the plume propagation

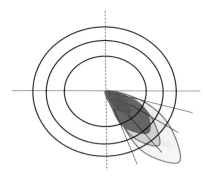

shows simplified illustration of atmospheric dispersion and propagation in polar grid format.

There are critiques of the Gaussian dispersion model, and this is one of the reasons for many advanced models that are used, for example, the puff diffusion model. The available literature shows many studies related to benchmarking of the Gaussian plum model. One of the inferences regarding the accuracy of the Gaussian model is that calculations in this model heavily rely on the accuracy of the stability class parameters. In non-idealized situations, for example, where the topology is complex or when stability classes are incorrectly applied, the results might be at variance with reality.

## 6.10.7   Exposure Pathways

So far, we have discussed the propagation of the radionuclide as a result of release from the source in the public domain, referred to as atmospheric dispersion. One of the major objectives of Level 3 PRA is assessment of health effects or fatalities induced due to exposure to nuclear radiation. There are six ways members of the public can be exposed to radiation hazard as can be seen in Fig. 6.19.

**Fig. 6.19** Exposure pathways for dose evaluation [27]

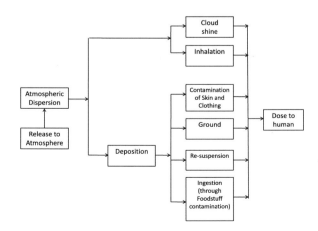

The exposure can be broadly divided into external and internal exposures. The cloud shine (exposure from passing cloud due to reflection of radiation), contamination of skin and clothing (surface absorption), ground release and re-suspension (due to leakage from containment building or release at ground level from any other source) also referred to as ground shine will cause external exposure, while the inhalation of radioactive air and consumption of contaminated food through ingestion rout will cause internal exposure. The dosimetry model is used to convert the concentration of radionuclide in the atmosphere, on the ground, on food stuff, and on the skin and clothing to the dose in humans.

### 6.10.8    Health Effects

This requires assessment of detrimental effects of ionizing radiation on human health mainly in respect of deterministic effects associated with high radiation doses immediately after the accident and stochastic effects associated with low chronic doses for extended periods of time. In deterministic effects, the severity of the consequences increases with dose and has a threshold below which they do not occur, whereas stochastic affects have no dose threshold and the probability of health effect increases with dose. Deterministic effects have dose threshold below which it does not occur. The health effects associated with individual radionuclide need to be evaluated; for example, early exposure say 24 h to mainly radioactive iodine or tellurium affects the bone marrow and associated cancer risk, also iodine has an affinity for thyroid gland, and a threshold is prescribed for humans, particularly for children. Exposure to cesium radioactivity and other isotopes causes latent cancer deaths. The health effects are evaluated for individuals and social groups at large in terms of fatal deaths due to early exposure or chronic cancer risks over extended exposure or latent cancer risks.

### 6.10.9    Counter Measures

To reduce the impact of an accidental release on the environment, public countermeasures or protective actions are envisaged. Depending on when the countermeasures are implemented, they can broadly be categorized as short-term or emergency response and long-term countermeasures. The aim of these countermeasures is to reduce the internal and external exposures of individuals and society from ionizing radiation. The short-term countermeasures are required to prevent the deterministic effects, while long-term countermeasures are implemented to reduce chronic exposure. The short-term countermeasures include sheltering, evacuation, distribution of stable iodine tablets, and decontamination of people. Sheltering of the population means asking people to stay indoors and close the doors and windows to protect from radiation cloud shine and ground shine until the radiation level drops to a predetermined low level. Evacuation involves transporting people to safe distances from the source of radiation based on certain guidelines or dose criteria. Iodine has the affinity to deposit in the thyroid of humans and causes detrimental health effects, and hence, stable iodine tablet distribution in the potentially affected areas, either before or after the release, forms part of emergency plans. Decontamination of humans is carried out to clear off loose contamination on skin and clothing.

Long-term countermeasures include relocation of affected population to outside the affected zone to prevent exposure and are carried out based on dose limit prescribed in the emergency plan, land decontamination, and control or total ban of food and agricultural produce.

### 6.10.10   Economic Losses and Public Conscience

There are economic losses associated with an accident; however, there is no formal framework or methodology developed for evaluation of economic losses. The major factors considered are loss of working hours, cost of relocation and evacuation, distribution of stable iodine tablets, and other health care support immediately after the accident. Long-term factors include loss to the agricultural forms and food products due to contamination and the cost of decontamination of large areas (e.g., desilting of land). Losses associated with damaging effects on the nuclear plant and loss of revenue and other property around the plant are also major economic considerations.

Even though the nuclear industry has maintained and demonstrates very high standards of safety, the major accidents do have an implication for the collective conscience. Public conscience or public opinion on nuclear systems affects government policy toward nuclear or the industry involved in the accident and can arguably be a more important factor than economic losses. Hence, this requires public outreach to educate the public about the potential risks and the enormous benefits associated with nuclear plant operations.

### 6.10.11   Results and Applications

The presentation of results of the Level 3 PRA requires development of a risk metrics that is commensurate with the objectives and scope of the study. Hence, development of a risk metrics is an integral part of a Level 3 PRA study which should take into account the interest of the stakeholders, state of the art in technical evaluation, availability of the data, and associated uncertainty in data and models. Nevertheless, there are certain general features that need to be considered while presenting the results of any Level 3 PRA. The three primary aspects of the risk metrics are the health effects on individuals and society, environmental effects and parameters, and economic consequences.

Level 3 PRA requires a special focus on addressing uncertainty in input data, particularly that associated with source term quantification, weather, and health factors. The evaluation of uncertainty associated with input parameters and the results of the Level 3 PRA study should be an integral part of the study. Also, sensitivity analysis for various assumptions of the study should be conducted to bring out the parameters that have a high impact on the main results.

One of the major objectives of Level 3 PRA is assessment of the risk in terms of number of fatalities and associated frequency of occurrence to individuals and society and comparison of the same with the available national/international safety goals as prescribed by the regulatory authority [82, 84]. For example, Fig. 6.20 shows a general example of safety goals in terms of frequency–consequence diagram.

The presentation of recommendations and use of insights in the development of off-site emergency plans forms a part of Level 3 PRA. The presentation should revisit parameters and requirements of the exclusion zone, the sterilization zone, and the emergency planning zone, which are based on deterministic considerations and can be a vital part of Level 3 analysis. The definition of core damage in Level 1 PRA is intuitively based on the potential consequence of damage to the core on the individual members of the public and society. Ideally, this information should come only after performance of Level 3 PRA. However, it is a vicious cycle. Now that a Level 3 PRA result in terms of risk to the public is available, there is a feedback to fine-tune the definition of core damage frequency and Level 2 release frequency. Although this is a complex recursive iteration, it is a useful exercise that brings us closure to the reality of accident scenarios.

**Fig. 6.20** Safety goals: frequency/consequence representation

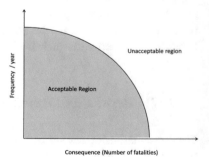

# 6.11 Conclusions and Final Remark

This chapter while reviewing PRA methods and tools focusses on Level 1 limited-scope model in rather detail and later provides an overview for aspects like low-power and shutdown and internal and external events also. The treatment to the whole subject was keeping in view the IRBE requirements.

# References

1. U.S. Nuclear Regulatory Commission, *PRA Procedure Guide (vol. 1 and vol. 2), NUREG/CR-2300* (USNRC, Washington, D.C., 1983)
2. NASA, *Probabilistic Risk Assessment Procedures Guide for NASA Managers and Practitioners*. NASA/SP-2011-3421, 2nd edn. (2011)
3. R.R. Fullwood, *Probabilistic Safety Assessment in Chemical and Nuclear Industries* (M/s Butterworth-Heinemann, 2000)
4. Environment Protection Agency, *Risk Assessment Forum: Probabilistic Risk Assessment Methods and Case Studies, EPA/100/R-14/004* (U.S. Environmental Risk Protection Agency, Washington D.C., 2014)
5. J. Rocha, A. Henriques, R. Calcada, Probabilistic safety assessment of short span high-speed railway bridge. Eng. Struct. **71**, 99–111 (2014)
6. M.G. Stewart, R.E. Melchers, *Probabilistic Risk Assessment of Engineering Systems*, 1st edn. (Chapman & Hall, 1997)
7. A. Papanikolaou (ed), *Risk-Based Ship Design—Methods, Tools and Applications* (Springer, 2009)
8. T. Bedford, R. Cook, *Probabilistic Safety Analysis: Foundation and Methods* (Cambridge University Press, 2001)
9. International Atomic Energy Agency, *IAEA Safety Glossary—Terminology Used in Nuclear Safety And Radiation Protection* (IAEA, Vienna, 2007)
10. International Atomic Energy Agency, *Applications of Probabilistic Risk Assessment (PRA) for Nuclear Power Plants, IAEA-TECDOC-1200* (IAEA, Vienna, 2001)
11. International Atomic Energy Agency, *Procedure for Conducting Probabilistic Safety Assessment of Nuclear Power Plants (Level 1), Safety Series No. 50-P-4* (IAEA, Vienna, 1992)
12. R.L. Winkler, Uncertainty in probabilistic risk assessment. Reliab. Eng. Syst. Safety **54**, 127–132 (1996)
13. G. Tokmachev, Application of PRA to Resolve Design and Operational Issues for VVER in Russia, in *Report of IAEA-Technical Committee Meeting*, vol. II, pp. 104–110 (1998)
14. Canadian Nuclear Safety Commission, *Probabilistic Safety Assessment for Nuclear Power Plants, Regulatory Standard S-294* (Canada, 2005)
15. T. Fuketa, *How PRA Results are to be Utilized in New Nuclear Regulations in Japan* (Nuclear Regulatory Authority, 2013)
16. Health Safety Executive, *Step 3 Probabilistic Safety Analysis of the Westinghouse ap1000, Division 6 Assessment Report No. AR 09/017-P* (HSE Nuclear Directorate, UK)
17. International Atomic Energy Agency, *Determining the Quality of Probabilistic Safety Assessment (PRA) for Applications in Nuclear Power Plant, IAEA-TECDOC-1511* (IAEA, Vienna, 2006)

18. International Atomic Energy Agency, *A Framework for Quality assurance Program for PRA, IAEA-TECDOC-1101* (IAEA, Vienna, 1999)
19. International Atomic Energy Agency, *Development and Application of Level 1 Probabilistic Safety Assessment for Nuclear Power Plants, IAEA Safety Standards Series No. SSG-3* (IAEA, Vienna, 2010)
20. International Atomic Energy Agency, *Development and Application of Level 2 Probabilistic Safety Assessment for Nuclear Power Plants', IAEA-Safety Standard, Specific Safety Guide No. SSG-4* (IAEA, Vienna, 2010)
21. ASME/ANS, *Addenda to ASME/ANS RA-S-2008—Standard for Level 1/Large Early Release Frequency Probabilistic Risk Assessment for Nuclear Power Plant Applications, ASME/ANS RA-Sa-2009* (The American Society of Mechanical Engineers, USA, 2009)
22. USNRC, *PRA Quality Expectations*
23. ASME/ANS RA-S-1.3 Standard, *Radiological Accident Offsite Consequence Analysis (Level 3 PRA) to Support Nuclear Installation Applications*
24. USNRC, *An Approach for Determining the Technical Adequacy of PRA Results for Risk-Informed Activities, Regulatory Guide 1.200* (USNRC, 2009)
25. Nuclear Energy Institute, *Process for Performing Internal Events PRA Peer Reviews Using the ASME/ANS PRA Standard, Revision 2, NIE 05-04* (Nuclear Energy Institute, 2008)
26. Brookhaven National Laboratory, *Probabilistic Safety Analysis: Procedure Guide, Rep. NUREG 2815 BNL-NUREG51559, Rev. 1 (2 vols.)* (United State Nuclear Regulatory Commission, Washington, D.C., 1985)
27. International Atomic Energy Agency, *Procedure for Conducting Probabilistic Safety Assessment for Nuclear Power Plants (Level 3), IAEA Safety Series No. 50-P-12* (IAEA, Vienna, 1996)
28. International Atomic Energy Agency, *Defining Initiating Events for Purposes of Probabilistic Safety Assessment, IAEA-TECDOC-719* (IAEA, Vienna, 1993)
29. US Department of Defense, *American Military Handbook, Reliability Prediction of Electronic Equipments, MIL-HDBK-217 Version F* (US Department of Defense, Washington, D.C., 1991)
30. International Atomic Energy Agency, *Component Reliability Data for use in Probabilistic Safety Assessment, IAEA-TECDOC-478* (IAEA, Vienna, 1988)
31. U.S. Nuclear Regulatory Commission, *Industry-Average Performance for Components and Initiating Events at U.S. Commercial Nuclear Power Plants', Idaho National Laboratory, NUREG/CR-6268, INL/EXT-06-11119* (USNRC, Washington, D.C., 2007)
32. D. Nicholls, An introduction to the RIAC 217Plus component failure rate models. J. Reliab. Inf. Anal. Cent. (2007)
33. Electronic Parts Reliability Data EPRDA, Reliability Information Analysis Center, NY, USA (2014)
34. TELCORDIA, Reliability Prediction Procedure for Electronic Equipment, SR-332, Issue 1-4. Telcordia, May 2006
35. Bell Communications Research, BELLCORE-TR-332 Issue 6 (1997)
36. Institute of Electrical and Electronics Engineers, *IEEE Guidelines to the Collection and Presentation of Electrical, Electronic, Sensing Component and Mechanical Equipment Reliability Data for Nuclear Power Generating Station, IEEE-Std 500* (IEEE, 1984)
37. International Electrotechnical Commission, *Reliability Data Handbook—Universal Model for Reliability Prediction of Electronic Components, PCBs and Equipment, IEC-TR-62380, 1st edn.* (International Electrotechnical Commission, Geneva, 2004)
38. Naval Surface Warfare Center, *Handbook of Reliability Prediction Procedure for Mechanical Equipment, CARDEROCKDIV, NSWC-98* (National Surface Warfare Center, Maryland, USA, 1998)
39. OREDA, Offshore and Onshore Reliability Data—OREDA Handbook 2015, vol. I and vol. II, 6th edn.

40. F. Nash, *Estimating Device Reliability: Assessment of Credibility* (AT&T Bell Labs/Kluwer Publishing, MA, 1993)
41. M. Pecht, *Why the Traditional Reliability Prediction Models Do Not Work—Is There an Alternative?*, vol. 2 (Electronics Cooling, 1996)
42. Human Reliability Review Paper
43. A. Swain, H. Guttmann, *Technique for Human Error Rate Prediction (THERP), NUREG-1278* (USNRC, 1983)
44. Human Cognitive Reliability (HCR)/Operator Reliability Experiments (ORE) Method, EPRI TR-100259 (1983)
45. J. Forester, et al., *A Technique for Human Event Analysis Error Rate Prediction (ATHENA), NUREG-1624* (USNRC, 2000)
46. U.S. Nuclear Regulatory Commission, *Shutdown and Low-Power Operation at Commercial Nuclear Power Plants in the United States, NUREG-1449, Office of Nuclear Reactor Regulation* (USNRC, Washington, D.C.)
47. International Atomic Energy Agency, *Low Power and Shutdown PRA, IAEA Training in Level 1 PRA and PRA Applications* (IAEA, Vienna)
48. Westinghouse Electric Company, in *Low Power and Shutdown Probabilistic Risk Assessment Modeling*. Nuclear Services/Engineering Services (2011)
49. International Atomic Energy Agency, *PRA for the Shutdown Mode for Nuclear Power Plants, IAEA-TECDOC-751* (IAEA, Vienna, 1994)
50. International Atomic Energy Agency, *Probabilistic Safety Assessments Nuclear Power Plants for Low Power and Shutdown Modes, IAEA-TECDOC-1144* (IAEA, Vienna, 2000)
51. M.-K. Kim, Risk monitor during shutdown of Candu NPPs. *PRAM 9*, pp. 18–23 (2008)
52. O. HaeCheol, K. K. Myung, R. S. Mi, S. C. Bag, Y. H. Sung, Development of Simplified Shutdown PRA Model for CANDU Plant, in *Transactions of the Korean Nuclear Society Spring Meeting* (Chuncheon, Korea)
53. T.-L. Chu, Z. Musicki, P. Kohut, *Results and Insights of Internal Fire and Flood Analysis of the Surry Unit 1 Nuclear Power Plant during Mid-Loop Operations, Conf-9511153, BNL-NUREG-61792* (Brookhaven National Laboratory/U.S. Nuclear Regulatory Commission, 1995)
54. D. J. Wakefield, Status of Low Power and Shutdown PRA Methodology Standard, ANSI/ANS-58.22, in *20th International Conference on Nuclear Engineering and the ASME 2012 Power Conference* (2012)
55. Nuclear Energy Agency, *Improving Low Power and Shutdown PRA Methods and Data to Permit Better Risk Comparison and Trade-Off Decision-Making, Volume 1: Summary Of COOPRA and WGRISK Surveys* (2005)
56. International Atomic Energy Agency, *Safety Assessment for Spent Fuel Storage Facilities, Safety Series No. 118* (IAEA, Vienna, 1994)
57. Electric Power Research Institute, Spent Fuel Pool Accident Characteristics, EPRI-Product-ID:3002000499 (2013)
58. United States Nuclear Regulatory Commission, *Spent Fuel Storage in Pools and Dry Casks—Key Points and Question and Answers*
59. E. Burns, L. Lee, A. Duvall, J. Sursock, *Integrating the Assessment of Spent Fuel Pool Risks into a Probabilistic Risk Assessment (Mark I and Mark II BWRs)* (Electric Power Research Institute, Palo Alto, USA, 2013)
60. Westinghouse Electric Company, *Spent Fuel Pool Probabilistic Risk Assessment Modeling, NS-ES-0231* (Nuclear Services/Engineering Services, 2012)
61. S. Y. Choi, J.-E. Yang, Flooding PRA by considering the operating experience data of Korean PWRs. Nucl. Eng. Technol. **39** (2007)
62. J. Mattei, E. Vial, V. Rebour, H. Liemersdorf, M. Turschmann, *Generic Results and Conclusions of Re-evaluating the Flooding Protection in French and German Nuclear Power Plants* (IPSN, GRS)

63. B. Heinz-Peter, R. Marina, *Current Status of Fire Risk Assessment For Nuclear Power Plants*
64. USNRC/EPRI, Nuclear power plant fire modeling analysis guidelines (NPP FIRE MAG), NUREG-1934/EPRI 1023259. Nuclear Regulatory Commission, Palo Alto (2012)
65. P. Schairer, Fire modeling examples from the nuclear power industry—an engineering planning and management (EPM) presentation, in *DOE Nuclear Facility Safety Programs Workshop*, Las Vegas, Nevada (2014)
66. EPRI/NRC-RES, *Fire PRA Methodology for Nuclear Power Facilities: Volume 1: Summary and Overview, EPRI-1011989 and NUREG/CR-6850*. Electric Power Research Institute (EPRI), Palo Alto, CA, and U.S. Nuclear Regulatory Commission, Rockville, MD (2005)
67. National Fire Protection Association, National Fire Protection Standard, NFPA 557: Standard for Determination of Fire Loads for Use in Structural Fire Protection Design (2016)
68. P. Sharma, *Computational Fluid Dynamic Fire Modeling for Nuclear Power Plants* (Bhabha Atomic Research Centre, Mumbai, 2014)
69. T. Kuromoto, Advancement of PRA and risk evaluation methodologies for external hazards, in *PRAM Topical Conference in Tokyo—In the Light of Fukushima Dai-ichi Accident*, Tokyo, Japan (2013)
70. International Atomic Energy Agency, *Extreme External Events in the Design and Assessment of Nuclear Power Plants, IAEA-TECDOC-1134* (IAEA, Vienna, 2003)
71. International Atomic Energy Agency, Internal and External Hazard Analysis in PRA, in *IAEA Workshop/Training Course on Safety Assessment of NPPs to Assist in Decision Making*, Vienna
72. United State Nuclear Regulatory Commission, *Fact Sheet on Seismic Issues for Nuclear Power Plants*
73. STUK, Seismic Events and Nuclear Power Plants, Guide YVL 2.6/, Finland (2001)
74. International Atomic Energy Agency, *Probabilistic Safety Assessment for Seismic Events, IAEA-TECDOC-724* (IAEA, Vienna)
75. T. Takada, On seismic design qualification of NPPs after Fukushima event in Japan, in *15th WCEE, LISBOA* (2012)
76. Y. Narumiya, Outline of Seismic PRA method and examples in Japan, in *1st Kashiwazaki International Symposium in Seismic Safety of Nuclear Facilities*, Niigata, Japan (2010)
77. JNES, Outline of Seismic PRA, in *1st Kashiwazai International Symposium on Seismic Safety of Nuclear Installation* (2010)
78. M. Hirano, T. Nakamura, K. Edisawa, Outline of Seismic PRA implementation standards on the atomic energy society of Japan, in *14th World Conference on Earthquake Engineering*, Beijing, China (2008)
79. J. Johnson, Fragility evaluation and seismic probabilistic safety assessment, in *Joint ICTP/ IAEA Advanced Workshop on Earthquake Engineering for Nuclear Facilities*, Italy (2009)
80. E. Raimond, Advanced safety assessment PRA2 (ASMPRA2—Best Practice Guidelines For L2PRA Development and Applications, Vol. 1—General/Technical Report ASAMPRA2/ WP2-3-4/D3.3/2013-35 (2013)
81. American Society of Mechanical Engineers/American Nuclear Society, Severe Accident Progression and Radiological Release (Level 2) PRA Standard for Nuclear Power Plant Applications for Light Water Reactors (LWRs). ASME/ANS RA-S-1.2-2004, New York, USA
82. Organization for Economic Co-operation and Development, *CSNI Technical Opinion Papers No. 9: Level 2 PRA for Nuclear Power Plants, Nuclear Energy Agency No. 5352* (OCED, 2007)
83. O. Sutton, A theory of eddy diffusion in the atmosphere. Proceedings of the Royal Society of London **135**, 143–165 (1932)

84. J. Preston, Overview of Level 3 PRA—assessment of off-site consequences, in *ANSN regional Workshop on Integrated Deterministic Safety Analysis (DSA) and Probabilistic Safety Analysis (PRA) for Risk Management in Nuclear Power Plant*, Manila, Philippines (2013)
85. A. Cladwell, *Addressing Off-site Consequence Criteria Using Level 3 Probabilistic Safety Assessment—A Review of Methods, Criteria, and Practices, Master of Science Thesis.* Department of Nuclear Power Safety, KTH Royal Institute of Technology, Stockholm, Sweden (2012)

# Chapter 7
# Risk-Based Design

> Design creates culture, culture shapes values, values determine the future.
>
> Deniels – Adedu Darlintine uche, boardofwisdom.com

## 7.1 Introduction

The traditional approach to design is conservative and rule based (i.e., prescriptive) and uses point estimates of design parameters while employing factors of safety to account for uncertainty in data, models, and other sources of knowledge. This approach, even though it is time-tested and proven, does not provide any quantified statement on design reliability or safety or, conversely speaking, the risks associated with failures. The other drawback of this approach is that it tends to result in relatively overdesigned parameters in terms of volume/weight or size ratio and further over-burdens the connected systems and, at times, performance. This approach is basically conservative and prescriptive in nature which makes the whole design process rigid in nature. The principles of defense in depth, redundancy, diversity, and single-failure criteria are employed in the design to incorporate safety. However, it may be noted that "risk," be it at qualitative level, remains the main driver for system designs.

The risk-based design approach enables a design process that considers risk as the major parameter driver or focus of the design. This focus is achieved by applying a probabilistic approach to the risk-based design, where input variables that are considered random enable characterization of uncertainty in the output for the associated failure probability estimates of the structures, systems, and components (SSCs). However, the quality of the results has great bearing on the accuracy of the input data and applicability of the probability distributions associated with each input variable.

A risk-based or probabilistic design approach treats the input variables as random variables while using the traditional deterministic models for estimating the risk associated with the design, performance degradation, or failure. Evaluation of the uncertainty associated with the estimates also often forms part of the risk-based design.

© Springer Nature Singapore Pte Ltd. 2018
P. V. Varde and M. G. Pecht, *Risk-Based Engineering*, Springer Series in Reliability Engineering, https://doi.org/10.1007/978-981-13-0090-5_7

The quantitative methods [i.e., quantitative risk assessment (QRA) or probabilistic risk assessment (PRA)] and structural reliability methods are central to the risk-based approach. The QRA/PRA or system modeling approaches are employed at the plant- or system-level modeling, whereas the structural reliability modeling approach provides an effective framework for component-level or passive structure modeling.

This chapter provides a holistic framework for risk design for modeling at the component as well as system level. Discussion of the state of the art and the advantages and limitations of the risk-based approach have also been summarized.

The risk-based design approach provides an effective and rational framework where the estimates of risk provide the basis for design optimization using quantified risk/reliability criteria. The quantified risk goals enable comparison of the estimated risk with the set risk goals and criteria and hence provide a rational-based approach to design. The probabilistic models allow systematic evaluation of the major risk contributors and hence identification of areas for modifications and changes. This approach provides quantified estimates of risk associated with the estimates or results of the analysis. This approach is effective in integrating human factors with the risk model of the plant, hence, it is possible to optimize hardware requirements against available human factors or operational and maintenance procedures.

## 7.2  Evolution of Risk-Based Design Approach—A Review

The traditional approach to design is deterministic in nature employs defense-in-depth principles involving multiple barriers and more than one level of protection [1]. Redundancy, diversity, and fail-safe criteria form the cornerstones of the traditional approach, which is conservative and prescriptive. If we look closely at the fundamental tenets of this approach, it can be argued that even in the deterministic approach, risk or, conversely speaking, "safety" is the main driver for design and operation of the systems, to the extent that conservative risk (or safety) and rule-based philosophy drive this approach. Although risk is the major component, the overconservatism built into the traditional approach and its prescriptive nature need to be revisited given the availability of accumulated design and operating experience, lessons learnt from accidents, the improved understanding of materials, and advances in engineering analysis. The competitive environment at the national and international level requires that this overconservatism is addressed, of course without compromising safety.

The net effect has been a move from a deterministic to a risk-informed approach to design changes/plant retrofits and modifications. Insights from a PRA study play a key role as one of the additional inputs along with engineering judgment in respect of maintenance of defense in depth, which includes maintenance of safety margins, meeting regulatory stipulations, and monitoring performance. With this constitution, the risk-informed approach is being extensively encouraged because it provides the needed flexibility in decisions and addresses issues related to over-conservatism without compromising safety [2, 3].

However, in recent times there has been a definite trend toward applying a risk-based framework in systems design. Here, the present form of the risk-based approach deals with use of the PRA framework, primarily in support of design evaluation and not exactly in designing the system.

The risk-based approach is increasingly being considered in manufacturing applications where risk is one of the major drivers, such as the design of ships and casing [4, 5], civil structures (e.g., dams) [6], bridges [7], off-shore drilling operations [8], space [9], and aviation, particularly for design of future/advanced systems [10, 11], process systems [12, 13], and manufacturing [14]. In the nuclear industry, this approach is used extensively as part of risk-informed applications in support of regulatory review.

There are a number of applications of the risk-based approach in social sectors, including financial management [15], hydraulic engineering [16], fire safety management [17], mitigation of dust explosions [18], environmental applications [19], and groundwater contamination from waste management [20].

The available literature shows that for complex engineering systems such as nuclear plant modeling, there is a growing interest in the risk-based approach [21]. In this context, even though the nuclear industry has come to terms with the risk-informed approach (a paradigm shift from conservative deterministic methods), there is a growing interest in exploring the application of the risk-based approach for nuclear reactors, particularly for advanced and new reactor systems. The available publications also show that the risk-based approach is being used in many cases involving design for extreme external events, namely the IAEA initiative on the development of methodologies for complementary assessment of nuclear plants against extreme events [22] and the risk-based design of flood defense systems [23]. Some of the studies on nuclear plants are focused on addressing challenges, for example, consolidation and not dilution of defense in depth by characterization of target risk and uncertainty metrics toward making a case for risk-based design for regulatory review [24]. There are specific studies investigating the role of defense in depth for risk-informed applications by arguing that the weaknesses of traditional defense in depth can be dealt by use of PRA as part of a risk-informed framework [25, 26].

Other studies have performed integrated deterministic and probabilistic safety analysis to demonstrate the safety of design [27]. Even though publications on the development of risk-based regulations are limited, the framework proposed by Jourdan [24] provides an effective case for regulatory considerations. Jourdan proposed risk goals which are in line with a conservative approach along with acceptable uncertainty levels for advanced reactors. Peterson and Fensling [28] provide good practices to be followed as part of regulatory reviews. As mentioned earlier, there are many applications of the risk-based approach to new reactors or new designs. Ritterbusch [29] developed a new design and regulatory process employing a risk-based approach for a nuclear plant. Burdick et al. [30] discuss the development of a risk-based approach for advanced reactor design.

Implementation of a risk-based approach requires advanced techniques such as discrete-time Bayesian networks [12], modern methods in reliability engineering for structural systems [31, 32] (including fundamental probabilistic tools and associated expertise [32]), and prognostics and health management tools and methods for prediction of degradation/incipient failures a priori, particularly in support of life extension and management [33, 34]. Apart from these techniques, should the probability for human error in design form part of the risk-based approach? As design error left or goes unnoticed could become root cause of failures during operational phase of the plant.

Experience and insights on the implementation of the risk-informed approach to nuclear plants can be adapted for risk-based applications. The critiques of defense in depth and the models and methods proposed employing probabilistic or rationalist approach can be investigated for risk-based application. For example, the risk-informed design guidelines proposed by Delaney et al. [35] could provide an effective input for the risk-based approach. Similarly, the studies dealing with investigation of role of defense in depth in risk-informed regulation [25, 26] could be extended to include the risk-based approach with further stringent criteria and risk goals.

The above literature review which includes critique of traditional defense in depth provides a positive motivation as well as tools and techniques for proposing a risk-based design approach employing basic tenets of integrated risk-based engineering (IRBE), such as conservative goals and criteria, prognostics and monitoring, and application of a structural reliability approach for detailed investigation of component failure.

## 7.3  The Approach

The risk-based design approach proposed in this chapter is based on an IRBE framework. The probabilistic as well as deterministic methodologies are integrated such that each one complements the other. Whereas the deterministic methodology provides the framework for defense in depth, redundancy, diversity, and single-failure criteria, the probabilistic approach enables plant model integration at all levels to quantitatively estimate risk.

Consolidation of defense-in-depth principles, and not dilution, is at the core of the risk-based design approach proposed here. Risk targets or risk criteria are fundamental to the implementation of risk-based design and are in line with IRBE philosophy. Further, evaluation of acceptable and unacceptable zones is intrinsic to the proposed risk-based framework.

## 7.4  Salient Features of Risk-Based Design

The risk-based design approach proposed in the chapter has the following salient features:

- Plant SSC reliability data and target risk criteria/goals form the basic inputs for design.
- The deterministic and probabilistic elements that drive the IRBE approach are an integral part of risk assessment and complement each other.
- The approach consolidates and demonstrates defense-in-depth principles and ensures their implementation by employing a rationalist approach to determine realistic margins using a science-based approach to uncertainty characterization.
- The deterministic models and procedures form the fundamentals. The design rules and factor of safety used in the traditional approach, i.e., ASME rules and methods, are critically evaluated when required to rule out over conservatism using probabilistic methods.
- Plant-level modeling and system-level modeling are performed by the PRA approach, whereas component-level modeling is performed by employing physics-of-failure or structural engineering methods that treat input variables as random variables.
- The component failure probabilities are determined based on the operating experience, generic data, or life testing experiments.
- The root cause analysis and failure mode and effect analysis are integral to critical evaluation of causes of failure such that design qualifies against these weaknesses.
- The target core damage, large early release, and risk frequency in respect of Level 1, Level 2, and Level 3 along with uncertainty bounds form the acceptance criteria for the plant.
- Risk criteria down to the system, component, and structural level are derived from the plant-level criteria such that there is no bias from single or few events.
- To account for random phenomena and lack of data, knowledge, and experience, aleatory and epistemic uncertainty modeling, respectively, are envisaged.
- The confidence bound obtained from statistical analysis characterizes the aleatory uncertainty. Other experimental, structural modeling approaches or configurational arrangements, such as redundancy and diversity, are used to address epistemic uncertainty.
- Plant-level modeling and system-level modeling are carried out using the PRA method.
- The detailed modeling for component failures or root cause failure analysis is performed using a physics-of-failure or structural analysis approach that also includes probabilistic fracture mechanics or life testing data.
- Since no standards or codes are currently available for risk-based design, the available standards and technical documents on PSA and deterministic procedures from ASME codes, along with guideline documents on the risk-informed approach, are used for the integrated risk-based procedure.

Generally, a nuclear power station or a site has more than one nuclear power plant (NPP) or unit and even up to 8 units. In such cases, the risk to members of the public is from all the units at a site. After the Fukushima event, there has been increased interest in multi-unit site risk frequency evaluation. In case the organization has the site frequency estimate, and then the core damage frequency (CDF) for each plant will be derived from the site risk frequency. Hence, the CDF target may be seen in the context of risk apportionment from each plant at a given site.

The other chapters in this book on PRA, life testing, probabilistic fracture mechanics, and uncertainty modeling provide the supporting information for integrated risk-based engineering. This chapter focuses on risk-based design aspects.

## 7.5  Major Elements of Risk-Based Design

Before we go into detail on risk-based design, we need to distinguish between two situations. The first situation is that a plant has already been built or is in an operational stage, and the study is performed to analyze design weakness such that retrofits/modifications can be incorporated. The second situation involves use of risk-based approach for evaluating the design or optimizing the system configuration through conceptual stages, preliminary design, commissioning stages.

This chapter proposes a risk-based framework for design evaluation be it during the design stage or operation stage. Initially, a conceptual deterministic design should be available that considers the expected performance of the plant. The design should include power-level ratings in milliwatts and expected broader configurations, such as fueling modes either online or during shut down. In short, the risk-based procedure discussed here starts with the availability of a skeleton or preliminary design. The risk-based procedure involves two major steps: (a) plant- and system-level modeling employing PRA procedures and (b) component- or process-level modeling using the applicable methods, like structural or physics-of-failure modeling to understand failure modes and mechanism, and finally the failure rates along with associated uncertainties; application-specific modeling such as nuclear, chemical, space, and aviation employing the domain-specific models. For example, the neutronic and reactivity aspects, decay heat removal aspects, and effects of ionizing radiation are modeled in the nuclear system as part of the loss of regulation (neutronic phenomenon), loss of coolant (primarily decay heat removal phenomenon) and loss of off-site power (power decay aspects) initiating events. In this chapter, the major modeling has been shown using stress–strength modeling for mechanical aspects of design. However, this concept is extended to operational and environmental stress and the capability of the design for electrical, electronic, thermal hydraulics, and neutronic phenomena.

Even though a framework has been proposed for safety-critical applications, such as nuclear plants, aviation systems, railway and transport systems, process and

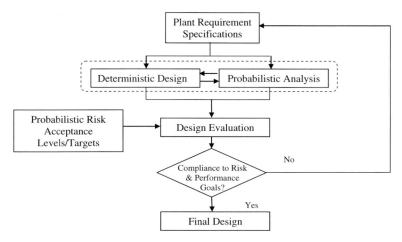

**Fig. 7.1** Risk-based design approach

industrial systems, it can be adopted with few modifications for other systems such as software, domestic appliances, and other systems where cost forms the governing criteria. Figure 7.1 shows the major elements of the risk-based design evaluation framework.

In this chapter, the design problems have been addressed in two ways, viz., design evaluation and designing the system for defined safety/reliability goals. At the system level the focus of this chapter is on design evaluation, whereas at the structural level, the design for a given reliability has also been addressed. The major procedural elements of integrated risk-based design are discussed in the following subsections.

## 7.5.1   Identification of Safety and Functional Objectives

This step envisages the identification of the system to be evaluated. The activities that need to be performed include: (a) defining objectives and scope of the evaluation, (b) system familiarization, and identification of system boundary and associated interfaces, (c) identification of system safety and performance requirements, (d) identification of major constraints and resources, and (f) validation and verification criteria in respect of safety and functional requirements.

The aim of this analysis is to sensitize the team involved and the management to the nature and expected output of the work. It goes without saying that the main objective function is to optimize "risk" so that overall risk is acceptable while optimizing the performance parameters. Often, certain applications may require a trade-off between risks and cost, hence, these aspects should be identified in the beginning of the project. However, optimization for cost does not form part of scope of this chapter.

## 7.5.2   Quality Assurance Program

There are two aspects of quality assurance (QA) in risk-based design. The first one concerns the realization of QA procedures and design rules during the design (e.g., conformance to ASME Section III for design of nuclear components). Here the objective is to follow the quality procedure by the designers. Noncompliance with these procedures could lead to unacceptable design and often leads to construction/fabrication of the components/structures that might have safety implications. The other aspect of QA is non-compliance of QA during construction and commission that leaves a plant vulnerable to certain risks. The PRA modeling experience has shown that many times QA-related aspects dominate the list of major contributors to risk.

For a component to qualify for a risk reliability or risk analysis modeling, the quality attributes, e.g., ASME or IAEA quality criteria for PRA modeling, need to be fulfilled. For example, it is not possible to qualify a system/component for reliability analysis or modeling in PRA if it does not conform to ASME design rules. The QA aspects related to risk modeling and how it contributes to enhanced uncertainty in risk modeling are addressed in chapter on PRA scope of this book; however, detailed for details of QA aspect, readers might refer to the literature on quality standards/technical documents for PRA. The QA aspects of risk-based design should be worked out at an organizational level for adequate assurance that quality attributes are incorporated in all sphere activities of the project.

## 7.5.3   Postulation of Events and Hazards

Even though SSCs are primarily designed to cater to the stresses during normal operations, deviations, threats, and emergency conditions, the design should also qualify against identified hazards. The design should also meet the life expectancy criteria and further provisions for aging-related aspects. This procedure is adequately laid out in the traditional deterministic design approach and hence does need not to be discussed in detail here. The aspects that need to be strengthened as part of IRBE, such as life prediction, prognosis of failures or degradations, and uncertainty modeling, are discussed in detail in other chapters in this book.

Hazard identification deals with the nature of the threats that need to be considered. The potential hazards are flood, fire, seismic, nuclear radiation, industrial health hazard, road or rail traffic, and chemical. However, it should be possible from the problem definition that what type of hazard against which the risk modeling need to be performed. For example, in the case of a nuclear plant, the main hazard against which the risk analysis is performed is "release of nuclear radiation." However, the scope of risk assessment forms the basis for consideration of internal hazards only or inclusion of external hazards, such as seismic, flood, or fire. The hazard identification sets in motion the procedure for PRA, the next step in design evaluation. The identification of applicable hazard sets the ball in motion for identification of applicable initiating events for a given design evaluation project/problem. For example, a case study in Chap. 15 in this book is presented on evaluation of design margin for a plant where the applicable internal initiating events are loss of off-site power (LOOP), loss of coolant accident (LOCA), and loss of regulation accident (LORA). These three events were evaluated considering reactor core as the source of hazard for full power conditions.

## 7.5.4   Formulation of an Integrated Design Framework

As mentioned earlier, the PRA framework is used for modeling plants and systems. The basic components include hardware, software, and human action or recovery mechanisms. The inputs required for estimating reliability of the hardware components are failure rate, mission time, repair time, and test intervals. These data are obtained from generic sources when the plant is in the design stages; however, when the PRA modeling is performed at operational stages, often the data obtained from the plant directly are used as input for evaluating component reliability. There are situations when the data are either not available or adequate. In such situations, the available data are updated using a Bayesian approach. As can be seen, the PRA framework provides an elegant approach to model a plant using the failure rate data of components to arrive at a risk level or CDF (when the scope of the study covers Level 1 PRA only)/r-y. However, for the case of design stage PRA, we know that the study is based on generic data, which could be the source of significant uncertainty. Even though it is a challenge, the efforts may be made so that, at least for critical components, the generic data are updated by accelerated life testing methods or a physics- or mechanics-based approach.

The proposed risk-based approach further estimates the probability of component failure of by using an integrated approach wherein the deterministic as well as probabilistic methodologies work together to evaluate not only the probability but the root cause of failure. This approach is particularly vital for new components or a new design for which no data is available. For example, the mechanical aspects are modeled using probabilistic structural analysis, software component modeling is performed using a software reliability approach, human error modeling is performed using a human reliability approach, and failure criteria definitions at the system and component level is derived using thermal hydraulic, neutronic, and reactor physics requirements.

The component reliability data are not available during the design stage of the system. Hence, the PRA is performed using generic data available from the open literature or from a similar plant. Often the data is not available for a certain new component or new design. In this case, accelerated life testing is carried out to generate the data on these new designs.

For some advanced designs, the safety features are based on innovative concepts, such as passive process and safety features and passive main coolant as well as shut down cooling systems. For such features, the available data and knowledge are of little use for risk analysis. These features require extensive thermal-hydraulic modeling and thermal-hydraulic testing using integral loops. The situation becomes complex because there are fewer safety margins in passive systems compared to active systems and there is no motive driving force (e.g., pump) available in passive systems. Hence, extensive experiment and data collection are imperative for characterizing the reliability of these systems.

Similarly, take the case of digital systems. There is no acceptable approach available for quantifying software reliability. Hence, the robustness of software is mainly dependent on validation and verification models and methods.

A comprehensive PRA procedure may not be required in many applications. For many applications, the scope of risk analysis requires fault tree or event tree analysis or even just failure mode and effect analysis (FMEA). The reliability estimates in such cases are directly utilized as an indicator of risk. In the case of FMEA, the risk priority numbers are used for identifying and prioritizing safety issues.

## 7.5.5    Evaluation of SSCs Failure Criteria

Formulation of failure criteria forms a critical and vital step in IRBE. The failure criteria are generally derived from the deterministic analysis. This might involve neutronic analysis, thermal-hydraulic analysis, or structural analysis. Neutronic analysis is required to understand reactor core physics and forms the basis for control of criticality, safe reactor shut down, and decay heat removal requirements. Suppose there are 15 shutdown devices that have been provided in the reactor for safe shut down. The requirement of minimum number of shutdown devices for safe shutdown of the plant is derived (e.g., failure to insert 2-out-of-15 shutdown assemblies in the reactor core) is arrived at by neutronic or reactor physics analysis. In this context, the failure criteria for shutdown system will be 2-out-of-15. The physics requirement of reactivity removal rate has been translated into the component failure criteria. That is, 2-out-of-15 is the failure criteria, which means the shutdown system can tolerate only one device failure out of 15 and the moment two devices fail at the same time/demand the system is declared failure.

Similarly, the reactor cooling flow and pressure requirements in emergency core cooling requirements are also translated into the number of injection valves and pumps needed to actuate out of a group of redundant components. For example, when a minimum of 2-out-of-4 pumps is required for ensuring cooling, then the failure criteria are 3-out-of-4 pump failures.

The structural criteria are related to damage to the fuel, reactor, or support structure such that the reactor reaches a point of no return (i.e., liable for being written off). This condition includes rupture of the reactor vessel, cooling lines, support structures, and damage to containment. For example, permanent and severe damage to the end shield of the reactor could be categorized as a condition for core damage.

These failure criteria are derived from the deterministic conditions, such as maximum tolerable temperature for fuel or cladding or even a condition that leads to boiling on a conservative basis. However, the IRBE approach seeks to reevaluate these criteria employing the best estimate approach using each input parameter as a probabilistic variable such that unnecessary conservatism can be avoided. The idea is to assess realistic margins. Advanced neutronic, thermal hydraulic, and structural codes are employed to arrive at an estimate along with uncertainty bounds.

### 7.5.6   Characterization of Risk and Uncertainty Goals/ Targets

Chapter 2 in this book on risk characterization presents the design of risk metrics (i.e., at the plant level, the target CDF, or large early release frequency (LERF)/r-y). These targets are also accompanied by uncertainty bounds. For new reactor designs, in line with conservative principles, for example CDF and LERF target could be more stringent compared to existing plants. Similarly, the unavailability targets at system level also need to be derived either from CDF targets or employing any optimization procedure. The basic approach to define these targets, as discussed by Jourdan [24] could be based on classification of unacceptable, tolerable, acceptable, and negligible regions, which is in line with the traditional deterministic approach.

### 7.5.7   Structural Safety Margin Assessment

The structural design approach has evolved over the years from the traditional deterministic approach that is referred to as the Level 1 approach where design rules and criteria including the safety factors are followed to realize postulated performance functions; through Level 2 methods where the state-of-the-art methods in structural reliability are used to provide approximate solutions for identified locations in the structure; to Level 3 methods, which attempt to achieve an exact solution.

Most of the systems are designed employing the Level 1 approach. The probabilistic approach could be used for not only component fragility evaluation but also to design the component or structural support.

This aspect forms an important area to address design risk; hence, it is covered in detail in the following section on probabilistic or reliability-based design.

## 7.5.8  Evaluation of Defense in Depth

The principle of defense in depth forms the cornerstone of the deterministic approach. The major elements of defense in depth include multiple barriers, and more than multiple levels of protection to protect the public and the environment. Implementation of redundancy and diversity and fail-safe design further forms part of defense in depth. Even though these principles are well tested and proven, the history of nuclear accidents in general and the three major accidents, viz. Three Mile Island, USA (1976); Chernobyl, erstwhile Russian Federation (1986), and the recent one in Fukushima, Japan (2011), arguably requires strengthening and focusing the safety philosophy to be based on a risk-informed/risk-based approach to design and operations. Further, critical reviews/critiques of the defense in depth in the literature suggest a move from the traditional approach to the risk-informed approach [25, 26].

Although the risk-based approach to design works toward removing overconservatism and the prescriptive nature of the traditional deterministic approach, it is conceived to be based on the (a) sound rationales supported by quantified estimates of failure, (b) improved understanding of physical processes including the evaluation of uncertainties that determine safety margins, (c) considerations of potential common cause failures that can fail redundant and independent systems, (d) built on the improved foundation provided that considers that any component that forms an SSC can fail, and (e) provides an integrated model of the plant to enable sensitivity analysis for postulation of any one or combination of component failure.

Hence, when we look at the framework of IRBE, it can be argued that while the proposed approach to design tends to remove overconservatism, it provides an effective risk-based framework to ensure safety.

## 7.5.9  Surveillance/Prognostics and Health Management Program

The genesis of the many surveillance and prognostics and health monitoring (PHM) programs is aging evaluation life extension. For existing NPPs [36, 37], the same is emphasized for new and advanced designs as an integral part of surveillance and monitoring such that consequences of failure either can be avoided or reduced. The ASME international document [40] observes that risk-based programs offer significant benefits in terms of prioritizing inspection, and maintenance for equipment life management, while improving plant economics and without comprising safety and environmental risk standards. This document provides training/guidance on risk-based methods for equipment life management primarily focused on fossil plants.

Monitoring and feedback provisions in design are an integral part of the risk-based approach proposed in this book. The objective is to monitor performance to detect any deviation or degradation from reference conditions and predict incipient failure well in advance so that consequences of failure can be avoided. The risk-based approach requires that design should cater to implementation of reliability, availability, maintainability, and inspectability (RAMI) provisions. For example, for a piping system there should be provision for leaks before a break to ensure reliability and availability, layout such that any maintenance function can be implemented with ease and access to the area such that an in-service inspection program can be performed on a periodic basis.

There are two approaches in developing and implementing monitoring and health management: a conventional surveillance program that includes monitoring, testing and maintenance (preventive and condition-based) and advanced methods that involve degradation monitoring employing management PHM program. The traditional methods can be used where periodic surveillance such as in-service inspection, condition monitoring, and traditional maintenance approaches caters to fulfill the safety as well as availability objectives. However, PHM approaches are required where failures lead to high consequences in terms of safety or availability implications. The traditional deterministic approach involving maintenance of barrier and safety provisions can be effectively enhanced by implementation of a PHM program on structural systems.

## 7.5.10   FMEA and Root Cause Analysis

Failure mode and effect analysis (FMEA) is performed to document the critical failure mode and its effect on the local and system level. In many PRA studies, a first-step FMEA or its variation is performed to capture major system failure modes as input for the risk modeling. It may be noted that the reliability or PRA study provides only part of the story associated with the failure (i.e., probability of failure). The other part (i.e., cause(s) of failure), which might be more important, is also performed, particularly for critical components and failure modes, as part of the risk-based approach. Once the cause/mode of failure is understood, system design can be revisited to affect the modifications/changes. Root cause analysis is also performed as part of the development of a physics-of-failure model of the components/structures. This aspect is discussed in Chap. 12 on physics-of-failure in this book.

## 7.5.11   Human Factor Considerations

The PRA model of the plant provides a list of human actions or human factors important to safety. The design should be commensurate with these human actions and work toward reduction of probability of human error. As mentioned earlier, the design itself should reduce the chances of human error. This requires postulation of normal operations and emergency conditions (e.g., provision for manually opening a damper if it is not opening on remote action). It could require a fail-safe feature such that if the motive power, (e.g., pneumatic pressure) is lost then the damper occupying a—conditions could be "open" or "closed" due to spring action.

Human error could also be reduced by imparting an adequate time window for human action particularly during emergency conditions. This could be done by enhancing the coping capability of the design for a given scenario such that adequate time is available for human action. In case the adequate time window is not available then the possibility of automation, or provision such that adequate conditions are created for performing intended human actions. The IRBE approach in this book envisages development of a human error database and human factors with due consideration and detailed modeling of human behavior. Chapter 10 is devoted for human reliability aspects in this book.

## 7.5.12   Identification and Prioritization of Design Issues

When the PRA is used to create the risk model of the plant or system, the results of this study mainly comprise following: (a) an integrated statement of risk, for example, for nuclear plants, the CDF statement forms the statement of risk; (b) a list of minimal cut-sets that provide information on the combination of components/ systems that result in risk to the plant (i.e., CDF); (c) a list of "risk importance measures" for systems and components; and (d) uncertainty associated with individual estimates of risk.

The above input is vital to understand and identify design weakness of the systems. Level 1 PRA and Level 2 PRA provide a vital indicator for the safety level of the plant (i.e., CDF and LERF, respectively). The design evaluation is performed by optimizing the CDF/LERF. The information on single or a combination of component failures, called cut-sets at the system level and accident sequences at the plant level, provides information on the ways these components lead to plant system failure. This information is critical in evaluating the ways in which plants can fail. It is seen that common cause failure (CCF) groups in the accident sequence provide information on the vulnerability of standby/redundant groups of component/systems. The importance measure of components enables prioritizing the safety issues/components list based on their safety significance. Similarly, the human error appearing in the cut-sets provides information on design issues related to human factors. The traditional deterministic methods are not adequate when it comes to systematic identification and prioritization the safety issues based on quantified parameters, be it arranging the component or human action based on its safety significance.

Once the safety issues are identified, the designers need to address these weaknesses depending on the nature of the challenges. The specific analysis enables the designer to incorporate the provisions that reduce the potential for failures and more particularly the CCF of redundant systems. There are well-established approaches to address the CCF, such as physical separation and independent power supply. The other information that is available from these results is the contribution of human error to the risk level of the system. Here also the issues related to human error can be further prioritized such that issues ranking high on risk ranking can be addressed on a priority basis compared to the issues that are not safety significant in relative terms.

Some of the weaknesses identified might require detailed structural analysis to assess the required safety margins using structural engineering approaches. The proposed risk-based design framework identifies the following modes to reduce the risks.

## 7.5.13   Evaluation of Plant/System Configuration

Often the identified issues require a revisit to assess the available redundancy. These requirements are derived from the basic safety functional criteria derived from deterministic assessment, which includes thermal-hydraulic criteria in terms of minimum flow and pressure in a safety injection train, or requirements of emergency loads in the plant to cater to operational preparedness in respect of a safety system. The PRA provides an effective tool to evaluate the provision of redundancy and diversity in the plant at the system as well as plant level. In case a design weaknesses is discovered, the designers should consider enhancement of redundancy or, if required, diverse provision. It may be noted that performance of PRA right through the conceptual stage and later toward finalization of design is very effective because once the design is finalized and the plant enters into construction or further into operational stages, the changes/modifications in plant configuration become more difficult.

## 7.5.14   Documentation: Safety Reports and Formulation of Technical Specifications

Traditionally, the safety report demonstrates and documents the safety analysis of postulated initiating events considering the response of the plant for given scenario based on the design provisions, particularly the safety systems and safety support systems. Based on the safety report, the plant technical specifications are derived that layout conditions or rules for operation of the facility. These conditions include safety limits, limiting conditions for operations, limiting safety system settings, and surveillance requirements and schedules. Often the criteria laid out for surveillance, for example, periodicity of inspection and monitoring are based on engineering judgment, which tends to be arbitrary. Similarly, allowable outage time estimates are also governed by conservative estimates of time.

The risk-based design involves the use of deterministic and probabilistic approach for safety evaluation. Hence, it is expected that for plant level an independent PRA is available while the physical/structural modeling is performed by deterministic models and methods with the considerations of random parameters. Hence, uncertainty characterization forms part of the results of physical quantity as well as probabilistic estimates.

The CDF, LERF, and risk estimate along with associated uncertainty bounds form the major metrics at Level 1, Level 2, and Level 3 PRA, respectively. Apart from this, safety system unavailability forms the lower level of metrics. The classification of SSCs based on their safety importance derived from the "importance measure" forms one of the major results.

The conditions in technical specifications are based on rationales/engineering basis and are expected to be flexible and dynamic compared to traditional technical specification rules, which are prescriptive. For example, the level of impact of an event on CDF might be used in a technical specification to derive decisions or a rule to support operational decisions. Further, the surveillance test interval and allowable outage time, instead of being prescriptive, might be determined based on the PRA studies performed against the regulatory criteria on changes in CDF/LERF. Hence, it is expected that the technical specifications derived from the risk-based approach will be dynamic and tend to remover overconservatism and rule-based operations.

*Implementation IRBE*

The risk-based design approach is implemented at two levels—higher-level modeling involving the plant and system levels employing a PRA framework and lower level or detailed modeling for analyzing safety-critical structures and components. PRA modeling is discussed in detail in Chap. 6. The following sections provide a brief description of PRA in support of design and the lower-level integrated modeling procedure involving application of deterministic-probabilistic procedure for design evaluation of components.

## 7.6   Higher-Level Modeling: Probabilistic Risk Assessment

The detailed PRA steps are as follows: (a) selection and grouping of initiating events based on the hazard for which the analysis is being performed; (b) identification of safety functions/safety systems; (c) event sequence modeling; (d) data collection and analysis, (e) safety system modeling; (f) quantification of the models; (g) uncertainty and sensitivity analyses; and (h) presentation of the results. These steps are discussed in Chap. 6, Sect. 6.1. In the PRA study, the event tree modeling for accident sequence analysis and fault tree analysis for system modeling are employed, and it can be argued that this exercise forms a major part of PRA modeling.

In this section, we will take a simple example of fault tree modeling for Class III power supply or emergency power supply system, as shown in Fig. 7.2, to show the framework and boundary of higher-level modeling and lower-level modeling. The system comprises two redundant diesel generator sets along with associated circuit breakers and the power supply bus. For the purpose of this example, it is assumed that the components in train 1 and train 2 are independent and there is no common cause failure (i.e., failure of identical components in two redundant trains).

The fault tree output forms the input to header events in the event tree along with other safety functions such as recovery actions or human error probabilities.

It is assumed that all the component failure data follows normal distribution and therefore the distributions are characterized by $\mu_i$ and $\sigma_i$. This data for each component either comes from a generic database or from generic sources. However, certain components which are critical to the success of systems are modeled in detailed to understand its failure mechanism. For example, let us assume that the diesel generator (DG) failures are dominant. Further, root cause analysis revealed that the failure of the jacket cooling water system of the faulty DG was tripping the DG. Further, the pump tripping was identified due to overload due to bearing failure. This information forms input to arrive at the DG failure rate along with other causes of failure and the root cause of the DG failure (i.e., bearing inner race failure). The failure rate estimation and the RCA fall in the scope of lower-level analysis in the risk-based design approach.

As can be seen, the mean value and the associated uncertainty from the component level have been propagated to the system level and further as part of the accident sequence analysis to the plant level as part of CDF.

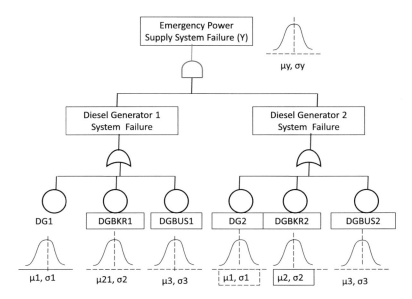

**Fig. 7.2** Scope of higher-level (PRA) and lower-level (detailed failure modeling)

## 7.7   Lower-Level Modeling: Structural Probabilistic Methods

From the above discussion, it emerges that structural failure, wear, fatigue, corrosion, and primary and secondary design, including thermal loads and nuclear characteristics particularly for passive structures, form an important aspect of risk-based design. As mentioned earlier, the PRA approach only identifies the SSCs that contribute to risk; however, further tools and methods are required to evaluate the required safety margins based on, for example, the stress/strength relationship. In this, we will consider details of structural reliability modeling. The following sections cover the reliability-based approach that predicts reliability from structural (stress/strength) modeling and also provides a mechanism for root cause analysis of mechanical failures as part of the risk-based approach. The fatigue-induced failure degradations are also discussed in Chap. 8 on probabilistic fracture mechanism in this book. For other degradation mechanisms, such as corrosion, thermal, creep, and mechanical wear, readers are referred to the relevant literature.

### 7.7.1   Structural Reliability: Stress/Strength Concept

The genesis of structural reliability resides in considerations of stress/strength factors as random in nature. Here, the failure is defined as stress increasing the strength. Hence, mathematically, the probability of failure can be expressed as follows:

$$P_{\mathrm{f}} = P(S < s)$$
$$\text{reliability}, R = P(S > s) \tag{7.1}$$

That is, for structure to be reliable, strength has to be more than $s$. The stress ($s$) and strength ($S$) are random variables. Let $M$ represent the safety margin as the difference between strength and stress as follows:

$$M = S - s \tag{7.2}$$

If $S$ and $s$ are assumed to follow normal distribution, then $M$ tends to follow the normal distribution. Hence, the probability distribution function (pdf) of $M$ can be given as follows:

$$f_M = \frac{1}{\sigma_M \sqrt{2\pi}} \exp\left[-\frac{1}{2}\left\{\frac{M - \mu_M}{\sigma_M}\right\}^2\right] \tag{7.3}$$

where $f_M(M)$ is the pdf of the safety margin $M$ and $\mu_M$ is the mean of the margin where

$$\mu_M = \mu_S - \mu_s \tag{7.4}$$

and

$$\sigma_M = \sqrt{\sigma_S^2 + \sigma_s^2} \tag{7.5}$$

If $S$ and $s$ are correlated, then

$$\sigma_M = \sqrt{\sigma_S^2 + \sigma_s^2 - 2\rho_{Ss}\sigma_S\sigma_s} \tag{7.6}$$

The pdf of $M$ is shown in Fig. 7.3 as follows. As can be seen from $M = S - s$, the design is safe as long as the strength is more than the stress, which means $M > 0$. Figure 7.3 represents the failure probability region for $M < 0$, and the same is shown in the diagram.

The mathematical expression for failure probability $P_f$ can be given as follows:

$$P_f = P(M < 0) \tag{7.7}$$

$$P_f = \int_{-\infty}^{0} f_M(M)dM \tag{7.8}$$

$$P_f = \int_{-\infty}^{0} f_M dM = \int_{-\infty}^{0} \frac{1}{\sigma_M \sqrt{2\pi}} \exp\left[-\frac{1}{2}\left\{\frac{M - \mu_M}{\sigma_M}\right\}^2\right] dM \tag{7.9}$$

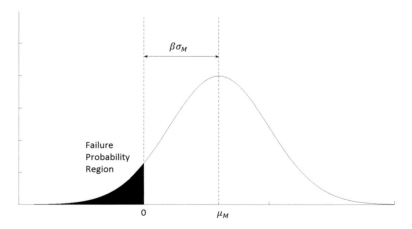

**Fig. 7.3**  Probability density function for safety margin ($M$)

Let $Z$ be a normalizing variable such that

$$Z = \frac{M - \mu_M}{\sigma_M}$$

$$dz = \frac{dM}{\sigma_M}$$

Therefore, $dM = \sigma_M dz$. Also, when $M = \infty$, then $Z = \infty$; and when $M = 0$,

$$Z = -\frac{\mu_M}{\sigma_M} \tag{7.10}$$

The ratio of mean to standard deviation is called the reliability index and is shown by notation $\beta$. Hence,

$$\beta = \frac{\mu_M}{\sigma_M} \tag{7.11}$$

After the substitutions, the equation of failure probability $P_f$ can be written as follows:

$$P_f = \frac{1}{\sqrt{2\pi}} \int_{-\infty}^{-\beta} \exp\left[-\left\{\frac{z^2}{2}\right\}\right] dz \tag{7.12}$$

In terms of the normal standard format $(z)$, the equation can be written as

$$\Phi(z) = \frac{1}{\sqrt{2\pi}} \int_{-\infty}^{-\beta} \exp\left[-\left\{\frac{z^2}{2}\right\}\right] dz \tag{7.13}$$

The standard normal cumulative distribution function $\Phi(z)$ and its value can be obtained from the standard normal chart. Hence,

$$P_f = |\Phi(z)|_0^{-\beta} \tag{7.14}$$

$$P_f = \Phi(-\beta) \tag{7.15}$$

## 7.7.2  Derivation of Reliability Expression for Stress/Strength Interference

For reliability, the stress should remain less than the strength of the structure, or, conversely speaking, the strength should be more than the stress during the service period under all the conditions. Therefore, the mathematical formulation is as follows [38, 39]:

$$\text{reliability}, R = P(S > s) \tag{7.16}$$

$$R = P(S - s > 0) \tag{7.17}$$

Since, the stress ($s$) and strength ($S$) are random variables, the pdf for these two variables can be denoted as $f_s(s)$ and $f_S(S)$, respectively. Figure 7.3 shows the stress/strength interference.

The following section describes the integration procedure to arrive at definition structural reliability in terms of the stress/strength relationship. Figure 7.4 shows the details at the intersection of the stress/strength distribution curve. As can be seen, a strip at a distance $s_o$ having a thickness d$s$ and height described by pdf of stress/strength is assumed to derive the expression of reliability.

The probability that the stress parameter $s$ will lie in a small interval d$s$ can be written as follows:

$$P\left(s_o - \frac{\mathrm{d}s}{2} \le s \ge s_o - \frac{\mathrm{d}s}{2}\right) = f_s(s_o)\mathrm{d}s \tag{7.18}$$

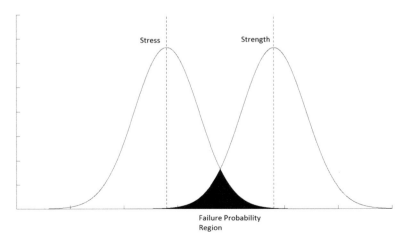

**Fig. 7.4**  Stress/strength interference

Further, the probability that strength will be more than a given value of stress $s_o$ can be written as follows:

$$P(S > s_o) = \int\limits_{s_o}^{\infty} f(S)\mathrm{d}S \tag{7.19}$$

The above two expressions enable us to find out the probability that the stress parameter $s$ will lie in a small interval $\mathrm{d}s$ when the probability that strength will be more than a given value of stress $s_o$, which is the product of the above two equations as

$$f_s(s_o)\mathrm{d}s \times \int\limits_{s_o}^{\infty} f_S(S)\mathrm{d}S \tag{7.20}$$

Now the reliability $R$ of a structural element is defined as the probability that strength will be more than the stress for all the possible values of stress in the structure, accordingly, the mathematical expression for reliability can be written as follows:

$$R = \int\limits_{-\infty}^{\infty} f_s(s) \cdot \left[ \int\limits_{s}^{\infty} f_S(S)\mathrm{d}S \right] \mathrm{d}s \tag{7.21}$$

Conversely, the reliability for an structural element can be defined as the probability that stress will be less than strength for all the possible values of strength and can be expressed as follows:

$$R = \int\limits_{-\infty}^{\infty} f_S(S) \cdot \left[ \int\limits_{-\infty}^{S} f_s(s)\mathrm{d}s \right] \mathrm{d}S \tag{7.22}$$

This procedure for computing reliability is called the procedure of integration. The limitation of this procedure is that often the closed formed solution is not available. This procedure is implemented employing numerical approaches where the approximate solutions can be obtained.

The next section introduces the major approaches for reliability structural reliability computation.

## 7.7.3 Selected Structural-Based Methods

There are four major structural reliability approaches, namely the first-order reliability method, first-order second moment method, Monte Carlo simulation, and the response surface method. It will be shown that each of the approaches/methodologies has benefits as well as limitations. Designers make a choice considering the characteristics and requirements of the problem on hand and the power and limitations of the structural reliability approach.

### 7.7.3.1 First-Order Reliability Method (FORM)

This approach comprises formulating a limit state function. This limit state is represented as the function of a set of variables. Before we discuss the definition of the limit state, we need to consider the nature of input variables. In any structural design problems, even though the traditional approach considers them as deterministic and accounts for the variability in data by considering the safety margins, a close review will reveal that most of the inputs are random and therefore require probabilistic treatment. Further, it is the designer's choice as to which variables are to be treated as probabilistic and which need deterministic considerations. A general heuristic is that the parameters that have little randomness associated can be treated as deterministic variables (i.e., by using the point value of these parameters). Mathematically, the coefficients of variation can be used as indicators for deciding upon the categorization as deterministic or random variable. If the coefficient of variation (COV) is very small, then the variable can be categorized as deterministic, whereas if the COV is relatively large then the parameter can be treated as probabilistic.

Let these basic variables be represented by $X_1$, $X_2$, $X_3$, $X_4$, ..., $X_n$. The performance function of a structural element can be formulated as follows:

$$g(X_1, X_2, X_3, \ldots, X_n) \begin{cases} > 0; & \text{Safe State} \\ = 0; & \text{Limit State} \\ < 0; & \text{Failed State} \end{cases} \tag{7.23}$$

Accordingly, the limit state equation will be

$$g(X_1, X_2, X_3, \ldots, X_n) = 0 \tag{7.24}$$

The function represents the failure surface or limit state. Suppose the limit state function is comprised of two variables, i.e.,

Then the concept of limit state is represented in Figs. 7.5 and 7.6.

If there are $n$ random variables, then the integration approach involves formulation of the problem in terms of joint pdf. Let $f_X(x)$ represent the joint probability

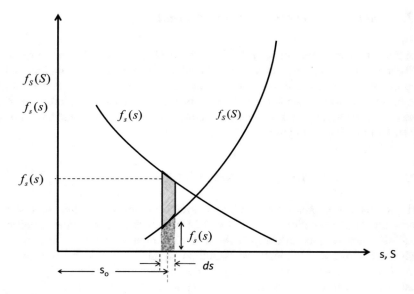

**Fig. 7.5**  Details of intersection on a stress/strength diagram

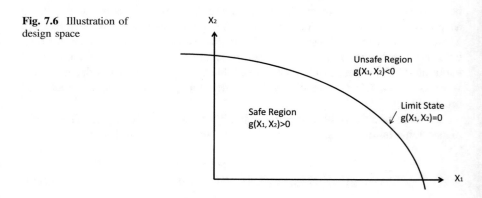

**Fig. 7.6**  Illustration of design space

function of the random variables $X_1, X_2,\ldots, X_n$. Then the failure probability can be expressed as

$$\iiint_{g<0} \cdots \int f_X(x)\mathrm{d}x \tag{7.25}$$

The solution is arrived by numerical integration over $g < 0$.

### 7.7.3.2 First-Order Second Moment Method (FOSM)

In this method, the performance function is also defined by $g(X_i)$ with usual meaning, while the limit state as $g = 0$ and failure space by $g < 0$ and success space as $g > 0$. However, there are additional considerations, namely (a) linear approximation of limit state, (b) input random variables are normal as is also the limit state, and (c) reliability index is the minimum distance of the mean of $g$ from the origin (measured in terms of standard deviations). As the name suggests, in this method the input variables are presented as first and second moment. To clarify this in case of normal random variables, the first moment is mean (i.e., $\mu$ and the second moment is standard deviation $\sigma$).

Consider the tailor series expansion of the function $g(X)$ as

$$g(X) = g(\mu_X) + \nabla g(\mu_X)^T (X - \mu_X) \tag{7.26}$$

As we noted here, the first two terms of Taylor series expansion have been considered here, ignoring the remaining terms for the purpose of generating the reliability index. Also, note that both terms are first-order terms. This is the reason this method is called the first-order second-moment method.

From the above, the mean and standard deviation of $g$ can be shown as

$$\mu_g = g(\mu_X)$$

$$\sigma_g = \left[ \sum_{i=1}^{n} \left( \frac{\partial g}{\partial X_i} \right)^2 \sigma_{Xi}^2 \right]^{1/2} \tag{7.27}$$

Accordingly, the reliability index can be defined as

$$\beta = \frac{\mu_g}{\sigma_g} \tag{7.28}$$

and the probability of failure as

$$P_f = \Phi(-\beta) \tag{7.29}$$

For further studies on the FORM, the readers may refer to the relevant literature.

### 7.7.3.3 Monte Carlo Simulation

The Monte Carlo method is a mathematical simulation approach where computers are used to perform simulations. This approach is different compared to the analytical approach. In the analytical approach, we have a limit state function or an equation where we input the value and obtain the result as a point estimate of the response variable, such as a deterministic calculation. The Monte Carlo methodology takes probabilistic input from distributions and iteratively works out uncertainty in the estimates of the response. This methodology is extensively used in many science disciplines such as engineering, nuclear physics, financial, and economics. In structural reliability analysis, it is used to find the distribution of the response variable by propagating uncertainty in the input parameter random variable. Monte Carlo simulation is used extensively when the problem is complex in terms of number of inputs that need to be handled, or where data either from experiments or other sources are prohibitory in nature. In respect of risk-based decisions, this method allows quantitative evaluation of risk.

The major steps involved in Monte Carlo simulation are as follows:

1. Identify the input variables and characterize the distribution parameters associated with each parameter. For example, if a parameter is following normal distribution then the parameter is characterized by mean and standardization.
2. Define the performance function or limit function for the response variable. For example, if the objective is to estimate the reactor CDF along with 90% confidence, then the parameter CDF is a response variable, whereas cut-sets along with their frequency and uncertainty form the input.
3. Perform random sampling of the value from this given distribution for each of the input variables and compute the value of the response variable.
4. Repeat the iteration, as given in steps 2 and 3 until an adequate number of estimates of response variable is obtained.
5. Construct the distribution using the values of the response variables.
6. Characterize the distribution with the parameters of the distribution.
7. Estimate the probability of failure.

Monte Carlo simulation forms an integral part of fault tree analysis when characterization of uncertainty, as mentioned earlier, forms part of the evaluation. The following example illustrates the salient features of Monte Carlo simulation as applied to fault tree analysis.

Generation of a random number for each iteration and estimation of distribution parameters are the two important aspects of the Monte Carlo distribution. In Monte Carlo simulation, the prerequisites are (a) characterizing the distribution of each input variable, (b) writing or using a random number generator because it is a computer simulation approach, (c) devising a scheme for analyzing the output of the simulation, and (d) working out the number of simulations so that iterations can be performed because the accuracy of the result of the simulation is sensitive to the number of iterations Monte Carlo simulations are regarded as a resource-consuming technique in terms of time and computations that need to be performed; hence, it is important that the CDF simulation is efficient in terms of memory as well as the code.

At the outset, the Monte Carlo technique is regarded as a simple and elegant approach to finding solutions when the application of an analytical technique becomes prohibitive. There are five main steps in Monte Carlo simulation:

1. Create a parametric model or limit state equation as $Y = (X_1, X_2, X_3, \ldots X_n)$;
2. Generate set of random inputs $x_1, x_2, x_3, \ldots x_n$;
3. Evaluate the response and store the results;
4. Perform the number of iterations for $i = 1$ to $M$ for steps 2 and 3;
5. Analyze the probability of failure and associated uncertainty.

The efficient approach to random sampling of the probability distribution of the input variable $x$ is toward building the distribution of the response variable $Y$ is the most important step in Monte Carlo simulation.

There are many random sampling techniques available in the literature which has been employed for various problems. In this section, we introduce only a few important techniques used in reliability and risk characterization of engineering systems. These techniques are: inversion approach to sampling, importance sampling plan, and Latin hypercube sampling.

**Inversion approach to sampling**

In this approach to sampling, the inverse of cumulative distribution function (CDF) of the $X_s$ is employed to generate the random probability number. We can understand this technique of sampling by considering the example of exponential distribution. Assume that the random number generator is set to generate a random number $M$. We express $M = F(x)$; where $F(x)$ is the cumulative distribution function of variable $x$. We know that $M$, and hence, $F(x)$, is uniformly distributed between 0 and 1. This implies that the probability will fall between $F(x)$ and $F(x) + dF(x)$ is $dF(x)$. We also express $f(x) = dF(x)$.

Hence, $x = F^{-1}(M)$ distributed as per $f(x)$. In case of exponential distribution

$$F(x) = 1 - \exp(x) \tag{7.30}$$

or

$$M = 1 - \exp(-x)$$

$$\exp(-x) = 1 - M$$

Taking the log of LHS and RHS,

$$x = -\ln(1 - M) \tag{7.31}$$

or given that the term $(1 - M)$ is uniformly distributed, we can have

$$x = -\ln(M) \tag{7.32}$$

which is nothing but an inversion for exponential distribution, and the same can be adopted for Monte Carlo simulation sampling.

### 7.7.3.4   Probabilistic Response Surface Method (PRSM)

The response surface method is comprised of developing an equation or limit state function for a complex structural risk evaluation problem where a well-defined analytical performance function or equation is not available. In the context of this book, the approach being covered in this section can be termed as probabilistic or stochastic response surface (PRSM) because it differs in the sense that the input variables are all probabilistic and application of the design of experiment and Monte Carlo simulation often forms part of PRSM. This approach can be considered an approximate approach but at the same time vital for complex problems where the traditional approach might not provide a satisfactory solution. The PRSM described in this section is in line with the subject covered in reference [39].

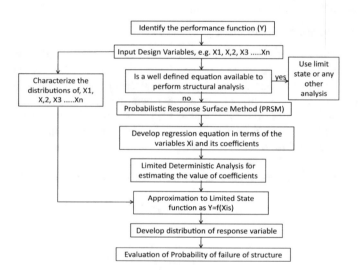

The main steps involved in PRSM are as follows:

1. Identify the input variables and performance function;
2. Develop the response surface;
3. Conduct deterministic structural analysis of the structure;
4. Develop a performance;
5. Develop the probability density function of the response variable;
6. Evaluate failure probability and associated confidence interval.

### Step 1: Identification of the input variables and response function

The PRSM is employed only in those situations where the limit state function or the equation relating the response variable and input and associated coefficients and constants could not be defined. Hence, the first step is to identify the response variable and the input variables. Let the response function be $Y$ and the input variables are defined as $X_1$, $X_2$, $X_3$, ...$X_n$.

**Step 2: Design of experiment to develop regression function**

The design of experiments is performed by assigning the levels, low, nominal, and high, to each variable. The number of vectors that need to be formulated is governed by the heuristics, $2^{2n} + 1$. For example, if three input variables are involved then the number of test runs required will be 13.

**Step 3: Performance deterministic analysis**

Deterministic analysis such as finite element analysis (FEM) is performed to work out the estimates of response variables. For example, for the 13 combinations or patterns the response variables will be estimated using FEM. Commercial software is available to perform these test runs. The output of this exercise will be formulation of a regression equation. The example of the first-order regression is given as

$$Y = A + BX_1 + CX_2 + DX_3 \tag{7.33}$$

where $A$ is a constant and $B$, $C$, and $D$ are the coefficients of variables $X_1$, $X_2$, and $X_3$.

**Step 4: Performance of Monte Carlo simulation**

Once the performance equation is identified, the next step is to perform the Monte Carlo simulation by characterizing the individual input variables with probability distribution. The result of the Monte Carlo simulation will be in the form of distribution of the performance variable, in this case, $Y$. The minimum number of iterations that need to be performed is governed by the nature of problem and the prediction accuracy.

**Step 5: Evaluation of failure probability**

Depending on the nature of the response variable(s), stress, strength, or safety margin as the case may be the probability of failure can be estimated. If the response variable is a stress for the structure then superimposing the failure criteria, (i.e., maximum limit of stress beyond which the structure is considered as failed) on the distribution of stress will provide the failure probability of the structure. The area under the failure reason can be numerically evaluated or can be analytically estimated.

## 7.8 Major Supporting Tools

Even though the static PRA techniques and the approach presented in this chapter might manage to be adequate for application of risk-based design, to make the process effective, it is desirable to have dynamic PRA (DPRA) because most of the fracture problems, thermal-hydraulic aspects, and structural aspects should be analyzed for their behavior in the time domain. Dynamic fault tree and event tree methods are discussed in the Chap. 4 on system modeling. Apart from this, dynamic modeling might require application of Markov analysis for analyzing complex scenarios. Other tools that for man integral part of risk-based design are the intelligent methods for machine learning such that degradation in structure can be captured more effectively. Apart from this, stochastic finite element modeling, particularly for fracture and wear analysis, forms tools for routine design analysis. An integrated life management design tool which facilitates sensitivity analysis for any change in the design on the affected and other systems should be part of the risk-based approach.

## 7.9 Codes and Standards

A survey of the available literature for a regulatory position on risk-based regulation shows that there is almost no literature available on the subject. However, Peterson et al.'s published [28] work on risk-based regulations bringing good practices and lessons learned that formed the proceedings of a regulatory conference held in 2011 in Melbourne. Certainly, it can be concluded that risk-based applications have not formed part of the regulatory position. In this situation, the existing design codes such as ASME codes, including ASME Section I and Section III for NPPs, continue to form general design rules and the upper bound of factor of safety and safety margin assessment. Further, for critical SSCs, reliability-based methods can be used to evaluate the margins [40]. The first requirements for application of the risk-based approach are availability of regulatory framework. At this point it can be argued that even a risk-informed approach is being applied using regulatory guides, codes, and standards available for PRA. Similarly, the same codes with stronger goals and criteria can be used for risk-based applications. As for the reliability of structural systems, the ISO standard ISO13822 [41] and ISO 2394 [42] can be adopted for developing reliability goals and criteria. The ASME/ANS standard RA-Sa-2009 [43] for design stage PRA can provide an effective quality assurance framework. The discussion on the subject is not exhaustive; however, for effective implementation of the risk-based approach, a regulatory framework is a must. The supporting codes and guides can enable limited applications where the traditional approach does not provide adequate answers. At present, the international community has just accepted risk-informed applications; therefore, the risk-based approach might be realized only in specific areas, as mentioned earlier.

## 7.10   Case Study: Use of Available Safety Margins to Demonstrate Reactor Safety

The traditional approach to safety in the reference plant does not take credit for availability of moderator dumping to shut down the reactor because this system is slow-acting. Therefore, the reactivity removal rate is also relatively slow compared to the primary shutdown system, i.e., shutoff rods which drop by gravity and spring action on sensing a completed reactor trip signal.

The genesis of this case study is insights developed from the Level 1 + PRA of a reference nuclear reactor on modeling on reactor protection system and further observation on review of the main results of the analysis. In short, this is a classical case of using the available safety margins in the plant. Even though the material presented in this chapter heavily discusses evaluation safety margin in terms of mechanical aspects of stress, strength, the case study deals with reactor and thermal-hydraulic aspects to investigate the available safety margin in terms of the availability of redundant and independent systems. The major features of this case study are (a) assessing the performance of moderator dumping against each initiating event, such as loss of offsite power (LOOP), major LOCA, minor LOCA, and loss of regulation accident (LORA); (b) revisiting the reactor physics and thermal-hydraulic calculations using best estimate approach such that the available margins in the parameters (which appear to be conservative during initial calculations); and (c) learning from over 30 years' experience to fine-tune the parameters and reevaluate the CDF and compare the same with CDF goals and targets [44]. This analysis shows that credit for moderator dumping can be taken for moderator dumping for LOOP, LORA, and minor LOCA. This case study recommends that structural analysis of major structural components, such as plenum and associated piping, to be performed to demonstrate that the probability of catastrophic failure of these structural systems is $<10^{-7}$/year to be categorized as beyond the design basis phenomenon. Chapter 15 provides details of this case study.

## 7.11   Summary

This chapter proposes a new approach to risk-based design as part of the IRBE approach. Why a new approach? Most of the risk-based approaches reviewed in the literature have risk rather simplified framework where risk assessment tools ranging from FMEA, HAZOP, to complex tools such as PRA are employed for quantified estimates of risk. The insight developed from these assessments can only be used in a limited sense to support decisions which are mostly based on deterministic approach. The risk-based approach to design proposed in this chapter uses both risk insights from deterministic as well as probabilistic assessments supported by advanced tools such as PHM for failure/degradation prediction and elaborate modeling of human factors, where risk and uncertainty characterization is performed with stringent targets/goals. These features make the approach more robust. Even though the framework of the proposed approach might appear similar to the risk-informed approach, the salient features of this approach provide a more robust basis for engineering decisions. Although this approach is expected to work well with the available state of the art in monitoring and surveillance, for more effective application, the advantages available through following methods will certainly provide the needed robustness to the proposed method: (a) PHM; (b) advances in human factor modeling; (c) uncertainty and risk targets; (d) advances in dynamic modeling, including methods employing deterministic and probabilistic methods, such as probabilistic fracture mechanics and structural reliability; and (e) passive system reliability and physics-of-failure based methods for electronics in general and digital systems in particular.

A case study has been performed on a reference reactor to demonstrate the applicability of this approach. It may be noted that in this chapter most of the aspects of the proposed risk-based approach to design have been discussed through a structural reliability engineering approach. However, depending on the application, other aspects involving thermal hydraulics, nuclear reactor physics, electrical/electronics, and other modes of degradation such as wear, corrosion, fatigue, creep, for mechanical systems and drop in change resistance, capacitance, thermal conductivity, and thermal expansion for electrical systems might be the subject of modeling. Nevertheless, the core features of the proposed risk-based approach remain same. For example, the case study deals with thermal hydraulics and neutronic aspects of the reactor to arrive at a decision.

# References

1. International Atomic Energy Agency, *IAEA Safety Standards: Safety of Nuclear Power Plants: Design—Specific Safety Requirements No SSR-2/1* (IAEA, Vienna, 2012)
2. United States Nuclear Regulatory Commission, Use of Probabilistic Risk Assessment Methods in Nuclear Activities: Final Policy Statement. Fed. Reg. **60**, 42622 (1995)
3. United States Nuclear Regulatory Commission, *An Approach for Using Probabilistic Risk Assessment in Risk-Informed Decisions on Plant Specific Changes to the Licensing Basis, Regul. Guide 1.174, Rev. 1,.* (USNRC, 1998)
4. P.D. Apost (ed.), *Risk-Based Ship Design: Methods, Tools and Applications.* Springer
5. M. Maes, K. Gulati, D. McKenna, Risk-Based Casing Design. J. Energy Res. Technol. **117**(2), 93–100 (1995)
6. Anhalt, M., & Meon, G. Risk-Based Procedure for Design and Verification of Dam Safety, in *4th International Symposium on Flood Defence: Managing Flood Risk, Reliability and Vulnerability* (Toronto, Canada)
7. A. Sahrapeyma, A. Hosseini, M. Marefat, Life cycle prediction of steel bridges using reliability-based deterioration profile: case study of Neka Bridge. Int. J. Steel Struct. **13**(2), 229–242 (2013)
8. S. Lee, B. Chu, D. Chang, Risk-based design of dolly assembly control system of drilling top drive. Int. J. Precis. Eng. Manuf. **15**(2), 331–337 (2014)
9. I. Tumer, F. Barrientos, L. Meshkat, *Towards Risk Based Design for NASA's Missions.* NASA

10. US Federal Aviation Administration (1999) *Probabilistic Design Methodology for Composite Aircraft Structures,* (Office of the Aviation Research, Washington D.C.)
11. J. Gruenwald, *Risk-Based Structural Design: Designing for Future Aircraft.* AE 440 Individual Technical Report (2008)
12. N. Khakzed, F. Khan, P. Amyotte, Risk-based design of process systems using discrete-time bayesian networks. Reliab. Eng. Syst. Safety **109**, 5–17 (2013)
13. DNV. *Risk-Based Design.* DNV Announcement
14. D. Burlington et al. *A Risk-Based Approach to current Good Manufacturing Practices (cGMPs).* PHRMA
15. FATF/OECD. *Guidance for Risk-Based Approach—Prepaid Cards, Mobile Payments and Internet-Based Payment Services.* (Financial Action Task Force (FATF), Organization for Economic Development and Cooperation, Paris)
16. F. Schoustra, I. Mockett, P. Gelder, J. Simm, A new risk-based design approach for hydraulic engineering. J. Risk Res. **7**(6), 581 (2004)
17. G. Sanctis, M. Fontana, Risk-based optimization of fire safety egress provisions based on the LQI acceptance criterion. Reliab. Eng. Syst. Safety **152**, 339–350 (2016)
18. Z. Yuant, N. Khakzad, F. Khan, P. Amyotte, G. Reniers, Risk-based design of safety measures to prevent and mitigate dust explosion. Industr. Eng. Chem. Res. **52**(50), 18095–18108 (2013)
19. O.S. Oladokun, Risk based design for safe development of reliable and environmentally. Science (2013)
20. J. Massman, R. Freeze, Groundwater contamination from waste management sites: the interaction between risk-based engineering design and regulatory policy: 1 methodology. Water Resour Res **23**(2), 351–367 (1989)
21. H. Golbayani, K. Kazerounian, On risk-based design of complex engineering systems: an analytical extreme event framework. *ASCE-ASME J. Risk Uncertainty Eng. Syst, Part-B.* Mech. Eng. **1**(1), 011002 (2015)
22. International Atomic Energy Agency, *Technical Meeting on Developing Methodologies for Complementary Assessment of Nuclear Power Plants' Robustness Against the Impact of Extreme Events* (Meeting Announcement, Vienna, 2014)
23. Y. Tung, *Risk-Based Design of Flood Defense Systems*, ed. Wu et al., (Science Press, New York, 2002)
24. G. Jourdan, *Using Risk-based Regulations for Licensing Nuclear Power Plants: Case Study of the Gas Cooled Fast Reactor.* (MS Thesis, Massachusetts Institute of Technology, USA) (2005)
25. J. Sorenson, G. Apostolakis, T. Kress, D. Powers. (1999) *On The Role of Defense in Depth in a Risk-Informed Regulation, PSA'99.* Washington DC
26. K.N. Fleming, F.A. Silady, A risk informed defence in depth framework for existing and advanced framework. Reliab. Eng. Syst. Safety **78**(3), 205–225 (2002)
27. F., Maio., Zio, E., Smith, C., Rychkov, V. (eds.), *Integrated Deterministic and Probabilistic Safety Analysis for Safety Assessment of Nuclear Power Plants*
28. D. Peterson, S. Fensling, Risk-Based Regulation: Good Practice and Lessons for the Victorian Context, in *Regulatory Conference*, (R. Melbourne)
29. S. Ritterbusch, *A Complete New Design and Regulatory Process—A Risk-Based Approach for New Nuclear Power Plants.* XA 0201905 Annex 17
30. G. Burdick, D. Rasmuson, S. Derby, A risk-based approach to advanced reactor design. IEEE Trans. Reliab **26**(3), 198–202 (1977)
31. A. H. Ang, Structural risk analysis and reliability-based design. *J. Struct. Div.* (n.d.)
32. A. Haldar, S. Mahadevan, *Probability, Reliability and Statistical Methods in Engineering Design*, (Wiley, 1999)
33. L. Pradeep, R. Lowe, K. Goebel, *Use of Prognostics in Risk-based Decision Making for BGAs Under Shock and Vibration Loads*, (IEEE, 2010)

34. P. Tipping, *Understanding and Mitigating Ageing in Nuclear Power Plants: Material and Operational Aspects of the Plant Life Management (PLiM)* (Woodhead publishing Series in Energy, Oxford, Woodhead, 2010)
35. M. Dalaney, G. Apostolakis, M. Driscoll, Risk-informed design guidance for future reactor systems. Nucl. Eng. Des. (accepted) (n.d.)
36. J. Coble, P. Ramuhalli, L. Bond, B. Upadhyay, *Prognostics and Health Management in Nuclear Plants: A Review of Technologies and Applications, PNNL-21515*, (Pacific Northwest National Laboratory, U.S. Department of Energy, 2012)
37. P. Varde, M. Pecht, Role of prognostics and health management in risk-based applications. Int. J. Prognostics Health Manage. (2008)
38. K. Kapur, L. Lamberson, *Reliability in Engineering Design*, (Wiley, 2009)
39. R. Ranganathan, *Structural Reliability Analysis and Design*, (Jaico Publishing House, 1999)
40. J. Mishra, V. Balasubramanian, P. Chellapandi, Reliability based code calibration of fatigue criteria of nuclear class I piping. SRESA Int. J. Life Cycle Reliab. Safety Eng. Soc. Reliab. Safety **5**(1), 8–17 (2016)
41. International Organization for Standardization. *Bases for Design of Structures—Assessment of Existing Structures, ISO 13822:2010(E)*, 2nd edn. (2010)
42. International Organization for Standardization. *General Principles on Reliability for Structures, ISO 2394:2015(E)*, 2th edn. (2015)
43. American Society of Mechanical Engineers/American Nuclear Society, *Addenda to ASME/ANS RA-S-2008: Standard for Level 1/Large Early Release Frequency Probabilistic Risk Assessment for Nuclear Power Plant Applications* (The American Society for Mechanical Engineers, USA, 2009)
44. P. Varde, T. Singh, T. Mazumdar, *Integrated Risk Based Design Engineering—Evaluation of available Safety Margin, BARC, E-004/2017* (BARC, Mumbai, 2017)

# Chapter 8
# Fatigue and Fracture Risk Assessment: A Probabilistic Framework

> Risk comes from not knowing what you are doing
> Warren Buffet

## 8.1 Introduction

More than 80% of the components in the modern industrial world are subjected to fluctuating loads and consequently fail through fatigue. Over 50–60 years ago, the fracture mechanics approach to predict the growth of a crack was not available, and therefore, consideration of a higher factor of safety was required to account for unforeseen factors [1]. On the other hand, in the nuclear industry, pressure vessels are designed by considering the conservative safety margins as per Sect. 8.3, Appendix G of the ASME code. Of course, a high standard of safety has been achieved, and no catastrophic failure of a reactor pressure vessel has been reported for pressurized water reactor (PWR) or boiling water reactor (BWR). However, a probabilistic evaluation of conservatisms used has been evaluated employing a probabilistic fracture mechanics approach. It was observed that the safety margins are considerable. Depending upon the conditions considered to be realistic in practice, the margins could be upwards of 10 or more orders of magnitude [2]. In fact, for any accident to be categorized as beyond design basis, it is required to establish that the likelihood of catastrophic failure is less than $10^{-6}$ per year [3]. Here, the probabilistic fracture mechanics approaches are being used to establish the beyond design basis accident (BDBA) criteria because even with the accumulated operating experience from 430 operating reactors world over, the statistical approach is not an adequate input on BDBA criteria.

Further, the traditional fracture mechanics tools and methods are not capable of characterizing the reliability or failure frequency of passive structural components in general and reactor vessels, primary coolant system piping, and pressure tubes in particular with an acceptable level of uncertainty. Deterministic as well as

© Springer Nature Singapore Pte Ltd. 2018
P. V. Varde and M. G. Pecht, *Risk-Based Engineering*, Springer Series in Reliability Engineering, https://doi.org/10.1007/978-981-13-0090-5_8

probabilistic analysis requires that the failure frequency of the reactor pressure vessel should be demonstrated to be $<10^{-6}$ per year to categorize the reactor pressure vessel failure in a BWR and PWR. The methodology followed to estimate the failure frequency of the components from operating experience, as applied for other components, such as pumps, valves, and electrical systems, is not adequate due to the lack of operating experience on passive systems such as reactor pressure vessel. Even if we consider the accumulated operating experience of around 400 PWRs and BWRs, the failure frequency of the order of $10^{-4}$ per reactor-years is not adequate to demonstrate the criteria for beyond a design basis event. The catastrophic failure or rupture of a reactor vessel leads to severe safety implications; therefore, the structural reliability of the reactor pressure vessel must be high deterministically and probabilistically.

Even though the available generic sources of component methods provide failure frequency data along with uncertainty bounds, it has been observed that the associated uncertainty levels are higher for structural systems in general and piping systems in particular (e.g., for piping failure an error factor 10). Application of these data to the risk model of the plant (e.g., estimation of major and minor loss of coolant accident (LOCA) frequency) introduces relatively higher uncertainties in LOCA frequency estimates. Even though there is excellent operating experience for primary coolant, steam line, and feed water piping systems, analysts have always felt the need for an approach that provides improved understanding of piping system degradation due to corrosion, fatigue, and induced degradation, so that the nondestructive examination approach can be improved to reduce the probability of catastrophic failure.

Fatigue-induced degradation is one of the major concerns that require attention during the design phase as well as the operational phases of the plant. The traditional fracture mechanics approach to fatigue and fracture is not adequate because this approach does not account for randomness in the parameters associated with estimation of failure or crack initiation and propagation.

Fracture risk assessment is an integrated approach that combines the results of preservice inspection performed during the preoperational phase (or in-service inspection during the operational phase of the plant), (in terms of the crack size) with a probabilistic fracture mechanics approach toward estimating the risk associated with the defect and thereby predict the remaining useful life of the SSCs. This is basically the simulation-based approach.

This chapter has two objectives, one, to provide an overview of the fatigue and fracture risk assessment requirements in a holistic sense and two, present the salient features of probabilistic tools and methods, particularly the probabilistic fracture mechanics approach and its application to nuclear plants.

## 8.2  Fatigue and Fracture: Background

The fatigue-induced failure phenomenon can broadly be divided into three phases—crack initiation, propagation or growth, and fracture or failure, as shown in Fig. 8.1.

From the analytical point of view, the fatigue phenomenon can be divided into low-cycle and high-cycle fatigue. The dividing line depends upon the material being considered; however, it usually falls between 10 and $10^5$ cycles. For example, considering that the vibration amplitudes falls within an elastic range, the life span of transport bridges, off-shore structures, and transmission towers estimated to exceed $10^8$ cycles, and they belong to high-cycle fatigue structures. There is a trend in fatigue research toward focusing on situations that require more than 104 cycles to failure in general and cases where stress is low and deformation is largely elastic (Fig. 8.2).

Fatigue failures are associated with degradation of material under cyclic mechanical, thermal, or corrosion stresses involving two distinct phenomena, crack initiation and crack propagation. There are good studies on characterization of crack propagation; however, there is no consensual work on crack initiation. Nevertheless, fracture mechanics is based on the implicit assumption that there is a crack in a work component, either inherent material or a defect developed during service period, and therefore, the fracture mechanics deals with modeling for growth of a known crack under certain loading conditions [1].

Even though the available literature shows a host of probabilistic approaches to evaluate fatigue and fracture reliability, from the nuclear system application, it was considered appropriate to deal with *P-S-N* (probability, stress, and number of cycles to failure) and probabilistic fracture mechanics approach to crack growth. Often these two approaches are integrated as a unified approach [5].

Surface finish, material microstructure, and crack orientation play a significant role in crack initiation and propagation. For example, generally, coarse colony microstructures are considered to have better crack growth resistance than fine colony microstructures in α/β titanium alloys. The coarser (Widmanstatten) microstructures exhibit higher closure levels and more torturous fracture modes [6].

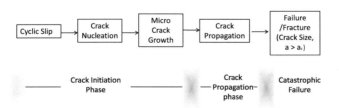

**Fig. 8.1**  Three phases in structural systems [4]

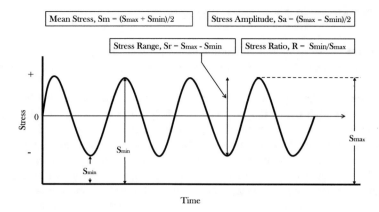

**Fig. 8.2** Typical representation of cyclic stress during fatigue test and associated terms and respective mathematical expressions

Cameron et al. [7] provide a statistical account of operating performance of nuclear and nonnuclear pressure vessels and piping. As mentioned earlier, the accumulated operating experience on nuclear pressure vessels is not adequate to give confidence that the failure frequency of the pressure vessel is lower than $10^{-6}$ per reactor per year. It is also shown that no failures have been reported in the primary coolant or the steam line. In nonnuclear vessels, this work provides an elaborate account while concluding that weld or welding-related failures are some of the few dominant causes of failure. The major causes include material in the heat-affected zone found to contain an initiation site, a manufacturing defect that escaped detection, lack of or inadequate post-weld heat treatment, and, finally, improper weld repair. This is why the focus of the fracture risk assessment is normally the weld and the affected parent material of the piping or nozzles [7].

There are three methods to determine the fatigue life of a material: the stress-life, strain-life, and linear-elastic fracture mechanics methods. The stress amplitude below which the material will not fail for any number of cycles is called the fatigue strength of material [8, 9].

## 8.3 Deterministic Approach

The deterministic approach lays a scientific and engineering foundation for the probabilistic approach. These models are further deployed by considering the randomness in the parameters while applying probabilistic models and methods to obtain the uncertainty in the estimates. This section discusses selected approaches; however, details of all available approaches are not in the scope of this book (Figs. 8.3 and 8.4).

**Fig. 8.3** ESH universal testing machines for the smallest machines universal specimen grips are available [10] *Courtesy* Soet Laboratory, Universiteit Gent, www.soetelaboratory.urgent. be/05_a_esh.shtml

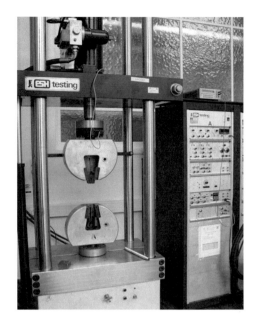

**Fig. 8.4** Some of the fatigue test specimens, **a** part through crack, **b** rotating bending, **c** three-point bend, and **d** compact tension [11]

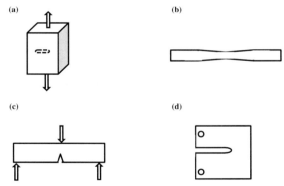

## 8.3.1 S-N *Approach*

The *S-N* diagram was introduced during the late nineteenth century when locomotive component failures led to many accidents. The pioneering study by Wholers in 1860 led to the development of the *S-N* diagram and consequent understanding of fatigue failure of the locomotive axle due to cycling stresses. The *S-N* approach comprised testing a few samples, the number that is considered adequate to give an idea of fatigue strength of the material, by applying cyclic loadings until the sample fails. The stress applied (*S*) and the number of cycles to failure are recorded for each sample. The graphical representation of *S-N* data or typical Whoeler *S-N* curve can be seen in Fig. 8.5.

**Fig. 8.5** Stress-number of
cycles to failure (*S-N*) curve

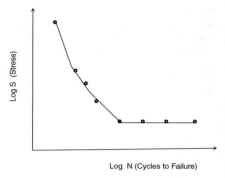

The endurance limit obtained by constant stress ($S_{en}$) line on the *S-N* diagram is modified to account for design considerations taking into account the effect of various factors, such as surface finish, surface treatment, residual stress, notches, temperature, and environmental factors. The equivalent endurance stress limit ($S_e$) is obtained as follows:

$$S_e = (S_{en})(K_s)(K_{size})(K_{load})(K_t)(K_{sc})(K_e) \tag{8.1}$$

where $S_{en}$ is the stress endurance limit (obtained from the *S-N* curve), $K_s$ is the modifying factor for the surface finish, $K_{size}$ is the modifying factor for projected load cycles, $K_t$ is the modifying factor temperature, $K_{sc}$ is the modifying factor stress concentration, and $K_e$ is the modifying factor for the environmental condition.

As indicated above, the modifying factors essentially show an approach to account for a factor of safety considering the real-time conditions. Hence, the applicable factors are used for a given problem in hand, although this also means that some factors may be added, whereas certain factors may not be applicable.

### 8.3.2  *Fracture Mechanics Approaches*

#### 8.3.2.1  Paris' Approach to Crack Growth

Suppose during the initiation and further growth of a fatigue crack, the crack size is represented by *a* and the number of cycles by *N*, then d*a*/d*N* represents the fatigue crack growth—an important observation that lays the foundation for fracture mechanics. Figure 8.6 shows the fatigue crack growth rate d*a*/d*N* as a function of the change in the stress intensity factor $\Delta K$.

Figure 8.6 shows three regions that characterize the crack growth phenomenon as a function of the stress intensity range. In region I, the stress intensity is very small. Region II is characterized by a linear relationship of crack growth with stress intensity range on a log scale. In this region, the crack is stable and change in the

**Fig. 8.6** Typical crack growth rate curve

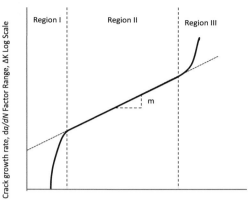

stress range $\Delta K$ has not reached the $\Delta K_c$ value region. In region III, the $K_c$ value is approaching and the crack becomes unstable and leads to fracture.

Paris law derives this relationship between crack growth rate and stress intensity range as follows [12]:

$$\frac{da}{dN} = C(\Delta K)^m \tag{8.2}$$

Where $C$ and $m$ in the Paris model are the material constants. The limitation of the Paris model is that it represents the case for experiment performed in a controlled laboratory environment. As mentioned earlier, there is a need to predict the fatigue growth by considering variability in loading, component specific parameters, and environmental conditions.

### 8.3.2.2 Minor's Rule (or Palmgren-Minor's Law)

Even though in the previous section we discussed sinusoidal representation of stress cycles, in real-life situations, often the stress amplitudes show a variable pattern as shown in Fig. 8.7. It can be seen that there are instances where the stress amplitude is more than the endurance limit with a frequency that forms a spectrum for magnitude in a certain range.

In 1945, Minor proposed a model wherein the fatigue life of an item experiencing variable stress cycles can be estimated employing the following model:

$$\frac{n_1}{N_1} + \frac{n_2}{N_2} + \frac{n_3}{N_3} \cdots \frac{n_k}{N_k} = 1 \tag{8.3}$$

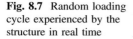

**Fig. 8.7** Random loading cycle experienced by the structure in real time

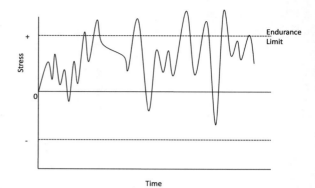

$$\sum_{i=1}^{k} \frac{n_i}{N_i} = 1 \tag{8.4}$$

where $n_i$ is the number of cycles for a given stress magnitude above the endurance limit and $N_i$ is the median number of cycles to failure for this given magnitude.

For example, we have three spectrum frequencies above the endurance limit, viz. $x_1$, $x_2$, and $x_3$ N/m$^2$. The number of cycles to failure for these $x_1$, $x_2$, and $x_3$ loadings is to $y_1$, $y_2$, and $y_3$ cycles. The frequency of loading or stress cycles, as obtained from the $S$-$N$ diagram, correspond to $z_1$, $z_2$, and $z_3$ cycles.

For further details on the existing approaches to fracture mechanics, readers are referred to the literature on the subject. Minor's rule is able to account for the random loads in a way; however, considerations of various random parameters employing the probabilistic method provide an improved framework.

## 8.4   Probabilistic Approaches

We have seen that most of the parameters in fracture mechanics are random; hence, the researchers in advanced laboratories are looking for solutions using probabilistic models. The probabilistic approach enables development of tools and methods for giving reliability and risk estimates, which can be used in design, and further in overall asset management.

There are many approaches to fatigue reliability prediction/evaluation. The weakest concept approach has two line fatigue modeling phenomena. The first one deals with the cases where the weakest spot is not significantly weaker than the gross material strength, while the second one deals with cases where the weakest spot is significantly weaker than the bulk material or structural component, such as a crack or some material flaws or physical shape that gives rise to stress concentration in the component [12].

The following section deals with the probabilistic background required in support of fatigue life risk assessment followed by a presentation of modeling of fatigue employing the *P-S-N* approach. This approach models cases where the weak spot is not significantly weaker than the bulk material and then discusses the PFM approach that models cases where the weak spot is significantly weaker than the bulk material in the structure.

## 8.4.1 Probabilistic Tools and Methods

Some of the essential probabilistic tools and methods needed for fatigue and fracture modeling have already been introduced in Chap. 3. For fatigue and fracture modeling, what we need at this stage is fundamentals of reliability and associated axioms, definitions of central tendencies, probability distributions, estimation of mean and standard deviations, application of the method of moments, major steps in the Monte Carlo simulation method, or a regression approach for data analysis. Hence, review of these chapters is a prerequisite to better appreciate the fatigue modeling concepts discussed in the following section. However, the following section provides an overview of the probabilistic models directly related to fatigue modeling.

## 8.4.2 An Overview of Probabilistic Fatigue Reliability Models

The tests and experimentation, as discussed in the previous section, form an integral part of fatigue and fracture analysis. It is observed that the data inherently show variability due to uncertainty in material properties, shape and size, and variables. These variabilities are captured by probability distributions. As mentioned earlier, the probability distributions have been discussed in detail in Chap. 3; however, this section comments on the relevant features of most distributions.

### Exponential Distribution

Due to its simplicity, this distribution is extensively used in life prediction and risk analysis of engineering systems. In fact, the compete fabric of the probabilistic risk assessment rests on the foundation of this distribution. This distribution is called memory less because the prediction is based on the current state of variables. Why is this distribution popular? The reason is that the constant hazard characteristic of this distribution can be realized through the real-time plant scenarios where a routine test and maintenance program provides corrective mechanisms and thereby tends to keep the failure/hazard rate constant. This distribution captures random or catastrophic failures and not the wear or aging phenomenon. Exponential distribution is one of the forms of Weibull distribution for the case when the shape parameter $\beta$ acquires a value $\sim 1.0$.

**Normal Distribution**

When the data show variability around the mean value such that the distribution is symmetrical about the mean, then this distribution is considered to be representing data. This distribution is very popular in representing the variability in scientific data analysis, production environment, quality controls, and engineering structural analysis. In the context of fatigue failure, when the number of cycles to failure $n$ is very large, then normal distribution can be approximated for estimating cycles to failure [13]. Fatigue design analyses have been carried out by accounting for variability in working stresses, yield strengths, and brittle fracture strength employing normal distribution [14].

**Log-Normal Distribution**

A random variable a (e.g., fatigue crack size) is considered to follow log-normal distribution when the log of the variable (i.e., log a) follows normal distribution. A review of the literature shows that log-normal distribution has been used to describe many fatigue and fracture studies.

**Weibull Distribution**

This can be said to be one of the most versatile distributions that have found maximum application in fatigue or wear modeling. This distribution can be used to model entire life cycles of the components. When the shape parameter $\beta$ has a value ranging between 3 and 4, then it represents an increasing failure rate that suits modeling for wear or aging phenomena. In fact, this distribution began in 1949 when Weibull proposed a probability distribution function (pdf) for the interpretation of the fatigue data [13]. This is the most widely used distribution in fatigue studies, and the literature shows many applications of this distribution in the study of fatigue as well as creep-induced failures.

There are many other distributions, such as gamma and uniform, which have been applied in aging/degradation studies; however, the scope of this chapter is limited to a brief overview of the distributions that are relevant for fatigue and fracture studies.

### 8.4.3   P-S-N *Approach*

Until 1939, the research community only had normal distributions to model engineering test data, including representations of the strength of the material. In 1939, Weibull proposed a distribution in his work on statistical theory of the strength of materials, which was later named after him [15, 16]. Further, he worked on fatigue data analysis and proposed a modification of the *S-N* curve to include the probability component for giving the statement/probability of failure risk or reliability [17]. He represented the fatigue data number of cycles to failure against the median value of probability of failure, where he proposed a model for median rank as follows:

$$P = \frac{m}{n+1}$$

The probability of failure is estimated by arranging the data in assenting order, where $m$ is the rank of the data and $n$ is the sample size that represents the population. The probability of failure is obtained using the equation. Further, the plot of a number of cycles to failure ($N$) for each specimen along with the associated probability of failure ($P$) has been shown by Provan as a $P$-$N$ graph as shown in Figs. 8.8 and 8.9.

### 8.4.4 Probabilistic Fracture Mechanics Approach

The major features for employing probabilistic fracture mechanics (PFM) for risk and reliability evaluation involve the following tasks:

(a) Definition of the problem;
(b) Application of deterministic fracture mechanic models as the fundamental to develop objective function;
(c) Characterization of input variables with probability distributions;
(d) Identification of in-service inspection parameters such as detection probabilities for various crack sizes and locations;

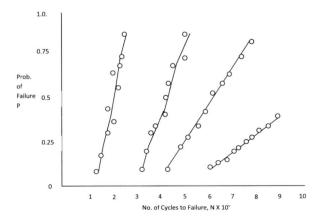

**Fig. 8.8** $P$-$N$ curves for different stress levels

**Fig. 8.9** A set of *P-S-N* curves

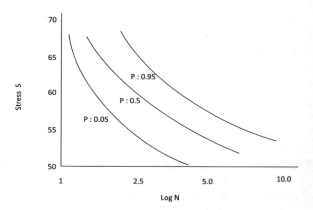

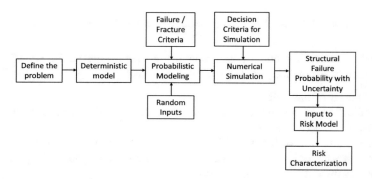

**Fig. 8.10** Generic framework for PFM for risk characterization

(e) Application of numerical analysis procedures employing Monte Carlo simulation to obtain the distribution of cycles or times to failure;

(f) Optimization of maintenance or in-service inspection strategies and tools to realize acceptable risk by performing sensitivity analysis.

Figure 8.10 shows the basic framework of the flow of the major activities.

The following section discusses the intimate and broad aspects of modeling and utilizing the PFM approach to risk characterization for specific structural risk analysis requirements that can be applied to the whole system or even a plant.

### Definition of the Problem

Concisely define the objective, scope, and associated criteria for application of the PFM approach along with the associated constraints. For example, the problem definition could be "Evaluation of weld failure probability in primary coolant system at common header location considering the vibration values recorded for the location in plant ISI program by considering subsurface elliptical crack in support of the development of vibration-induced degradation factor for this location and evaluation of risk over the next 15 years. Evaluation of uncertainty should be part of this analysis."

**Deterministic Modeling**

The first step in deterministic modeling is to state and record the assumptions of the analysis and the constraints in a clear and concise manner. This is crucial because later once the PFM modeling is completed, the simulation can be performed to analyze the impact of these assumptions. In case safety cannot be demonstrated against these assumptions, then engineering solutions are sought.

For instance, the analysts decide to use Parry's model of Eq. (1) for crack growth rate as follows:

$$\frac{da}{dN} = C(\Delta K)^m$$

Here, the required inputs are material and associated properties. If the analysis is being performed for the weld location, then weld configuration and heat-affected zone parameters; values of material constant $C$ and $m$; the initial crack size parameters are the required information for the fracture analysis.

The next step in crack growth modeling is to evaluate and find the model for stress intensity factor $K$ and then find the applicable value of the material constant $C$ and $m$. For the purpose of this problem, we assume a plate-type geometry having an elliptical crack as shown in Fig. 8.11 [1]:

The stress intensity factor for this geometry is given as [1]

$$K_I = \frac{\sigma(\pi a)^{1/2}}{I_2}\left[\sin^2\theta + \left(\frac{a}{c}\right)^2\cos^2\theta\right]^{1/4} \tag{8.5}$$

**Fig. 8.11** Large plate with an embedded initial elliptical crack

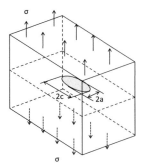

**Table 8.1** Value of $I_2$ as a function of $a/c$

| $a/c$ | 0.0 | 0.1 | 0.2 | 0.3 | 0.4 | 0.5 | 0.6 | 0.7 | 0.8 | 0.9 | 1.0 |
|---|---|---|---|---|---|---|---|---|---|---|---|
| $I_2$ | 1.000 | 1.016 | 1.1051 | 1.097 | 1.151 | 1.211 | 1.277 | 1.345 | 1.418 | 1.493 | $\pi/2$ |

where $I_2$ is the elliptical integral of a second kind that depends on $a/c$ (ratio of minor to major axis) as defined in Fig. 8.11 and is given by

$$I_2 = \int_0^{\pi/2} \left( 1 - \frac{c^2 - a^2}{c^2} \sin^2 \alpha \right)^{1/2} d\alpha \qquad (8.6)$$

The value of $I_2$ for ready use for the designer as a function of $a/c$ is provided in Table 8.1.

The value of parameters $m$ and $C$ that can be obtained from the generic sources is evaluated from regression analysis. For the purpose of this illustration, the value of $m = 3.4$ and $C = 3.4 \times 10^{-4}$ for austenitic steel is assumed.

The intensity factor, $K_{IC}$, for a crack depends on several factors, mainly the heat treatment, the speed of the crack, the temperature of the specimen, the manufacturing process, the orientation of the crack in respect to the grain orientation at the crack tip, and the test methods. However, the representative values of $K_{IC}$ for mild steel, medium carbon steel, alloy steel, nuclear reactor steel, and stainless steel are given as 220 (for a yield stress value of 240 MPa), 54 (for a yield stress value of 54 MPa), 99 (for a yield stress value of 99 MPa), and 190 (for a yield stress value of 190 MPa, 80–150 MPa√m).

### Definition of Failure Criteria

Failure in terms of stress intensity factor is defined, for the purpose of this illustration, as the condition when the stress intensity factor for a structure increases more than the critical value of the stress intensity factor, i.e., $K_I \geq K_{IC}$. This definition requires assessment of $K_{IC}$. For the purpose of this illustration, an approximate approach is presented. An alternate approach is to use the critical crack length; i.e., when the crack size ($a$) reaches a critical size, then the crack become unstable, $a \geq a_c$.

### Probabilistic Modeling

The fundamental or basis of PFM is that even though most of the variables should be treated as random for achieving acceptable uncertainty in the final results, there are some variables, such as crack size and probability of detection, as part of risk assessment, that require to be defined as random variables and not as deterministic variables [13]. The deterministic equation obtained in the previous step forms the basis for probabilistic analysis.

Probabilistic analysis can be divided into two main tasks—characterization of the distribution for the identified variables in the deterministic equation and creation

**Table 8.2** Identification of probability distribution for some random variables in PFM

| Random variable | Distribution | Mathematical form |
|---|---|---|
| Crack size, $a$ | Weibull | $a \sim W(\alpha, \beta)$; a has Weibull distribution with parameter and scale parameter $\alpha$ and shape parameter $\beta$ |
| Far-field stress, $\sigma_f$ | Log-normal | $\sigma_f \sim \log N(\mu - \sigma)$; far-field stress $\sigma_f$ has log-normal distribution with mean $\mu$ and $\sigma$ standard deviation of far-field stresses |
| Constant, $C$ | Normal | $C \sim N(\mu, \sigma)$; $C$ has normal distribution with mean $\mu$ and standard deviation $\sigma$ |
| Constant, $m$ | Normal | $C \sim N(\mu)$; $C$ has normal distribution with mean $\mu$ and standard deviation $\sigma$ |

of probabilistic distribution of variables such as detection probabilities. It is not required to transform all the deterministic variables to probabilistic variables in the governing deterministic equation. The analysts choose only those variables for probabilistic characterization where uncertainty is considered to be higher. Often the requirements identified for the analysis (e.g., risk characterization metrics and the available resources) form the governing considerations for probabilistic analysis.

For example, for the given equation, it might be required to use far-field stress $\sigma$ as the random variable following log-normal distribution, the constant $C$ following normal distribution, and the crack size $a$ following Weibull distribution. The non-detection probability might use a Poisson or log-normal distribution. Table 8.2 shows how the variables can be stated.

The identification of the distribution involves two steps. Usually, in the first step, the data are plotted on a versatile distribution, for example a Weibull graph, and based on the $\beta$ obtained (slope of the line), the distribution is approximated. For example, if $\beta = 1$, then the data follow the exponential distribution, which means the rate function is constant; when $\beta = 2$, the data follow log-normal distribution and the rate function could be increasing/decreasing; when $\beta = 3$, the data follow normal distribution and the rate function is increasing. Further, regression analysis is performed to assess the adequacy of how well a distribution represents the data set. For details of the evaluation of applicable distribution, readers should refer to the relevant literature.

**Numerical Analysis**

Numerical analysis is extensively used for implementing the probabilistic approach for complex systems. The core model for solutions in this approach remains deterministic; however, the variables are presented as probabilistic variables. In fact, the numerical approach is required as part of the deterministic approach because, for example, the crack length should be treated as a random variable, which is also the case with the stress intensity factor. Hence, the numerical approach is required even for the deterministic modeling. As for probabilistic modeling, the conventional methods, such as stress/strength interference theory, variance/covariance, and Monte Carlo simulation, are some of the tools employed for probabilistic fracture/failure analysis. In all these cases, finding the applicable distribution of the input parameters (e.g., stress, strength) is crucial. Keeping in view the scope of this section, we provide the major features and procedural steps in the following section. For details on numerical analysis and Monte Carlo simulation, readers are advised to refer to the literature, for example reference [13].

Monte Carlo simulation is one of the numerical techniques that employ random number generation for sampling a probability value from given distributions. Computer-assisted simulation in an iterative manner forms the characteristic feature of Monte Carlo simulations. Why is this required? In the context of reliability engineering in real life, a series of tests are performed on products or systems, particularly the new designs, to collect data to gain adequate confidence in predicting its performance for postulated environmental and operational loads. However, for many situations/systems, testing is an expensive, prohibitory proposition because it is not practical to perform testing on an integrated system such as a space, aviation, or nuclear pressure vessel to predict its reliability. In such scenarios, the simulation in general and Monte Carlo simulation in particular provide an effective solution.

Because the quality of the results will have great bearing on these aspects, some basic questions need to be answered for employing Monte Carlo simulation [18]. For example, How is a distribution for the random variables selected? How are random variables generated? How many iterations or simulation runs are required to predict the parameters of the final distribution?

The characterization of the random parameters by appropriate distribution was discussed in previous section where, apart from the graphical, analytical approach regression analysis was discussed to select a distribution.

To generate random variables, random number generators are used that generate a random number between 0 and 1. The algorithm for generating random parameters in Monte Carlo simulation, from a distribution, is as follows [19].

At this point, we understand that the probability distribution for a random variable has been identified. Suppose that $X$ is a random variable, then the probability that $X$ will occupy a value less than or equal to $x$ is written mathematically as the cumulative distribution function (CDF) of $x$ as $F(x) = P(X \leq x)$. The value of $F(x)$ lies between 0 and 1. However, here, we want to know the value of $F(x)$ for a given $x$. We define a reverse function $G(F(x)) = x$. How does it works? The random number generator algorithm generates a random number $r$ between 0 and 1. Say this value is 0.4. The corresponding value of the parameter $x = 610$ units can be seen from Fig. 8.12.

For further details on sample techniques to improve Monte Carlo process efficiency, readers can refer to the literature on the subject, including reference [9, 20].

**Fig. 8.12** Generation of reverse function

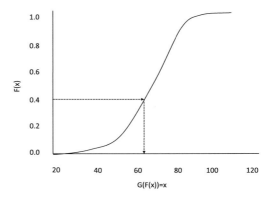

The major procedural steps in Monte Carlo simulation are as follows:

1. Create a deterministic parametric function in the form of $y = f(x_1, x_2, x_3, \ldots x_k)$.
2. Assign probability distribution to each of the $k$-dependent random variables.
3. Generate a set of $k$ random parameters.
4. Evaluate the value $y_i$ using the function given in step 1 and the random set of $k$ parameters.
5. Repeat steps 3 and 4 iteratively for the predefined sampling number $n$.
6. Analyze the outcome of the $n$ iterations by creating histograms to evaluate the distribution, mean/median value, and uncertainty bounds.

Determining the number of iterations that need to be performed is crucial to the outcome of a Monte Carlo simulation. There are some heuristics, for example, as many iterations as possible should be performed. However, in Monte Carlo methods, the number of iterations to be performed is limited by another set of guiding factors. In this case, we are dealing with an outcome probability in the range of $10^{-5}$, and the number of iterations should be $\sim 100{,}000$. By the same observation, say we have a system expected to have a probability of failure to be $\sim 10^{-3}$, e.g., a Class III power supply or ECCS system. The number of iterations could be around $10^3$–$10^4$. Hence, it is left to the analysts to judge the number of iterations to be performed for a given analysis.

## 8.4.5 Risk Assessment and Impact Analysis

The crack growth data can be used as input in support of an in-service inspection (ISI) program where periodic monitoring supported by probabilistic fracture analysis can form part of the prognosis and maintenance/repair approach. The PHM supports the application of a risk-informed approach to plant structural health management. However, we have learnt in the previous section that some aspects such as the probability of non-detection need to be considered while evaluating the risk impact of a failure.

There are a host of techniques that are employed during ISI depending on the requirements of the ISI program. For example, the most used technique is visual examination. This technique is commonly used by expert/qualified ISI staff to inspect structures such as the piping for surface defects. Another technique is the liquid penetration test. A liquid and a developer are applied to the surface to locate any crack or defect. The most used technique is the ultrasonic test for locating a defect/crack in the bulk material. Ultrasonic testing is also used for thickness reduction measurement. The detection probability is a function of the technique used, the adequacy of coverage during the inspection, the staff's expertise/qualification level of the, and the location or accessibility of the area being inspected. Expert opinion is one of the common methods for evaluating non-detection probability, which is an important parameter in identifying the potential leakage location.

The second situation involves the analysis of catastrophic failure irrespective of the availability of an ISI-assisted probabilistic fracture analysis approach. This situation is addressed by a risk-informed ISI approach. In this approach, the ISI program is structured keeping in view the likelihood and consequences of the failure of pipe segments in the system. This analysis links the individual pipe segment failure to the net risk levels for the plant in terms of conditional core damage probability. It may be noted that development of risk-informed ISI requires a Level 1 PRA model of the plant. Here, the PFM results are used to predict the likelihood of loss of coolant accident (LOCA) frequency. Further, a piping section failure that leads to leakage or double-ended pipe break and results in loss of coolant from the primary system provided the consequences of the failure, while the conditional probability of a pipe that has ruptured provides the likelihood of core damage. The likelihood coupled with consequences provides the estimate of risk levels posed by the pipe segment in question. The categorization of piping for identified risk levels helps structure the ISI program such that risk forms the driver for the ISI program.

The developmental work by Kurz et al. [19] on assessment of fracture risk deals with the integration of quantitative nondestructive inspection or ISI and probabilistic fracture mechanics. Development of a failure assessment diagram ($S_r$, $V_s$, $K_r$) is facilitated, like any other risk assessment diagram, by having two regions, acceptable and unacceptable, separated by a continuous failure assessment boundary. In this procedure, $S_r$ is expressed in terms of stress as follows:

$$S_r = \frac{\sigma_{ref}}{\sigma_f} \tag{8.7}$$

and

$$K_r = \frac{K_I}{K_{IC}} \tag{8.8}$$

where $\sigma_{ref}$ is the reference stress

$$\sigma_f = \frac{\sigma_y + \sigma_{UTS}}{2}$$

where $K_I$ is the stress intensity factor; $K_{IC}$ is the material fracture toughness; $\sigma_{ref}$ is the reference stress; $\sigma_f$ is the nominal stress; $\sigma_y$ is the yield stress; and $\sigma_{UTS}$ is the tensile strength.

The salient feature of this approach is that it can evaluate the reliability non-destructive inspection by integration of FAD and probability of detection (POD) parameters. For details of this approach, readers are advised to refer to the literature.

## 8.5 Conclusion and Remarks

The probabilistic approach to risk assessment in general and probabilistic fracture mechanics in particular is crucial to demonstrating that the likelihood of failure of vital components such as reactor pressure vessels and piping is within the acceptable limit. The deterministic model and methods were discussed because they provide the required foundation for implementing the probabilistic methods. The role of the Monte Carlo technique for probabilistic analysis was discussed. The evaluation of crack growth parameters in probabilistic fracture mechanics was carried out using the Monte Carlo technique. It has been discussed that the results of probabilistic fracture mechanics in conjunction with ISI and probabilistic risk assessment of the plant provide the required risk-informed framework for risk impact assessment. In the background, there is an understanding that the non-detection probability associated with ISI needs to be accounted for in risk-impact assessment.

## References

1. P. Kumar, *Elements of Fracture Mechanics* (Tata McGraw Hill Education Pvt. Ltd., New York, 2009)
2. G. Jouris, Probabilistic evaluation of conservatisms used in section III, appendix G of the ASME code, in *Probabilistic Fracture Mechanics and Fatigue Methods—Application for Structural Design and Maintenanc*, ed. by B.J., E.J.C. (1983)

3. International Atomic Energy Agency, *Deterministic safety analysis for Nuclear Power Plants, Specific Safety Guide no. SSG-2* (IAEA, Vienna, 2009)
4. A. Azeez, Fatigue Failure and Testing Methods, Bachelor's Thesis (HAMK University of Applied Sciences)
5. P. Darcis, D. Santarosa, N. Recho, T. Lassen, *A Fracture Mechanics Approach for the Crack Growth in Welded Joints with Reference to BS 7910*, ECF15
6. A. Soboyejo, S. Shademan, V. Sinha, W. Soyejo, Probabilistic methods in fatigue and fracture, in *Statistical Modeling of Microstructural Effects on Fatigue Behavior of α/β Titanium Alloys*, ed. by A.B.O. Soboyejo (Trans Tech Publications, 2001)
7. G. Cameron, G. Johnstone, The reliability of pressurized water reactor vessels, in *Probabilistic Fracture Mechanics and Reliability*, ed. by J. Provan (Martinus Nijhoff Publishers, Dordrecht, 1987)
8. C. Bathias, There is no infinite fatigue life in metallic materials. Fatigue Fract. Eng. Mater. Struct. **22**(7), 559–565 (1999)
9. J.E.S. Jump, M.R. Charles, B.G. Richard, *Mechanical Engineering Design*, 7th edn. (McGraw Hill Higher Education, New York)
10. Universitate Gent, Soete Laboratory, www.soetlaboratory.urgent.be/0s_a-esh.shtml
11. P. Paris, M. Gomez, W. Anderson, A rational analytic theory of fatigue. Trends Eng. **13**, 9–14 (1961)
12. J. Provan (ed.), *Probabilistic Fracture Mechanics and Reliability* (Martinus Nijhoff Publishers, Dordrecht)
13. K. Mishra, *Reliability Analysis and Prediction—A Methodology Oriented Treatment* (Elsevier, Armsterdam, 1992)
14. C. Osdood, Fatigue Design, 2nd edn. (Pargamon Press, 1982)
15. W. Weibull, A Statistical Representation of Fatigue Failure in Solid, *Acta Polytech. Mec. Engrg. Serv.* **1**(9) (1949)
16. D. Datta, L. Guneshwor, in *Heat Conduction Using Monte-Carlo Simulation, Uncertainty Modeling and Analysis*, ed. by H. Kushwaha, (Bhabha Atomic Research Centre, Mumbai)
17. W. Soboyejo, W. Shen, A. Soboyejo, Probabilistic Modeling of Fatigue Growth in Ti-6Al-4V, in *Probabilistic Methods in Fatigue and Fracture*, ed. by A.O.I., S.W. Soboyejo (Trans Tech Publications, 2001)
18. R.K. Durga, Treatment of Aleatory and Epistemic Uncertainty in Safety Assessment, in *Uncertainty Modeling and Analysis*, ed. by H. Kushwaha (Bhabha Atomic Research Centre, Mumbai, 2009)
19. J. Kurz, D. Cioclov, G. Dobmann, C. Boller, Quantitative NDI Integration with Probabilistic Fracture Mechanics for the Assessment of Fracture Risk in Pipelines, in *Proceedings of the 4th European–American Workshop on Reliability of NDE–Fr.1.A.3*
20. H. Shen, T. Nocolas, Reliability High Cycle Fatigue Design of Gas Turbine Blading System Using Probabilistic Goodman Diagram, in *Probabilistic Methods in Fatigue and Fracture*, ed. by A.O.I., S.W. Soboyejo (Trans Tech Publications, 2001)
21. http://www.barringer1.com/wa.htm

# Chapter 9
# Uncertainty Modeling

*In this world of uncertainty, death is certain and the fear is apparent.*

Srimad Bhagavada Gita

## 9.1 Introduction

It is not exaggeration to say that the adequacy and applicability of risk-based methods to a large extent depends on the accurate characterization of not only risk metrics but also associated uncertainties. In the context of nuclear plant safety evaluation, risk-based engineering deals with the evaluation of safety cases using probabilistic risk assessment (PRA) where uncertainty analysis forms one of the components to create a model of the plant [1], while component reliability and associated uncertainties form the input for the model of the plant. The risk-based approach need not necessarily consider that probabilistic and deterministic methods are in two explicit domains. Here, modeling requires considerations of deterministic as well as probabilistic methods together. Otherwise, the solution may not be adequate and complete. Deterministic variables, such as design, process, and nuclear parameters, are often random and require probabilistic treatment; therefore, uncertainty analysis also forms part of the design analysis to improve understanding of safety margins. Figure 9.1 illustrates the advantages of employing probability distributions to understand the safety margins through an example considering the stress/strain relationship. The safety factor, which is based on experience and conservative assumptions without a sound scientific basis supported by data, tends to result in overdesign without any quantified information on the reliability or safety

This chapter is a revised and updated version of paper entitled "Uncertainty Analysis in Support of Risk-based engineering" by P. V. Varde, published in Proceedings of the International Symposium on Engineering under Uncertainty: Safety Assessment and Management, Edited by Chakraborty S. & Bhattacharya G., (Editors), (ISEUSAM-2012), Springer, 2012.

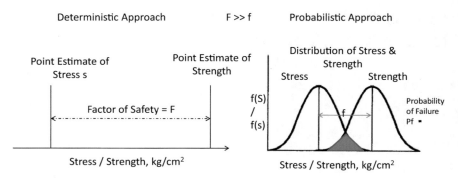

**Fig. 9.1** Simplified perception of the factor of safety and uncertainty for the deterministic approach (left) and probabilistic approach (right)

of the design. The probabilistic approach reflecting uncertainty due to randomness in the data is able to reduce the safety margin while giving the probability of failure.

The reliability of various protection barriers needs to be characterized, which forms the basic instrument of defense in depth. Similarly, probabilistic methods cannot work in isolation and require deterministic input in terms of plant configurations, failure criteria, and design inputs. Hence, it can be argued that a holistic approach is required where deterministic and probabilistic methods have to work in an integrated manner in support of decisions related to design, operation, and regulatory review of nuclear plants. The objective should be to remove over conservatism and the prescriptive nature of the current approach, introduce rationales, and make the overall process of safety evaluation scientific, systematic, effective, and integrated. The key issues that need to be considered for applications are characterization of uncertainty, assessment of safety margins, and requirements of dynamic models for assessment of accident sequence evolution in the time domain.

This chapter provides a brief overview of uncertainty evaluation methods relevant to risk-based applications. It introduces types of uncertainty that include, apart from epistemic and aleatory uncertainty, the subjective category of uncertainty that deals with characterization of cognition, conscience, and consciousness summed up as ethical/moral components of uncertainty. The subject is treated in a philosophical manner but not all of the specifics are covered because of the volume and complexity that would be required. However, these aspects can be found in the referred or available literature.

## 9.2   Treatment of Uncertainty: A Historical Perspective

Most of the cases evaluated as part of safety assessment employ deterministic models and methods. However, if we look at the assumptions, boundary conditions, factors of safety, data, and models of a traditional safety analysis, it can be argued that many of the elements are probabilistic. These elements or variables have qualitative notions for bounding situations and often provide comparative or relative aspects of two or more prepositions. To understand this point further, let us review the traditional safety analysis report(s) and take a fresh look at the broader aspect of this methodology. The major feature of the traditional safety analysis approach is based on the maximum credible accident, and for nuclear plants, these scenarios are mainly loss of coolant accidents (LOCAs) and loss of regulation accidents. It was assumed that the plant design should consider LOCAs and other scenarios, such as station blackout, to demonstrate that the plant is safe enough in terms of regulatory acceptance criteria. However, now the safety community considers a list of postulated initiating events ranging from anticipated, design basis, beyond design basis, and severe accident event conditions.

A reference to a typical safety analysis report or, to be precise, a deterministic analysis report will make it clear that there is an element of probability in a qualitative manner. These insights can include statements such as: (a) the possibility of two-out-of-three train failures is *very low*, (b) the possibility of a particular scenario involving multiple failures is *very unlikely or low*, or (c) the series of assumptions that certain conditions are assumed to be *not probable or possible*, such as failure of three components/systems in series particularly at a single instant. These assumptions or qualitative insights form the boundary conditions for the safety evaluation and show that uncertainty characterization employing qualitative probabilistic aspects is part of the deterministic methods. These aspects are qualitative in nature.

Keeping in view the above, and considerations of factors of safety in the design as part of the deterministic methods, brings out the fact that characterization of uncertainty is integral to safety. Integrated risk-based engineering (IRBE) seeks to (a) be more rational-based and not prescriptive, (b) remove over conservatism, (c) be holistic, (d) allow realistic safety margins, and (e) provide a framework for dynamic aspects of the accident scenario. Therefore, the role of uncertainty evaluation becomes integral to the IRBE approach. Concerning uncertainty characterization, IRBE requires that random phenomena are addressed through aleatory uncertainty and the model and data-related uncertainty are represented as epistemic uncertainty. There are a host of issues in risk analysis where handling of the uncertainty is more important than quantification of uncertainty. The complex issues related to human behavior through the subjective treatment of cognition, consciousness, and conscience are summed up as ethical modeling to provide a holistic approach. These rather qualitative aspects of uncertainty or cognitive uncertainty need to be addressed by having safety provisions in the plant. For application of the integrated risk-based framework, it is vital to address these aspects, perhaps through automation, checklist procedures, or incorporation of an operator support system, particularly in a control room environment.

## 9.3   Risk-Based Methods and Uncertainty Characterization

This section outlines the major features or elements of the integrated risk-based approach that are relevant to uncertainty characterization. Most of the aspects are listed for modeling that employs the PRA approach; however, the same observations can be adopted for any other approaches. Some challenges are inherent to risk assessment; however, they also need an evaluation from the point of view of uncertainty characterization.

### 9.3.1   Major Features of Risk-Based Approach Relevant to Uncertainty Characterization

To characterize uncertainty for the risk-based approach, one must understand the nature of major issues that need to be addressed in risk-based characterization and accordingly look for the appropriate approach. At the outset, there appears to be a general consensus that on a case-to-case basis, most of the above-listed approaches may provide efficient solutions for a specific domain. However, here the aim is to focus on the most appropriate approach that suits the risk-based applications. The PRA in general and Level 1 PRA in particular, as part of risk-based approaches, have the following major features [2–5]:

- The probabilistic models for complex systems, such as nuclear plants, process, and aviation systems, are basically complex and relatively large compared to the models developed for other engineering applications. Hence, they require computerized approaches to analysis that are capable of connecting structures, systems, and components (SSCs) to the plant level where the interrelationship among the component forms the format of the model.
- The major part of probabilistic modeling is performed using fault trees and event trees; hence, the uncertainty modeling approach should be effective for these models.
- Because there is a growing interest in dynamic probabilistic risk assessment (DPRA), the method chosen for uncertainty modeling should be commensurate with the simulation aspects in the dynamic environment.
- In the probabilistic models for risk assessment, randomness in the data inherent to the model at an integrated level requires aleatory uncertainty characterization.
- Epistemic uncertainty analysis is a major requirement to address the aspects related to completeness of knowledge in terms of adequacy of data, models, and assumptions.
- The uncertainty characterization for PRA models requires simulation tools/ methods that are efficient and effective in terms of optimizing/conserving resources.

- Confidence intervals for the components and human errors, estimated using statistical analysis form the input for the probabilistic models.
- There should be provision to integrate the prior knowledge about the events for getting the posteriori estimates, i.e., the approach should be able to handle subjective probabilities. For example, the Bayesian updating should be part of the risk model and uncertainty characterization.
- Often, instead of quantitative estimates the analysts come across situations where it becomes necessary to derive quantitative estimates through "linguistic" inputs. Hence, the framework should enable estimation of variables based on qualitative inputs, for example, application of fuzzy logic to deal with imprecise data.
- Evaluation of deterministic variables forms part of risk assessment. Hence, provision should exist to characterize uncertainty in structural, thermal, hydraulic, and neutronic modeling, particularly in evaluation of failure criteria.
- Sensitivity analysis for verifying the impact of variability in data and assumptions forms the fundamental requirements.
- The probabilistic methods offer an improved framework for assessment of safety margins—a basic requirement for risk-based applications.
- Although PRA provides an improved methodology for assessment of common cause failure (CCF) and human factor data, keeping in view the requirements of risk-based applications, further consolidation of data and model is required. Here, more than quantification, the deterministic evaluation of defense against CCF is vital for a robust approach that addresses CCF-induced events.

There are some domain-specific requirements, such as chemical, environmental, geological, and radiological dose evaluation, that also need to be modeled. As discussed, the uncertainty characterization for risk assessment is a complex issue and requires expertise and modern methods for enhancing the accuracy of the results.

## 9.3.2   Major Issues for Implementation of Integrated Risk-Based Engineering Approach

One of the major issues that form a bottleneck to successful application of risk-based engineering is characterization of the uncertainty associated with the data and the model. Even though reasonable data are available for internal events, characterization of external events poses major challenges. Apart from this, the non-availability of probabilistic goals/criteria along with uncertainty bounds and safety margins as regulatory policy also poses a challenge. Issues related to the required knowledge base for new and advanced features of the plants such as passive system modeling, digital system reliability in general and software system reliability poses special challenges. Relatively large uncertainties associated with common cause failures of hardware systems and human action considerations particularly with accident scenarios are one of the major issues.

It can be argued that reduction of uncertainty associated with the data and the model in characterization of uncertainties, particularly for rare events where the data and the model are either not available or inadequate, is one of the major challenges in implementation of integrated risk-based applications.

## 9.4  Uncertainty Analysis in Support of IRBE: A Brief Overview

### 9.4.1  Risk and Uncertainty

Even though many definitions of uncertainty are given in the literature, the one that suits risk-informed/risk-based decisions has been given by ASME as "representation of the confidence in the state of knowledge about the parameter values and models used in constructing the PRA" [6]. Uncertainty characterization in the form of qualitative assessment and assumptions is an inherent part of risk assessment. However, as the data, tools, and statistical methods developed over the years, the quantitative methods for risk assessment became more accurate and effective in terms of reflection of real-time scenario. The uncertainties in estimates have been recognized as an inherent part of any analysis results. The actual need for uncertainty characterization was felt while addressing many real-time decisions related to the assessment of realistic safety margins.

Uncertainty characterization as part of IRBE requires conservative target goals for core damage frequency (CDF) along with upper confidence bounds, compared to Level 1 PRA or as part of a risk-informed approach by at least an order. The limit on increase in CDF is also worked out on a conservative basis because conservatism forms the major feature of deterministic analysis and needs to be reflected IRBE-based decisions.

Risk and uncertainty are discussed in further detail the Chap. 2 on Risk Characterization.

### 9.4.2  Taxonomy of Uncertainty

Tannert et al. [7] in their article "The Ethics of Uncertainty" have presented an uncertainty taxonomy that considers the moral and ethical components of uncertainty other than epistemic and other natural phenomena. The taxonomy of uncertainty as presented in this section has been adopted from the taxonomy presented by Tannert et al. [7], with small variations to account for the human factor. Evolving socioeconomical conditions require consideration of human factors based on, apart from cognition, the conscience and consciousness-related factors that might play a role in determining human behavior. Apart from this, there was a need to modify the semantics to suit the risk assessment requirements.

The proposed taxonomy as shown in Fig. 9.2 has two major categories of uncertainty, viz., objective and subjective/cognitive uncertainty. The objective uncertainty sub-categorization is in line with the conventional approach in risk assessment, i.e., the epistemic and aleatory components. The lack of knowledge in the data, models, and parameters contributes to epistemic uncertainty, while inherent randomness in the data contributes to aleatory uncertainty. Epistemic uncertainty can be reduced by improving the data, models, and consolidation of parameters; however, aleatory uncertainty cannot be reduced because of the inherent randomness in the data [8].

Figure 9.3 depicts the relationship of epistemic and aleatory uncertainty to overall uncertainty. There are two major observations offered by this illustration:

(a) Initially, due to lack of knowledge and randomness, there is a variability in the epistemic and aleatory components of uncertainty in a given SSC. The first three stages in this figure show that there is uncertainty in the mean, shown by three probability distributions each having its own mean. It can be argued that this can be due to lack of knowledge because as the knowledge base grows the mean keeps changing along with reduction in variability around the mean. However, the fourth probability distribution has one mean. It need not be the case; however, for the purpose of this illustration, an attempt has been made to

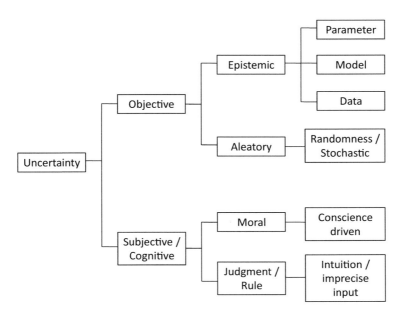

**Fig. 9.2** Taxonomy of uncertainty adopted for IRBE

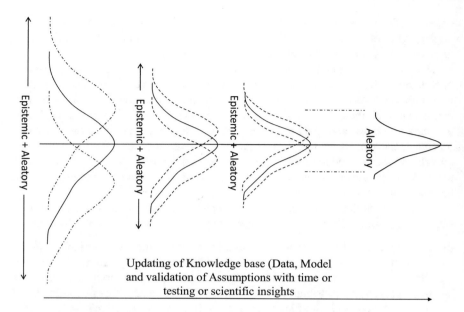

**Fig. 9.3** An illustration of epistemic and aleatory contributions

show that due to randomness the likelihood of variability will be observed around the mean, or there could also be a small shift.

(b) As the knowledge base improves, the epistemic content reduces while the aleatory content still plays its part in the overall uncertainty estimates. Even though it cannot be claimed, practically that knowledge component is perfect; however, for this illustration, consider that the knowledge component is complete; hence, in the last stage, we would see only an aleatory component of uncertainty as shown in the final stage of Fig. 9.3.

(c) Even though the epistemic and aleatory uncertainty has been shown together in Fig. 9.3, however, in practice these two uncertainties should be treated separately. As mentioned, the objective is to reduce the uncertainty; hence, if the uncertainty contribution is due to epistemic content, the improving knowledge base will work to reduce the uncertainty; however, if the aleatory content is predominant, then reduction of the uncertainty is not possible [9].

Up to this point, the proposed taxonomy is more or less similar or in line with risk assessment parlance. The second category (i.e., subjective uncertainty) is included in the IRBE model. The human factor in IRBE involves application of the Cognition, Consciousness, Conscience and Brian ($C^3B$) approach for development of the human model, which is one of the salient features of IRBE human factor modeling. From the subjective considerations, there are two subcomponents, viz., moral or ethical behavior modeling of human aspects. This type of uncertainty can be due to erosion in ethical or moral provisions, such as service oaths/pledges, behavior guidelines, or certain provisions that contribute to unbecoming ethical situations. To reduce this type of uncertainty, a situation should be created for moral or ethical behavior by the institutional provisions and encouraging and creating humans with high conscience that leads to ethical behavior. The second sub-category (i.e., intuition/imprecise uncertainty) can be attributed to fuzzy or imprecise knowledge, such that the additional procedures, rules, and knowledge-base might not reduce uncertainty. In this situation, intuition-guided decisions provide the available options.

It may be noted that subjective uncertainty is not limited to man/machine interface modeling; it might extend to institutional behavior and further to national and cross-national boundaries. Further, it was also observed that the above discussion tends to cross the scientific boundary to model human behavior. Given the fact that modeling of human behavior is complex and often incomprehensible, the modeling approach also should explore the boundary to include aspects such as intuition, ethics, and morality, which tend to be driven by cognition (which has improved scientific acceptance), consciousness, and conscience (which are acceptable at a philosophical level but scientifically need to be further explored experiments dealing with ethical aspects).

The above subject aspects of human behavior are further detailed in Chap. 10 on human reliability in this book.

### 9.4.3   Overview of Approaches for Uncertainty Modeling

If we look at the modeling and analysis methods in statistical distributions, we will note that probability distributions provide one of the important and fundamental mechanisms to characterize uncertainty in data by estimating the upper and lower bounds of the data [10, 11]. Hence, various probability distributions are central to uncertainty characterization. The fuzzy logic approach also provides an important tool to handle uncertainty where the information is imprecise and the probability approach is not adequate to address the issues [12]. Like in many situations the performance data on the system and components is not adequate, and the only input that is available is the opinion of the domain experts. There are many situations where linguistic variables are required as an input. The fuzzy logic approach suits these requirements.

There are many approaches for characterization/modeling of uncertainty. Keeping in mind the nature of the problem being solved, a judicious selection of applicable methods has to be made [9]. The following methods can be considered as possible candidates for uncertainty modeling:

1. Probabilistic approach

   – Frequentist approach,
   – Bayesian approach.

2. Evidence theory—imprecise probability approach

   - Dempster–Shafer theory,
   - Possibility theory—fuzzy approach.

3. Structural reliability approach (application-oriented)

   - First-order reliability method,
   - Stochastic response surface method.

4. Other nonparametric approaches (application-specific)

   - Wilk's method,
   - Bootstrap method.

Each of the above methods has some merits and limitations. The available literature shows that the probabilistic approach is the general practice for uncertainty modeling in PRA uses [13–15]. The application of probabilistic distributions to address aleatory as well as epistemic uncertainty forms the fundamentals of this approach. There are two major basic models—the classical model, which is also referred to as the frequentist model, and the subjective model (e.g., Bayesian approach). The frequentist model tends to characterize uncertainty using probability bounds at component level. This approach has some limitations such as a lack of information on characterization of distribution and non-availability of data and information for tail ends. Hence, it is required to have some way to use the probabilistic approach where issues like lack of data can be addressed. In this situation, the Bayesian approach provides an effective mechanism to update the existing data for uncertainty characterization. The Bayesian approach enables updating of the opinion of the analysts as "prior" knowledge to be integrated with the data or evidence that is available to provide the estimate of the event called a "posteriori" estimate [16]. Even though this approach provides an improved framework for uncertainty characterization in PRA modeling compared to the frequentist approach, there are arguments against this approach. The subjectivity of this approach has become a topic of debate in respect of regulatory decisions.

Hence, there are arguments in favor of methods that use evidence theory, which addresses "imprecise probabilities" to characterize uncertainty [17, 18]. Among the existing approaches for imprecise probability, the one involving "coherent" imprecise probability and which provides upper- and lower-bound reliability estimates has been favored by many researchers [19].

Amongst the other methods listed above, each one has its own merits for specific applications. For example, the response surface method and the first-order reliability method (FORM) are generally used for structural reliability modeling [20]. There are some application-specific requirements, such as problems involving nonparametric tests where it is not possible to assume any particular distribution (as is the case with probabilistic methods). In such cases, the bootstrap nonparametric approach is employed [21]. Even though this method has certain advantages (e.g., it can draw inference even from small samples and estimate standard error), it is computationally intensive and may become prohibitive for complex problems that are encountered in risk-based applications. Other nonparametric methods that find only limited application in risk-based engineering are not included in the scope of this paper. From the above, it could be concluded that the probabilistic methods form the major approaches for uncertainty characterization in the risk-based approach.

At times, the fuzzy logic-based approach is used where the situation involves information comprising imprecise input such as linguistic variables. Often for such cases, the fuzzy logic approach provides an effective mechanism. However, fuzzy logic applications need to be scrutinized for a methodology that is used to design the membership functions because these functions have been found to introduce subjectivity in final estimates. Further, selection and qualification of experts in fuzzy logic or, for that matter, in expert elicitation introduces an added component of uncertainty. Hence, design of experiment, while employing fuzzy logic to generate quantified outputs from imprecise data, plays a crucial role in arriving at most appropriate solutions. For example, calibration of experts to remove an individual expert's bias forms a tricky exercise. For further detail, readers may refer to the available literature on fuzzy logic and expert elicitation approaches for uncertainty analysis.

The scope of this section has not been to discuss all of the approaches listed above, but to present the possibility of using the candidate approach to risk-based approach. For further details, readers are advised to explore the interested topic in the open literature.

## 9.4.4  Uncertainty Propagation

Propagation of uncertainty is central to uncertainty analysis. The literature shows that the Monte Carlo simulation and Latin hypercube approaches form the most appropriate approaches for uncertainty propagation [22]. Even though these approaches are primarily used for probabilistic methods, there are applications where simulations have been performed in evidence theory or application where the priori has been presented as interval estimates [23]. The risk-based models are generally complex in terms of the size of the model, the interconnections of nodes and links, and the interpretation of results. The available literature shows that the Monte Carlo simulation approach is extensively being used in many applications and is considered computationally intensive and approximate. Even with these complexities for risk-based applications, both the Latin hypercube and Monte Carlo simulation approaches have been working well. Even though it is always expected that higher efficiency in uncertainty modeling is required for selected cases, for overall risk-based models these approaches are considered adequate. Generally, Monte Carlo simulation has been found to work well for uncertainty analysis in PRA and applications such as risk-based configuration system in which the uncertainty characterization for core damage frequency has been performed using MonteCarlo simulation [24]. However, efforts to develop a more efficient approach for uncertainty analysis are required because Monte Carlo simulation might become prohibitive or time- and resource-consuming when the model involves analysis employing physics of failure, passive system modeling, and software system modeling.

## 9.5  Decisions Under Uncertainty

At this point, it is important to understand that the uncertainty in engineering systems stems basically from two sources, viz., non-cognitive, generally referred to as quantitative uncertainty and subjective or cognitive, generally referred to as qualitative uncertainty [25]. Most of this chapter has so far dealt with the non-cognitive part of uncertainty, i.e., uncertainty due to inherent randomness (aleatory) and lack of knowledge (epistemic). Unless we address the sources of uncertainty due to cognitive/subjective uncertainty in the decision environment, the topic of uncertainty has not been fully addressed. The cognitive/subjective uncertainty is caused by an inadequate definition of parameters, such as structural performance, safety culture, deterioration/degradation in system functions, level of skill/knowledge base, and staff experience (design, construction, operation, and regulation) [26] or by moral and ethical aspects. Dealing with uncertainty using statistical modeling or any other evidence-based approach (including approaches that deal with precise or imprecise probabilities) cannot address issues involving vagueness of the problem arising from missing information and lack of intellectual abstraction of real-time scenarios, be it regulatory decisions, design-related issues, or operational issues. Traditional probabilistic and evidence-based methods mostly deal with subjectivities, perceptions, and assumptions that may not form part of the real-time scenarios and that do not require cognitive part of the uncertainty to be addressed. The following subsections discuss the various aspects of the cognitive/ subjective part of uncertainty and the methods to address these issues. The following sections do not discuss the ethical or moral components of uncertainty because we assume that in the decision environment being presented this aspect is not a subject to be explored. Hence, the majority of the discussion follows subjectivity/cognition components that affect the decision process.

### 9.5.1  Engineering Design and Analysis

The issues related to "uncertainty" are part of engineering design and analysis. The traditional working stress design (WSD)/allowable stress design (ASD) in civil engineering deals with uncertainty by defining "suitable factors." The same factors of safety are used for mechanical design to estimate the allowable stress (AS = yield stress/Safety Factor (SF)). This SF accounts for variation in material properties, quality-related issues, degradation during the design life, modeling issues, variation in life cycle loads, and lack of knowledge about the system being designed. The SF is essentially based on past experience but does not guarantee safety. This approach is also highly conservative.

It is expected that an effective design approach should facilitate a trade-off between maximizing safety and minimizing cost. Probabilistic or reliability-based design allows this optimization in an efficient manner. The design problems require treatment of both cognitive and non-cognitive sources of uncertainty. It should be recognized that the designer's personal preferences or subjective choices can be the source of uncertainties, which bring in a cognitive aspect of uncertainty. Statistical aspects such as variability in assessment of loads, variation in material properties, and extreme loading cycles are the source of non-cognitive uncertainties.

In the probabilistic-based design approach, considerations of uncertainty when modeled as the stress/strength relation for reliability-based design form an integral part of the design methodology. The load and resistance factor design (LFRD), first-order reliability methods (FORMs), and second-order reliability methods (SORMs) are some of the applications of probabilistic approach to structural design and analysis. Many civil engineering codes are based on probabilistic considerations. The available literature shows that design and analysis using a probabilistic-based structural reliability approach has matured into an "engineering discipline" [20] and new advances and research have further strengthened this area [26].

The Level 1 PRA models are often used in support of design evaluation. During the design stage, complete information and data are often not available. This leads to a higher level of uncertainty in estimates. On the other hand, the traditional approach using deterministic design methodology involves use of relatively higher safety factors to compensate for the lack of knowledge. The strength of the PRA framework is that it provides a systematic framework that allows capturing of uncertainties in the data and the model due to missing or fuzzy inputs. Be it probabilistic- or evidence-based tools and methods, it provides an improved framework for treatment of uncertainty. Another advantage of the PRA framework is that it allows propagation of uncertainty from the component level to the system level and further up to the plant level in terms of confidence bounds for system unavailability/initiating event frequency and core damage frequency, respectively.

## 9.5.2 Management of Operational Emergencies

If we take lessons from the history of nuclear accidents in general and the three major accidents, viz, Three Mile Island (TMI) in 1979, Chernobyl in 1986, and the recent one Fukushima in 2011, it is clear that real-time scenarios always require some emergency aids that respond to the actual plant parameters in a given "time window." One of the insights from Chernobyl as presented by Abotte et al. [27] is that the real challenge is due to uncertainty associated with information from the governmental sources and other management issues along with the risk posed by the accident itself. Further, in his research on "Managerial decision making under risk and uncertainty" Riabacke [28] discussed the risk attitude of plant managers. The study concludes that managers often make decisions based on intuition and gut feelings because probability estimates are based on inadequate information and imprecise data with large uncertainty. The unavailability of the computer-based operator support systems/tools and methods is also one of the contributing factors in a scenario involving risk and uncertainty.

Hence, dealing with uncertainty in decision making in design, operation, and regulatory setup in general and during emergency conditions is vital and requires potential uncertainties to be addressed either by deterministic provisions a priori or credible evaluation using probabilistic approaches. In this context, a probabilistic risk analysis framework may facilitate prediction and prioritization of these scenarios, but that only addresses the modeling part of the safety analysis. It is also required to consider the qualitative or cognitive uncertainty aspects and their characteristics for operational emergency scenarios.

The major characteristics of operational emergencies include:

- Deviation of plant condition from normal operations that require safety actions, e.g., plant shutdown, actuation of shutdown cooling.
- Overwhelming the control room staff with announcement of plant parameters which includes process parameters crossing their preset bounds, parameter trends, and indications, which might challenge the cognition capability of the operating staff;
- Constraint posed by the available time window for grasping the situation and correcting it;
- Plant response/feedback in terms of plant parameters regarding the improved/ deteriorated situations;
- Decisions regarding restoration of systems and equipment status if the situation is moving toward normalcy;
- Decision regarding declaration of emergency, which requires a good understanding of whether the situation requires declaration of plant, site, or off-site emergency;
- Interpretation of available safety margins in terms of the time window that can be used for designing the emergency operator aids;
- Communication within the plant with various agencies directly involved in routine operation, the plant authorities, and other agencies to deal with the deviation or emergency conditions.

The literature shows that responding to accidents/off-normal situations as characterized above calls for modeling that should have the following attributes:

- Modeling of the anticipated transients and accident conditions in advance such that the knowledge-based part is captured in terms of rules/heuristics as far as possible;
- Adequate provisions to detect and alert plant staff in advance of threats to safety functions;

- Unambiguous and automatic plant symptoms based on well-defined criteria such as plant safety limits and emergency procedures that guide the operators to take the needed action to arrest further degradation in plant condition;
- Consideration of the limits of plant parameters, assessment of actual time window that is available for applicable scenarios;
- The system for dealing with emergencies should take into account the plant-specific attributes, distribution of manpower, laid down line of communications, and, other than the standard provisions, the tools, methods, and procedures that can be applied for planned and long-term or extreme situations;
- Heuristics on system failure criteria using available safety margins;

Obviously, the ball is out of the "Uncertainty modeling" domain and requires the scenarios to be addressed from other side, i.e., making decisions such that actions as part of a real-time scenario compensate for the missing knowledge base and bring the plant to a safe state.

The answer to the above situation is development of knowledge-based systems that not only capture the available knowledge base but also provide advice to maintain plant safety under uncertain situations by maintaining plant safety functions. Note that we are not envisaging any role for a "risk-monitor" type of system. We are visualizing an operator support system that can fulfill the following requirements (the list is not exhaustive and only presents a few major requirements):

- The system should detect plant deviation based on plant symptoms.
- The system should exhibit intelligent behavior, such as reasoning, learning from the new patterns, conflict resolution capability, pattern recognition capability, and parallel processing of input and information.
- The system should be able to predict the time window that is available for the safety actions.
- The system should take into account operator training and optimize the graphical user interface (GUI)
- The system should be effective in assessment of plant transients—parallel processing of plant symptoms is required to present the correct plant deviation.
- The system should have adequate provision to deal with uncertainty and incomplete data.
- The results of the reasoning with confidence limits should be presented.
- The system should have an efficient diagnostics capability for capturing the basic cause(s) of the failures/transients.
- The advice should be presented with an adequate line of explanations.
- The system should be interactive and use graphics to present the results.
- The systems should provide provisions for presentation of results at various levels, such as abstract level advice (e.g., open MV-3001, start P-2) or advice with reasonable details (e.g., open ECCS Valve MV-3001 located in reactor basement area, start injection pump P-2, it can be started from control room L panel).

Even though there are many examples of R&D efforts to develop intelligent operator advisory systems for plant emergencies, readers may refer to paper by Varde et al. for further details [29]. Here the probabilistic safety assessment framework is used for knowledge representation. The fault tree models of PRA are used for generating the diagnostics, while the event tree models are used to generate procedure synthesis for evolving emergencies. Intelligent tools such as the artificial neural network approach are used for identification of transients, while the knowledge-based approach is used for performing diagnostics.

As can be seen above, the uncertainty scenarios can be modeled by capturing the lessons learnt from the past records for anticipated events. For even the rare events where uncertainty could be of higher levels, the symptom-based models, which focus on maintaining the plant safety functions, can be used to model the plant's knowledge base.

## 9.5.3  Regulatory Reviews

The available literature on decisions under uncertainty has often focused on the regulatory aspects [30]. One of the major differences between operational scenarios and regulatory reviews or risk-informed/risk-based decisions is that there generally is no preset/specified time window for decisions that directly affect plant safety. The second difference is that the regulatory setting requires collective decisions and a deliberative process, unlike operational emergencies wherein the decisions are often made by individuals or among a limited set of plant management staff where the available time window and resources are often the constraints. Expert elicitation and treatment of the same often forms part of the risk-informed decisions. Here, the major question is "What is the upper limit of the spread of confidence bounds that can be tolerated in the decision process?" In short, "How much uncertainty in the estimates can be absorbed in the decision process?" It may be noted that the decisions problem should be evaluated using an integrated approach where, apart from probabilistic variables, even deterministic variables should be subjected to uncertainty analysis. One major aspect of risk assessment from the uncertainty point of view is updating of the plant-specific estimates with generic prior data available either in the literature or from other plants. This updating introduces subjectivity to the posteriori estimates. Therefore, it is required to justify and document the prior inputs. The Bayesian method coupled with Monte Carlo simulation is the conventional approach for uncertainty analysis. The regulatory reviews often deal with inputs in the form of linguistic variables or "perceptions," which requires a perception-based theory of probabilistic reasoning with imprecise probabilities [17]. In such scenarios, the classical probabilistic approach alone does not work. The literature shows that application of fuzzy logic offers an effective tool to address qualitative and imprecise inputs [31].

The assumptions often form part of any risk assessment models. These assumptions should be validated by performing the sensitivity analysis. Here, apart from independent parameter assessment, sensitivity analysis should also be carried out for a set of variables. The formation of a set of variables requires a systematic study of the case under consideration.

The USNRC document NUREG-1855, on "Guidance on the treatment of uncertainties associated with PRAs in risk-informed decision making" deals with the subject in detail, and readers are recommended to refer to this document for details [24].

## 9.6  Codes, Guides, and References on Uncertainty Characterization

The scope and objectives of risk-based applications determine the major element of Level 1 PRA. However, for the purpose of this discussion let us consider development of base Level 1 PRA for regulatory review as the all-encompassing study. The scope of this study includes full-scope PRA, which means considerations of (a) internal events (including loss of off-site power and interfacing loss of coolant accident, internal floods, internal fire); (b) external events, such as seismic, events, external impacts, and flooding; (c) full power and shutdown PRA; and (d) reactor core as the source of radioactivity (fuel storage pool not included) [5].

The point to be remembered here is that uncertainty characterization should be performed keeping in view the nature of the applications [6, 29]. For example, if the application deals with the estimation of a surveillance test interval, then the focus will start right from the uncertainty including the initiating event that demands automatic action of a particular safety system, unavailability for safety significant component, human actions, deterministic parameters that determine failure/success criteria, and assumptions that determine the boundary conditions for the analysis.

An important reference that deals with uncertainty modeling is United States Nuclear Regulatory Commission (USNRC) document NUREG-1856 (USNRC, 2009) which provides guidance on the treatment of uncertainties in PRA as part of risk-informed decisions [29]. Although the scope of this document is limited to light-water reactors, the guidelines can be adopted with little modification for uncertainty modeling in either Canadian deuterium reactor (CANDUs), pressurized heavy water reactors (PHWRs), or any other nuclear plant. In fact, even though this document provides guidelines on risk-informed decisions, requirements related to risk-based applications can be modeled giving due considerations to the conservative approach used to arrive at the risk metrics used in the risk-based approach. Significant contribution of the ASME/ANS framework includes incorporation of the "state-of-knowledge correlation" [6].

Concerning the standardization of risk assessment procedures and dealing with uncertainty issues the PRA community finds itself in a relatively comfortable position. There is a consensus at the international level as to which uncertainty aspects need to be addressed to realize certain quality criteria in PRA applications. The major references that cover this aspect are: (a) American Society of Mechanical Engineers (ASME)/American Nuclear Society (ANS) Standard on PRA Applications [6], (b) International Atomic Energy Agency-Technical Document (IAEA-TECDOC-1120) on Quality Attributes of PRA applications [32], and (c) various Nuclear Energy Agency (NEA) documents on PRA [4]. Any PRA applications to qualify as "quality PRA" should conform to these quality attributes as laid out for various elements of PRA. For example, the ASME/ANS code provides a very structured framework, wherein there are higher-level attributes for an element of PRA, and then their specific attributes support the higher-level attributes, and so forth. These attributes formulate a program in the form of checklists that need to be fulfilled in terms of required attributes to achieve conformance quality level for PRAs.

There are some quality attributes that can be considered in support of uncertainty analysis. Following are some examples from ASME/ANS in respect of the PRA element, initiating event (IE) modeling. Examples of some lower-level specific attributes from ASME/ANS include: ASME/ANS attribute IE-C4: "When combining evidence from generic and plant-specific data, USE a Bayesian update process or equivalent statistical process. JUSTIFY the selection of any informative prior distribution used on the basis of industry experience." Similarly, ASME/ANS attribute IE-C3: "CALCULATE the initiating event frequency accounting for relevant generic and plant-specific data unless it is justified that there are adequate plant-specific data to characterize the parameter value and its uncertainty."

The USNRC guide as mentioned above summarizes in details the uncertainty-related supporting requirements of ASME/ANS documents systematically. These documents may be referred to for details. The availability of this ASME/ANS standard, the NEA documents, and IAEA-TECDOC is one of the important milestones for risk-based/risk-informed applications because these documents provide an important tool toward standardization of and harmonization of the risk assessment process in general and capturing of important uncertainty assessment aspects that impact the results and insights of risk assessment.

## 9.7  Summary and Conclusions

Uncertainty characterization is vital in analysis and interpretation of safety factors—a characteristic component of deterministic analysis. Uncertainty characterization helps deal with over conservatism in the defense-in-depth approach. The objective is to arrive at realistic safety margins such that effective solutions are arrived in design, operations, and regulation—one of the major objectives of the IRBE framework. This chapter provides an overview of uncertainty analysis while discussing the requirements of the risk-based approach in respect of uncertainty characterization. The state of the art in uncertainty analysis, particularly in respect of the implementation of IRBE, has been discussed in brief without going into the details of each technique and method. Even though it was suggested that Monte Carlo simulation and Latin hypercube approaches in uncertainty modeling and analysis have been working well for PRA analysis, the work to develop efficient algorithms and methods is emphasized. The available literature shows that there is an increasing trend towards the use of risk assessment or PRA insights in support of decisions.

There is a general suggestion that uncertainty evaluation must be strengthened for realizing risk-based applications. It is expected that the IRBE approach will provide the required framework to implement decisions in support of design, operations, and regulations. It is also argued that apart from probabilistic methods, evidence-based approaches and fuzzy logic-based approaches should be further developed to deal with "imprecise variables or probabilities," which often form important input for the risk-based decisions.

There are subjective or cognitive issues that need to be addressed by incorporating management tools for handling real-time situations. This is true for both operational and regulatory applications. The taxonomy adopted from the available literature in this chapter is rather over encompassing to include not only the cognition but ethical or conscience components for modeling uncertainty as part of human factors development. It can be argued that the PRA community needs to explore this aspect, particularly in the twenty-first century, to reflect the social and economic conditions.

Finally, this chapter drives the point that both quantitative as well as quantitative aspects need to be addressed to realize more holistic solutions. Further research is needed to deal with imprecise probability, whereas conscience/ethical and cognitive aspects form the cornerstone of uncertainty evaluation.

# References

1. J. Chapman, et al., Challenges in using a probabilistic safety assessment in risk-informed process. Reliab. Eng. Syst. Saf. **63**, 251–255 (1999)
2. R.L. Winkler, Uncertainty in probabilistic risk assessment. Reliab. Eng. Syst. Saf. **54**, 127–132 (1996)
3. A. Daneshkhah, Uncertainty in probabilistic risk assessment: a review
4. Nuclear Energy Agency, *Use and Development of Probabilistic Safety Assessment, NEA/CSNI/R(2007)12* (Committee on Safety of Nuclear Installations, 2007)
5. IAEA, *Development and Application of Level 1 Probabilistic Safety Assessment for Nuclear Power Plants, IAEA Safety Standards—Specific Safety Guide No. SSG-3* (IAEA, Vienna, 2010)

6. American Society of Mechanical Engineers/American Nuclear Society, *Standards for Level 1 Large Early Release Frequency in Probabilistic Risk Assessment for Nuclear Power Plant Applications, ASME/ANS RA-Sa-2009* (2009)
7. C. Tannert, H. Elvers, B. Jandrig, *The Ethics of Uncertainty, EMBO Reports*, vol. 8, no. 10 (2007)
8. G.W. Parry, The characterization of uncertainty in probabilistic risk assessments of complex systems. Reliab. Eng. Syst. Saf. **54**, 119–126 (1996)
9. H. Kushwaha (ed.), *Uncertainty Modeling and Analysis* (Bhabha Atomic Research Centre, 2009)
10. M. Modarres, *Risk Analysis in Engineering—Techniques, Tools and Trends* (CRC-Taylor & Francis Publication, 2006)
11. M. Modarres, M. Kaminskiy, V. Krivtsov, *Reliability Engineering and Risk Analysis—A Practical Guide* (CRC Press Taylor & Francis Group, 2010)
12. K.B. Mishra, G.G. Weber, Use of fuzzy set theory for level 1 studies in probabilistic risk assessment. Fuzzy Sets Syst. **32**(2), 139–160 (1990)
13. J. Weisman, Uncertainty and risk in nuclear power plant. Nucl. Eng. Des. **21**, 396–405 (1972)
14. M. Pate-Cornell, Probability and uncertainty in nuclear safety decisions. Nucl. Eng. Des. **93**, 319–327 (1986)
15. T. Nilsen, T. Aven, Models and model uncertainty in the context of risk analysis. Reliab. Eng. Syst. Saf. **79**, 309–317 (2003)
16. N. Siu, D.L. Kelly, Bayesian parameter estimation in probabilistic risk assessment. Reliab. Eng. Syst. Saf. **62**, 89–116 (1998)
17. L. Zadeh, Towards a perception-based theory of probabilistic reasoning with imprecise peobabilities. J. Stat. Plann. Inf. **105**, 233–264 (2002)
18. F. Caselton, W. Luo, Decision making with imprecise probabilities: Dempster-Shafer theory and application. AGU. Water Resour. Res. **28**(12), 3071–3083 (1992)
19. I. Kozine, Y. Filimonov, Imprecise reliabilities: experience and advances. Reliab. Eng. Syst. Saf. **67**(1), 75–83 (2000)
20. R. Ranganathan, *Structural Reliability Analysis and Design* (Jaico Publishing House, 1999)
21. A. Davison, et al., *Bootstrap Methods and Their Applications* (Cambridge University Press, 1997)
22. Atomic Energy Regulatory Board, *Probabilistic Safety Assessment for Nuclear Power Plants and Research Reactors, AERB Draft Manual, AERB/NF/SM/O-1(R-1)* (AERB, 2005)
23. K. Weichselberger, The theory of interval-probability as a unifying concept for uncertainty. Int. J. Approximate Reasoning **24**, 149–170 (2000)
24. M. Agarwal, P. Varde, Risk-informed asset management approach for nuclear plants, in *21st International Conference Structural Mechanics in Reactor Technology (SMiRT-21), India* (2011)
25. I. Assakkof, *Modeling for Uncertainty: ENCE-627 Decision Analysis for Engineering, Making Hard Decisions* (Department of Civil Engineering, University of Maryland, USA)
26. A. Haldar, S. Mahadevan, *Probability, Reliability and Statistical Methods in Engineering Design* (Wiley, New York, 2000)
27. P. Abbott, C. Wallace, M. Beck, Chernobyl: living with risk and uncertainty. Health Risk Soc. **8**(2), 105–121 (2006)
28. A. Riabacke, Managerial decision making under risk and uncertainty. IAENG Int. J. Comput. Sci. **32**(4) (2006)
29. P. Varde, et al., An integrated approach for development of operator support system for research reactor operations and fault diagnosis. Reliab. Eng. Syst. Saf. **56** (1996)

30. M. Drouin, G. Parry, J. Lehner, G. Martinez-Guridi, J. LaChance, T. Wheeler, *Guidance on the Treatment of Uncertainties Associated with PRAs in Risk-informed Decision Making (Main Report), NUREG-1855*, vol. 1 (Office of Nuclear Regulatory Research Office of Nuclear Reactor Regulation, USNRC, USA, 2009)
31. I. Karimi, Hullermeier, Risk assessment system of natural hazards: a new approach based on fuzzy probability. Fuzzy Sets Syst. **158**, 987–999 (2007)
32. International Atomic Energy Agency, *Quality Attributes for PRA Applications, IAEA-TECDOC-1120* (IAEA, Vienna)

# Chapter 10
# Human Reliability

*...The Earthquake and Tsunami of March 11, 2011 were
natural disasters of a magnitude that shocked the entire world.
Although triggered by these cataclysmic events, the subsequent
accident at Fukushima Daiichi Nuclear Power Plant cannot be
regarded as a natural disaster. It was profoundly a manmade
disaster—that could and should have been foreseen and
prevented. And its effect could have been mitigated by a more
effective human response. ...What must be admitted—very
painfully—is that this was a disaster "made in Japan". Its
fundamental causes are to be found in the ingrained
conventions of Japanese Culture: our reflexive obedience: our
reluctance to question authority: our devotion to 'sticking' with
the program; poor grouping and our insularity. Had other
Japanese been there in the shoes of those who bear
responsibility for the accident, the results may well have been
same....
Evaluating operational problems, the commission concludes
that there were organizational problems within TEPCO. Had
there been a higher level of knowledge, training and equipment
inspection related to severe accidents, and had there been
specific instructions given to the on-site workers concerning the
state of emergency within necessary time frame, a more effective
accident response would have been possible ....*

Kiyoshi Kurokawa
Chairman, Fukushima Nuclear Accident Independent
Investigation Commission

The National Diet of Japan [1]

## 10.1 Introduction

During a three-day program at an *Adhyatmic* or spiritual center in India, I observed
that all the activities of the center were well organized and managed. The schedules
were met, the services were timely, and the *sewaks*, or volunteers, not only served
the participants meticulously but, more importantly, happily. The whole ambiance
was filled with positive energy, and nobody seemed tired or depressed. I was

© Springer Nature Singapore Pte Ltd. 2018
P. V. Varde and M. G. Pecht, *Risk-Based Engineering*, Springer Series in Reliability
Engineering, https://doi.org/10.1007/978-981-13-0090-5_10

wondering, what is so unique about this organization that it can easily surpass the standard of any professionally managed organization. Nevertheless, we underwent the three-day program which included yoga sessions, meditation, spiritual discourses, and visit to places in this center.

The final program was marked by a spiritual discourse by the *head acharya*, the head teacher. We assembled in the central hall and were surprised when she began by saying, You all must be wondering since your entry into this center until today as to "What keeps this place not only so well organized and managed but a happy surrounding?" and "Why do these people want to serve you so happily and welcome you with intimacy often more than your family?". She made it a point to mention that there is no code or guide or "do and don't do" prescription for any of the volunteers or even for the participants. She further said, "You must be wondering what is the secret of this center that everything surpasses the overall ambiences and working culture of a professionally managed organization and where the happiness quotient is an aspect left to individual's observation, but certainly could not be called as one of the happiest place on the earth—while this aspect can easily be attributed to this center". Then, she further asserted that the secret is "awakened consciousness and conscience" which makes this place not only happy but emits the positive vibes that connect everyone with positive energy.

From the above discussion, we can certainly agree that cognition, consciousness, along with conscience, work at the level of the human brain and this provides a fundamental requirement as also inspiration and motivation to develop a human model in support of human factor development and human reliability evaluation.

The traditional approach to human factor modeling to a large extent deals with safety, whereas international and national organizations are working extensively to develop technical and management tools and methods to address these situations involving modeling for design basis threat from inside as well as outside elements, particularly human elements. However, in the last 5–10 years, we find that apart from safety, the security aspects involving humans and organizations are posing challenges. This observation motivates us to think about a parameter that can capture the "ethical" quotient in a human model. Hence, this chapter also proposes, apart from cognition and consciousness, and the brain, a new parameter called "human conscience" or simply "conscience".

Most of the existing approaches and techniques for human reliability analysis represent the human model through "cognitive" aspects of the task requirements, and it can be argued that none of the approaches for human reliability modeling have a robust human model—a fundamental requirement for human reliability analysis. The data collected from plants, expert opinion, and simulator experiments might provide a manageable condition, but this condition is not sufficient for human reliability modeling. Without a human model, uncertainty in estimates will be derived only from data, which itself may not be adequate.

In the engineering context, when we talk about human performance, we need to consider the human model, human performance modeling, human resource management, and decision-making. This chapter presents an Indian system of organizational/system management that can provide an effective framework for management. Human factors and plant/system management can be improved by addressing the root causes of human and institutional failures through analysis of consciousness, cognition, and conscience working at a common ground, i.e., human brain. The history of accidents in engineering systems shows that the human factor is one of the major contributory factors to accidents; e.g., 70–80% of accidents are attributed to human error [2]. Even in the nuclear industry, the history of accidents shows that the human factor is one of the contributors to accident sequences.

This chapter proposes a new approach to human reliability modeling by incorporating cognition, consciousness, and conscience at the brain level and modeling human behavior in general and human reliability in particular.

In the current human reliability approaches, the time window available and time it takes for actually performing a given task along with other empirical constants is considered for evaluating the human reliability. However, the current literature shows that cognitive and psychological components of the human stress can be measured. For example, there are physiological techniques like saliva sample analysis and conventional techniques like blood pressure, pulse rate, body langue which can provide a direct approach to stress measurement. Apart from this, the electroencephalography (EEG) technique is also being investigated to evaluate cognitive and conscience phenomenon associated with human. In respect of psychological component of stress, the checklist procedure is used to evaluate the psychological stress and this approach can also be effectively employed for stress assessment. Hence, the proposed approach explores direct measurement of stress at physiological level referred as intrinsic component, along with psychological component through the existing approaches in human reliability.

This approach is referred to as the $C^3B$-based or the CQB-based approach for human reliability modeling and is designed to meet the requirements of complex engineering systems. This chapter presents the CQB approach in the context of the nuclear plant environment; however, its fundamentals and frameworks can also be adopted for other safety-critical systems.

## 10.2 A Brief Overview of Human Reliability Techniques

The human reliability approaches or methodologies, as discussed in the available open literature, are divided into three general categories—first-generation techniques, second-generation techniques, and the present, i.e., third-generation, techniques. A Health and Safety Report, UK, reports that over 72 reliability modeling techniques have been developed. Of these techniques, 17 were considered for detailed review by applying the following criteria: the origin of the technique, a description of the tool, the validation level, the application domain, availability of the tool, advantages and limitations, and the relationship of the given approach to other methods [3]. Readers interested in a detailed review of human reliability can refer to the Health and Safety Report or other documents available on the subject. This chapter reviews only a few of the techniques. The considerations for selection of these techniques in this chapter are: (a) the human reliability methodology or technique should be available in open literature with adequate input for critical review, (b) the methodology should have application in the nuclear domains, or it should have relevance to nuclear environment, (c) either the technique provides a complete technology for human reliability or it provides specific considerations such as cognition aspects in second-generation methodology, and (d) the technique provides input for development of the proposed CQB-based approach to human reliability.

The literature shows that the first human reliability study was carried out for weapon systems by Sandia National Labs in 1952 [4]. The first version of United States Nuclear Regulatory Commission (USNRC)-sponsored technique for human error rate prediction (THERP) was presented in 1963 [5], while a later version was published in the early 1980s [6]. The first comprehensive PRA study entitled "Reactor Safety Study," popularly known as WASH-1400, developed by Rasmussen and his team, was published in 1975. In this work, a human reliability model was developed for application as part of probabilistic risk assessment for pressurized water reactors (PWRs) and boiling water reactors (BWRs) [7]. The Accident Sequence Evaluation Program (ASEP) was developed to evaluate the impact of human error on risk, i.e., core damage frequency of the plant [8]. The first-generation techniques WASH-1400, THERP human cognitive reliability (HCR) [9], success likelihood index [10], accident sequence evaluation program (ASEP) [8], human error assessment and reduction technique (HEART) [11] and time reliability correlation (TRC) [12], and systematic human action reliability procedure (SHARP) [9] along with other techniques were developed in the early to mid-1980s. Most of these techniques were based on expert elicitation due to the scarcity of human error data. Techniques that were developed in the early to late 1990s were categorized as second-generation techniques. In these techniques, apart from cognitive aspects, there is an attempt on modeling dynamic aspects of human reliability. The cognitive reliability evaluation and assessment methodology (CREAM) [13] and technique for human error analysis (ATHENA), which employ cognitive as well as dynamic aspects of human reliability [14] and success plant risk analysis-human reliability (SPAR-H) [15] can be considered as second-generation methods. The most recent technique, i.e., Nuclear Action Reliability Assessment (NARA) published in 2005 [16], is referred to as a third-generation human reliability assessment (HRA) method, which is based on HRA data collected from plant experience and is a modification of the Human Error Assessment and Reduction Technique (HEART), a first-generation technique.

In fact, there are many publications that present a review of either human error or its application to incident investigations. Among these reviews, one notable publication is the USNRC document on review findings for human error contribution to risk in operating events in identified nuclear plants in the USA [17]. The study indicates that human error contributed significantly to risk in nearly all events analyzed, and latent errors were present in every event and more predominant than active errors. This study concluded that design changes and work practice errors were present in 81% of the events and maintenance-related errors were present in 76% of the events analyzed.

For further details, readers can refer to these review reports available online. The literature shows that there are extensive reviews on human reliability techniques [18]. Hence, this section reviews the capability of the nine most discussed and used techniques. The broad criteria considered for this review are as follows: the basic model, pre- and post-initiators, data and quantification, performance-shaping factors (PSFs), dependency, uncertainty, context and general remarks on applicability, use, and major strengths and limitations. Table 10.1 presents the capabilities and applications of these techniques.

The three, major nuclear power plant accidents, viz. Three Mile Island in 1976, Chernobyl in 1986, and Fukushima in 2011, demonstrated that human error along with institutional failures contributed significantly to these accidents [19]. Further, the above review and the insights from the published literature indicate that the present techniques of human reliability analysis are not adequate to model human error effectively. The insights from review of PRA studies underline the fact that human error contribution to risk is significant, particularly when human error induces common cause failures that result in failure of redundant safety provisions and recovery actions [20].

The above discussion suggests two-pronged approaches/options: (a) to reduce human intervention in nuclear systems and (b) to address the issue of human reliability in a more holistic manner than the existing approaches. The first option is being addressed by the nuclear industry, which is incorporating safety systems such that human intervention, particularly in accident conditions, is not anticipated during the initial period ranging from hours to days. However, this aspect is not within the scope of this chapter. This option is futuristic for advanced reactor systems, with inherent and passive safety features such that there will not be a threat to safety even if operator intervention is not there for extended duration from the initiation of the accident. Because a majority of the more than 400 nuclear power plants operating worldwide are first-or second-generation systems, we need to realize the second option. The second option requires that we heed the lessons from the accidents and incident report reviews to identify the weaknesses in the existing modeling human reliability techniques. The areas for further research must be determined such that the human reliability analysis integrates with error reduction techniques toward demonstrating the impact of human error in risk assessment. These human error reduction systems should effectively address real-time scenarios —for example, application of an intelligent operator system for emergency conditions in a control room environment.

**Table 10.1** Overview of major features of selected human reliability techniques

| S. No. | Technique (Year) | Specific features | Present use/applications/gap areas |
|---|---|---|---|
| 1. | THERP (USNRC), 1983 [10.42] | Models preinitiators as well as post-initiators and PSFs. Uses an HRA event tree for representing the procedural part. The model is quantified using nominal HEPs derived from judgment and experience-based data mostly from 1960s non-nuclear plants. Diagnosis and execution model the scenario. Time reliability correlation is used to model diagnosis. Consideration of uncertainty employing error factors. Seven PSFs subdivided into three categories—external, internal, and stress. Added features include five levels of a dependency model | Most comprehensive technique targeted for nuclear power plants. However, the data and uncertainty need to be considered while giving final results. This technique is very resource-consuming and often requires a human reliability expert for exploiting its potential |
| 2. | WASH-1400, 1975 [10.43] | The first comprehensive PRA study, where the human model is comprised of skill-, rule-, and knowledge-based actions | Developed as part of WASH-1400 study. Provides the human model that is popular among the risk assessment community |
| 3. | ASEP, 1987 [10.48] | Simplification of THERP model that proposes conservative values of nominal probability, considers recovery probability, provides improved guidance on preinitiators, lookup tables, and curves employed for quantification | Easy to use and targeted for PRA application having a simplified dependency model |
| 4. | SLIM, 1983 [10.53] | Success-likelihood model that employs expert judgment for relative importance of weights and PSFs to arrive at success index for a given task. The calibration values form a crucial aspect of this approach. No model for dependency; however, the approach includes a model for estimation of epistemic uncertainty | Even though the methodology focuses on post-initiators, it can be applied to any event including the post-initiators |

(continued)

**Table 10.1**  (continued)

| S. No. | Technique (Year) | Specific features | Present use/applications/gap areas |
|---|---|---|---|
| 5. | HCR, 1984 [10.49] | In this technique, the time available for completion of the task mainly considers diagnostic aspects. There is more clarity required for action part of the task. Data are obtained mainly from simulator experiments and expert judgments. Major steps are: (a) task classification, (b) estimation of nominal value of median time $T_{1/2}$, (c) estimation of median time by applying PSFs, (d) assessment of time window available for the task, and (e) estimation of non-response probability | Heavy dependence on the simulator data in this approach leaves questions on applicability to plant-specific evaluation of human reliability. There is no method to correlate the simulation data and plant-specific data |
| 6. | CREAM, 1998 [10.47] | Developed for performance prediction and accident analysis. The concept of cognition is implemented through four basis control modes, viz. scramble, opportunistic, tactical, and strategic control. The complete methodology can be summarized in four major steps: (a) identification of the work, (b) identification of the context, (c) evaluation, and (d) recommendations for reducing error producing conditions | The basic version is used for screening analysis, whereas the extended version is intended for human reliability prediction |
| 7. | SHARP, 1984 [10.51] | Mainly focuses on methodological or procedural aspects while discussing the dependency levels in detail. Even though this methodology discusses quantification approaches, it does not provide the data needed for quantification | SHARP is considered as an HRA guide by the analysts. An analyst is required to choose among the available options of HR requirements |

(continued)

**Table 10.1**  (continued)

| S. No. | Technique (Year) | Specific features | Present use/applications/gap areas |
|--------|------------------|-------------------|-------------------------------------|
| 8. | SPAR-H, 2005 [10.28] | Employs a psychological model of human performance. The diagnosis (nominal HEP:0.01) and action parts (nominal HEP:0.001) are evaluated by applying 8 PSFs to give human error probability for the task. Treats dependency in line with the THERP approach. The organizational factors are covered in the work process | The approach has been rated as simple and less resource-consuming; however, when realistic or more accurate prediction of human reliability is required this approach may not be the ideal option |
| 9. | ATHENA, 1996 [10.45] | The human model is derived from behavior science taxonomy. However, the major source of data is based on expert opinion and judgment. Training of plant operator and staff is part of the expert elicitation. This second-generation approach is characterized by context-driven mechanisms and has provisions for assessment of aleatory uncertainty estimation | The available literature shows that there is very limited application of this technique |

*Legend* THERP: technique for human error rate prediction; ASEP: accident sequence error prediction; HCR: human cognitive reliability; TRC: time reliability correlation; CREAM: cognitive reliability; SHARP: systematic human action reliability procedure; SLIM: success-likelihood index method; ATHENA: a technique for human error analysis; CREAM: cognitive reliability evaluation and assessment methodology; SPAR-H: standardized plant analysis risk-HRA

## 10.3  Motivation for the CQB-Based Human Reliability Approach

This section discusses why a new reliability analysis approach is needed, what is new in the proposed approach to human reliability (here onwards will be referred to as the CQB approach), and how this approach works.

### 10.3.1  Why We Need a New Approach

The history of accidents and the recent ones in particular show that human and organizational factors are crucial to improving the safety of engineering plants in general and nuclear plants in particular. Talk about relatively recent paradigms such as "safety culture" is not adequate to effectively address the issues associated with human factor improvement. There is a need is to go beyond a procedural or slogan-based approach to develop a system that works at the fundamental levels of human performance. The following are some of the major factors that motivate development of a new paradigm for human reliability modeling.

- The human model is central to human reliability and in this context a comprehensive and integrated and robust human model is required.
- Human failure is a major contributor to accidents, and human reliability, arguably, is the most ignored and uncertain area in safety research.
- There is no consensus among the practitioners of human reliability regarding the applicability of available techniques.
- Most of the models and methods in the existing approaches were developed mostly 1980s–1990s, and majority of them were based on expert elicitation. The models and methods need to be revisited in the context of available plant-specific and simulator data.
- Although the philosophy of safety culture was developed for real-time implementation, it appears to have not made the required impact and remains more of a slogan than a real-time tool.
- Most of the existing approaches have focused on ensuring safety and reliability, whereas existing scenarios require a special focus on the security aspects of plants.
- In the recent past, there has been an increasing interest among highly qualified practitioners to perform R&D related to human reliability and to provide new insights that can be adopted for developing a new approach.

The proposed approach developed in this chapter considers an improved and robust human model to develop a human reliability approach referred to as CQB approach.

## 10.3.2  What Is New in the CQB Approach

The reasons why this approach has improved potential to evaluate human factors are as follows:

- A comprehensive CQB-based human model, that extends cognition to include consciousness (like attention, awareness, learning) and conscience (e.g., ethics).
- This approach proposes a human reliability model where stress values obtained from direct measurements of physiological parameter and other psychological stress parameters interpreted from psychological data.
- The CQB concept operates from at and within the human level and extends to the whole organization and beyond up to national level and further higher levels and has potential to operate at a universal level.
- The CQB concept facilitates extension of the "safety culture" philosophy to development of "conscious-safety culture" factors that can be used in human factor modeling.

- The consciousness concept in CQB is more effective to model institutional aspects of human failure because consciousness does not remain limited to an individual, like cognition, but connects the individuals at higher levels in an integrated manner.
- The CQB concept provides a mechanism for connecting cognition to action, For example, in the cognition approach one of the stimuli is "seeing," whereas in the conscious-based approach the stimuli become "conscious-seeing", i.e., grasping power associated with the act of seeing. The same argument also applies to other stimuli.
- Human error modeling deals with identifying the intended actions and what goes wrong. The cognition remains limited to assessing the capability of human faculties for the correct or intended actions, whereas the CQB model connects motivation and ethics to perform the correct or intended actions.
- In the present scenario, as security is becoming important for organizational and individual safety, the conscience concept provides the suitable framework for improving human as well as organizational security factors.
- While the cognition capability provides a framework for logical reasoning, the human consciousness and conscience complements the decision-making by providing a basis holistic view of the subject by considering associated aspects, like tangible or intangible, direct or indirect, short-term and long-term consequences and ethical quotient associated with various options, in decision-making.
- The human factor has been found to be one of the major inputs in probabilistic risk assessment; hence, a conscious consideration for employing stress as direct parameter for emergency scenario modeling provides an effective approach to incorporate human factor at system level as well as at accident sequence level.

## 10.3.3    Proposed Human Model for CQB

The literature review shows that most of the approaches discussed in Sect. 10.2 remain limited up to cognitive aspects derived from the branch of psychology. However, it is well-accepted fact that physiological aspects along with cognition psychological phenomenology facilitate direct measurement of stress, a vital component for modeling human reliability. In the proposed human model, as shown in Fig. 10.1, the consciousness, cognition, and conscience phenomenon work in human brain. In this model, consciousness has been considered to work at not only human brain level but extends and connects with the outer environment. This consideration is relevant as this effectively explains the context of connection and communication among humans at plant, organization, and national level, for example, collective safety consciousness.

Hence, the human model in CQB depicts human interaction with the outside world through external and internal objects through stimulus (e.g., seeing, smelling, touching, listening, sensing, and cognitive functions like diagnosis, analysis). In

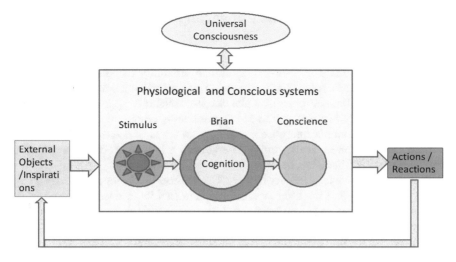

**Fig. 10.1**   Basic human model in CQB

this model, human consciousness (life force or awareness) communicates with the outside world on perpetual basis and provides basis for phenomenon beyond awareness and explains communication with the surroundings. The cognitive activities are performed in brain with sensation through nervous system of the body. Finally, the conscience element in CQB presents conditioning of human personality and provides the basis for assessing integrity, ethical quotient, and other attitude and personality aspects associated with not at human and at organizational level. This conscience element enables modeling of one more aspect, i.e., security aspects, that has not been addressed in other human reliability techniques.

## 10.3.4   How the CQB-Based Human Reliability Approach Works

The CQB-driven philosophy shapes the present system of training and qualification, revisits the safety culture skeleton and institutional aspects that impact safety, and integrates the human factor into the plant risk model. Once the risk areas are prioritized through risk studies, the CQB-driven approach provides the broad guidelines on the factors that have the potential to improve human performance, such as quality attributes of operational procedures and actions, cognitive stresses, the role of communication, and man–machine interfaces. In fact, the CQB model can effectively be used to develop an operator support system where intelligent tools such as artificial neural network and knowledge-based systems can be used to improve the human factor during normal operations as well as during emergency conditions [21].

## 10.4    Basic Philosophy and Supporting Input for the CQB Model

Even the traditional and state-of-the-art human reliability models available in the literature do not have a robust human model. This one missing component, which is so fundamental to human reliability, makes these reliability approaches stand on a weak ground. Many factors, such as training, learning, communication, lapse, error of omission, and commission, pose a challenge in terms of associating root causes to these factors. Hence, in the proposed approach a human model is proposed to provide a basis for human reliability.

There are four basic elements in the CQB approach, viz. consciousness, cognition, conscience, and brain. How are these elements interpreted and used in the CQB approach to model human reliability? First let us look at the definition of these elements in terms of the functional requirements.

### 10.4.1    Consciousness

The dictionary definition of consciousness is: "(a) The state of being conscious—aware of one's own feeling; what is happening around one; and (b) the totality of one's thoughts, feelings and impressions; consciousness mind" [22]. The attribute "attention" is normally used interchangeably, informally, or even unintentionally in place of "consciousness". This broad general attribute is associated with "selection" or even "identification" of a subject or activity for further conscious evaluation. For the purpose of CQB modeling, "attention" will be discussed as part of conscious activity.

The CQB approach assumes that consciousness is related to the life (awareness quotient) for cognition; hence, consideration of consciousness is a prerequisite for modeling cognition. Furthermore, consciousness does not remain limited to living individuals but is a part of existence and nature. However, there could be a debate whether consciousness is a subset of cognition or cognition is a subset of consciousness. Therefore, apart from being a personal attribute consciousness is communication that connects individuals in organizations and between nations. Given this background, cognition is a subset of consciousness. This means that consciousness is the basis for cognition. In the developed approach, consciousness has two major connotations, here referred to as type I and type II consciousness.

#### 10.4.1.1    Psychological: Type I Consciousness

The attributes associated with type I consciousness are the level of the awakened state, awareness, attention based on sensing and determined by response quotient to the environment, emotions through external physical stimuli (e.g., sound, vision), and internal processes such as feelings or thoughts. This type of consciousness is referred to as a "lower state of consciousness" and is related to the level of functioning of various faculties (or of the physical and mental body, or sense bases). This type of consciousness has its basis in science and has been accepted as a scientific phenomenon over the 10 years. Even though the essential/salient features of physiological consciousness will be touched upon in this chapter, for further details, readers are advised to refer to the available literature. One of the major references is the handbook by Bernard and Gage entitled Cognition, Brain and Consciousness—Introduction to Cognitive Neuroscience [23]. In fact, the cognitive stimuli will be conditioned with consciousness quotient and will be formulated as conscious-seeing, conscious testing, conscious-attention (or concentration), conscious feeling, and conscious hearing. For the purpose of the proposed approach, type I consciousness is defined as "ability of an individual to receive the sensory input and respond to the surrounding environment in the interest of safety."

### 10.4.1.2   Spiritual: Type II Consciousness

It is more appropriate to discuss type II consciousness by an evidence-based approach and the recent insights available from the experiments that have been carried out. Realization that one's awareness, existence, thoughts, impressions, awareness, and communication are a part of the universe and connection with other individuals and systems fall in the category of type II consciousness. This type of consciousness can be termed a "higher state of consciousness," and it has philosophical/spiritual basis [24, 25]. However, it is also related to behavioral, cultural, or attitudinal evidence at the individual, plant, and national level and it can be seen in the universal phenomena, for example, the findings of the National Diet of Japan in regard to the Fukushima accident. The National Diet of Japan relates this accident to the cultural/safety-conscious aspects of plant and regulatory body. The report emphasized that the culture of not questioning and prevalent regulatory and utility protocol made the accident a man-made event [1]. Therefore, in the way we accept the theory related to electrons in the atomic model (no one has ever seen the electrons), wherein we validate the presence of electrons by looking at the evidence, type II consciousness can also be validated by the cultural and behavioral evidence that affects human performance.

The fundamental considerations, directly relevant to human reliability, in the consciousness model are as follows [24]:

- The six internal sense bases are the eyes, ears, nose, tongue, body, and mind.
- The external sense bases are visible forms, sounds, odors, flavors, touch, and mental objects.
- Sense-specific consciousness is dependent on an internal and external sense bases.
- Contact is the meeting of an internal sense base, external sense base, and consciousness.

- Feeling is dependent on contact.
- Craving/aversion is dependent on feeling.

In the CQB approach, it is assumed that type I consciousness and type II consciousness together provide a basis for cognitive functioning and further physical or internal actions in humans. This also includes aspects related to human conscience.

## 10.4.2   Cognition

The Webster Dictionary definition of cognition is "(a) The process of *knowing* in the broadest sense, including perception, memory, judgment and (b) The result of such a process, perception, *conception*, etc." [22]. The capability of having good perception, memory, and judgment can form the basic parameter to characterize cognition capability for human model. It is possible to evaluate these parameters qualitatively and even quantitatively in simulator experiments or through qualification program.

In fact, the initial assessment can be obtained from the performance of individuals during training and simulator experiment data. The striking observation here is that human cognition is characterized by modeling human performance in training and qualification (in line with hardware reliability where qualification and testing form an integral part of component qualification procedure) and by verifying these insights during operation (again in line with hardware performance evaluation procedure). As can be seen, there are two ways cognitive aspects can be modeled, either by collecting data on human performance behavior for various contextual conditions or by monitoring and evaluating bodily or mental responses employing techniques such as EEG and ECG, and mapping this to data to external and/or internal phenomena. What is the difference between the CQB approach and other approaches? In traditional approaches, the tasks are characterized and categorized as skill, rule, or knowledge-based, whereas in the CQB approach it is assumed that for a given task the response is based on an individual or a crew depending on their cognitive capabilities. The aim of the CQB approach is to evaluate human factors or human error probability, which is related to an individual's stability to varying stress levels. The traditional approaches lack the ability to evaluate responses considering individual attributes or implicitly assume that all the individuals will respond to a simulation alike. If we take a leaf from hardware system modeling, we know that there is an attribute "quality" associated with each component, and upkeep of the components determines the probability of success and failure.

Hence, we can infer the definition of cognition reliability as "the probability that an individual has the capability for correct, perception, inference, and recall for a given application/scenario and time window".

How to assess these constituent factors, (a) capability for correct perception, (b) capability for correct inference and capability to recall correctly certain procedural steps or particular information? Simulators, training, and checklist procedures provide an effective mechanism for these attributes of cognition. The fact that human error is one of the dominant factors itself suggests that our qualification and training program should be based on certain criteria and guidelines such that human behavior, particularly during postulated emergency conditions, can be evaluated.

### 10.4.3  Conscience

The Webster Dictionary meaning of conscience is "(a) (conscientia , consciousness), ethical principles, moral sense, to know, knowledge within; and (b) a knowledge or sense of right and wrong, with an urge to do right; moral judgment, that opposes the violation of previously recognized ethical principles and that leads to feelings of guilt, if one violates such a principle" [22]. IAEA in its nuclear energy series NG-T-1.2 has published guidelines/manuals for establishing a code of ethics for staff at all the levels in nuclear operating organizations [26]. Further, it has been emphasized that values and ethics of the individuals and organizational units play an important role in safety and security. It has been emphasized that the organizations that have a code of ethics benefit in terms of improved bottom line. This is contrary to the general feeling that a code of ethics might affect the organization adversely. The document provides guidelines for the development and implementation of code of ethics. For the purpose of the CQB technique, the conscience aspects have been incorporated in the model as the conscience-quotient for individuals. Accordingly, the conscience-quotient is defined as the "probability that an individual or an organization follow ethical practices that are in the interest of public at large." Details related to modeling of the conscience-quotient are a matter of subject and problem on hand; hence, it is out of the scope of this chapter.

### 10.4.4  Brain and Brain Waves

Molecules are the fundamental entities that make up genes, and genes are the fundamental entities that make up the brain. The physiological faculties of the brain are composed of trillions of neurons and associated links that enable stimuli and memory to perform the given function and give rise to perception, feeling, emotions, and motor actions. For the purposes of the CQB approach, the brain is the central processing unit, in the sense that all the cognitive processes, consciousness phenomena, and conscience come into being only at the level of the brain. Figure 10.2 shows the major faculties that play a key role in encoding information in memory and the various areas responsible for sensing vision, speech, and sound.

These are directly responsible for human performance/activities in an engineering setup, be it design, fabrication, operation or regulation, particularly in support of collecting input from the subject and decision-making. The six major brain functions are creative visualization, memory and learning, executive planning, language and

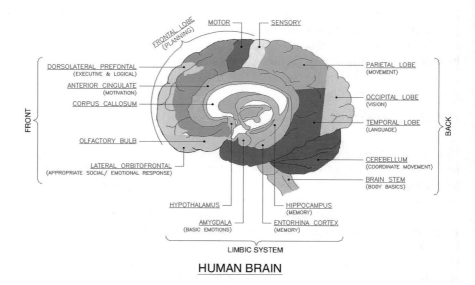

**Fig. 10.2** Simple schematic showing a human brain and constituent faculties and their functions

math (i.e., logical reasoning, emotional response, and social interaction [27]. Further, the brain's memory function is comprised of three major memory systems, each having a role for life services in general and relevant for consolidating the human model. *Working memory* is a short-term memory system, such as online holding of information (equivalent to cache memory in a compute). This memory system is predominantly active for knowledge-based job; however, it is always active irrespective of skill-, knowledge-, or rule-based actions. *Procedural memory* is used for skill-based and rule-based jobs and has a small component of knowledge-based activity. *Declarative memory* deals with conscious recollection of facts and events, also known as explicit memory. Declarative memory differs from procedural memory as the procedural memory facilitates repetitive tasks that involves skill, e.g., movement of objects and body parts that are intuitive and embedded such that tasks can be performed without being much conscious about the function being performed. Procedural memory facilitates the routine and repetitive jobs, like routine inspection and sampling in a lot containing number of components or taking down readings from the control panel on hourly basis. However, inter-pretation of any change in trends or significant observation requires knowledge- or rule-based actions (Table 10.2).

The neurons are the basic building blocks of the brain, and neurotransmitters are responsible for communications between neurons. Brain waves are produced by synchronized electrical pulses generated when neurons communicate with each other—the fundamental phenomenon associated with thoughts, emotions, and behaviors [28]. The advances in understanding of brain waves have revealed new avenues to study the implication of brain processes. Scientific experiments confirm

**Table 10.2** Brain waves, associated activities, and relevance to the CQB approach [28]

| Brain wave | Frequency (Hz) | Description | CQB relevance/remarks |
|---|---|---|---|
| Infra-low | <0.5 | These waves are responsible for timing and connective network functions between neurons in the brain | These brain waves are subtle and pose a challenge for detection and monitoring. They are not considered for the CQB approach |
| δ (Delta) | 0.5–3 | Associated with the deepest meditation and dreamless sleep stages. These brain waves are slow and loud and are associated with healing and regeneration. Delta stage requires isolation from the external world | Since these brain waves are generated during meditation states that stimulate healing and regeneration, for CQB this state could be considered as the ideal and healthiest state of mind |
| θ (Theta) | 3–8 | Sleep and deep meditation are the two stages when these brain waves have been recorded. It is a state when all the senses are withdrawn from the external world | Deep restorative sleep is a prerequisite to performance of human functions, such as learning, memory, imagination, and perception, which are some of the major components that characterize training and the performance quotient in operational scenarios |
| α (Alpha) | 8–12 | Characterizes calmness of mind and a feeling of comfort and living in the present | Higher alertness, mental coordination, mind/body integration, and learning are the attributes that are associated with this brain wave and are a required condition in the operational scenario |
| β (Beta) there are three bands as follows | 12–38 | Normal waking consciousness is characterized by this range of frequencies | Directly relevant to cognitive functions and associated with alertness and attention; involved is routine to complex problem solving in a typical operational environment |
| β-1 | 12–15 | Idle and restful state | Cognitive load is minimal/lowest—normal plant operation—routine tasks, normal heuristics are adequate to cater to task requirements, without many resource constraints. The knowledge-based quotient is low or minimal, the |

(continued)

**Table 10.2**  (continued)

| Brain wave | Frequency (Hz) | Description | CQB relevance/remarks |
|---|---|---|---|
| | | | procedural task quotient is dominant, while the skill-based quotient is lower than rule-based tasks |
| β-2 | 15–22 | Conscious tasks require focus and attention, often in a team environment | Decision-making in typical operational scenario. Sequential processing diagnostics. Review and deliberation in regulatory environment. Knowledge-based quotient is more than beta-1 tasks, such as shutdown jobs management, coordination factor associated with rule-based task is predominant with skill-based quotient higher then beta-1 category |
| β-3 | 22–38 | Conscious level that takes cognitive loading when resources as constraints (or stressful situations) often aggravated by new insights or inadequate input | Forced reactor trip or plant shutdown or transient conditions require a higher quotient of the knowledge-based approach and very little of the skill-based type of activity |
| γ (Gamma) | 38–42 | Fastest brain waves, often employing multiple faculties for parallel processing of information | Reactor transient having potential to enter into reactor threat or accident conditions. These types of situation pose a high dependence on adequacy of parallel processing of information involving multiple faculties of the brain. Like LOCA situations where the operator is ensuring ECCS function trying to confirm the location and quantum of threat perception |

the evidence of ranges of frequency associated with (a) infra-low waves <0.5 Hz, (b) δ (delta) waves 0.5–3 Hz, (c) θ (theta) waves (3–8 Hz), (d) α (alpha) waves (8–12 Hz), (e) β (beta) waves (12–38 Hz), and (f) γ (gamma) waves (38–42 Hz). The frequency of brain waves changes depending on how and which activities the brain is engaged in [28] (Fig. 10.3).

**Fig. 10.3** Brain waves and consciousness, showing the dominance of gamma waves [29]

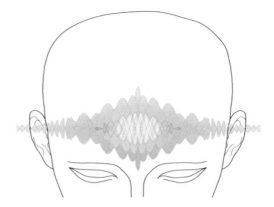

In CQB approach, the model of brain as discussed above has been employed to model cognition (knowledge and perception) and consciousness (like awareness, communication). The stress–strength model is central to the proposed approach. Here, the stress has usual meaning, i.e., the context which affects human stress level and to a large extent it is measured using inputs like saliva testing, blood pressure, heart rate that affects a person's cognitive and motor capabilities. However, cognition does not work in isolation and depends on the consciousness levels of human. Hence, in CQB approach the human brain is central to model consciousness and cognition.

## 10.4.5   Inter-relationship of Basic CQB Elements

For the purposes of the CQB approach, it is assumed that the brain is central to the cognitive, consciousness, and conscience process and the major aspects of the developed model in this chapter. The available literature shows that these three phenomena particularly are not independent; rather, they overlap and are interdependent. For example, the stimulus operates at the cognitive level; however, unless seeing becomes "conscious-seeing" or reading becomes "conscious-reading," the process may be of more use in cognitive modeling. Here, both cognition and consciousness are operating at brain level. Conscience is dictated by conditioning of the mind–brain, while the information is processed at the cognitive as well consciousness level. Genes assisted by conditioning can be argued to play a role in determining the conscience. Hence, it can be concluded that some functions are independent, while other functions are interdependent, and their inter-relationship is shown in Fig. 10.4.

The reference human model discussed in this section provides the necessary base for developing the CQB framework, discussed in the following section.

**Fig. 10.4** Representation of
the functional
interrelationship among three
Cs and Brain in CQB

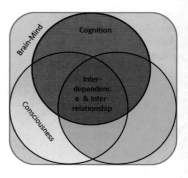

## 10.5   CQB-Based Framework

### 10.5.1   *Fundamental Tenets of CQB Approach*

The fundamental tenets of the proposed integrate framework are:

- Human reliability is a subject not limited to man–machine interaction but includes national-level administrative arrangement and policy percolating down to regulatory and design and operating institutions, and further down to man–machine interactions, as shown in Fig. 10.5.
- Human reliability should use a suitable "human model" that facilitates considerations of functioning of biological brain, cognition, consciousness, and conscience aspects in an integrated manner.

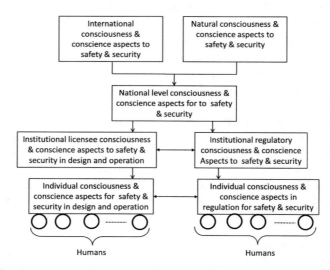

**Fig. 10.5** Over-encompassing levels and role of consciousness and conscience to safety and security in the CQB approach

- The human model for human reliability prediction is primarily based on stress–strength model where stress is considered as measurable parameter, viz. physiological as well as psychological parameters are considered as measurable parameters, directly or indirectly.
- Training and education enable simplification of complexity involved in development of human model.
- The scope of human reliability should extend beyond error rate prediction to include the ways and means of reducing human errors, not merely as recommendations but also in terms of technology applications.
- The proposed framework should facilitate integration of new insights in the forms of data, factors, and models from R&D or plant-specific sources such that uncertainty in human factors can be reduced on perpetual basis.

In many traditional approaches, the scope of human reliability is seen remains limited to operation and maintenance aspects of the plant. However, it is also known that the institutional-level aspects of further higher national and international level, as shown in Fig. 10.5, including the government policies influence human factors and might affect motivation, design, recruitment, training, safety culture, and, of course, the operational and training aspects of human reliability. Hence, in CQB approach apart from plant-level man–machine interaction, there are higher-level considerations that have significant bearing on human reliability during operations.

Level 1    Even though the national-level policy and safety-conscious culture plays a vital role, the input available from international developments provides insights on safety aspects that becomes a critical part of national program.
Level 2    National considerations that have the potential to affect human reliability, e.g., availability of codes and guides that reflect human reliability considerations, attributes that have bearing on plant safety-conscious culture, human resource policies, and the regulatory framework and its interaction with operators.
Level 3    Institutional aspects that include recruitment policies, qualification procedures, plant design, operational and maintenance safety-conscious culture.
Level 4    Plant-level man–machine interaction during operation, maintenance, services that relates to direct actions/non-actions that impact the plant directly

Many of the human factors discussed in the open literature deal with operational environment at Level 4, i.e., man–machine direct plant interactions, and remain limited to either the control room scenario the plant areas outside of the control room. Here, a broad categorization of human error events, specific to plant operational environment been made depending on the types of human errors/actions and time of occurrence, are as follows [30]:

Type 1   *Preinitiators*: Human error committed during normal operations (that remain dormant) and before an initiating event, e.g., human errors associated with maintenance, like mis-calibrations, misalignment, blocking of a safety system channel.

Type 2   *Initiators*: Human error that causes an initiating event. The action of plant staff that results in an initiating event can be categorized as error of commission, for example, maintenance errors that cause the main supply feeders to trip and result in Class IV power failure.

Type 3   *Post-initiators-A (Procedural)*: Error while executing the identified/ established recovery procedure, as part of response in a given time window after the initiating event, has occurred, e.g., probability to start a standby system/equipment in a given time as part of an identified procedure. It can either be error of omission or error of commission or slip.

Type 4   *Post-initiators-B*: Human error in recovering faulty component due to error of judgment, slip error of commission, or error of omission, after an initiating event and further aggravating plant condition.

Type 5   *Post-initiators-C*: Human interaction that recovers a failed standby system or equipment either by following the recovery procedure or any improvised action, for example, during an off-site power failure scenario, manual starting of an on-site diesel generator that failed to start on demand.

As can be seen, the above categorization is formulated for risk assessment studies and seeks to model human behavior for various plant scenarios including emergency conditions. Even though the procedural steps discussed below have been tailored to model all three levels and five types of human actions, the scope of this section is modeling in support of probabilistic risk assessment studies only.

The salient feature of the CQB approach is that it treats the complete human reliability to apply consciousness-based management principles that go beyond the man/material management objectives to include consciousness aspects related to each of the activities of the project [31]. The terms "attention," "communication," and "seeing" in the CQB approach become "conscious-attention", "conscious-communication," and "conscious-seeing," respectively. This conscious attribute enhances the quality of the output while keeping the atmosphere of the project environment harmonious. The aim is also to embody the attributes of conscience such that the credibility of the work can be elevated.

Characterization of national policies and framework are a prerequisite to the CQB methodology. These aspects are discussed as follows:

National-level policies or acts that govern the design, operation, and regulatory framework for the facility might have a bearing on human performance and safety of the subject plant. For example, NUREG-1478 provides the basic framework for a training and qualification program [32]. A broad characterization of social systems and behavioral attitude is required as this may have a bearing on the human factor development. This step deals with familiarization and development of common minimum attributes that can be used as normalizing parameters in human reliability studies. These parameters will relate to general consciousness and conscience of the society. For example, one of the major observations from the Fukushima accident was that even if different sets of people were in-charge of the plant, the consequences of the accident could not have been avoided. This points toward a general conscience and consciousness than can be attributed to the national level. How these factors can be incorporated in the human factor studies is not within the scope of this section. The point being made here is that to have more credible results of from the human reliability study this factor cannot be ignored. Further, the insurance coverage for the plant and the employees as a matter of policy provides an added element of security and works at a consciousness and conscience level of the site. This aspect can also be addressed at a qualitative or quantitative level as the scope of a given study. To address this aspect requires motivation for a very high level of national consciousness-based safety culture and conscience such that there are transparency at all levels.

Further, the hierarchy presented in Fig. 10.4 relates to various levels of conscience and conscience factors that need consideration as a prerequisite to implementation of the CQB methodology.

The interorganizational- and institutional-level safety framework, particularly the safety culture involving the utilities and regulatory body, influences the safety of not only the man–machine interface but overall human behavior, which determines plant safety. In fact, the safety consciousness at the institutional-level percolates down to the working staff in the regulatory body as well as the utility. Direct indicators of this culture are the number of and effectiveness of regulatory oversights, pending safety recommendations, provision of multi-tier reviews, and regulatory inspections. Other important indicators are staff qualification criteria, adequacy of staff and its deployment, effectiveness of the qualification and training program, and the adequacy and effectiveness of plant operating, emergency operating, and maintenance procedures.

The conscious quotient of the organization is reflected in the quality of the training and qualification program, staff motivation, availability of the revised procedures, periodic review of the safety documents (e.g., periodic safety review reports (PSARs), technical specifications, and emergency operating procedures), and direct involvement of higher management in resolving the plant issues.

The national and institutional characterization should form part of the human reliability evaluation. It has been a general observation that the safety culture changes from plant to plant and nation to nation. In case any weakness is revealed that has potential for latent procedural or regulatory faults, it can be argued that gap areas at the higher level can have higher penalties than at the man–machine level of human interactions.

Even though it might not be possible to quantify these gap areas, however, understanding the weakness and working toward overcoming them or having a defense against them is more important than quantification of these indicators.

A conscious approach to safety and security is driven by input from the international environment to national policy that drives the organizational conscious and conscience levels and further down to design, operational, and regulation through the prevailing conscious-safety culture approach as shown in Fig. 10.5. To reinforce this point, the Chairman of the Fukushima Accident Investigation Committee stated

> What must be admitted – very painfully – is that this was a disaster 'made in Japan'. Its fundamental causes are to be found in the ingrained conventions of Japanese culture…there were organizational problems within TEPCO. Had there been a higher level of knowledge, training and equipment inspection related to severe accident, …*a more effective accident response would have been possible…* [1].

This is because unlike cognition, which remains limited to individuals, human consciousness and conscience run through the nation to organizations and further down to individuals. Here the consciousness connects humans. Hence, the broad framework should include all the elements from a global level because international awareness of human reliability aspects plays an important role in framing national policies and these policies in turn affect administrative policy and the individual consciousness.

## 10.5.2  The Integrated Human Model in CQB

Traditionally it has been accepted that it is a challenge to model human behavior particularly during testing or emergency conditions. At the same time, it has been witnessed that an over-simplified model of human derived from evidence- or symptom-based approach forms a major source uncertainty. One of the basic requirements that should satisfy a reference human model is that human behavior motivation is governed not only by interactions with the outside world, but also with internal processes/factors. This particular aspect makes prediction of human behavior more challenging. Hence, let us have a just model with appropriate assumptions and boundary conditions, even if calls for further R&D efforts such that the estimates of human reliability cater to the safety requirements. For example, we can consider a reference human that is characterized by training and qualification attributes for a given job. Development of a system where this reference human model modified with factors associated with jobs of performing knowledge-based actions like analysis which requires, say reasoning, learning, capability, etc., provides the robust approach for predicting human reliability factors associated with individual staff in the plant. These attributes specific to individuals can be derived from simulator experiments as also from in-plant performance review of individual.

Before we look at the major factors/assumptions of the human model, let us see how Buddhist philosophy defines consciousness.

> It asserts Why we call it consciousness? ..because it cognizes thus it is called consciousness .. What does it cognize? It cognizes what is sour what is bitter, sweet, pungent, …Because it cognizes it is called cognition [24].

Here we are discussing tongue-consciousness. If we talk about mind-consciousness, then we can talk about awareness. When this information is processed by various faculties in the brain, it leads to development of feelings, perception, and knowledge about the present. This knowledge is weighted or processed at a conscience level—about right or wrong, employing ethical conditioning of the mind–brain. Once a decision is made at the conscience level, knowledge or perception leads to mental or physical action based on the conscience processing. This process leads to conditioning or formation or *sanskar* at the conscious level. The formation is stored in the brain whether it is declarative information or long-term perception. This processing is fundamental to learning and enhancing the experience.

This model has been developed so that it provides a basis/mechanism for understanding aspects vital to modeling human behavior in a complex environment (e.g., in a nuclear plant operational scenario), namely learning, perception, recognition, identification, recall, logical processing, multitasking, motivation, mental and physical stress, and ethical behavior. Figure 10.6 presents the human model considered in the CQB framework.

The major considerations/assumptions of the proposed human model are as follows:

(a) Consciousness forms the basis of the human model because human actions, whether physical or cognitive, are realized through required level of consciousness.
(b) Internal sense bases communicate with external sense objects and generate stimuli; e.g., seeing is generated when eyes focus on an external object. Depending on the conscious level, a perception is developed.

**Fig. 10.6** Integrated human model in the CQB approach

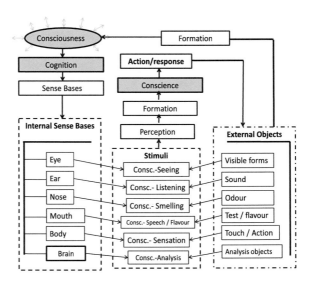

(c) The mechanism is defined by six sextets as discussed in Sect. 10.4.1.2 as part of type II consciousness.
(d) Based on the six sense bases, a number of mental factors arise, including the six types of consciousness, viz. (a) eye-consciousness, (b) ear-consciousness, (c) nose-consciousness, (d) tongue-consciousness, (e) body-consciousness, and (f) mind-consciousness.
(e) Consciousness is one of the five classically defined experimental aggregates. The other four aggregates are the material form, sensation, perception, and volitional formations or fabrications.

### 10.5.3   Stress–Strength Model

The proposed model for human reliability is in-line stress–strength model in structural reliability. Hans Selye in 1936 borrowed the term "stress" and performed experiment to determine stress response or strain in human. The fundamental considerations which have in favor of this model are (a) the available state of the art in human stress characterization suggests that both physiological and psychological stresses can be measured [33]; (b) the traditional models in human reliability consider the available time for action to model stress; however, the fact is that time is an indirect indicator of stress; and (c) available time for action is just one of the many indicators, like context, individual's capability, or tolerance to stressful situations.

Figure 10.7 shows the stress–strength interference model. In the context of human reliability in this model stress can be interpreted as stress experienced by the individual and the strength can be characterized by design provision in the plant to cope with the scenario under consideration and here it means strength of plant provision to reduce psychological stress on the human operator or crew.

The other modification which the proposed model may be considered to have achieved is that stress in this model is considered as specific phenomenon; for example for a given emergency/abnormal situation, there can be varying responses, from "no effect' facilitation," i.e., small amount of stress improves performance, varying degree of degraded response to chocking or outright panic [34]. In traditional approach, there is one curve that determines non-response probability such that even the uncertainty bound may not be able to capture the actual uncertainty that remains latent till actual situation occur. The proposed model attempts to highlight this sensitivity to human reliability modeling.

In this definition, failure may occur when human stress is more than the strength. The mathematical expression of human reliability in the context of stress and strength can be formulated for human reliability as follows:

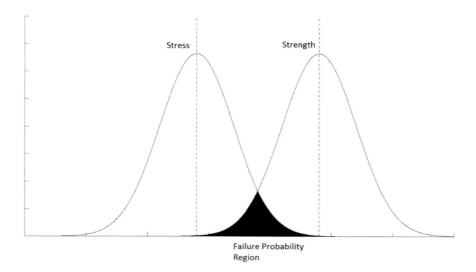

**Fig. 10.7**  Stress–strength representation to model human reliability

$$R = \Pr(S_D > S_A) \tag{10.1}$$

$$R = \Pr\left(\frac{S_D}{S_A} > 1\right)$$

$R$    Human reliability as a function of $S_A$ actual human stress;
$S_A$   Human physiological stress;
$S_D$   Plant design safety strength for potential psychological stress response to a situation or

Context will be referred as design strength or just "strength."

Human non-performance or failure probability, $P$, can be given as

$$P = 1 - R \tag{10.2}$$

This concept has been discussed in detail in the following Sect. 10.5.5, considering the reference non-performance probability $P_0$, i.e., failure probability during normal condition and impact of other factors.

The premise of the proposed model is "stress can be measured" either directly through physiological parameters or for a given context, indirectly through psychological assessments employing checklist procedure. Centre for Studies on Human Stress in its report published in 2007 explained the mechanism how body responds to stress and the end product cortisol and catecholamine can be measured to assess physiological stress by analyzing blood, urine, and saliva sample collected from human experienced stressful situations. The design safety strength, i.e., the 'the parameter $S$' for strength, on the other hand can be measured by following a psychological response checklist, e.g., the design of systems to address base the knowledge-, rule-, and skill-based tasks. The expert elicitation technique has been used to infer the potential amount of psychological robustness or strength of the plant against a given emergency scenario.

The above formulation is based on the design and operational approaches where adequate factor of safety provision exists keeping in view the potential design basis as well as design extended or even beyond design basis accidents.

It is important to note that the above model integrates consciousness, cognition, and conscience. The first term in the model accounts for conscience of human. The intrinsic component of this model contributes to model cognition, in terms of human as direct measurable parameter "$s$" in the model. The psychological aspects derived from the robustness of the context or the plant systems contribute to "$S$" the strength. For arriving at value of strength, i.e., inherent capability of the plant that reduces psychological stress on the human, expert elicitation approach is employed where input from simulator experiments along with psychological checklists are used to derive the value of plant strength. The unit of both $s$ and $S$ are same, like for stress of saliva only difference is human stress is measured using direct experiment data like quantity of cortisol or catecholamine measured in RIA (radioimmunoassay), while for strength this parameter is determined based on expert judgment, i.e., for a given context what will be the value of, say, cortisol.

## 10.5.4   Human Performance Influencing Factors (HPIS) in CQB

Keeping in view the unique and integrated features of CQB, it is required that human performance influence factors are developed such that we have a structured approach for human reliability analysis. As mentioned earlier, the major feature of this approach has been to develop a human model that accounts for basic blocks, viz. consciousness, cognition and conscience and brain. Apart from human model, the other factors which are vital that determine human performance are task characteristics, system, and environment. These factors affect the human performance. The taxonomy proposed here is that, for each subelement of the task can be evaluated in respect of assessing the stress generation potential and availability of safety provision or human element that provides input for strength considerations.

The term consciousness in the CQB approach has two connotations, viz. physiological that has a considering of awareness levels of human, and second one is in terms of connecting people in the organization at communication level. The second aspect can also be related to consciousness at national, organizational and plant as also at regulatory level. As mentioned in previous section, the EEG can determine the brain wave generated during awareness and activity level. The consideration of conscience is exclusively from security considerations. It may be noted that the list of HPIF is not exhaustive and is indicative only. Table 10.3 presents the HPIF considered appropriate for CQB approach.

**Table 10.3** Human reliability influencing functions (HRIF) in CQB approach

| Main element | Group element | Subelement (parameter and unit) | Measured as part of **stress** characterization | Design/expectation for **strength** considerations |
|---|---|---|---|---|
| **Human** | *Consciousness* | Consciousness—Type I | | |
| | | Sleep quality levels—brain wave range δ and θ (Hz) | Values assessed/ measured simulator environment | Corresponding to δ and θ range |
| | | Calmness of mind during normal condition—brain waves α range (Hz) | Values assessed/ measured simulator environment | Corresponding to α range |
| | | Normal waking consciousness— brain wave for complete β range (Hz) | Values assessed/ measured simulator environment | Corresponding to β range |
| | | Normal restful state (e.g., normal steady state reactor operation) —brain waves β-1 range (Hz) | Values assessed/ measured simulator environment | Corresponding to β-1 range |
| | | Normal procedural task —β-2 range (Hz) | Values assessed/ measured simulator environment | Corresponding to β-2 range |
| | | Stressful situations β-3 range (Hz) | Values assessed/ measured simulator environment | Corresponding to β-3 range |
| | | Stressful situation employing coordination of multiple brain and faculties γ range | Values assessed/ measured simulator environment | Corresponding to β-3 range |
| | | Awareness quotient | Evaluated in simulator training/ requalification program | Expected awareness quotient level |
| | | Psychological health quotient | Evaluated in training/ requalification program | Expected reference psychological attributes |

(continued)

**Table 10.3** (continued)

| Main element | Group element | Subelement (parameter and unit) | Measured as part of **stress** characterization | Design/expectation for **strength** considerations |
|---|---|---|---|---|
| | | Sleep quality factors | Assessment during training/ requalification program | Free of sleep disorders, like sleep apnea |
| | | Concentration | Assessment of concentration levels | Expected concentration for routine and emergency procedural tasks |
| | | Age | Mental agility assessment | Administrative provisions for qualified staff |
| | | Consciousness— Type II | Evaluation of communication skills and capability to work in a team during training and qualification | Expected communication skills and capability to work in a team |
| | | | Evaluation of attitude toward system, organization, co-workers and subordinates | Expected personal attributes like attitude toward system, organization, co-workers and subordinates |
| | | | Psychological checklist evaluation grading related to consciousness parameters | Psychological checklist criteria |
| | *Cognition* | Cognitive or (physiological factors in psychology checklists | Evaluation of physiological aspects in checklist | Plant provisions to strengthen cognitive support, like quality and mode of operation of emergency operating procedure (EOP), like manual, computerized, or as part of operator support systems, hard/soft mimic, training provisions like availability of simulator, period of training, level of details in training |

(continued)

**Table 10.3** (continued)

| Main element | Group element | Subelement (parameter and unit) | Measured as part of **stress** characterization | Design/expectation for **strength** considerations |
|---|---|---|---|---|
| | | Learning capability | Evaluation through the training program followed by qualification written exam and interview. Annual performance review grading | Quality and strength of licensing and qualification procedure and periodic internal and regulatory review findings. Administrative provisions like annual performance review |
| | | Memory | Same as above | Same as above |
| | | Intelligence | Same as above | Same as above |
| | | Diagnosis capability | Same as above | Same as above |
| | | Comprehension capability | Same as above | Same as above |
| | | Interpretation of information and analysis capability | Same as above | Same as above |
| | | Decision-making | Simulator training and in-plant performance | Provisions for simulator sessions and control room real-time training on making decisions on normal operations, deviations from normal operation, accident conditions and beyond design basis scenario. Availability of operator support systems |
| | | Physiological health | Periodic monitoring of general health of staff | Free from diseases like diabetes, hypertension, blood pressure, sleep apnea, and any psychological |

(continued)

**Table 10.3**  (continued)

| Main element | Group element | Subelement (parameter and unit) | Measured as part of **stress** characterization | Design/expectation for **strength** considerations |
|---|---|---|---|---|
| | *Conscience* | Integrity | Assessment of general behavioral from ethical point of view | Organizational activities are covered to ensure security considerations |
| | | Commitment | General observation on meeting the target work assigned considering quality and completeness | Periodic performance monitoring and grading systems |
| | | Attitude | Assessment of attitude toward work, environment and with colleagues, particularly in team work | Provision for psychological monitoring and tracking |
| | | Motivation | Assessment of general behavioral elements | Provision for psychological monitoring and tracking |
| **Task** | *Task characteristics* *Task type* | Normal/routine task | Evaluation of performance for normal surveillance and monitoring which also includes testing and maintenance and coordination activities and compliance to technical specification requirements | Plant technical specifications, good operating practices, normal and anticipated operational occurrence procedures. Reporting of reporting of events and significant events and the safety review and feedback |
| | | Transient conditions | Evaluation for handling anticipated operational and other transients. Statistics on human error contribution to incidents and transients | Redundant trip and protection and shutdown cooling system and an effective man–machine interface. Transient handling procedures. Technical specification on |

(continued)

**Table 10.3**   (continued)

| Main element | Group element | Subelement (parameter and unit) | Measured as part of **stress** characterization | Design/expectation for **strength** considerations |
|---|---|---|---|---|
| | | | | minimum staffing that ensures adequate back up |
| | | Accident conditions | Evaluation for handling of deviation, transients and finally to accident on simulator. Statistics on human error contribution to incidents and transients and accident response | Redundant trip and protection and shutdown cooling system and an effective man–machine interface. Emergency handling provisions like emergency operating procedures, Technical specification on minimum staffing that ensures adequate back up |
| | | Severe accident management (during initial hours into the accident management) | Performance during simulator sessions on severe accidents | Training of staff and periodic refresher program on severe accident management. Training on using the available safety margins in the plant. Mock drill for site and off-site emergency. Radiation protection provision for on-site as well as off-site conditions. Provision of exclusion zones |
| | | Static task | Performance during normal operation and shutdown conditions | Same as normal operations |
| | | Dynamic task | Same as transient and accident conditions | Same as transient and accident conditions |

(continued)

**Table 10.3** (continued)

| Main element | Group element | Subelement (parameter and unit) | Measured as part of **stress** characterization | Design/expectation for **strength** considerations |
|---|---|---|---|---|
| **Organization framework** | *National level* | National safety policies and act | Periodic review of the qualification, licensing and authorization policy | National safety, reliability and quality policy. Reporting and appraisal system. Qualification, licensing and authorization policy. Regulatory provision for periodic review |
| | *Operating organizational/ plant level* | Human resource development provision | Periodic evaluation of adequacy of human resource quality, safety consciousness and ethical practices | Technical specification requirements on training and qualification. Provision for review and analysis of human error events. Availability of training and qualification infrastructure, e.g., availability of simulator and on the job training. Risk-informed policy—technical and administrative provisions. Safety consciousness and code of ethics |
| | *Regulatory framework* | Plant training and qualification overseeing | Organizational hierarchy and Safety and quality policy. Regulatory strength in terms of standards, codes and guides. Review and updating based on changes in organization chart. Adequacy of trained and qualified staff | Regulatory strength in terms of (a) safety codes and standards, (b) multi-tier review. Protocol for regulatory review on safety and quality including training and qualification of staff and review of human error related root cause and follow up action or |

(continued)

**Table 10.3** (continued)

| Main element | Group element | Subelement (parameter and unit) | Measured as part of **stress** characterization | Design/expectation for **strength** considerations |
|---|---|---|---|---|
| | | | | feedback mechanisms. Risk-informed policy—technical and administrative provisions. Safety consciousness and code of ethics |
| | | Training of regulatory staff | Adequacy of training of regulatory staff | Availability of trained and qualified staff |
| **System** | *System characterization* | System design safety philosophy and provisions | Evaluation of safety in design by deterministic and probabilistic methods and identification of human error contribution and further changes and modification to reduce contribution of human error to potential accident scenario | Compliance to defense-in-depth philosophy in design and implementation of principles of redundancy, diversity, fail-safe logic and strict quality specifications. Strict compliance to codes and standards |
| | | System operation safety philosophy and provisions | Evaluation of operational safety employing deterministic and probabilistic safety assessment. Feedback on human errors considered in design stage analysis | Compliance to defense-in-depth philosophy in operation and maintenance of spirit of principles of redundancy, diversity, fail-safe logic and strict quality specifications. Strict compliance to technical specification |
| | | Man–machine interface | Human error events related to in control room and ex-control room | Organizational framework to identify, correct, and monitor effectiveness of modification, change in operating policy and performance |

(continued)

**Table 10.3** (continued)

| Main element | Group element | Subelement (parameter and unit) | Measured as part of **stress** characterization | Design/expectation for **strength** considerations |
|---|---|---|---|---|
| | | Plant safety and reliability level | Periodic performance review to ensure that plant safety goals and reliability targets are met | Objective safety goals and policy. Reliability and safety targets for plant and systems. Provisions to evaluate and account for uncertainty or gap area that need to be narrowed down to address safety. Root cause analysis program to avoid recurrence of an event |
| | | Level of automation | Evaluation of reliability of interlocks electrical as well as mechanical and automation systems | A just automation in the plant such that human cognitive load can be reduced. The administrative provisions are closely followed and monitored |
| | | Emergency operating procedures | Regulatory or plant-level provisions to ensure maintenance and periodic updating of emergency provisions | Availability of emergency operating procedures, operator advisory systems, periodic emergency drills and corrections based on feedbacks |
| | | Complexity level | Periodic evaluation of performance of systems to ensure that complex scenario should not pose challenges, particularly during emergency | System design should be as simple as possible, e.g. provision of inherent and passive features, automatic actuations determines the strength against complexity |

(continued)

**Table 10.3** (continued)

| Main element | Group element | Subelement (parameter and unit) | Measured as part of **stress** characterization | Design/expectation for **strength** considerations |
|---|---|---|---|---|
| **Environment** | *Location* | Control room | Qualification of control room keeping in view the safety requirements ranging from normal operations to design basis accident and beyond design basis conditions | Plant ergonomically designed control room providing ground benign conditions (temperature 22 °C and relative humidity 60%), and having digital control, mimic displays and control, control console for power maneuvering, operator information/operator support systems. The safety protection and control logics is design is based on standard control room requirements. Meets standards and code requirements for control room design. Provision of plant wide announcement and communication systems |
| | | Ex-control room | The layout and working conditions including the access to repair locations in the area facilitate proper environment for maintenance staff. Review and modifications required from safety including industrial safety and maintainability requirements should be an | Design of system-based RAMS (reliability, availability, maintainability conditions) and the working environment is maintained such that maintenance and operation of the equipment do not pose challenge to human comfort. Availability of |

(continued)

**Table 10.3**   (continued)

| Main element | Group element | Subelement (parameter and unit) | Measured as part of **stress** characterization | Design/expectation for **strength** considerations |
|---|---|---|---|---|
| | | | indicator of level of design for a given area | equipment identification tags, window forms, information on fire loads, safe entry and exit markings. The area meets all industrial safety requirements. Provision of area wise announcement and communication systems |

## 10.5.5   CQB Mathematical Model

As mentioned in Sect. 10.5.3, the stress–strength relationship forms the basic model for human reliability in CQB.

Accordingly, the following axioms or considerations governed the formulation of human reliability model:

- The model should be able to model the human performance taking consciousness, cognition and conscience parameters into consideration.
- The model should account for the data collected from the plant or simulator records on human performance during normal condition (with no or minimal stress) as reference probability. These data provide statement of non-performance probability $P_0$ at normal stress.
- It should relate stress ratio level to non-performance probability.
- It should take into considerations, the extrinsic factors, like system and task characteristics, environment and any other factor that might affect human reliability estimate.
- Non-performance probability should increase with stress ratio.
- At stress ratio one, the non-performance probability should be equal to the reference non-performance probability, $P_0$.
- At infinite stress ratio, the non-performance probability should be 1.
- Effect of operator training must be incorporated in the model.

**Proposed Model**

Non-performance probability is expressed using an exponential function. This is in line with the previous available models of human reliability.

Non-performance probability under stress ratio $S_R$ is given as:

$$P(S_R) = \left[1 - (1 - P_0)\exp\left(-\frac{(S_R - 1)}{B}\right)\right]$$ (10.3)

where

$P_0$ is the non-performance probability under reference stress condition. This is representation of random errors that an operator commit even in absence of any external stress.

$S_R$ is the stress ratio. It is the ratio of actual stress acting on an operator in any situation to the reference stress. It quantifies the cognitive load acting on the operator. Multiple measurable parameters are available to quantify stress. These include techniques like saliva analysis, EEG. Parameters like heart rate, pulse rate, blood oxygen level, blood pressure can also be used for stress determination. Ratio of time required for any action to time available can also be used for stress quantification as is used in HCR model. Reference stress refers to the stress acting on the operator in normal working scenario. Reference stress depends on quality of system design, man–machine interface, operator information and support systems, and other plant parameters.

In presence of multiple parameters, the weighted linear combination can be used to determine stress level.

$$S_A = \sum W_i p_i$$ (10.4)

where $W_i$ is the weighting factor and $p_i$ is the measured value of stress parameter.

$B$ is the stress normalizing parameter. It quantifies the operator training level and can be determined from the experimental data available on human reliability using the relation:

$$\ln(1 - P) = \ln(P_0) - \frac{S_R}{B}$$ (10.5)

Hence, slope of the straight line plot between $\ln(1 - P)$ and $S_R$ is $-\frac{1}{B}$. This plot is generated from the experimental or plant data. Training level of an operator reduces the probability of error. Higher value of $B$ is expected for better training.

The nature of job (skill-, rule- or knowledge-based) also affects the reliability of an operator. A skill-based job will have higher reliability as compared to that of a knowledge-based job. In order to incorporate this effect, parameter $A$ is introduced in the model. Higher value of $A$ signifies higher difficulty in taking any action. The difficulty sequence starting from least difficult is:

Skill based $\rightarrow$ Rule based $\rightarrow$ Knowledge based

Incorporating this, the model becomes:

$$P(S_R) = \left[ 1 - (1 - P_0) \exp\left( -\frac{S_R}{B} A \right) \right] \tag{10.6}$$

The variation of non-performance probability with stress ratio is shown graphically in Fig. 10.8.

Conscience takes into account the intent of an operator. A person with malice intent will always commit error, this can be considered as a non-performance probability of 1. Let, parameter $C_{sc}$ defines the conscience that can either take value of one (good intent) or zero (bad intent). The model thus becomes:

$$P(S_R) = 1 - C_{sc}(1 - P_0) \exp\left( -\frac{S_R}{B} A \right) \tag{10.7}$$

This model takes into account contributions of work stress, operator training, nature of job and conscience through parameters $S_R$, $B, A$ and $C_{SC}$ respectively. Stress ratio can be quantified using various measurable parameters available as discussed earlier. $A$ quantifies the difficulty of job taking into account the nature of job to be done. Wise judgment is required to determine the value of $A$. $P_0$ is the non-performance probability at stress ratio one. $P_0$ and $B$ can be determined from the available plant or experimental data. The experimental data can be generated using simulators. Methodology for determination of $P_0$ and $B$ is discussed in next section.

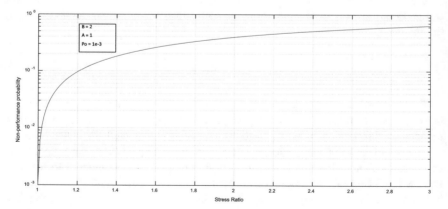

**Fig. 10.8** Non-performance probability versus stress ratio

**Table 10.4** Results of human reliability experiment performed on simulator

| Stress ratio | Non-performance probability |
|---|---|
| 1.0 | 0.001 |
| 1.2 | 0.096 |
| 1.4 | 0.182 |
| 1.6 | 0.160 |
| 1.8 | 0.330 |
| 2.0 | 0.394 |

### Determination of $P_0$ and B from Simulator Experiments

Consider a hypothetical experiment conducted on an operator to determine the non-performance probability under varying levels of stress. The results are given in Table 10.4.

From above data we get, $P_0 = 10^{-3}$ (Non-performance probability at stress ratio one).

The plot between $\ln(1 - P)$ and $S_R$ is shown in Fig. 10.9.

Slope of above plot is $-\frac{1}{2}$. Hence, we get,

$$B = 2$$

Thus, the human reliability model becomes:

$$P(S_R) = 1 - C_{\text{sc}}(1 - 0.001) \exp\left(-\frac{S_R}{2} A\right)$$

Value of A depends on the nature of job to be performed.

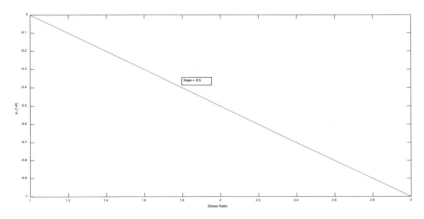

**Fig. 10.9** $\ln(1 - P)$ versus stress ratio

## 10.6  The CQB Methodology

As mentioned earlier, the proposed human reliability technique called the CQB approach or methodology has been developed with an objective to provide holistic solutions in terms of assessment as well as improvement of human factors through the implementation of CQB attributes and management principles. The human reliability projects are generally complex, in terms of the activities involved in each of the procedural steps. In fact, it is required to have a consciousness- and conscience-based management approach to human reliability projects to ensure acceptability and credibility of the results and insights developed through these technical activities [31]. Figure 10.10 depicts the major steps involved of this technique where the CQB aspects form part of each of the given activities.

The major steps as shown in Fig. 10.10 illustrate the applicability of the CQB methodology in support of risk assessment. Each step will the objective and scope, a description of the activities and the CQB attributes that are applicable for the subject step. It will be interesting to note that this methodology will be at par with other approaches if the CQB factors are not addressed explicitly. In short, we can say that provision exists to apply the approach given in Fig. 10.6 to evaluate human reliability as part of either an independent investigation or as a probabilistic risk assessment activity. The following sections bring out the salient features of each procedural step of the CQB framework.

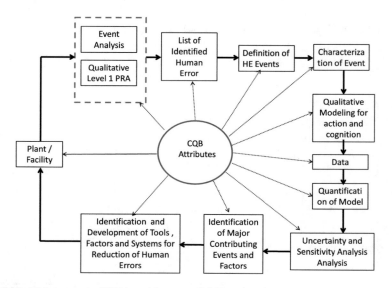

**Fig. 10.10**  Major steps in CQB-based human reliability procedure

### 10.6.1 Plant/Facility Familiarization

Familiarization with the plant or system for which the human reliability analysis is being performed forms an integral part of this methodology.

**Objective**: To intimately know the organizational structure and various administrative and technical provisions of the plant such that plant-specific inputs are utilized effectively to realistically assess human reliability aspects.

**Scope**: Covers the study of administrative setup, including human resource management systems, documentation, communications, safety documentation, changes or modifications, and overall regulatory framework relevant to human reliability analysis.

**Activities**: The major activities involved in familiarization of the plant/facility include:

- Analysis of organizational chart of the plant. This chart should provide information on plant hierarchy or the management structure of the plant. The line of authority and line of communication provide insights into the administrative as well as regulatory process applicable to the plant. The study of regulatory reviews for over a period of say 10 years can provide vital input on human factors, plant safety culture, and, most importantly, the consciousness and conscience-quotient of the organization. How to reflect these factors is a matter of individual organization/facility.
- Information on provision of qualification levels and actual status of qualification levels. The staffing status provides insights into human resource management aspects of the facility. It is important to know the policy on academic or educational qualification of the staff. For example, the eligibility criteria for a shift-in-charge of the plant could be a graduate degree or diploma in engineering, minimum experience criteria in the preceding or entry level. These aspects in a way enable determination of cognition and conscious capability of the individual who gets certified through the qualification exams.
- Evaluation of training and qualification program to assess its impact on PSFs. Study of the examination patterns and their coverage, including written tests, walkthroughs, and interviews, forms part of this evaluation. It has been found that most of the training programs are built around practical training focused toward understanding the equipment and system functioning toward smooth operation of the plant. The evaluation should review the training and qualification in respect of its impact on risk mitigation and risk information. In short, the program should have a risk-informed characterization along with practical aspects of training.

The factors that are relevant to modeling of human PSFs are the communication framework between one agency/individual to other or many agencies and requirements on communication among the facility staff and further with the regulator, as directed by the facility technical specification or limiting conditions for operation document.

The availability of a training simulator and the quality of training staff for these facilities has been attributed as part of a good safety-conscious culture of the plant and has been recognized as one of the major quality attributes at the organizational level.

Documentation, including the major procedural elements, safety analysis reports, technical specifications, emergency operating procedures, event reporting systems, design manuals and operating procedures, standing instructions, quality assurance program, staffing levels, and overall human resource management system of the plant; should be reviewed for gathering input on organization safety management of the plant.

The plant walk down to collect data and information to facilitate development of a human factor study forms an integral part of the plant/facility familiarization. The objective is to understand the gap areas in respect of human factors that have been credited in the safety analysis and provisions that are available in the plant. This aspect will be discussed in the section on plant-specific data collection.

The insights developed and data collected in this step provide a reference for the detailed analysis as given in the following steps.

**CQB Aspects**: Safety of the plant to a large extent depends on the quality of recruitment, training, evaluation, and qualification and finally criteria used for authorization of the staff for a given posting and responsibility. The conscious-safety culture can be determined based on the quality attributes derived based on the study of the above aspects. The *consciousness* factors that are important to this step are characterizing the organizational values, learning quotient, and sensory consciousness developed during the training. Evaluation should also focus on development of cognitive ability for knowledge-based talks. Developing an insight on the collective and individual conscience of the staff is an important part of this approach.

## 10.6.2   Identification of Human Error Events

**Objective**: To prepare a list of human error events (HEEs) as complete as possible such that there is a reasonable confidence that at least a loss of the safety-critical events have been identified.

**Scope**: Identification of events of Types I-VI as per the requirements of the project—in this case, risk assessment study. Review of incidences forms the plant as well as the generic source; qualitative PRA model of the plant and review of similar other PRA models. Apart from this, expert opinion and operating experience are two important sources of data and information on human error events.

**Activity**: Preparation of an exhaustive list of human error events (HEEs) in a risk assessment study is crucial as considerations of human error have been observed to have impact on the final results of the analysis. The six major sources for human error data, which include human error events, precursor to human error, and events having the potential for plant safety, are as follows:

(a) **Plant Reports**: A review of incident reports of the plant is performed to identify those events where human error was directly or indirectly responsible for the incidents/accidents. This review should also identify the latent human errors which otherwise were not recorded as human error but as a component failure or a procedural failure. A calibration error or error introduced during the testing or maintenance that resulted in unsafe failure of the component should be classified as latent human error. Similarly, procedural errors need to be identified as institutional errors. In this step, significant efforts and debate are required to identify events that can be attributed to organizational/institutional errors. These types of events have much a wider impact than the events at lower, viz. man–machine direct interactions.

(b) **Plant Procedures and Operating Experience**: Normal and emergency operating procedures form one of the sources of safety-significant human error events. Expert opinion has traditionally been a useful source of data. Certain assumptions in operating procedures related to the availability and adequacy of cooling provisions its, such as connecting a hose from a service water system to meet cooling requirements, issues related to completion of a given task in a given time, and considerations that do not take into account modifications in the system, need to be reviewed in the context of direct or latent human errors. Apart from this, operating experience and expert opinions can provide vital information that is otherwise not available in other sources of data on human errors.

(c) **Simulator Data**: The training and R&D activities performed on the plant simulator provide input on human factors relevant to accident management scenarios as well as an opportunity to reflect modification in the plant, which might include provision of redundancy or enhanced automation.

(d) **Expert Opinion**: For rare scenarios such as LOCA, not much insight is available apriori; hence, expert opinion based on the safety and operational insights can be vital input for identifying the human error events.

(e) **PRA Study**: Human error events directly impacting safety can be identified while performing PRA [35]. Human errors form either header recovery events as safety functions in the event tree or basic human actions, e.g., actuation of redundant systems, in the fault tree. These human error events can be identified in the qualitative analysis stage of the PRA modeling. References to other similar PRA studies can also be a good source of human error events.

(f) **Generic Sources**: Data collected from similar plants or from generic literature can be used to prepare a list of human errors after validating the applicability of the data for the subject plant.

**Outcome**: An exhaustive list of human error events and factors that is as complete as possible.

**CQB Aspects**: There should be conscious efforts to figure out all the human actions right from Type I to Type V such that the list of human error events is as complete as possible and credit can be taken for these events, particularly during emergency scenarios. This requires a conscious revisit of all postulated scenarios and emergency operating procedures (EOPs) to look for human factors that can adversely affect the implementation of an EOP during an actual emergency scenario. For example, procedural steps may require alternate safety provisions which are not part of standard provisions. However, these provisions may be required to mitigate an accident scenario, like connection of service water or fire water to the emergency cooling system as last mitigation feature through a temporary line or hose. Even though the plant emergency procedure requires this mitigation fire/service system to be connected, the adequacy of the ad hoc provision determines the success of connecting the mitigation measures, e.g., absence of a tapping or nozzle (hardware) to connect a hose to the cooling system. Here, the human error to connect a hose cannot be credited as the adequate provision may not be realized during emergency.

### 10.6.3   Screening

In the previous step, an exhaustive list of human errors has been generated that ranges from a very impact on safety to a low impact. It is prudent that the resources for risk analysis are utilized in an effective manner such that risk benefits are maximized. Hence, a screening analysis is performed.

**Objective**: To produce a list of important human errors having direct bearing on safety by screening out those not important to safety such that efforts on safety-significant human actions can be focused by performing an iterative process.

**Scope**: Covers screening of all the events that form a part of the fault tree and the event tree by employing screening criteria such that safety-significant events are screened for detailed modeling. This covers all Level I to Level III events and Level I and II for higher-level aspects that might have on impact on Level 3, i.e., procedural and man–machine aspects.

**Activities**: There are two situations that need to be considered. Here, it is understood that the human reliability analysis is being performed as part of probabilistic risk assessment and the process for preparing the qualitative model has been completed. Also, the generic screening value from the generic data source is available.

There are two major approaches for screening—qualitative and analytical or quantitative. The qualitative approach involves operational/engineering judgment and use of expert opinion. The insight on organizational/institutional aspects related to command and control, procedures, and communication need to be screened employing engineering and executive judgment. For example, the plant insights on potential procedural lapses or communication errors can be better screened-in/screened out based on administrative or executive deliberations. In line with the existing approaches, the human error events directly related to plant and the man–machine interface, particularly for control room scenarios, are screened using mainly an analytical or quantitative approach as these aspects form part of the PRA model and there exists the screening values on human error events in generic data sources. The available literature suggests that an upper confidence bound on the estimates should be used for screening. A review of the list of cut-set provides the information on potential for human error induced common cause failure as well as independent human errors. The human errors that cause common cause failure (CCF) are classified as the most significant events, particularly when these CCFs cause knocking of a safety function all together or significantly contribute to core damage. Performing a sensitivity analysis using fault tree and event tree model provides an effective mechanism for screening.

**Outcome**: The output of the screening activity is a list of human error events that need to be analyzed in detail as part of the PRA program. Some additional events, which even though they do not form part of the list, but are based on engineering judgment, will need attention in the final analysis. For these events, a sensitivity analysis will be required when a quantified model of the plant is made available.

**CQB Aspects**: Human reliability analysis can easily be manipulated to demonstrate low CDF by considering certain human actions or using a very low value of human error probability. Hence, sound arguments and data such that it qualifies the regulatory guidelines should be provided giving due considerations to organizational ethics/values. Conscious-attention is required to ensure that no human events can be credited in risk modeling, if these events are not covered in plant emergency procedures. Finally, a conscience-based criteria should be used to screen-in or screen-out human error events. If a quantified screening criterion is used, the rationales and aspects related to it should be validated and documented.

## 10.6.4 Definition of the Event

**Objective**: To provide a crisp and objective definition of the human action that includes cognitive as well as physical action required to manipulate systems or components in the plant to bring the plant to a safe state. The definition should also include a definition of success in terms of plant parameters.

**Scope**: The scope includes identification of the system boundary, the plant status before and after the human actions, the means of recovery and definition of success and failure.

**Activities**: Drafting of the definition of human action requires input on plant systems being affected and familiarization with operating policies and procedures, resources, and provisions in the plant to affect recovery. The only criterion in defining the event is that the definition should be crisp and clear without any ambiguity. Let us discuss this aspect through an example of a definition as follows:

> Failure of human action in recovering failed shutdown cooling system by starting the standby shutdown pump SD-P-Y to cater to shutdown cooling requirement within the stipulated time after failure of selected shutdown cooling pump SD-P-X.

This definition should be supported by a definition of failure of a human action as well as the system. For example, the failure of the system could be not able to deliver required flow as per the core cooling requirements. Failure of human action can be those "actions or non-actions" that blocked the system/component manipulation, such that the core cooling function could not be recovered. Here, the failure could relate to diagnosis, planning, and analysis, i.e., cognitive failure and physical action such as wrong component manipulation. As mentioned earlier, the human reliability analysis should explicitly define criteria for success/failure in terms of plant parameters, such as minimum flow or clearing of certain alarms or trips. These aspects should be clearly documented.

**Outcome**: Definition of identified human action that is clear and crisp, that needs to be analyzed in detail in support of a given project, in this case a probabilistic risk assessment study.

**CQB Aspects**: Drafting a correct and objective definition might require a conscious-attention, definition of failure success criteria, and understanding of cognitive and action part of the human reliability analysis. This will required an upfront and clear understanding of the provisions and resources in the plant. Hence, the supporting analysis should be documented while drafting the definition of "human action" associated success and failure criteria.

### 10.6.5   Event Characterization

**Objective**: Often the definition of the events formulated in previous steps requires supporting information in terms of type of human error category, definition of success and failure criteria, and plant characterization. The aim of this task is to evaluate the factors that must be considered to facilitate the detailed modeling at a qualitative and quantitative level in the following steps, keeping in view the focus of risk assessment.

**Scope**: Includes considerations of all those PSFs that directly or indirectly cover the direct human error including, latent as well as institutional human errors.

**Action**: Each human error event requires its description such that the available mechanism including the nominal quantified parameters, various conditions, and the constraints can be stated. The major parameters required for characterizing a human error event include:

- Human action being analyzed as part of organizational/procedural event tree/fault tree;
- Human action type: action/recovery;
- Time required to complete the task ($T$);
- Time window available ($T_w$);
- Type of system or man–machine interface;
- Stress levels: normal/emergency;
- Type of recovery/checks that are available;
- Type of crew and training level;
- Applicable PSFs;
- Cognitive characterization: skill-, rule-, and knowledge-based.

Depending on the analysis for every event, all the factors may not be required or at the same time some new factors may be required; however, the point to be noted is that the event characterization should be as complete as possible. There should be adequate input to address the conscious, cognition, and action part of the event analysis.

**Outcome**: Identification of major parameters that need to be considered to facilitate qualitative as well as quantitative evaluation in the following steps.

**CQB Aspects**: This step is crucial in terms of characterization of cognition, consciousness, and conscience factors as follows:

- Consciousness factors: Deals with characterization of organizational/institutional consciousness, level of conscious-safety culture, characterization of training and qualification program, characterization sensory consciousness/awareness quotient.
- Cognition levels: Evaluation of the capability of the average individual for knowledge, rule-, and skill-based events, through simulator sessions and psychological mapping.
- Conscience factors: This factor is derived from national as well as organizational values systems. The degree of adherence to a code of ethics prescribed by the management also provides input in determining the ethical quotient. In an operating plant the incidences/observations related to lack of commitment and overall attitude toward safety are some of the indicators that can be used to characterize the conscience level of the staff as well as individuals.

A close look at the above factors indicates that the role of plant management is vital for overall improvement in consciousness, conscience, and cognition levels. The role of spiritual conditioning of mind and yoga practices can be crucial in improving CQB factors.

## 10.6.6 Qualitative Assessment

**Objective**: To define the chronological as also parallel order of events that includes cognitive as well as action components, keeping in view the available safety options and resources available in terms of time and human resources in context of the scenario being analyzed/modeled. The major aim is to prepare a framework or model for quantitative analysis.

**Scope**: To connect the information/input available in previous steps and pass on the graphical model to the next step for quantification.

**Action**: A qualitative analysis is performed to get the holistic logical model for the analysis. There are a host of techniques developed for human reliability analysis; however, this section will discuss some of the graphical techniques as follows:

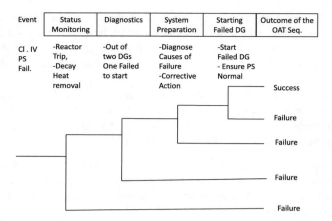

**Fig. 10.11**  Simplified operator action tree (OAT) for class power supply failure

### 10.6.6.1   Operator Action Tree

This graphical model is similar to the event tree diagram employed in PRA [12]. The analysis process involves induction of an event in the organization/plant or system and propagation of the scenario based on the human response in chronological or logical order. The header events can be classified into two parts, related to cognition (i.e., observation and diagnostics) and physical action. The aspects related to observation and analysis are governed by conscious and cognition capability or factors while the action parts are governed by only awareness-consciousness or mind-consciousness. The consequences of this analysis are recorded in the right-hand column of the action tree as shown in Fig. 10.11.

This part of the analysis deals with generation of sequences of human errors, and quantification is performed in the following steps using the quantified estimates on human error probabilities.

### 10.6.6.2   HRA Tree

The HRA tree is another graphical representation for human reliability analysis. The procedural part of the human reliability analysis is generally performed using an HRA tree. This analysis investigates the alternative or mirror image of the action envisaged in the procedure. Figure 10.12 demonstrates the applicability of this tool for a four-step procedure. This is a binary representation, and each node represents a procedural step. The left side shows the procedural steps and associated probability of successful completion of each task while the right side shows the probability of failure of each step. The two types of error that form part of the procedural steps, assuming that the procedure is written correctly, are error of commission and error of omission. This representation is used for quantification using probability for

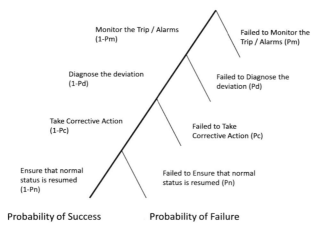

Fig. 10.12   HRA event tree

correct action as well as probability of error in each step. An arithmetic summation of all probability, i.e., (Pm + Pd + Pc + Pn), estimates the probability of error associated with the procedure. The error in preparing the error can be apportioned in institutional or organization faults. An HRA tree could be one of the auditing tools to investigate the error in preparing procedures.

#### 10.6.6.3   Confusion Matrix

This technique employs tabular representation for misdiagnosis of an event or error in identifying the correct event [36]. The top row and the left column show the events that can be misdiagnosed for other events. The probability for misdiagnosis is indicated in the respective position in the metrics. Figure 10.10 shows the confusion metrics for power supply failure.

There are many approaches for graphical modeling in HRA, readers may refer to the available literature on human reliability analysis methodologies/techniques.

### 10.6.7   Data Collection and Analysis

This step is vital in human reliability analysis, as the general observation is that normally human reliability estimates have relatively larger uncertainties. One of the major factors responsible for this is scarcity of data particularly for accident conditions. As pointed out in this chapter, often the available approaches do not provide an adequate framework for human reliability modeling due to adequate representation of human model.

**Objective**: To collect data and information such that the analysis uses, as far as possible, realistic data for quantification of the human reliability model. In the CQB approach, information is required in respect of cognition modeling in respect of the nature of the job.

**Scope**: To collect relevant and applicable data, including PSFs from various sources. This includes plant-specific data, generic data, expert opinion, and simulator data.

**Activities**: At the outset, human reliability data collection and analysis requires a data collection approach or methodology. To make the data collection process more effective and consistent, the data collection form should spell out the expectation of the analysts such that the staff dealing in data collection are clear about what is required and what kind of flexibility is available to them. Many data collection projects do not provide expected results because the staff involved are not clear about what type of data is required and the flexibility available to them. Reference to various documents on the data collection and analysis like IAEA-TEDDOC-1048 on Collection and Classification of Human Reliability Data for Use in Probabilistic Safety Assessments [30] helps in consolidating the data collection program. An initial assessment on meeting the data requirement for the analysis helps in identifying the sources for the data; for example, as it is encountered in most HR studies, all the data from plant-specific sources are not available; hence, the existing sources such as THERP, ASEP, SPAR-H, and HCR documents can be used for meeting the data requirements.

### 10.6.7.1   Plant-Specific Data

Many organizations have a system of data collection on regulatory reviews, quality audits, incidences, test and maintenance records, operational records, deliberations of plant-level safety committees, and procedures that includes operating manuals and emergency procedures [37]. These records provide information on human error events that are precursor events and that have the potential for human error. The operating procedures provide information on the opportunity that is available for human errors. Even though it is challenge to derive information on PSFs on plants as well as human aspects, efforts should be made to use plant insights such that PSFs are more realistic. That standard procedures are recommended that include computerized systems and formats that also facilitate classification of human actions into various categories and computation of the reliability estimates.

### 10.6.7.2   Generic Data

The available literature on human reliability analysis shows that many approaches, such as THERP, ASEP, SPAR-H, and HEART, provide data and information on nominal or base failure rates for various human errors as also associated PSFs. These documents form a generic source for human reliability data. The information/

data available from the generic sources can be divided into time reliability models/curves, events (i.e., human error data), uncertainty bounds of data, and dependency aspects. The scope of this chapter is not to go into details and benefits of one study over other. However, certain observations can be made before we use these sources for human reliability analysis. Most of these approaches are known as Handbook approaches. The limitation in respect of application data and information from these sources is that they provide a black box solution in respect of considerations of data and PSFs. What one comes across is the outcome of certain experiments and data analysis while a little or no information is available on basis of design of experiments and details of the experiments.

However, these approaches are still being employed because many institutions involved in risk assessment do not have their own approaches. The limitation of these approaches is that they are, to a large extent, targeted toward control room scenarios while not addressing the aspects related to institutional errors. The field actions are modeled only for some limited events. Hence, the results obtained from generic approaches tend to carry relatively large uncertainty in estimates.

### 10.6.7.3   Simulator Data

In the CQB approach, simulator experiments are performed to evaluate the stress component associated with individual tasks. For example, electroencephalogram (EEG) brain wave patterns and associated frequencies can relate the stress level for a given task. Even though at this point of time having correlation coefficients to relate observations from simulator experiments with actual plant conditions, particularly emergency scenarios, is a matter of research, the insights available from these experiments can be utilized employing conservative criteria. For example, if a simulator experiment records that the time taken for a particular diagnosis is 10 min, for plant conditions a conservative estimate of >10 min, say, 15 min, can be considered. In the human reliability field, it is expected that the data available from the operating plants should be compared with simulator experiments toward developing insights into correlation coefficients for a given task. Virtual models also form part of simulator exercises toward development of human factors. In fact, there are many publications, and one of the important sources of information is from Halden Man-Machine Laboratory on Simulator experiments for development of human factors [38].

### 10.6.7.4   Expert Elicitation

Expert elicitation approach is used when the data from plant records are not adequate to estimate human reliability data or PSFs [39]. Design of experiment plays a crucial role in deriving these estimates. Qualification of experts is another major step in expert elicitation. One of the major limitations of this approach is the "subjectivity" that is associated with these estimates. Hence, the calibration of the experts forms an important aspect of this approach. Often the large uncertainty bound associated with these estimates due to imprecise nature of input handled in this approach poses a question on applicability of the data. Hence, it is recommended that advanced methods such as a fuzzy approach which is effective in handling ambiguous or imprecise linguistic variables need to be employed in expert elicitation.

### 10.6.7.5   Dependency

Dependency evaluation is required because the dependency of one task upon another arises due to many factors. Some of the major factors have been addressed in the THERP approach. For example, same crews inducing errors in performing the same maintenance job on three redundant pumps adversely affect the redundant provision in the plant. Similarly, there is the possibility of error in operating a switch correctly when the second switch is located side by side on the control console. For example, the Spar-H approach considers dependency due to crew, time, location, and crew assignments. Here, the dependency is "complete" when the scenario involves "same crew, close in time, or same location with or without additional cues." The dependency is "high" when the case is "same crew, close in time, different locations, and with or without additional cues" or for "same crew, not close in time, same location, and no additional cues". There are other lower-level ratings such as "moderate" and "low" for other combinations of the same set of crew, time, location, and cue assignments. Readers can refer to SPAR-H NUREG-CR-6883 for details [15].

## *10.6.8   Quantitative Analysis*

**Objective**: To estimate the human reliability by considering the CQB aspects in mental processes as well as in action parts by making the results more accurate though application of appropriate performance-shaping factors (PSFs).

**Scope**: Estimation of probability human error and modeling using logical or graphical models. Often uncertainty and sensitivity evaluation also form part of this step.

**Activities**: The major activities in this step are: Formalize the qualitative graphical representation (e.g., AOT, HR Tree); acquire the probability or frequency of the initiating event from previous steps; estimate the human error probability associated with the header event; and estimate the human error probability for the task.

The graphical model has two types of header events, error in observation, analysis/diagnosis (related to cognitive part) and second action part, like error in starting of an equipment. The OAT of Fig. 10.4 can be referred to as an example. The event probabilities are calculated by modifying the nominal value employing

PSFs. After quantifying the human error probability for the header as well as initiating event, the event sequences are evaluated and the results are presented in the right-hand column of the OAT as shown in Fig. 10.6. A summation of all the accident sequences provides the human error probability for the scenario being analyzed.

The OAT can be analyzed manually or by using a computerized software environment.

## 10.6.9   Uncertainty Analysis

Ideally speaking, every parameter involved in human reliability analysis has uncertainty, be it from a generic source or a plant-specific source. It may be noted that human reliability itself is associated with the larger part of uncertainty in the risk assessment model. This is due to the inherent nature and complexity that is involved in the human model, poor understanding of the contextual scenario, and severe shortage and lack of accuracy of human reliability data. This is evident in the large error factor that is part of any data available in the generic source. The error factor ranges from a minimum of 3 to 30 or more.

There are three major types of uncertainty that need to be addressed: the aleatory uncertainty due to inherent randomness in the parameter; epistemic uncertainty due to lack of knowledge about the scenario/parameter; and the uncertainty associated due to imprecision and ambiguity about the parameter, generally referred to as "fuzzy" uncertainty. The uncertainty of analysis is first carried out at a task level where uncertainty is characterized in diagnosis part considering the conscious component of the human model along with associated PSFs and then at the execution level, which involves characterizing uncertainty in data. The uncertainty from the task level to the procedural or safety function level is propagated in the fault tree and event tree through well-established methods, for example, Monte Carlo simulation.

The resources for uncertainty evaluation depend on the source of data. In case the human reliability analysis is carried out using plant-specific data then it becomes necessary to estimate the uncertainty associated with each parameter. If the data are used from generic sources for diagnosis as well as execution, then the error factor or upper bound and lower bound associated with the nominal or median value characterize the associated uncertainty. There is a separate chapter on uncertainty analysis; however, this section presents the specific aspects associated with uncertainty characterization in human error data.

In this step, procedural guidelines are developed for uncertainty evaluation for plant-specific input dealing with the randomness of the parameter (i.e., aleatory uncertainty); the lack of knowledge (i.e., epistemic); and the imprecise langue variable obtained from experts (i.e., fuzzy uncertainty). The scope of this discussion is limited to what should be done and not how it is done because there are adequate details available to evaluate specific parameters. To have improved understanding, let us see these steps with examples as follows:

**Aleatory Uncertainty Analysis**: Let us assume that data collected from plant-specific sources on a human interaction deal with starting a failed diesel generator set during a Class IV Power failure scenario. There are two components to this evaluation: (a) diagnostic, i.e., cognitive or conscious analysis to first identify and analyze which DG out of two DGs has failed and the reason why it has failed; and (b) execution component, i.e., take corrective action and recover the failed DG and normalize plant load for Class III failure condition. Let us assume that there are 10 records available in the past 10 years where we have two sets of 10 data point each for diagnosis and execution. The 10 data points when plotted on a log-normal paper show a straight line; i.e., the data follow log-normal distribution. The upper bound and lower bound estimate show the aleatory uncertainty in the data.

**Epistemic Uncertainty Analysis**: As mentioned earlier, this uncertainty is due to lack of knowledge or data. Suppose this lack of data or knowledge is reflected in only two data points in the above exercise, let us say in diagnosis. If we characterize the upper bound and lower bound using only two data points, then the uncertainty band will be much wider, let us say to the extent that this estimate of uncertainty may not useful or as good as "no information". Let us assume that on the same parameter, i.e., diagnostic time, information is available from generic sources, i.e., the median value of the parameter and associated upper and lower bound. It is natural that we would like to use this information to consolidate our plant-specific data. This can be achieved by using Bayesian updating. In Bayesian updating, the information from generic source is called the "priori input", the plant-specific limited data act as "evidence", and these two together can provide the posteriori estimate. Posteriori estimate provides information on time required to perform diagnostic with upper and lower bound, which is nothing but epistemic uncertainty.

**Fuzzy Uncertainty**: The available literature shows that most of the techniques on human reliability use simulator data, expert opinion, or judgment approach to generate human reliability as well as performance-shaping factors. Most of the techniques lack characterization of the data because it is a challenge to deal with the linguistic variable, which an expert or plant employee uses to expresses his or her perception on a subject question [40]. The fuzzy logic approach provides an effective mechanism to deal with imprecise or ambiguous linguistic response for a question like "high", "medium," or "low", for a probability value, or for an estimate of time taken for a task. Fuzzy elicitation techniques are used for converting expert opinion, which is treated as imprecise input, to arrive at quantitative estimates of human reliability [41]. For example, seven experts were requested to provide their opinion on "time taken to perform diagnostic", during the DG failure scenario as discussed in the preceding section. The expert may provide answers, like high, medium, or low. The membership functions used in the fuzzy approach provide an elegant approach to estimate the nominal time based on expert opinion. Of course, the accuracy of the prediction depends on the design of the experiment, which includes the criteria used in selection and calibration of experts.

## 10.6.10  Identification of Major Human Risk Contributors

In the preceding sections, we discussed how human reliability estimates along with its uncertainty bounds are obtained. We are aware that this section forms part of the probabilistic risk assessment where apart from hardware, human reliability data are required toward predicting the core damage frequency for the plant. The human reliability estimates can be divided into four major categories: (a) human failure frequency that initiates an incident/accident; (b) estimates of human actions plugged in the system fault tree as basic events; (c) estimates of human reliability, which forms the header event in the event tree as safety functions; and (d) recovery probability estimates for failed systems in the fault tree or event tree analysis (as dictated by plant EOPs).

The estimates obtained using this procedure are plugged into the system PRA model. The PRA simulation will produce the estimate of core damage frequency, the accident sequences with its frequency, and the importance measure for each accident sequence.

The impact analysis is comprised of:

- A review of the list of the cut sets identifying the cut set having human error as the common cause contributor.
- A single-order cut set leading to a safety system failure or for that matter (hypothetical case) leading to core damage acquires the highest attention and remedial measures.
- A general check should be performed to justify credit taken for human action and documenting this fact. (The PRA model should not take credit for human action while plant safety philosophy, resources and provision in the plant do not support this human action.) For every human action credited there should be EOPs apriori.
- A second-order cut set, i.e., upon failure of the hardware safety function the human action is required to recover the safety function that needs special attention in terms of time required for the recovery, resources available, and adequacy of the EOP.
- Some cut sets, irrespective of their order, will require review considering the probability ranking. High probability cut sets need attention.
- A general review of the cut set having low probability value that need attention. Here, these cut sets should be verified for lower probability associated with these sets such that these values are justified for exclusion.

The above exercise will result in a list of major human contributors to the risk.

## 10.6.11    Assessment of Impact of Human Error in PRA

Once the list of major human contributors is available, the next job is to evaluate the impact of these contributors by performing the sensitivity analysis. Sensitivity analysis is performed considering uncertainty in the data for the events identified in the preceding section. Assumptions made for that data and PSFs on a conservative basis, and the recorded justifications for crediting the human action for recovery or redundant actions.

If the insights available in the impact assessment are such that human reliability considerations are not in line with the plant's available resources, the plant's technical specifications, or the EOP requirements, then there should be an in-depth review for crediting/discrediting these human actions.

## 10.7    Tools and Approaches to Reduce Human Error Probability

Human reliability should not stop at the impact assessment, but continue to address/ overcome the limitations revealed by performance-shaping factors. Be it the limitations imposed by organizational factors such as communication, procedures, lines of authority, or organizational hierarchy; or availability of time for any diagnosis or execution, man–machine interface, safety upgrade or improvement program should address these limitations. How these improvements improve the performance-shaping factors should also be documented. Some of the techniques to reduce the probability of human error are discussed below.

## 10.7.1    Deployment of Operator Support Systems

There is an increasing interest in development of an operator support system in nuclear plant control rooms. The operator information system has already been deployed in many control rooms of NPPs. The advanced information system provides the operator status of the safety parameters such that any threat to safety functions can be identified and corrected during the inception itself. These information systems employ symptom-based identification where safety takes precedence over other aspects. Further, there is increasing interest in development of advanced operator support systems that employ intelligent techniques such as artificial neural networks for transient identification and a rule-based approach for diagnostics [21, 42]. In fact, the PRA framework is utilized for structuring the knowledge base in these systems. The objective here is to reduce the cognitive load on the operator during the plant deviation in general and emergencies in particular. Since a computer-based system provides online guidance support to the operator along with tracking mechanisms, the operator's burden for the analysis is expected to reduce significantly. However, before putting these tools in the control room, they need to be validated and verified either in simulated conditions or in proper simulator experiments.

## 10.7.2   Reduction in Human Action by Automation

The option of automating the processes or actions based on plant symptoms or feedback should be examined if the procedural part is causing a heavy cognitive load on the operator. Suppose a lengthy procedure or checklist to perform a job is becoming cumbersome, then a justified and balanced approach should be considered such that there is net improvement in safety.

## 10.7.3   Modification to Secondary Factors

The traditional approach to training is heavily focused on practical training; however, based on the insights available from impact analysis the training approach could be risk-informed. What does this mean? The training should focus on those aspects where monitoring of plant deviations, handling of emergency operating procedures. The qualification procedure should, as an institutional factor, recognize that at various stages of qualification, be it written tests, walk through, or interview should have a risk-informed orientation.

## 10.7.4   Simulator-Based Training

Requirement of simulator has become a regulatory stipulation for NPP commissioning and licensing. The reason is simulator-based training and qualifications has made significant impact not only for routine operations but also for evaluating response capability of emergency operating procedures and the crew for anticipated as well as rare emergency scenarios. Apart from this, simulators are extensively being employed for developing the human factors and human reliability data. Apart from this, the deployment of virtual tools forms one of the modern tools, such as a virtual environment that enables a simulation-based approach for training in emergency scenarios.

## 10.8   Conclusions and Remarks

Human factor has been found to be one of the major contributors to accidents in complex engineering systems. The experience in nuclear plants has not been different. The experience on development of PRA for nuclear plants shows that there is need to improve the existing human reliability methods such that uncertainty in data and model can be reduced to an acceptable level.

A new approach has been developed considering a human model captured through the three Cs, i.e., consciousness, cognition, and conscience and brain, referred as CQB approach. Another, modification in this approach compared to many contemporary approaches is the stress has been used as parameter to evaluate develop the human reliability model. It is recognized that human reliability evaluation requires a project management approach where the CQB philosophy works in all the sub tasks of human reliability assessment project. Further work is required to develop a human reliability handbook based on CQB approach. It is felt that simulator experiment and data available on plant operating experience will provide an effective mechanism to address issues related uncertainty in data and model.

## References

1. *The Fukushima Nuclear Accident Independent Investigation Commission* (The National Diet of Japan, 2010)
2. M. Stringfellow, *Accident Analysis and Hazard Analysis for Human and Organizational Factors*. Ph.D. Thesis, Massachusetts Institute of Technology, 2010
3. J. Bell, J. Holyroyd, *Review of Human Reliability Assessment Methods*. Research Report—RR679, Health and Safety Executive, 2009
4. A. Swain, Human reliability analysis: need, status, trends and limitations. Reliab. Eng. Syst. Saf. **29**, 301–313 (1990)
5. A. Swain, *A Method for Performing a Human Factors Reliability Analysis, Monograph SCR-685* (Sandia Laboratories, USA, 1963)
6. A. Swain, H. Guttmann, *Handbook of Human Reliability Analysis with Emphasis on Nuclear Power Plant Applications, NUREG-CR-1278* (U.S. Nuclear Regulatory Commission, Washington, 1983)
7. N. Rasmussen, *WASH-1400* (Nuclear Safety Study, 1975)
8. A. Swain, *Accident Sequence Evaluation Program-Human Reliability Analysis Procedure, NUREG/CR-4772* (US Nuclear Regulatory Commission, Washington, D.C., 1987)
9. G. Hannaman, A. Spurgin, Y. Lukie, *Human Cognitive Reliability (HCR) Model for PRA Analysis, EPRI NUS-4531* (Electric Power Research Institute, USA, 1984)
10. D.E.A. Embrey, *SLIM-MAUD (Success Likelihood Index Method—Multi Attribute Utily Decomposition): An Approach to Accessing Human Error Probability Using Structural Expert Judgment, NUREG/CR-3518* (1984)
11. J. Williams, HEART (human error assessment and reduction technique)—a proposed method for assessing and reducing human error, in *9th Advances in Reliability Technology Symposium, London, UK* (1984)
12. R. Hall, J. Fragola, J. Wreathall, *Post Event Human Decision Errors: Operator Action Tree/ Time Reliability Correlation, NUREG-CR-3010, BNL-NUREG-51601* (U.S. Nuclear Regulatory Commission, 1983)

13. E. Hollnagel, *Cognitive Reliability and Error Analysis Method: CREAM* (Elsevier, 1998)
14. S.E.A. Cooper, *A Technique for Human Error Analysis (ATHENA), NUREG/CR-6350* (1996)
15. G. Gertman, H. Blackman, J. Marble, J. Byres, C. Smith, *The SPAR-H Human Reliability Analysis Method, NUREG/CR-6883* (U.S. Nuclear Regulatory Commission, Washington, D.C., 2005)
16. B. Kirwan, Nuclear action reliability assessment (NARA)—a data-based HRA tool. Saf. Reliab. **25**(2) (2005)
17. U.S. Nuclear Regulatory Commission, *Review of Findings for Human Error Contribution to Risk in Operating Plants, Job Code E8238, INEEL/EXT/-01-01166, NUREG, Office of Nuclear Regulatory Research* (U.S. Regulatory Commission, Washington, D.C., 2001)
18. J. Forester, A. Kolaczkowski, E. Lois, D. Kelly, *Evaluation of Human Reliability Analysis Methods Against Good Practices, NUREG-1842* (U.S. Nuclear Regulatory Commission, Washington, D.C., 2006)
19. J. Vucicevic, *Human Error—Crucial Factor in Nuclear Accidents* (2016)
20. D. Gertman, B. Hallbert, M. Parrish, M. Sattision, D. Brownson, J. Tortorelli, *Review of Findings for Human Error Contribution to Risk in Operating Events, NUREG, INEEL/EXT-01-01166, Idaho National Engineering and Environmental Laboratory* (U.S. Nuclear Regulatory Commission, 2001)
21. P. Varde, S. Sankar, A.K. Verma, An operator support system for research reactor operations and fault diagnosis through a connectionist framework and PSA based knowledge based systems. Reliab. Eng. Syst. Saf. **60**(1), 53–69 (1998)
22. Webster Dictionary
23. B. Baars, N. Gage (eds.), *Cognition, Brain and Consciousness—Introduction to Cognitive Neuroscience* (Academic Press)
24. "Vignana," Wikipedia, http://en-wikipedia.org/Wiki
25. "Consciousness," Wikipedia
26. International Atomic Energy Agency, *Establishing a Code of Ethics for Nuclear Operating Organizations, IAEA Nuclear Energy Series No NG-T-1.2* (IAEA, Vienna, 2007)
27. http://www.brainwaves.com
28. www.brainworksneurotherapy.com/what-are-brainwaves
29. http://visualmeditation.co/gamma-brainwaves-facts-and-benefits/
30. International Atomic Energy Agency, *Collection and Classification of Human Reliability Data for Use in Probabilistic Safety Assessments, IAEA-TECDOC-1048* (IAEA, Vienna, 1998)
31. G. Gupta (ed.), *Management by Consciousness—A Spuirituo-Technical Approach* (SRI Aurobindo Institute of Research in Social Science, Pondicherry)
32. A. Swain, D. Gertman, *Handbook of Human Reliability Analysis with Emphasis on Nuclear Power Plant Applications, NUREG/CR-1478* (US Nuclear Regulatory Commission, USA, 1983)
33. Centre for studies on human stress, *How to Measure Stress in Humans? A Document Prepared by the Centre Studies on Human Stress* (Quebec, Canada, 2007)
34. L.E.J. Bourne, R.A. Yaroush, *Stress and Cognition: A Cognitive Psychological Perspective* (2003)
35. P. Varde, et al., *Level 1+ Probabilistic Safety Assessment for Dhruva* (BARC Internal Technical Report, 2002)
36. L. Potash, *Experience in Integrating the Operator Contributions in the PRA of Actual Operating Plants, ANS/ENS Topical Meeting on PRA* (American Nuclear Society, Port Chester, NY, 1981)
37. Nordic Liaison Committee for Atomic Energy, *Human Errors in Test and Maintenance in Nuclear Power Plants* (Nordic Project Work, NKA/LIT, 1985)
38. A. Skjerve, A. Bye, *Simulator-Based Human Factors Studies Across 25 Years—The History of the Halden Man-Machine Laboratory* (Springer, 2011)
39. Y. Donghan, Expert Opinion Elicitation Process Using a Fuzzy Probability. J. Koreas Nucl. Soc. **29**(1), 25–34 (1997)

40. S. Rivera, P. Baziuk, J. NunezMcleod, Fuzzy uncertainties in human reliability analysis, in *Proceedings of World Congress on Engineering*, London, UK, 2011
41. G. Ariavie, G. Ovuworie, Delphi fuzzy elicitation technique in the determination of third party failure probability of onshore transmission pipeliine in the Niger delta region of Nigeria. J. Appl. Sci. Environ. Manag. **16**(1), 95–101 (2012)
42. S. Lee, P. Seong, *Design of an Integrated Operator Support System for Advanced NPP MCRs: Issues and Perspectives*, ed. by H. Yoshikawa, Z. Zhang (Springer, Japan, 2014)

# Chapter 11
# Digital System Reliability

*Each business is a violation of digital Darwinism, the evolution
of consumer behavior when society and technology evolve
faster than the ability to exploit.*

Brain Solis

## 11.1 Introduction

In the 1950s and 1960s, control and protection systems for complex systems, say
for process and nuclear systems, were built around valve-based technology and later
employed electromagnetic relay logics. In the 1970s, solid-state technology changed the way protection and controls were built. Solid-state electronic technology
employed microchips to process the control and logics and provided an elegant
solution for the design of control and protection systems. These systems were
finding applications through the 1970s until the end of the millennium when digital
technology was used even in safety-critical systems.

Digital systems are emerging as a replacement for existing analog/solid-state
systems when they reach their end of life or become obsolete. They are also
increasingly being deployed in new plants due to their various advantages including
their compact design, efficient and faster information processing, and design flexibility. This flexibility leads to reconfigurable and self-diagnostic features that
facilitate distributed controls and make digital technology a designer's choice for
development of instrumentation and control (I&C) for complex safety-critical
systems.

The major feature of digital systems that distinguishes it from conventional
electronics is the use of the software component along with hardware electronics.
Designers perform extensive verification and validation (V&V) of software components to reduce the chances of software and common cause failures. However,
V&V alone is not adequate for postulating with high confidence the likelihood of
software-induced failures. Hence, a quantified estimate of reliability requires a
probabilistic approach along with a deterministic approach. Digital systems also use
complex boards and microchip designs that pose challenges toward evaluating the

© Springer Nature Singapore Pte Ltd. 2018
P. V. Varde and M. G. Pecht, *Risk-Based Engineering*, Springer Series in Reliability
Engineering, https://doi.org/10.1007/978-981-13-0090-5_11

safety of the controls and logic. For example, microchips in general and field-programmable gate arrays (FPGAs) in particular, multi-layer board designs, advanced connectors, and complex input/output communication at the board level all come with new failure modes. Reliability modeling that considers these failure modes requires application of physics-of-failure (PoF), root cause analysis, and development of failure mode taxonomy to achieve the required confidence in results.

The approach used for reliability evaluation of analog systems is not adequate for digital systems. Apart from hardware electronics, modeling the software component introduces additional challenges. Although software reliability as a field has matured, when it comes to reliability modeling of safety- or mission-critical systems, the existing approach to software reliability is not adequate because these systems require very high reliability and modeling for common cause failure among the redundant channels.

To a great extent, reliability analysis of digital systems is being performed with the same approach as was used for analog systems. However, with time there is evidence of accumulated operating experience on digital systems as well as technological developments in risk modeling. In spite of these developments and the state of the art in digital system reliability, there is still no consensus on an accepted methodology for digital system reliability modeling [1]. There are focused efforts to formalize the failure mode taxonomy, and although this is an important step toward improving the modeling techniques, further work is required in respect of a formal approach to digital system reliability. Further, digital system modeling requires a life cycle approach, particularly the hardware components.

Given this background, this chapter discusses the deterministic as well as probabilistic aspects of digital systems. It then proposes a simplified approach to support the regulatory case for safety demonstration on a conservative basis. The goal of this approach is to make the evaluation process risk-based by integrating deterministic and probabilistic aspects.

## 11.2   A Brief Overview

A deterministic philosophy provides the basis for qualification of safety-critical digital system design [2]. The deterministic approach is mainly characterized by conservative and prescriptive criteria in design of systems (e.g., defense in depth, redundancy, diversity, and fail-safe criteria), as well as guidelines on maintenance of adequate safety margins [3]. Along with qualification testing and verification and validation, these very principles form the basis for safety evaluation. Adherence to and compliance with elaborate quality assurance plans for vendor selection, formulation of requirement specifications, design, fabrication procurement, commissioning, and operation formed the major considerations. A general approach for qualifying these systems was evaluating the requirements of protection and controls against a set of postulated accident sequences, particularly considering the common cause failure potential of the digital redundant and diverse protection channels. Conservative designs, for example, incorporate redundant provision of the analog channel for tripping the reactor or postulate human action in tripping the plant such that safety consequences could either be eliminated or reduced to the level of mitigation and plant safety is ensured in the long run. Since software is one of the major features of digital technology, rigorous V&V formed the basis for qualifying the software [4, 5].

The deterministic approach has served well to establish digital I&C systems in nuclear plants. However, because digital technology is relatively new, there were concerns about its reliability. The new modes of failure in hardware as well as software components needed to be studied [6]; thus, varying approaches were employed to evaluate the reliability of digital systems. Initially, the approaches were not much different than the ones employed for analog systems, with some assumptions related to software component reliability modeling. For example, the modeling focused on hardware digital electronics configured in, say, four redundant channels, with another division poised for the eventuality that the primary division failed. Here, functional failure analysis formed the major approach at the module or electronic card level, whereas component-level analysis was conducted using data either from the vendor or from handbooks such as MIL-217-F [7] to determine the failure rate at the module/card and division/train level. Failure probability at the system level was evaluated by applying voting logic [8].

The software components were modeled very simplistically by attributing assumed value for independent as well as common cause failure level. Plant-specific experience was used later for estimating the system unavailability [9]. These initial efforts provided adequate confidence in digital system reliability; however, the safety community required more realistic modeling of the digital systems. For example, there was a requirement to account for the dynamics of interactions (particularly software components and software/hardware interactions), the availability of diagnostics and thereby online fault-tolerant features, and new/ additional modes of software/hardware failures. New technology, such as FPGAs, smart instrumentation, and other embedded features, became part of the digital technology and posed additional challenges to system reliability modeling.

This was a time when analysts used available resources and technology at their disposal to model digital systems. As a result, there were as many methodologies/ approaches as the number of digital reliability studies. Each study had its own way of looking at the system configurations, handling common cause failures, using expert judgment, and arriving at component failure rates, albeit in the absence of full knowledge of component failure modes. The level of detail of software modeling also varied significantly.

There is still no consensus on any one accepted methodology. However, at this point, the risk and reliability community feels necessary to develop a harmonized and standardized methodology approach for modeling digital system reliability.

This chapter discusses the deterministic aspects of design and operation that have a direct bearing on system reliability. However, at the end of the day you would like to have measurable or quantified aspects of system reliability; thus,

Sect. 11.4 discusses modeling for evaluating the impact of digital systems through the case of reactor protection systems in nuclear plants. Sections 11.5 and 11.6 focus on some special areas that now need attention.

## 11.3   Design for Reliability

Design for reliability (DfR) is an established approach to design of complex engineering systems [10]. In the context of digital systems, this approach deals with designing and developing techniques, methods, materials, and process that seek to make digital electronics reliable. This approach is essentially deterministic because the principles of fail-safe design, independence, redundancy and diversity, fault tolerance, and testing and validation are incorporated in the design. Hence, in this sense, this section covers the deterministic aspects of reliability in support of design of digital electronics. The major elements of the DfR process are: formulation of reliability requirements and specifications, materials selection, and a qualification program.

The other aspect of DfR deals with engineering and management and engineering of the design and to a large extent regulatory or code stipulations to ensure design and operational aspects for system implementation. These tasks impact electronic hardware reliability through the selection of materials, structural geometries and design tolerances, manufacturing processes and tolerances, assembly techniques, shipping and handling methods, operational conditions, and maintenance and maintainability guidelines [11]. The following section discusses in brief the major features of DfR.

### 11.3.1   Governing Design Considerations

The major element of the design approach for nuclear plants or any other safety-critical systems is governed by a defense-in-depth philosophy, which involves principles of independence, redundancy, diversity, and application of fail-safe and single-failure criteria. I&C systems cater to two broad classes of functions, viz. protection and control, or, in the context of reactors, regulation of reactor power [12]. The digital I&C design enables integration of the controls and logic of engineering safety features with protection systems. Maintenance of adequate design margins and quality assurance in all phases (i.e., design and operations) on a conservative basis provides for addressing uncertainty due to inherent randomness in design and operational parameters and lack of knowledge—and forms another governing design consideration. Experienced nuclear plant operations have suggested that often this conservative safety margin has worked to address certain unpredicted events or faults encountered in real time.

The objective, here, is to design highly reliability and fault-tolerant system. Once the parts, materials, processes, and stress conditions are identified, the aim is to design a product using parts and materials that have been sufficiently characterized in terms of performance over time when subjected to the manufacturing and application profile conditions. A reliable and safety-critical system can be designed only through a systematic and methodical design approach using PoF analysis, testing, and root cause analysis.

The following subsections will elaborate on the major attributes or governing considerations of design for reliability and risk:

### 11.3.1.1 Protective Architectures

For safety-critical systems, it is generally required to ensure that design likelihood of failure is as low as possible because the protection function must perform a safety function to avoid the undesirable consequences. If we look at the major design features of the reactor protection system architecture, the overall emphasis is on protecting the reactor in a timely and efficient manner. The protection functions should be designed to address the requirement of confining the hazard. For example, a reactor protection function is called for when an operating parameter (reactor trip parameter) reaches a limiting system setting. The reactor characteristics in terms of neutronic (reactivity) and thermal hydraulics will govern the demand and further dynamics of the protection function. For example, the protection systems are ideally designed to automatically shut down the reactor by terminating the neutron chain reaction and bringing the reactor to a subcritical level. In case the protection system fails to actuate, then what? Does the design have provision to automatically actuate the secondary reactor shutdown system? Or does the design permit manual action within a specified time limit to actuate the shutdown function followed by the mitigation function, say, by injecting boron into the coolant circuit. The point being made here is that the overall protection system architecture will be governed by the intrinsic behavior of the core in terms of the dynamics of neutronic and thermal-hydraulic parameters.

At a lower level, when it comes to protecting the safety function itself we see that even protection/safety features require protection. Most of the protection systems or safety systems remain dormant during normal operation of the plant and may pose challenges to locate faults which remain latent. Hence, provision for surveillance and monitoring is required to ensure healthiness of the system. Certain provisions like built-in test features, like Fine Impulse testing—that sends an electrical signal through the protection logic periodically and monitors health of the system based on input and output signal mapping. The diagnostic signal pulse generated is so small (of the order of ~ few m-seconds) duration such that it is not adequate to actuate the final electromagnetic relay.

### 11.3.1.2   Stress Margins

The design of the component and further qualification procedure should ensure that operating range of the component should such that apart from normal conditions the component also conforms to design basis and severe accident conditions also. The stresses generated, particularly during design basis and beyond design basis condition might induce stresses in the component that adversely affect the component to perform its intended function satisfactorily. The vendor generally provides the, apart from duty cycles, operating ranges, e.g., temperature range, humidity and vibration ranges for which the component has been demonstrated to be performing satisfactorily at vendors end. However, the qualification procedure at user end should verify that the component's performance for intended application. Here, it is important that subcomponent meet the accepted tolerances provided in the design. The objective here is to ensure that adequate safety margin has been established through statistical procedures. In case these are issues in respect assessment of margins, derating can be used as part of design philosophy. What it means is that components having higher stress capability that for example will keep actual applicable stresses like thermal, mechanical, electrical stresses to lower then the manufacturer's specified ratings.

### 11.3.1.3   Independence

The actuation of a safety function on demand in general and the protection function should not be dependent on some active source. For example, the functioning or actuation of the protection function should depend on availability of the power supply. If the power supply fails, then the system should acquire a fail-safe state. For example, the protection system function may be independent of the availability of a power supply; nevertheless, the failure of the power supply should result in safe actuation, i.e., tripping of the reactor. That is why we claim that the reactor protection function should be independent of the availability of a power supply. Further, many reactor systems have two independent shutdown systems, for example, pressurized heavy water reactors (PHWRs) have a primary shutdown system comprised of shutoff rod systems and a secondary liquid poison injection system. In case the primary shutdown system fails, the secondary shutdown system activates to bring the reactor to a safe shutdown state.

However, the coder guidelines require that these two systems should be physically and functionally independent. That is, no single conceivable common cause failure event can fail in both of the systems. The point we make here is that independence among the redundant I&C channel as also among two systems (e.g., the primary and secondary shutdown systems) is a requirement to ensure that common cause failure (CCF) does not affect the redundant provisions in the plant, be it four redundant channels or two redundant systems.

### 11.3.1.4   Single-Failure Criteria

The basic assumption here is that the hardware component can fail and the design should be such that this failure should not affect the safety function. Single-failure criteria means no single component failure should affect the availability of a safety function. The design of an I&C system should meet single-failure criteria, which means failure of a component on a board, failure of a module, failure at the I&C division level, and, finally, failure at the system level should not adversely affect the protection function.

### 11.3.1.5   Fail-Safe Design

Failure of any component, I&C module, or system should result in the safe state of the plant and the system itself. For example, failure of power supply to protection logic should result in a reactor trip. Power supply failure to any transmitter should take the reading of the instrumentation such that a trip signal is generated. In case the trip parameter is "low compressed air pressure," then the power supply failure to the compressed air switch operation should generate a low-pressure condition. Conversely, if the trip parameter is "high pressure in tank," then the power supply failure to the tank pressure transmitter should generate a high-pressure condition to fulfill the "fail-safe criterion." Even though it is impossible to meet this criterion in all aspects of the design, nevertheless, from a regulatory point of view this criterion should be met as far as possible. In case this criterion is not met then there should be justification or compensating provisions in the system to ensure safety.

   Failure mode and effect analysis (FMEA) should be performed to ensure that the design is "fail-safe" criteria-compliant. The observations/insights obtained from this systematic analysis should be documented for further transmission to operation and maintenance agencies. The input obtained from these safety studies should form an input to design of online diagnostics and prognostics in digital systems.

### 11.3.1.6   Redundancy

Provision of redundant components to have fault-tolerant features in the design is a safety requirement in design of safety- and mission-critical systems. Provision of more than minimum required component to cater to a safety function, ensures system normal operation, safety and availability even in the event of one or more failures. The provision of redundancy should be checked against common cause failure, like environmental harsh conditions, like increase in temperature, flooding, vibration adversely affecting redundant components channels. In fact, it is required that even supply of actuating provisions like power supply compressed air to redundant channels should be such that failure of one power source does not affect provision of redundant functions.

### 11.3.1.7   Diversity

When there is a requirement for negligible failure or failure probability to be lower than $10^{-6}$/demand, it is difficult to achieve this figure by a single protection system. The main reason is the probability that the common cause failure knock out all the redundant provision in the system. For example, for a protection system that has 4 parallel divisions, the failure criteria is 2-out-of-4 division failures, i.e., one failure system can absorb and two or more division failures take the system to a failed state. Here, any common cause failure event can knock out two or more divisions, e.g., any calibration error during maintenance, any software fault on redundant logic or processor units, or any combination of inputs in all the channels. Hence, to achieve very high reliability or, conversely, a lower demand failure probability, then an independent system called a "secondary system" capable of performing the same safety function is required such that common cause failure probability of these two system is negligible. That means these two systems should be independent of each other.

The above scenario is addressed in the nuclear industry by incorporating a secondary system that is working on different principles than the primary system. For example, the reactor protection function (i.e., fast shutdown of the reactor) is achieved by essentially a gravity fall of all the shutdown devices on demand such that the reactor becomes critical and achieves a safe shutdown state. To incorporate diversity, a secondary shutdown system is incorporated that works on a different principle and not by gravity action (i.e., injection of liquid poison into the reactor core) such that the secondary system is equally fast and effective as the first system in terms of making the reactor subcritical to the required margins and ensuring the thermal-hydraulic parameters to avoid any consequences of decay heat removal. As can be seen, if we use different sets of parameters for actuation of these two systems, then we have achieved diversity because system works by its own principles. These two systems together can provide a low probability of failure.

The principle of diversity is employed in many ways in many safety-critical systems. For example, a motor-operated valve and air-operated valve can work in parallel in a coolant water injection line such that if a power supply is not available then the air pressure opens the valve and achieves the water injection function. Conversely, in the eventuality the air supply is not available, the opening of the motor-operated valve can achieve the injection function.

For safety- and mission-critical systems, diversity is not an option but a requirement to achieve a function and in turn avoid undesirable consequences.

### 11.3.1.8   Level of Automation and Human Intervention

The protection function in a nuclear plant is completely automated. The monitoring of plant parameters is done in online mode and si continuous. The sampling frequency of the parameter is such that tracking a minor deviation or a small spurt in a parameter gets recorded in the control room and any deviation more than a preset limit of the parameter generates a trip. In case the voting logic senses that the majority of the channels validate that the parameter is reaching the set value, the reactor trip function is automatic and complete. For example, out of four divisions/ channels, A, B, C, and D, if only one division has registered in control room trip windows, then the operator has time to investigate whether the fault is genuine or spurious I and if it is genuine, then to determine what is causing the parameter to deviate. If the second division, let us say Channel B, also registers, then the protection system will actuate the shutdown devices and trip the reactor. However, if only Channel A remains registered, then the operator decides to shut down the reactor or keep the reactor operating. However, this decision should be based on well thought out rationales.

Normally, as a design rule human intervention is not expected during the first 30 min of an accident; however, the decision to postulate human action sometimes could be based on the reactor neutronic and thermal-hydraulic aspects. Further, maintenance actions are inevitable, whether they are surveillance or repair. However, these actions should not induce a fault in the system. In case the fault gets induced, then the diagnostic module should be able to detect the fault or deviation such that corrective action can be implemented.

The amount of autonomy that is passed on to the operator should be based on a well thought out strategy. For example, raising of shutdown devices from the reactor core, even in special circumstances involving testing, maintenance, and worth measurements, and even if required should be done with approved procedures to avoid reactor transients and unintended addition of reactivity. Apart from the operational hierarchy, these procedures should be authorized by the reactor physicist or the instrument maintenance agency, such that all the safety aspects are addressed.

### 11.3.1.9   Passive Features

Digital systems move the operational logic from active components (e.g., the traditional relay-based designs) to solid-state semiconductor logics and later to software logic operating on complex microchips. The basic idea for implementing passive systems, be they mechanical, thermal-hydraulic, electrical, or electronics, is to make systems after and more reliable. Hence, passive features should be central to the design of digital systems—for example, embedded architecture, wherein the whole system, including the logics, operates from an advanced chip such as an FPGA. Hence, deployment of passive features often requires new and advanced technology. However, the new technology brings along with it new failure modes. Hence, detailed FMMEA and qualification tests should be carried out for passive features or components particularly in the context of the potential for CCFs, before incorporating them into the system design.

## 11.3.2   Formulation of System Requirements and Constraints

System requirements and constraints are defined in terms of system design objective functions, scope of operation, safety policy (including applicable codes and guides), plant-specific operational and maintenance requirements, plant safety culture, and reliability and risk goals. There are two major risk components in requirements and constraint definitions: (a) inclusion of irrelevant requirements and (b) the omission of relevant requirements. Irrelevant requirements pose penalties in the form of unnecessary overheads and provision in the design and subsequent overheads when the plant goes into operation. On the other hand, the omission of relevant requirements makes the design inadequate and might pose safety issues. Even though system retrofitting/modifications can address these limitations to some extent, it has often been seen that the physical constraints may allow only limited upgrades.

The perpetual process of verifying requirements with the design process is necessary to avoid any gaps or to implement certain upgrades that were not covered in the specifications. Documentation of changes in requirement specifications should be considered as part of the safety culture in design.

## 11.3.3   Postulating the Life Cycle Environment

Postulating the life cycle scenario, starting from the conceptual, through the design, operation, aging management, and refurbishment stages, including disposal strategies, is vital for reliability-based designs. It may require quality plans that need to be followed during each stage of the system. For controls and electronics, generally the normal operating conditions are ground benign conditions, i.e., control room environment characterized by 22 °C temperature and 50% relative humidity for clean room or dust-free environment. Considerations of anticipated operational occurrences, which might occur with a frequency of $\sim 1$ year to design basis loads considering a frequency of $<10^{-3}$/year and further the beyond design basis loads with a very low frequency of $<10^{-6}$/year for severe accident management, provides a robust design. Environmental and operational testing forms part of the qualification process.

It is required to predict the life cycle performance requirements, in terms of continuous operating hours (e.g., power supply or protection logic cards), number of cycles (e.g., relays and switches), and number of demands (e.g., shutdown devices, safety valves). This facilitates right from material selection, design optimization, and plan for system qualification. In fact, projecting the life cycle environment for the I&C systems deployed in nuclear or any other process plants is relatively less complex compared to automotive or space missions; however, there are some strategies that help to implement and control reliability for projected life.

Rigorous market studies on expected component reliability or using quality components (e.g., MIL-grade components) may enable higher reliability in the system than use of commercial or industrial components. As an operational stage of the plant is concerned, in situ health monitoring of the systems can provide valuable information on any incipient generic degradation.

## 11.3.4   Supply Chain Management and Quality Assurance

The quality level of the component and decisions related to supply chain of components (including storage, screening, and final integration of these components into the system) requires a techno-management approach because the individual parts and the process of integration govern reliability to a large extent. Development of a procedure that is approved and qualified by management authorities for parts selection and screening is central to address reliability issues before and during the design stage. This procedure should identify the means and methods to identify counterfeit components, which are detrimental to system reliability. One of the approaches is burn-in tests or any other test that is considered appropriate. The process enables maximizing the operational safety and reliability while minimizing the maintenance resources when the product goes into operation.

The parts' selection criteria should be broadly based on the following three major steps as follows:

1. Supply Chain Evaluation: Identification and evaluation of complete supply chain nodes, mainly the raw materials supplier, the manufacturer, and the distributor. Before we do this, it is critical that the acceptance criteria developed by the plant are adequate to address the unsafe failure modes in the I&C system, particularly the CCF possibilities.
2. Performance Assessment: The component should meet the quality attributes and performance specifications. For a power supply module, it is power quality & reliability attributes; in terms of limits on voltage and current range, environmental conditions under which the module is supposed to perform, e.g., temperature limits, vibration levels; becomes the major criteria for performance assessment. The acceptance criteria should be applied for general conditions and extreme conditions such as fire and LOCA. This requires qualification of these components under rare events, viz. LOCA, radiation, and fire and scenarios related to flooding conditions.
3. Reliability Assessment: This step is central to DfR. In fact, in this step it should be ensured that the failure rate of the components, particularly the unsafe failure modes, should be so low that it is commensurate with the system reliability target, as specified by the regulator. In the digital sector, failure rates at the part or module level (e.g., processors, memory), and even at the electronic card level, are evaluated by the supplier. Often these estimates are directly used for system reliability evaluation. However, the process of integration and many factors that affect the workmanship may have an effect on system reliability. Hence, an

independent reliability assessment should be carried out at the system level where, apart from vendor data and life test data, any plant-specific experience should be used as input for system-level reliability evaluation. One of the techniques often used is virtual simulation employing a PoF approach or even a handbook approach. This aspect will be discussed in the following sections.

### 11.3.5   Failure Mode, Mechanism, and Effect Analysis (FMMEA)

FMMEA forms a critical component of digital system verification and validation. The more common approach to qualify the digital system is through functional failure mode and effect analysis. However, the performance of the FMMEA by the designer at conceptual or fabrication level enables improved coverage to assess postulation of system performance. The major challenge is to design testing vectors that provide improved coverage for software faults. However, the FMMEA has been designed to investigate the hardware failure mode and associated mechanisms and assessment of failure consequences. This aspect has been discussed in detailed in Chap. 12 on physics of failure of electronics.

### 11.3.6   Manufacturing Issues

Manufacturing and assembly processes can significantly impact the quality and reliability of hardware. Improper assembly and manufacturing techniques can introduce defects, flaws, and residual stresses that act as potential failure sites or stress raisers later in the life of the product.

Auditing the merits of the manufacturing process involves two crucial steps. First, qualification procedures are required, as in design qualification, to ensure that manufacturing specifications do not compromise the long-term reliability of the hardware. Second, lot-to-lot screening is required to ensure that the variability of all manufacturing-related parameters is within specified tolerances [13, 14]. In other words, screening ensures the quality of the product by identifying latent defects before they reach the field.

Before the components or electronics hardware board goes into regular operation, it is required that the process of manufacturing is qualified against applicable codes and standards. This also includes, apart from material tolerances, machine settings, requirement of trained manpower and calibration checks on quality checking instruments, maintenance of environmental conditions, and management and regulatory stipulations on acceptance criteria. In case the components are subjected to accelerated testing, then the accreditation of laboratories and environmental chambers is a prerequisite.

Even though the design team goes by the design specifications and rules based on the material capabilities, system configurations and performance criteria, however, for manufacturing a quality product on sustained basis the design team should have good exposure to manufacturing environment in terms of material processing and constraints in terms of configurations and operational parameters such as voltage and current in the context of the process capability. Very stringent tolerances may be prohibitive in terms of manufacturability. In short, the designers should have good understanding of process capability. The component specifications should be related to failure criteria either by scientific modeling or through field performance data. There should be a system of feedback on process quality control and performance of product or component in the field conditions. The defect tolerance capability of the product or component should be revisited in light of field data. The system root cause analysis of defect is a vital tool to ensure increasing the yield and improvement in quality. Here, the distinction should be made between random defect and systematic defects based on the analysis of quality control charts.

Often with the availability of POF models, it is possible to evaluate reliability at component, card or at system level, by performing simulation experiments, as part of virtual qualification during the design stage itself. Virtual qualification complemented by accelerated testing of real electronic card provides needed confidence in predicting performance of product in the field conditions.

## 11.3.7 Special Safety Issues

Some safety aspects overlap on the deterministic side in the context of hardware provision to achieve the required safety margins, whereas the adequacy of these provisions is evaluated employing a probabilistic approach. This section discusses three aspects, viz. requirements of defenses to protect the system from common cause failures, provision of monitoring and surveillance, and assurance on software reliability.

### 11.3.7.1 Defenses Against Common Cause Failures

In the deterministic approach, the principles of independence and diversity are intended to reduce the probability of failure. Hence, in the design stage the redundant equipment functional requirements are ensured such that one failure does not lead to knocking out off all the redundant provision. For example, the power supply to redundant pumps is provided from separate buses; similarly, the power supply to redundant protection channels is also provided from different control power supply buses such that one failure in the power supply bus affects only one bus and other redundant buses keep functioning. In the design stage, the location of the redundant channels is physically separated so that one common cause event, such as an increase in humidity, flooding in the area, or an abnormal increase in temperature, does not affect all the protection channels adversely. Hence, provision includes physical separation in terms of distance and physical barriers to avoid CCF events. Even during plant operation, the surveillance schedule is structured such that no single human error affects redundant channels. Examples of human error include calibration or maintenance-induced failure and installation or inadvertent removal of the wrong jumpers. Since CCFs have safety implications, be it software or hardware, a systematic evaluation is required. There is a need to design the software such that the probability of common software faults does not result in CCF of the protection function. However, even today there is no consensus on one state-of-the-art software reliability in general and analysis of the probability of CCF due to software fault. The CCFs need to be evaluated probabilistically for assessing their potential risk.

### 11.3.7.2   Monitoring and Surveillance

Monitoring and surveillance is at the core of defense in depth. The main function of the control and protection system is to keep a watch on plant parameters such that any deviation can be brought back to normal operation. In case the process parameters reach the limiting system settings, the protection system actuates plant trip. In case the plant crosses its operating limit, safety functions are actuated to mitigate the accident condition. However, there is a watchdog that monitors the health of an otherwise dormant safety system, such as a digital I&C system, through a provision called "fine impulse testing" or inbuilt fault diagnostics such that any degradation/failure in any I&C channel or more than one channel can be detected and corrected in time.

Apart from online monitoring, the periodic maintenance and surveillance checks ensure integrated testing of a channel one at a time, even during plant operations such that any latent fault can be eliminated. This also means calibration checks for root instrumentations.

In case of any incident/accident, there is an administrative provision for root cause analysis of the incident such that the recurrence of the incident can be avoided. The health of the system is monitored online so that any deviation can be arrested. System reliability needs to be ensured using a monitoring diagnostics and fault detection system. Data obtained from field failures, maintenance, inspection and testing, or health (condition) and usage monitoring methods can be used to perform timely maintenance for sustaining the product and preventing catastrophic failures.

### 11.3.7.3   Software V&V

The V&V program for hardware components is well established, and the available national and international codes and guides provide adequate assurance on hardware reliability. However, for a few new components and assemblies, additional analysis, testing, and quality checks are needed [4].

Given the current state of the art, the existing approach to assuring software reliability for digital electronics is deterministic in nature and, like hardware, involves rigorous testing as part of the V&V of the system [5]. This step involves two main types of test—positive and negative. The purpose of positive testing is to validate the software against set or intended functional requirements. The limitation of this test is that it can only validate that the software works for specified test vectors. In short, the coverage of the test vectors limits the testing effectiveness because a finite number of test cases may not provide full coverage by corresponding to scenarios that will be encountered during the operational phase. On the contrary, only one failed test is an indication that the software does not work. Negative testing refers to tests aimed at failing the software by projecting new or unlearned test vectors. Software should have sufficient exception handling capabilities.

Advanced methods have been developed in the commercial and industrial sectors for evaluating software reliability. Here, these methods cater to complete software life cycle in five major phases, viz. (a) software specifications are prepared in the analysis phase; (b) the software structure, including logics and interaction and information flow, is worked out in the software design phase, (c) the software is coded by different teams followed by integration in the coding phase, (d) the software is tested to improve quality and reliability in the testing phase, and V&V forms a major component of the testing; and (e) the software is put into operation in the operation phase. In fact, the earlier the software is put to test, the better it is for software reliability as fixing the software at later stages is resource-consuming. In fact in certain situations such as safety- and mission-critical applications the software failure consequences become prohibitively high.

However, software reliability for safety-critical systems requires protection of the software against CCFs. Hence, the techniques available in other domains may not work in safety-critical applications because even one software CCF disabling complete protection logic may pose prohibitory high consequences. Thus, the nuclear industry keeps addressing eventualities in case of digital channel failure, and independent safety studies are performed to demonstrate the safety of the plant, by either crediting human action or by keeping an independent analog channel that can independently trip the reactor.

## 11.4   Risk-Based Modeling for Design Evaluation: A Probabilistic Approach

Digital system modeling requires modeling of hardware and software failure to have required confidence in the quantified estimates of system reliability. Even though many PRAs are performed considering hardware reliability, there are a number of approaches that attempt modeling software reliability as part of digital system modeling. The following section provides the probabilistic approach to digital system modeling.

## 11.4.1 Background

The foregoing discussions have led to the consolidation of a design approach to digital systems that includes hardware as well as software aspects and occasional references to how the human factor affects operational aspects. The approach appears exhaustive and covers many aspects that the system designers have learnt through experience. Some of these aspects have been translated into requirements stipulated in relevant codes and guides. This knowledge base is vital to achieve confidence in the reliability of the digital system design, albeit qualitatively and not quantitatively.

To provide a quantitative estimate of reliability, it is imperative to employ a probabilistic approach to the reliability of the digital system itself and to a risk evaluation of the system this digital system is protecting or controlling. For safety- and mission-critical systems, the development of probabilistic risk assessment models is normally a regulatory requirement and this approach is increasingly being employed in support of risk-informed/risk-based applications.

The reliability assessment of digital systems poses many challenges mainly in terms of modeling system dynamics, fault-tolerant features, communication protocols, software modeling, new technology (e.g., FPGAs), and distributed features. The available literature suggests that a new approach is required to address these issues. A general understanding is that there is no consensus on a single approach for evaluating reliability of digital systems.

The Nuclear Energy Agency Working Group on Risk Assessment (NEA/ WGRISK) has develop a taxonomy for digital system reliability which has certainly given a push toward harmonizing the model and the procedure for categorizing failure modes for digital system reliability. These efforts are expected to improve the groundwork for digital system modeling [15].

## 11.4.2 Limitations of the Traditional Approach

Some issues still need to be fully resolved before a realistic model for digital systems can be developed, as follows:

- Digital systems are relatively new technology, and there is a need to understand fully various failure modes. For example, FPGAs, unlike microchips, are a new technology and further research employing a physics-of-failure approach is required to understand their performance and reliability.

- There is no consensus on software reliability modeling.
- The dynamics of interaction, particularly intermittent phenomenon capturing, remains a challenge.
- Digital boards are more complex in terms of hardware/software interactions and population of electronic components; hence, the understanding and documentation of failure modes of boards is a challenging task.
- Although there is new information on electronic component degradation, further rigorous developmental work is required to understand how degradation manifests and determine how to identify and associate these aspects with component and module failures. There are certain physical phenomena, such as electromigration, time-dependent dielectric breakdown, hot carrier effects, and negative bias temperature instability (NBTI) [16], need to be modeled to arrive at estimate of system reliability. However, the state of the art of digital systems has not reached the required level to accurately characterize the rate of degradation and model the instant of failure in general and new modes of digital system failure, in particular.
- Digital technology has brought some challenges, such as software CCF potentials, new possible failure modes, and system dynamics.
- The current approach to system modeling is essentially static in nature, while for digital system modeling ideally a dynamic approach is required.

At this point, we take note of the existing approaches for digital system modeling and the procedure adopted by the NEA/WGRISK taskforce to demonstrate the taxonomy [15]. Further, the protection system using digital technology has a higher safety impact than any other system using digital technology. The only undesirable mode of a protection system is "failure to actuate" (or failure to function) the shutdown devices. How this mode happens is a matter of root cause analysis and requires efforts to understand how software and/or hardware failures on the module/board will lead to the unsafe failure of the protection system (i.e., failure to actuate the shutdown devices). Some failure modes in the protection system might lead to spurious actuation of the shutdown; however, they are "safe failures." Of course, these modes may not be fully safe because they introduce transient and put demands on the safety and process systems whereby the performance of these systems determines the "safe or unsafe" state of the plant. However, for this demonstration let us assume that spurious shutdowns do not have safety implications.

## 11.4.3   Failure Mode Taxonomy for Digital Systems

In Sect. 11.3 on design for reliability, we discussed FMMEA; however, this technique is primarily targeted at hardware failure. The focus of the section was on understanding, along with the modes of component failure, the associated mechanism, and its effect on the local board and the system. However, FMEA or FMECA is not adequate in the context of digital I&C systems because the hardware/software combination introduces additional challenges in terms of additional or new failure modes. Further, the operating experience shows that the software failure needs to be understood in more detail, particularly the CCF potential of the software. The available literature shows that there are as approaches of analyzing the digital systems; however, there is no consensus on a single approach to be used in PRA. Hence, harmonization of the procedure for digital I&C was a requirement.

The objective of a failure mode taxonomy for digital systems is to build a platform for developing a digital system reliability model in the context of PRA. The available experience operating a digital system certainly provides the needed input. Even though the initiative for the taxonomy came from VTT, Finland [17], and Nordic Nuclear Safety Research [6], the task group formed by the NEA's Working Group [15] on risk wherein many countries shared the methodology and knowledge of failure modes of digital systems, provided the needed push toward harmonization and standardization of the approach for digital system modeling.

The salient features of the digital system failure mode taxonomy are:

- The taxonomy is meant to support modeling of digital protection and control systems in nuclear plants as part of the development of a PRA model of the plant.
- The taxonomy lays down criteria to provide guidance on identifying and characterizing the applicable failure modes, e.g., clear definitions, exhaustive lists, maintenance of hierarchy, exclusivity, and data/information to support the mode and PRA in general and CCFs in particular.
- A structural framework proposes a hierarchy with five levels—system, divisional/channel, I&C, module (hardware/software), and component.
- The framework proposes failure modes for hardware and software and the interaction (software/hardware) processes.
- The vendor or designer generally provides the hardware failure data for risk modeling. The root source could be based on handbook estimation or plant operating experience.
- For software failure, the taxonomy takes note of available approaches such as screening out, screening value, expert judgment approach, or operating experience approach. No accepted approach is available in the context of PRA, so a simple approach around the probability of software failure is used.

The available taxonomy provides guidance based on the modeling expertise generated over the years, as well as operating experience. The vital contribution of this work is that it brings to light the gap areas that still exist in digital system modeling, which leads to the goals for future R&D work and informs analysts about major assumptions that can be indicators to address the gap areas between the model and real-time behavior of the system.

What we have discussed so for is the state of the art in digital system modeling as part of risk evaluation or risk impact of digital systems; however, there are some design processes/guidelines that need to be followed to introduce reliability into the design. In fact, the design for reliability for digital systems is vital for ensuring that the system meets the highest criteria and requires a life cycle approach. The following section presents the design for reliability approach for digital hardware systems.

### 11.4.4   Reference System Description

Figure 11.1 shows the reference digital plant protection system (DPPS). This study, wherever possible, will use the digital system taxonomy such that this step goes toward harmonization or standardization of the procedure for reliability modeling of digital protection system. Accordingly, the complete system has been structured into four levels, viz. field, monitoring and detection, processing, and actuation levels. The DPPS has four redundant divisions. Each division can take analog as well as digital input from the field sensors. There are two types of parameters that form input to the DPPS, viz. digital and analog inputs. The digital inputs are fed directly to the digital input card, whereas analog parameters are fed first to the analog-to-digital conversion card and then to the digital card. The output of the digital card is fed to the processor unit along with signals from other divisions, viz. B, C, and D. Similarly, the output of the digital input card is also fed to other processors in other division, viz. B, C, and D. The output of the processor unit (PU) is fed to the voting module (VU). The output of the voting logic goes to the reactor trip circuit. If any parameter in at least two divisions reaches the trip setting, a reactor trip should be initiated. The reactor trip circuit after satisfying the condition of 2-out-of-4 majority voting actuates the n-shutdown devices and trips the reactor. Figure 11.2 shows the reference DPPS architecture.

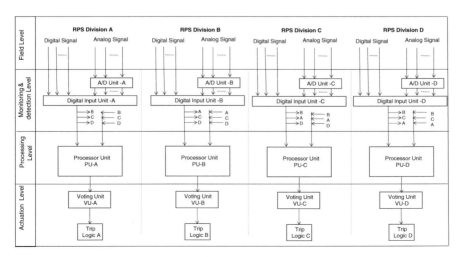

**Fig. 11.1** Simplified architecture of a digital reactor protection system

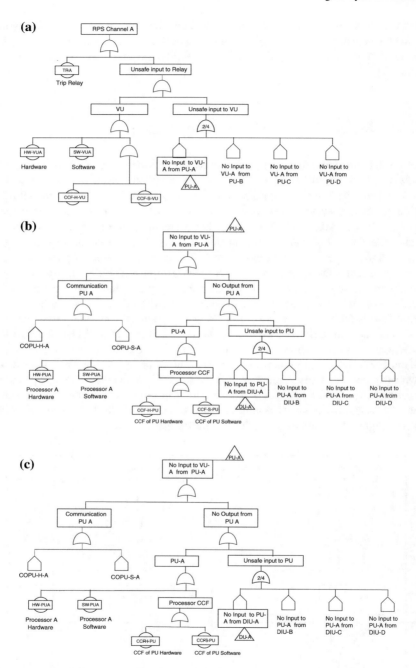

**Fig. 11.2  a** A simplified fault tree for a digital protection system depicting dynamic nodes, hardware and software failures, including CCF events. **b** A simplified fault tree for digital protection, 2/3. **c** A simplified fault tree for digital protection, 3/3. *Legend* RPS: reactor protection system; CCF: common cause failure; HW: hardware; SW: software; PU: processor unit; VU: voting unit; S: software; H: hardware; COPU: communication–processor unit; DIU: digital input unit

## 11.4.5   Technical Requirements in Probabilistic

This section discusses aspects that are directly related to improve the quality of a digital system reliability modeling. Any digital system modeling approach can be evaluated considering capability of the subject approach in respect of the following aspects.

### 11.4.5.1   Dynamic Environment

There are two types of interactions in digital logic operation, viz. type I and type II interactions which are dynamic in nature. Type I interactions are dynamic interactions between the various components and subsystems of I&C systems, and type II interactions represent the interaction between plant physical parameters and further between the I&C systems itself. There are concerns about the ability of static fault tree and event tree methods to capture these type I and type II interactions [18].

### 11.4.5.2   Input Data

The digital component and subsequent system reliability evaluation require data that can predict the system performance; the analysis should not be based only on historical data. Keeping in view this requirement, the hardware failure rate data should ideally be generated employing a PoF approach, as fundamentals of this methodology are based in science. Of course, the data collected from plant records are valuable input; however, to get the required confidence, these data should be updated using life-testing experiments or physics-based approaches. Further, conservative and realistic assumptions should be used to arrive at CCF probabilities. There should be explicit considerations of the availability of defenses against CCFs in the plant.

### 11.4.5.3   Software Reliability

Software-independent failure probability can be expected to be low; however, the concern is about the CCF potential of software systems in digital systems. In fact, there are new failure modes in digital systems due to dynamic interactions, dynamic memory handling efficacies, and online diagnostics. There should be enough justification for qualifying the digital system that software failures will not adversely affect the system.

### 11.4.5.4   Common Cause Failure

As discussed in earlier in Sects. 11.3.1.3 on Independence and 11.3.7.1 on Defenses against Common Cause Failure, there should be adequate assurance that digital system's unsafe failure does not pose safety implications. There should be a risk-informed approach to qualification of software or hardware to address CCF potentials of digital systems.

### 11.4.5.5   Intermittent Failure

Adequate provisions should be demonstrated to identify the source and causes of intermittent failure in general and intermittent unsafe failures. The online diagnostics provisions should have demonstrated the ability to capture and/or isolate intermittent failure and enable further maintenance action to remove the faulty component. For example, the fault category, such as single-event upsets (SEUs), needs to be either precluded by hardening approaches or should be captured by the online diagnostics. The requirements related to hardening of the digital systems against high-energy electromagnetic pulses and their potential for CCF should be evaluated.

### 11.4.5.6   Human Factor Considerations

Even though digital systems have been operating in many nuclear research and power reactors, this technology is still relatively new. In the context of challenges posed by new failure modes particularly related to software as well as new components such as FPGAs, a new culture of looking at system operation is required. For example, the dependency on self-diagnostics and the response of the maintenance crew are critical to safety and performance. Performance of operation and maintenance activities and their effect on the system needs special considerations. Further, digital technology is being implemented by employing standard cards/modules (e.g., digital input cards, digital output cards, processor modules). These cards are used in different systems by implementing application software for different implementations as also jumper configurations. While this approach has made the system development process very efficient and effective, the chances of human-induced error might pose challenge.

### 11.4.5.7   PRA Requirements and Meeting Reliability Targets and Goals

For any approach to qualify as a candidate for reliability modeling of digital systems, it should be possible to integrate this methodology with the PRA model of the plant. The results of the analysis should meet the regulatory goals and criteria. The results of digital system reliability should be used for computing the core damage frequency (CDF), and the regulatory goals for CDF should be satisfied not only for the point estimates but also for the uncertainty bounds. Further, a sensitivity analysis should be performed against these uncertainty bounds as well as major assumptions of the analysis.

### 11.4.5.8   Considerations of Security Aspects

Over the last two decades, computer or digital system security has acquired added dimension due to malicious attacks having potential to affect system safety. The exploitation of these vulnerabilities has been witnessed with growing frequency and impact [19]. The digital architecture employed for protection and regulation configures the network to operate with distributed and interconnected nodes, while the communication buses act as links between various nodes. Given this configuration, the system becomes more complex at least compared to its predecessor solid-state technology. Although the system has redundant architecture that makes it fault-tolerant, the cyber security-related issues pose a threat potential for normal operation of the protection and regulation function. Hence, the system needs to be resilient to the threats. In the current context, the security aspects form part of the system design, but they do not rule out the possibility of security issues. Hence, an approach is needed that protects or handles incoming threats by taking the system to a safe state while minimizing the probability of unsafe events. Hence, it is required that the security of computer-based system is demonstrated against identified design basis scenarios or design basis threats. Here, consideration should be given to incorporating into such scenarios threats of either stand-alone attacks using/against computer systems or coordinated attacks including the use of computer systems. Design basis threats are identified from either credible intelligence information or from the history of incidents and postulation of likely scenarios. The IEC Standard IEC-27005 [20] makes recommendations on information security systems, and one of the most relevant recommendations is that quantitatively or qualitatively assess (i.e., identify, analyze, and evaluate) relevant information risks, taking into account the information assets, threats, existing controls, and vulnerabilities to determine the likelihood of incidents or incident scenarios, and the predicted business consequences if they were to occur, to determine a "level of risk." In this context, it becomes necessary that the risk assessment, either qualitative or quantitative, should be performed and the design basis of the system should be demonstrated.

## 11.4.6   A Brief Overview of Modeling Approaches: State of the Art

In most of the present generation approaches to reliability of protection systems, electronics hardware modeling forms the major aspect, while the considerations of CCF and implications of human factors are also analyzed in detail. The procedure for hardware failure modeling either employs the handbook approach, such as MIL-217 [7], expert opinion for special/new components, and, in some cases, the limited plant-specific data that form the input for analysis. Software failure is treated depending on insights on the V&V of the software and using assumptions ranging from no software failure to expert opinion inputs. Generally, the static fault tree, event tree, and fault tree analysis approach, except for the cases discussed in NUREG-6901 [18], are used for system modeling. The available literature [21] shows that there is some sort of inference that the traditional static fault tree and event tree approach as part of PRA is not adequate to capture and model the digital system failure modes [21].

In this context, there are two major developments—the NEA/WGRISK document on development of digital system taxonomy [15] and NUREG-6901 on review of available methodologies for modeling digital systems [18]. While digital system taxonomy is certainly a step forward in understanding and modeling digital systems, the message from NUREG-6901 is that there is no consensus on one methodology as none of the reviewed work is meeting all the criteria discussed in the previous sections. The host of methodologies reviewed include Markov models, Bayesian network, Petri-net, dynamic flowgraph, dynamic event tree, test-based methods, software metrics, and black-box approaches. NUREG-6901 concludes that each approach has unique features and its own advantages, but none satisfies all the requirements of digital system modeling.

However, we need an approach that addresses the requirements of system modeling, such as dynamic scenario, software, communication, and CCF aspects. Markov [22] proposed a simple but elegant dynamic methodology for dynamic fault tree and its application, which can handle events evolving in the time domain. This approach enables modeling for digital systems. The Markov model captures the dynamics of voting redundancy and the fault-tolerant nature of the system by evaluating the degradation, detection, and recovery or repair coverage [23]. However, the limitation of the Markov approach is that it poses constraints in terms of model complexity and state-space explosion that challenge computer memory and CPU overloads. To this extent, analysts have to keep the complexity to a level that can be handled by a Markov model such that conservative assumptions truncate the boundary of the analysis.

Reliability data are another major issue. Given the critiques on the applicability of handbook approaches, particularly MIL-217 [24], operating experience backed up by a PoF approach to hardware components is a desirable option [16]. However, even though extensive R&D is being performed on development of a PoF model for electronic hardware components, there is no authentic document like a handbook that makes information readily available as input for PRA modeling. It can be argued that the available, accumulated operating experience may be used for PRA modeling backed up by PoF models available for microchips, capacitors, and connectors in the open literature.

The way forward is to take the advantage of the available models, methods, and current state of the art to have a pragmatic view of the situation and to develop an approach that attempts at least to fulfill the regulatory requirements, either employing deterministic or probabilistic arguments.

## 11.4.7   A Simplified Approach to Digital System Modeling

This section presents a simplified approach that is an extension of the existing reliability modeling approaches, in terms of limited aspects of dynamic modeling, addressing of data issues, and assumptions for software aspects employing conservative approach or through sensitivity analysis. Note that treatment of the subject remains limited to a philosophical level with illustration of concepts, and no full-scope analysis for the reference system is considered as the authors feel a qualitative treatment of the subject that is expected to be more effective than performing a full-scope case study.

### 11.4.7.1   Assumptions

The broad assumptions of this simplified procedure are as follows:

- Only the digital reactor protection system along with digital and analog forms part of this model; i.e., the hardware shutdown devices are excluded from this modeling.
- The simplified model of digital plant protection system, driven by conservative assumptions, such that it captures essential and key features of digital protection system having implication on safety, is adequate to demonstrate the simplified procedure for digital system reliability analysis.
- Even if the failure data at the board/module level are provided by the vendor/supplier, it needs to be verified by other available approaches, such as operating experience or life-testing/PoF approach.
- Given the function of the reference digital system, i.e., to shut down the reactor when the plant parameter generates a demand, the model captured only in limited sense, the operational and process dynamics enable by the present state of the art in dynamic fault tree and Markov model such that the model does not become too complex for inclusion in PRA model.
- The propagation of the fault occurs from the individual component(s) to board, from the board to the I&C, and from the I&C to the channel.
- Other aspects such as online diagnostics and repairs are intrinsic to digital systems, and communication forms a part of the model without the complexity associated with these aspects.

The following section is limited to specific modeling in the context of the reference protection system, as discussed in Sect. 11.4.4.

### 11.4.7.2   Objective and Scope of This Procedure

The objective is to discuss modeling aspects that address considerations of risk-based evaluation. Hence, the approach should be capable of modeling risk-important interactions in digital systems. In the context of digital systems, the minimum requirements and major aspects that need to be addressed adequately are safety-critical dynamic aspects, CCF modeling, and software integration with hardware [18]. The probabilistic method is limited to digital system modeling and presents a best estimate approach for data analysis while keeping event sequence modeling out of the scope of this section. Accordingly, the candidate major considerations are as follows.

### 11.4.7.3  Modeling for Dynamic Aspects of the System

Digital system operations, as discussed in earlier sections, are dynamic in nature. In this procedure, a limited-scope dynamic fault tree modeling is performed to model one division of the digital reactor protection system, while the Markov model is employed to evaluate system failure probability by considering the effects of repair or investigating voting redundancy. Different initiating events require a reactor to come to a safe state, for example, a class IV power failure or loss of regulation or even a low-probability event such as loss of coolant. This will require processing a different set of plant inputs. In some reactor systems, provision is kept for tripping the reactor; in case the automatic actuation fails due to CCF event, here the operator can trip the reactor manually by tripping the motor generator (MG) set supplying power to the actuating relays. Unlike static fault tree, a recovery provision can be incorporated into the dynamic fault tree for simulating the failure of the primary shutdown system while giving credit to human action. Modeling of voting logic involving k-out-of-n division failure can be modeled with the Markov diagram. The plant's technical specifications allow taking one division out of service for surveillance or maintenance after bringing this division to tripped state. In such situations, we have one more case for dynamic simulation where the division in maintenance was not rejected either due to human error or some hardware/software fault. There are safety demonstration requirements in respect of CCF induced by either software or hardware failure in all the redundant divisions such that the actuation of the redundant reactor protection divisions is blocked in an unsafe manner.

For demonstration, a simplified dynamic fault tree for the reference digital system shown in Fig. 11.2a–c has been developed. This procedure involves use of house events to model failure of various software and hardware components during the timeline of the scenario being modeled. Table 11.1 shows simplified example of matrix for generating more than one scenario in a fault tree, while Table 11.2 describes initiating events associated with matrix. Status of each house events for a given initiating event forms a vector for dynamic modeling.

The objective is not to demonstrate the procedural aspects for analyzing a dynamic scenario, but to model the hardware, software, and combined failures that can cause unsafe or fatal failure of the top event (i.e., protection system failure) and how different branches of the fault tree can be switched in/out to model a dynamic scenario. This is a simplified version of a complex fault tree, and all the nodes have not been modeled to their logical basic component level. Also, certain aspects have not been included as the exercise was meant to demonstrate how to capture critical aspects such as CCF and software and hardware failure related to I&C modules, including communication among the redundant channels. This fault approach dynamic facilitates: (a) one channel taken out for maintenance (as permitted by plant technical specification) and system operates 2/3 logic for a limited duration again as permitted by technical specification, (b) analyzing the effect of software and hardware failure, (c) failure of the communication bus that blocks the signals

**Table 11.1**  A simplified example of matrix for generating more than one scenario in a fault tree

| RPS-FT-Con | House event | | | | | | | |
|---|---|---|---|---|---|---|---|---|
| | RPS-1 | RPS-2 | RPS-3 | RPS-4 | RPS-5 | RPS-6 | RPS-7 | RPS-8 |
| PU-A | T | T | T | T | T | F | F | T |
| PU-B | T | T | T | T | T | T | T | T |
| PU-C | T | T | T | T | T | T | T | T |
| PU-D | T | T | T | T | T | T | T | T |
| COPU-H-A | T | T | T | T | F | T | T | T |
| COPU-S-A | T | T | T | T | F | T | T | T |
| DIU-A | T | T | T | T | T | T | F | T |
| DIU-B | T | T | T | T | T | T | T | T |
| DIU-C | T | T | T | T | T | T | T | T |
| DIU-D | T | T | T | T | T | T | T | T |
| CODIU-A | T | T | T | T | T | T | T | T |
| CODIU-B | T | T | T | T | T | T | T | T |
| LOOP-A | T | F | F | F | T | T | T | T |
| LORA-A | F | F | T | T | T | T | T | T |
| LOCA-A | F | T | F | F | T | T | T | T |
| HU-A | T | T | T | T | T | T | T | F |

**Table 11.2**  Description of events in matrix

| Case No. | RPS-status | Description |
|---|---|---|
| 1 | RPS-1 | Loss of off-site (LOOP) power condition is an anticipated transient, and the frequency range could be from $\sim 1$/year to 0.1/year |
| 2 | RPS-2 | Loss of coolant condition (LOCA) is a design basis scenario frequency around or less than $10^{-4}$/year |
| 3 | RPS-3 | Loss of regulation incidence (LORI) is a design basis scenario frequency that may range from $10^{-2}$ to $10^{-3}$/year |
| 4 | RPS-4 | Processor combination of initiating event: Initiating event is LOCA that also introduces LORI, i.e., reactivity transients |
| 5 | RPS-5 | Communication bus failure between Channel A with other three channels. This could be a hardware or software failure |
| 6 | RPS-6 | The processor in Channel A is taken out for maintenance and during this time a processor or in other channel |
| 7 | RPS-7 | The DIU in Channels A and B fail due to calibration fault on a demand generated by seismic sensors |
| 8 | RPS-8 | Human failure in tripping the reactor when RPS fails |

between two channels, and (d) modeling for failure of signals for more than one initiating event.

Certain aspects cannot be modeled even by employing a dynamic fault tree, as discussed in this section. Examples of these aspects include modeling more than two states of the system (e.g., apart from success and failure), a third state (e.g., degraded), or analysis of various aspects, such as repair/maintenance, system state identification, and fault coverage. The following section demonstrates modeling of the voting logic of a reactor protection system using the Markov approach.

### (A)  *Markov Modeling of 2/4 Voting Logic with One Repair*

Consider the same protection system for one division. If the requirement is to estimate the protection system failure probability that has four divisions A, B, C, and D, as discussed earlier. The failure criteria is 2 channel failures out of 4 channels; i.e., "2/4 channel failure" is the failure criteria.

The modeling inputs/configurations are as follows:

The objective here is to formulate a Markov model to assess the "availability" or, conversely speaking, the "unavailability" in this example. We assume that there is only one repair facility available. The data collected show that the failure rate $\lambda$/unit time and repair rate $\mu$/unit time and $\lambda$ and $\mu$ follow exponential distribution.

The Markov model for the system is shown in Fig. 11.3. In the Markov model, "1" indicates all the four channels are healthy and operational and "0" indicates reactor protection function has failed. The system configuration for various states is shown in the bracket.

The system has three states as follows:

State 1: All the three services are available (1111);
State 2: One channel out of four failed and three are operating (1110), (1101), (1011), (0111);
State 0: Two channels out of four failed, which takes the system to the failed state (1100), (1010), (1001), (0011): This is the failure criteria of the system, viz. 2/4-failure.

Suppose the probability of the system in states 1, 2, and F at time $t$ is given as $P_1(t)$, $P_2(t)$, and $P_F(t)$, respectively. Then the differential equation for the above systems can be written as:

$$\frac{dP_1(t)}{dt} = -4\lambda P_1(t) + \mu P_2(t) \tag{11.1}$$

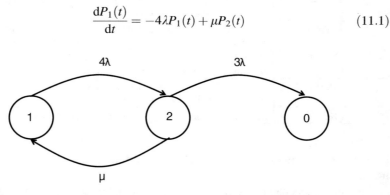

**Fig. 11.3** Markov diagram for 2-out-of-4(F) channels in a reactor protection system with one repair

$$\frac{dP_2(t)}{dt} = -(3\lambda + \mu)P_2(t) + 4\lambda P_1(t) \tag{11.2}$$

$$\frac{dP_0(t)}{dt} = 3\lambda P_2(t) \tag{11.3}$$

We assume that failure rates of the four divisions derived from unavailability estimates from the analysis of channel A. To do this, the failure frequency ($\omega$) was assumed equal to the failure rate ($\lambda$). It is also assumed that the failure rates of all the four channels are the same, i.e., $\lambda_A = \lambda_B = \lambda_C = \lambda_D = \lambda$. With these assumptions, these equations can be solved and the unavailability for these channels for 2/4 failure criteria and with one repair can be estimated. In this way, we have the qualitative model of the system and further quantification can be carried out based on the estimates of failure rates and test intervals as well as maintenance schedules or repair rates of the protection channels.

### (B) *Markov Modeling of 2/4 Voting Logic with Repair, Fault Diagnosis, and Recovery*

We consider one more demonstration, dealing with dynamic modeling of 2/4 (failure) voting logic of reactor protection system crediting online diagnostics and recovery or reconfiguration management system, in a reactor protection system. Once the online prognostic systems detect a deviation, they alert the system manager/operator about incipient fault so that recovery can be initiated and failure can be avoided.

The Markov model depicted in Fig. 11.4 incorporates provision for online detection and subsequent recovery. The reliability estimates obtained in this manner are closer to real-time conditions where the degradation is declared in the control

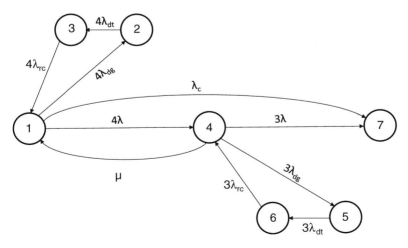

**Fig. 11.4** Markov model for 2/4-division reactor protection logic giving credit to recovery from degraded state of a channel

room and recovery actions are initiated to recover the system from a degraded state. These estimates while being more accurate enable enhancement of safety unlike a static fault tree where systems are either shown to be healthy or in a failed state that overestimates unavailability.

A description of the 7 states of this Markov model is as follows:

State 1 (1111): The reactor protection system is an operational state (all four units are healthy).
State 2 (1111): The system is operating in a degraded state and requires immediate action.
State 3 (1111): The degraded state and the fault has been detected, and recovery has been initiated.
State 4 (0111, 1011, 11011110): One out of four units has failed, but the system is operating with three channels in healthy condition.
State 5 (111): The system is operating with 3 channels; however, there is degradation of one of the three channels.
State 6 (111): The system is operating, the fault has been detected, and recovery has been initiated.
State 7 (110, 101, 011): The degraded state and the fault has been detected and recovery has initiated.

Further, the additional transition modes are: (a) The recovery through repair/maintenance on the failed unit in stage 4 is successfully performed to bring back a failed system; hence, an arrow depicts this condition from states 4 to 1 through a repair rate parameter $\mu$, and (b) from the healthy state (1), the system transits directly to a failed state due to CCF in the redundant channels, which fails more than one channel. The transition parameter $\lambda_c$, as depicted in the Markov model, is used as input for CCF probability.

The transition rates in the Markov model are:

$\lambda$      representative failure rate for all four channels, i.e., $\lambda = \lambda_1 = \lambda_1 = \lambda_1 = \lambda_1$
$\lambda_{dg}$   degradation rate of all four channels
$\lambda_{dt}$   detection rate of all four channels
$\lambda_{rc}$   recovery rate of all four channels
$\mu$     repair rate of all four channels

The differential equations for the above system are as follows:

$$\frac{dP_1(t)}{dt} = -(4\lambda + \lambda_c + \lambda_{dg})P_1(t) + \lambda_{rc}P_3(t) + \mu P_4(t) \tag{11.4}$$

$$\frac{dP_2(t)}{dt} = 4\lambda_{dg}P_1(t) - 4\lambda_{dt}P_2(t) \tag{11.5}$$

$$\frac{dP_3(t)}{dt} = 4\lambda_{dt}P_2(t) - 4\lambda_{rc}P_3(t) \tag{11.6}$$

$$\frac{dP_4(t)}{dt} = 4\lambda P_1(t) - (\mu + 3\lambda)P_4(t) + 3\lambda_{rc}P_6(t) \tag{11.7}$$

$$\frac{dP_5(t)}{dt} = 3\lambda_{dg}P_4(t) - 3\lambda_{dt}P_5(t) \tag{11.8}$$

$$\frac{dP_6(t)}{dt} = 3\lambda_{dt}P_5(t) - 3\lambda_{rc}P_6(t) \tag{11.9}$$

$$\frac{dP_7(t)}{dt} = \lambda_c P_1(t) + 3\lambda P_4(t) \tag{11.10}$$

We have 7 equations for the 7 states. As the number of states increases, the complexity of the problem increases and, beyond a limit, becomes computationally prohibitive. The solution of the above equations can be found analytically, but a numerical solution is obtained for complex scenarios.

Chapter 4 provides the procedure to analyze fault tree (i.e., qualitative analysis), which includes minimal cut-set analysis, estimation of system probability, and dynamic aspects of system modeling including dynamic fault tree analysis, and further procedures for Markov analysis. Of course, the material covered may not be adequate for specific and complex aspects of digital system model; however, it is expected that this input is adequate to model the core aspects, i.e., event sequences evolving in the time domain that requires modeling in simulation mode.

#### 11.4.7.4   Hardware Failure Data

As discussed earlier, in the proposed approach two sources of data are recommended, viz. operating experience and physics-of-failure approach. The following section explores the state of the art and adequacy of these two sources.

#### A.  *Operating Experience*

Given that many nuclear plants and facilities are presently operating with digital protection system functions or similar protection functions, there is need to explore the benefit of extracting reliability data from the accumulated operating experiences. Bickel [9] reviews the failure event reports on the first generation of digital reactor protection systems for US nuclear plants for the period from 1984 to 2006 and discusses that the total accumulated operating experience for this period is 145.5 years or 1.27 million hours. The important observation is that out of 141 reports, 26 could be attributed to CCFs in which 6 events were put in the category of series events. Among these six CCF events, it can also be inferred that the human factor was one of the major causes, specifically while loading of data to the digital systems. This work further explores CCF contributions from individual modules of core protection calculators or digital protection systems. A Westinghouse publication "Nuclear Automation—Core Protection Calculator System" [25] while mentioning the advantages of the core protection calculator incorporated in 1980 (e.g., like gain in margin, software flexibility, and other inherent advantages that come with computer technology) asserts that the core protection calculator system established the groundwork for digital protection technology in nuclear plants. As of now, this core protection calculator, which forms the heart of thermal-hydraulic parameters based on signals from core sensors, has been in operation in over 23 nuclear plants.

Given the fact that in each plant the digital system generally employs four redundant channels, the accumulated operating experiences on each module, viz. input signal/information processing module, logic processor and voting module, and further down to memory, multiplexer, board, and data link, are enhanced by a factor of four, and this experience can be a good starting point to estimate the reliability. Apart from this, these data provide a good insight on CCFs including the human factors that induce potential for CCF. Bickel's work can be referred to for the data extraction from first-generation plants in the USA. Hence, in this approach the operating data will be updated employing the Bayesian approach using new evidence [26].

In fact, through this work, the author would request an international organization such as the International Atomic Energy Agency or NEA/WGRISK to take up the project and develop a data library for digital hardware systems based on operating experience available from member countries. This becomes more important because many countries are operating plants that have digital protection or regulation systems.

### (B)  *Accelerated Testing and Physics-of-Failure Approach*

The accelerated life-testing approach is employed extensively for predicting the life and reliability of components and systems [27]. This approach is particularly useful for new designs or new generations of components, such as a new connector or FPGA [28]. The handbook approach and system-specific approach may not work for new design of components. Even generic components such as connectors require an accelerated life test approach to characterize the life and reliability of the component [29]. The life test models discussed earlier are used to estimate the life of the component.

The PoF approach is essentially a science-based approach to understand various failure mechanisms or degradation mechanisms and thereby provide an effective approach to formulate models toward predicting the remaining useful life and reliability of the components [30]. This approach deals with the application of first-principle models to understand the various failure mechanisms and thereby predict the remaining useful life and reliability of the components. The predictions in this approach are based on the component characteristics, such as material properties, geometrical attributes, and activation energy for applicable degradation processes for given operational and environmental stressors. Accelerated life testing is central to the PoF approach, which enables identification of dominant failure modes and mechanisms, and thereby the precursors for monitoring the health of the component. Failure mode, mechanism, and effect analysis (FMMEA) forms the cornerstone of this approach to identify and prioritize the applicable degradation mechanisms.

The PoF model can be expressed in a general form as:

$$t_{50} = f(x_1, x_2, x_3 \ldots) \tag{11.11}$$

where $t_{50}$ is the median life and $x_i$ are the model parameters. The commonly known PoF model for life prediction is the Arrhenius model, expressed as follows:

$$t_{50} = A\exp\left[\frac{E_a}{kT}\right] \tag{11.12}$$

where $A$ is a process constant, $E_a$ is the activation energy of the process in eV (electron volts), $k =$ Boltzmann's constant $= 8.617 \times 10^{-5}$ eV/K, and $T$ is temperature in Kelvin.

Current-generation advanced microchips such as FPGAs are complex and built using millions of transistors, complex layouts, and higher inter connect densities. As a result, even though these chips are built with extremely higher capability, the reliability of these components remains questionable and should be further researched. Traditional approaches may not provide satisfactory answers. Here, the PoF approach, where FMMEA and life testing are integral to the formulation of PoF models, enables degradation assessment life prediction of the component. This aspect will be discussed in greater detail in a separate Chap. 12 on PoF dedicated to electronics.

This section, in the context of the proposed approach, can be summarized as follows. The plant-specific data are preferred for risk assessment studies; however, if the hardware data are not adequate for certain components, then the life-testing and PoF approach will be explored for required assurance. Often in some situations analysts are required to use the vendor's data on hardware reliability. In these situations, it becomes necessary to validate at least a few sets of data by either operating experience or the PoF/life-testing approach. It may also be noted that, most of the phenomena responsible for CCFs that fall in the category of implicit CCFs can be evaluated using the PoF approach. For example, the potential CCF coupling mechanism, such as high vibration (low-cycle as well as high-cycle), flooding, high humidity, high temperature, low temperature, electromagnetic pulses, and insulation failure, can be evaluated using the PoF approach.

Appendix provides sample hardware data used in reliability modeling of a digital system for a pressurized water reactor PWR [8].

### 11.4.7.5 Software Failure Data

Software failure differs from hardware failure because software does not wear out. Another modeling aspect of the software is related to the type of software, e.g., the operating system or functional software and application software. Even though software reliability as a field has evolved over the years and there are many reliability growth models, Bayesian belief network, and testing-based techniques available in the open literature, for safety-critical systems these approaches may not be adequate due to stringent safety requirements, i.e., a low failure probability, particularly CCF reliability targets and ways to demonstrate through a test metrics that provide required assurance on software reliability. This is why there is no consensus on any one state-of-the-art approach to software reliability. There is, however, general consensus that software reliability quantification is a complex issue.

Certain aspects that are emerging as possible or broad guidelines for software reliability modeling can be summarized as follows:

- The software reliability evaluation should be initiated only after ensuring that a well-established code/standard has been used and adhered to with very high conformance level, for example, conformance to IEC-61508 Part 3 [31], or any other national code.
- For modeling, the software can broadly be divided into three major functional modules, application modules, communication modules, and other functions such as diagnostics and man–machine interface. Often the analysts or developer may not have access to proprietary software component; hence, a sensitivity analysis should be performed to assess the impact of this aspect of software.
- The failure mode taxonomy is a critical part of the software reliability that provides a functional as well as consequential component of the software analysis procedure.
- Software failure can broadly be summarized into fatal and non-fatal failures (fatal failures result in unsafe conditions), which could be due to no output from, say, the processor module, or unsafe output that causes failure of actuation of devices. Non-fatal failures might result in spurious actuation of the system.
- The software running on the redundant module has potential for CCFs. This aspect should primarily be addressed deterministically, while probabilistic aspects may be evaluated for quantification.
- The data and insight generated primarily based on operating experience, either from the same plant or from generic sources should be used for software reliability modeling.
- During the design stage of the plant, the operating experience from generic sources and data and insight gathered on testing performed as part of a V&V program should be used for reliability characterization.
- Human factor is related to software failure in three specific ways: (a) provision of human action modeling in the event of CCF of software/hardware for emergency operator action, (b) human error induced during software updates, and (c) calibration or parameter setting changes incorporated during operation and maintenance of the plant.

Even though extensive work has been reported on evaluation of software reliability in the open literature, there are some procedures like the NKS-341 methodology that takes the state-of-the-art one stage ahead [32]. The objective of NKS report is to develop a methodology for quantifying the critical modes of software failures in digital systems.

Based on the discussions in international meetings on risk modeling on software quantifications, one observation was made by an expert that independent failure probability software is generally significantly less compared to the typical hardware system and is normally inconsequential because it affects only one of the redundant channels. This argument also appears palatable because software does not wear out and the reliability growth that occurs during the software testing as part of verification and validation program that removes faults at least the one have potential for unsafe or fatal failure. In fact, this observation forms the basis for deterministic qualification. From the foregoing discussion, it can be assumed for this procedure that either a software-independent fault can be ignored or a screening value, which is a fraction of the hardware component node in the fault tree, can be assumed.

Regarding hardware failure, there are two observations: (a) use operating experience either from plant-specific sources or from generic sources for CCF evaluation at module levels [9]; (b) use a conservative value of 10% of hardware failure rates for explicit CCFs and a screening value of $10^{-6}$ for implicit CCF which accounts for new phenomena that lead to CCF events [8].

Even though this section discusses the approach for quantification of software components, the final safety aspects will be determined by the backup provision to protection system failure due to software-related CCF, which might include (a) operator action to trip the reactor (by tripping the MG set); (b) availability of independent and diverse protection channels; or (c) a fully independent hardwired (relay-based or solid-state logic-based) system whose functioning is independent of the digital system from the sensors to the shutdown devices.

### 11.4.7.6   Human Error Data

The design of a protection system is such that upon sensing that a parameter has reached its safety system setting values, an automatic actuation occurs and shuts down the reactor. The design ensures actuation of all the safety functions, such as reactor shutdown, and even actuation of emergency safety features to mitigate the consequences of accident are automatic and no operator intervention is expected.

However, in some reactor designs, human action is required in case the primary reactor protection system fails. In such cases, the operator is supposed to actuate a manual emergency reactor knob in the control room that interrupts power supply to the clutch coil of the control and drive mechanism (CDM) of the shutdown devices. This isolation of the power supply to the MG causes actuation of the CDM and thereby shuts down the reactor due to insertion negative reactivity by insertion of shutdown rods into the reactor core. Human error probability for this action has been estimated to be $8.0 \times 10^{-2}$/demand [8] to $2.8 \times 10^{-2}$/demand [33], if the cognition process was considered as knowledge based and the time available is $\sim 1.0$ min. The probability is high because (a) the time window available to the operator is around 1 min, whereas the action itself takes more than 0.5 min; (b) the situation is stressful because the operator needs to perform a diagnostic that reactor protection system has failed and the reactor needs to be tripped manually; and (c) even though the action appears to be a rule-based type actually it also involves a knowledge-based part.

Other errors that may adversely affect the reactor protection system are the ones that induce CCF, for example, a calibration error, such that error is on the unsafe side (i.e., it blocks the reactor trip action), a maintenance error, or an error while loading software updates. These aspects should be analyzed in a separate section on "human reliability."

### 11.4.7.7   Considerations for Common Cause Failure

The existing procedure for CCF evaluation for digital systems is not adequate. Further, the uncertainty associated with CCF in digital systems is posing challenges to sustainable implementation of digital systems. To improve understanding of CCFs, the Oak Ridge National Laboratory has developed a CCF taxonomy [34].

For digital systems, the broad definition of CCF is single failure or an undesirable event that results in unsafe failure of more than one redundant division instantly. This word "instantly" has different connotations for different safety actions; however, for the reactor protection action, it means in one instant or actuation of shutdown devices simultaneously to achieve safe shutdown.

If the protection configuration has four parallel and redundant channels and only one channel failure or isolation can be tolerated, then the definition of a failure protection system could be failure of at least two-out-of-four (2/4) divisions. Given the fact that digital systems use new design rules as well as new technology, including components such as field-programmable gate arrays (FPGAs) and complex programmable logic devices (CPLD), a typical observation is that the density of transistors and metallization on these devices is much higher than conventional microchips. Further, manufacturing technology, including foundry process, development of dies, and packaging, is also different; hence, these components are also expected to have new failure mode and need detailed FMMEA. This calls for a physics-based approach to understand the mode and expected frequency of failure. The priority here should be to analyze CCF of the existing components along with the new component by employing a physics-based approach. This can be done by postulating certain plant conditions that cause deviation from normal environmental as well as operational stresses such that it induces CCF of redundant components. Normally, to reduce CCF potential, a philosophy of independency, physical separation, and functional isolation are followed; however, the CCF can enter through some common features, for example, the same manufacturer, and maintenance team performing calibration for the redundant channels. It is be possible that jumper configuration on redundant modules might disable operation of the entire channel and induce CCF.

Keeping the above discussion in view, the procedure for DfR should have a provision for CCF whereby deterministic evaluation of CCF employing a physics-based approach and human factor assessment in design, fabrication, and operation should be evaluated, such that the potential for CCF can be demonstrated to be either reduced or eliminated altogether.

### 11.4.7.8  System Security

The current approach to deal with system design requires considerations of security along with safety [35]. The new draft documents from IAEA on computer security of I&C systems at nuclear facilities under the category "nuclear security series" further highlight the role of computer security in the future [36]. Given the fact that the digital systems are becoming part of nuclear system design, evaluation of the design basis threat forms an integral part of the risk modeling [37]. Further, cyber physical systems (CPSs) are also being increasingly being implemented in safety as well as public services systems. CPS are defined as physical and engineered systems whose operations are monitored, coordinated, controlled, and integrated by a computing and communication core [38]. CPSs can be a confluence of embedded systems, real-time systems, distributed sensor systems and controls. Given this background, there should be protection from malicious attacks either from the outside or from inside. Even though electromagnetic interference (EMI)/electromagnetic compatibility (EMC) qualifications form part of the I&C validation and verification program, there should be provision for monitoring and surveillance during normal operation of the plant as part of the technical specifications of the plant. Personal training and awareness at all levels should be integral to plant operations and maintenance practices. Any mechanism or initiating event, whether it is EMI or EMP (electromagnetic pulse), that might have implication for CCF should be investigated and plant safety should be demonstrated. PRA model of the plant should include evaluation of initiating event lists to have security aspects modeled as part of PSA.

## 11.5  Codes and Standards for Digital Protection Systems

To ensure the highest safety and reliability, the design of an I&C system should conform to the applicable national or international standards. The major goal is to follow the best practices in design operation and regulation of I&C aspects. For nuclear plants, there are many standards that can be utilized keeping in view design and national regulatory requirements. However, in this section we will discuss two international standards in brief. These standards are IAEA standard SSG-SSG-39 on "Design of Instrumentation and Control Systems for Nuclear Plants'" [35] and IEC standard IEC-61508 [31] on "Functional Safety of Electrical/Electronic/Programmable Electronic Safety-Related Systems." As the title indicates, the IAEA standard is specific to nuclear power plants, while the IEC standard is applicable to I&C for safety-critical applications.

The IAEA standard covers all the aspects specific to reactor protection and emergency safety feature designs, in detail.

The major feature of the IEC standard is its treatment of the subject in seven parts starting from general requirements, compliance, software requirements, methods for determining safety integrity levels (SILs), and overview of the techniques. The SIL categorization for systems having duty cycles as demand-related and continuous mode, along with associated frequency, is one of the major features that is of interest to safety-critical systems such as nuclear plants. There is general acceptance of these SIL levels in safety-critical design. For example, the reactor protection system uses a demand mode of operation and, keeping in view its impact factor on plant safety, it should go to the highest SIL category where the probability value as per this standard should fall between $10^{-5}$ and $10^{-4}$/demand. However, if we look at the target CDF of less than $10^{-5}$/reactor-year, for new and advanced plants, then the probability of failure of protection function in nuclear plants is expected to be even less than IEC targets. As a modest assumption, a Class IV power failure frequency of 1.0/year with IEC's lowest limit will take the CDF to $1.0 \times 10^{-5}$/reactor-year, whereas there are host of other events that contribute to CDF. Hence, in nuclear plants, the protection function failure is expected to be less than $10^{-5}$/demand. This is achieved by either a single protection system or another diverse system that gives a total failure probability even less than $10^{-6}$/demand, or a single system backed up by operator action to trip the reactor manually. Another interesting observation is that the SIL for continuous mode (i.e., the reactor operation is a continuous mode), for the highest level; i.e., SIL-4 are in line with CDF targets of $1.0 \times 10^{-5}$/reactor-year. This provides a target for new and advanced systems; with inherently safe design and passive features, a target for reactor protection function is between $10^{-6}$ and $10^{-7}$/demand, such that a CDF of at least $10^{-7}$/reactor-year can be realized.

## 11.6   Conclusions and Final Remarks

Digital systems are the integral part of design of controls and protection for new designs and a replacement option for the current operating plants due to end of service life or obsolescence. Digital technology has many advantages over traditional analog electronics. The advantages are many as highlighted in the chapter, for example, flexibility, drift-free operation, and embedded architecture, and this is the main drivers for increasing deployment of digital systems. However, digital systems have software components along with hardware, complex functional architecture, and new components and architecture (e.g., bus architecture, advanced ASICs like FPGAs/CPLD-based designs, and memory functions), while the operating experience is relatively limited compared to analog systems.

These aspects pose a challenge for modeling and analysis of these systems as part of PRA, and this is one reason why the regulatory approach for digital systems to a large extent is deterministic in nature. The available literature shows that extensive work is being performed, and in some cases, there has been progress, for example, the digital system taxonomy, where the NEA itself has supported the efforts. However, the larger picture is that the PRA analysts still follow either the same procedure employed for analog systems or a variation of it because there is no consensus on any one methodology for digital system reliability assessment. This raises the question "What should the utilities do to present their safety case?" Regulators would like a safety demonstration, with major considerations being the conceivable failure modes in relation to both software and hardware and their complex interaction. This safety demonstration should be set against challenges posed by CCF and human factors, particularly for emergency scenario modeling.

The risk analysis challenges for modeling are: (a) the dynamic aspects of digital systems; (b) fault identification, coverage, and repairs; (c) software reliability; and (d) failure characteristics of new aspects of the design. At the outset, it is concluded that further R&D efforts are required to consolidate the state of the art and develop a methodology that addresses all the issues discussed, such that there is a larger consensus on the approach.

In this chapter, a simplified approach has been proposed that is expected to address major issues such as dynamic aspects in a limited sense, CCF of software/hardware on a conservative basis, qualitative evaluation of security concerns, and reliability data. It is assumed that there is accumulated operating experience available within the utilities. This methodology is a minor departure from the standard methodology followed traditionally; however, it attempts to address the issues and take advantage of recent developments.

Through this chapter, the authors would like to make an appeal that efforts should be started to create a database of operating experience of digital systems that also include CCF entries, such that this experience provides the required insights into the complex issues (as they look now). This will further motivate the required R&D toward developing an improved understanding of the risk potential from digital systems.

## Appendix: Sample Data of Digital System Failure Rates [8]

| S. No. | Component name | Unit | Failure mode | Failure rate/ Probability | Error factor |
|--------|----------------|------|--------------|---------------------------|--------------|
| 1 | Processor module [Advent 645C and primary rack] | /h | Fail to generate trip output | $3.24 \times 10^{-6}$ | 3 |
| 2 | Digital input/output card | /h | Fail to generate trip output | $8.96 \times 10^{-7}$ | 3 |
| 3 | A/D module | /h | Fail to generate trip output | $2.0 \times 10^{-6}$ | 3 |
| 4 | Fiber-optic transmitter | /h | Fail to actuate | $4.4 \times 10^{-6}$ | 3 |
| 5 | Watchdog timer | /d | Fail to open | $8.21 \times 10^{-8}$ | 3 |

(continued)

(continued)

| S. No. | Component name | Unit | Failure mode | Failure rate/ Probability | Error factor |
|---|---|---|---|---|---|
| 6 | Shunt trip device | /d | Fail to energize | $1.2 \times 10^{-4}$ | 3 |
| 7 | U/V trip device | /d | Fail to energize | $1.7 \times 10^{-3}$ | 3 |
| 8 | Reactor trip circuit breaker | /d | Fail to open | $4.5 \times 10^{-5}$ | 3 |
| 9 | Push button switch | /d | Fail to function | $1.5 \times 10^{-5}$ | 3 |
| 10 | Interposing relay | /d | Fail to de-energize | $6.2 \times 10^{-6}$ | 3 |
| 11 | Instrument power supply | /d | Fail to supply | $1.6 \times 10^{-3}$ | 3 |
| 12 | 125 VDC bus supply | /d | Fail to supply | $1.8 \times 10^{-8}$ | 3 |
| 13 | Pressure transmitter | /h | Fail to provide signal | $4.4 \times 10^{-6}$ | 3 |

# References

1. S. Authen, J.-E. Holmberg, Reliability analysis of digital systems in a probabilistic risk analysis for nuclear power plants. Nucl. Eng. Technol. **44**(5) (2012)
2. K. Coyan, *Digital I&C PRA Research*. http://www.nrc.gov/docs/ML102700629.pdf. Accessed 25 Apr 2017
3. International Atomic Energy Agency, *Deterministic Safety Analysis for Nuclear Power Plants, Specific Safety Series No SSG-2* (IAEA, Vienna, 2009)
4. International Atomic Energy Agency, *Verification & Validation of Software Related to Nuclear Power Plants Instrumentation and Control, Technical Report Series No. 384* (IAEA, Vienna, 1999)
5. International Atomic Energy Agency, *Validation Procedures of Software Applied in Nuclear Instrumentations, IAEA-TECDOC-1565* (IAEA, Vienna, 2007)
6. S. Authen, J.-E. Holmberg, *Nordic Nuclear Safety Research, Guidelines for Reliability Assessment of Digital System in the Context of PSA—Phase 3 Status Report* (2013)
7. American Military Standard—MIL-HDBK-217F, *Reliability Prediction of Electronic Components* (RIAC, Washington, D.C., 1991)
8. P. Varde, J. Choi, D. Lee, J. Han, *Reliability analysis of advanced Pressusized Water Reactor-APR-1400* (KAERI, Daejeon, 2003)
9. J.H. Bickel, Risk implications of digital reactor protection system operating experience. Reliab. Eng. Syst. Saf. **93**, 107–124 (2008)
10. D.G. Raheja, *Design for Reliability*, ed. by G.J. Louis (Wiley, 2012)
11. M. Pecht, *Integrated Circuit, Hybrid, and Multichip Module Package Design Guidelines—A Focus on Reliability* (Wiley, New York, 2008)
12. International Atomic Energy Agency, *Defense in Depth in Nuclear Safety, INSAG-10* (IAEA, Vienna, 1996)
13. M. Pecht, A. Dasgupta, J. Evans, J. Evans, *Quality Conformance and Qualification of Microelectronics Packagies and Interconnects* (Wiley, New York, 1994)
14. K. Upadhyay, A. Dasgupta, *Guidelines for Physics-of-Failure Based Accelerated Stress Testing* (New York, 1998)
15. Nuclear Energy Agency, *Failure Modes Taxonomy for Reliability Assessment of Digital I&C Systems for PRA* (NEA, Paris, France, 2015)

16. M. White, J. Bernstein, *Microelectronics Reliability: Physics of Failure Based Modelling and Life Time Evaluation*
17. J.-E. Holmberg, Failure modes taxonomy for digital I&C systems—common framework for PSA and I&C experts, in *Submitted for the Nordic PSA Conference—Castle Meeting 2011, Johannesberg Castle, Sweden*, 2011
18. T. Aldemir, D. Miller, M. Stovsky, J. Kirschenbaurr, P. Bucci, A. Fentiman, L. Mangan, *Current State of Reliability Modeling Methodologies for Digital Systems and Their Acceptance Criteria for Nuclear Power Plants, NUREG/CR-6901* (USNRC, Washington, D.C., 2006)
19. International Atomic Energy Agency, *Computer Security at Nuclear Facilities—A Reference Manual, IAEA Nuclear Security Series No.17* (IAEA, Vienna, 2011)
20. British Standard ISO/IEC, *Information Technology—Security Technique-Information Security Risk Management, ISO/IEC 27005* (ISO/IEC, 2008)
21. T. Chu, G. Martinex-Guridi, M. Yue, J. Lehner, P. Samanta, *Traditional Probabilistic Risk Assessment Methods for Digital Systems, NUREG/CR-6962/BNL-NUREG-80141-2008* (USNRC, Washington, D.C., 2008)
22. M. Cepin, B. Mavlo, A Dynamic fault tree, in *Reliability Engineering & System Safety* (2002), pp. 83–91
23. R.W. Butler, S.C. Johnson, *Techniques for Modeling the Reliability of Fault Tolerant Systems with the Marov State-Space Approach* (NASA, Hampton, Verginia, 1995)
24. P. Anto, M.G. Pecht, D. Das, *Reliability Growth: Enhancing Defence System Reliability* (The National Academies Press, 2015)
25. Westinghouse, *Core Protection Calculator System*, March 2013. www.westinghousenuclear. com/portals/0/operating%20plants%services/automation/protection%20systems/NA0094%20cpcs.pdf. Accessed 19 Apr 2017
26. P. Badoux, R. Sander, *Bayesian Methods in Reliability* (Springer Sceince+Business Media, B.V., Netherlands, 1991)
27. V. Naikan, *Reliability Engineering and Lifetesting* (PHI Learning, New Delhi, 2009)
28. L. Srivani, B. Kumar, S. Swaminatan, P. Satyamurty, Accelerated life testing of field programmable gate arrays, in *ICRESH-2010, Mumbai, India*, 2010
29. P. Varde, M. Agarwal, P. Marathe, U. Mohapatra, R. Sharma, V. Naikan, Reliability and life prediction of electronic connectors for control applications, in *ICRESH-2010, Mumbai, India*, 2010
30. M. While, J. Bernstein, *Microelectronics Reliability: Physics-of-Failure Based Modeling and Life Time Evaluation* (National Aeronautical Space Administration, 2008)
31. International Electrotechnical Commission, *International Standard for Functional Safety of Electrical/Electronic/Programmable Electronic Safety Related Systems, IEC-61508* (IEC, Switzerland, 2010)
32. O. Backstorm, J.-E. Holmmberg, M. Jockenhovel-Barttfeld, M. Porthin, A. Taurines, T. Tyrvainen, *Software Reliability Analysis for PSA: Failure Mode and Data Analysis, NKS-341* (Nordic Nuclear Safety Research, Roskilde, 2015)
33. International Atomic Energy Agency, *Case Study on the Use of PSA Methods: Human Reliability Analysis, IAEA-TECDOC-592* (IAEA, Vienna, 1991)
34. R. Wood, K. Korsah, J. Mullens, L. Pullum, *Taxonomy for Common-Cause Failure Vulnerability and Mitigation, ORNL/SPR-2015/209* (Oak Ridge National Laboratory, 2015)
35. International Atomc Energy Agency, *Design of Instrumentation and Control System for Nuclear Poweer Plants, SSG-39 IAEA Safety Standard Series* (IAEA, Vienna, 2016)
36. International Atomic Energy Agency, *Computer Security of Instrumentation and Control Systems at Nuclear Facilities', Draft Technical Guidance, NST036* (IAEA, Vienna, 2014)
37. International Atomic Energy Agency, *Development, Use and Maintenance of Design Basis Threat* (Vienna, 2008)
38. R. Rajkumar, I. Lee, L. Sha, J. Stankovic, Cyber-physical systems: the next computing revolution, in *Design Automation Conference 2010, California*, 2010

# Chapter 12
# Physics-of-Failure Approach for Electronics

*Ask the right questions and nature will open the doors to her secrets.*

Sir C. V. Raman

## 12.1   Introduction

Physics-of-failure (PoF) approach is integral part of IRBE as in this approach reliability of the component and systems is predicted based on scientific models for identification of applicable failure mechanism and evaluation degradation to arrive at time to failure. The traditional statistical approach for reliability evaluation can predict the probability of failure but incapable of providing information on instant of failure. The physics-of-failure approach as part of prognostics framework can predict failure in advance with acceptable level of uncertainty. PoF as an approach is extensively being applied for electronics reliability; hence, this chapter deals with PoF approach to electronics.

"Electronic packages" refers to all electronic technologies from an integrated chip to a fully assembled electronic system and the bridge technologies between them, such as passive components and circuit cards. The evolution of electronic package design is trending towards integrating these separate packages into single or fewer packages, making them smaller, multifunctional, and energy efficient. Some of the technologies that are already in production to achieve these design goals include direct chip attach flip-chip assembly, multichip module, and chip-scale packages. The electronics industry is blurring the boundaries between different components of electronic packages by developing new materials and processes that can produce high levels of integration, such as system-on-chip and system-in-package configurations [1]. The drive towards smaller and more complex packages poses challenges in terms of dealing with higher current densities and hence higher thermal loads on a smaller package and smaller structures, making them vulnerable to mechanical loads.

The reliability of the new designs, materials, and processes that enable these sophisticated packages must be evaluated under life cycle conditions. Because the

© Springer Nature Singapore Pte Ltd. 2018
P. V. Varde and M. G. Pecht, *Risk-Based Engineering*, Springer Series in Reliability
Engineering, https://doi.org/10.1007/978-981-13-0090-5_12

electronics industry is striving to meaningfully go beyond Moore's law for enabling new and advanced technologies [2], many large companies are still employing traditional reliability estimation techniques. Traditionally, the reliability estimates for electronics components are obtained by one of the following three approaches: using a standards handbook; using field failure statistics; or performing life-testing experiments. Some of the prominent handbooks used to estimate reliability are MIL-HDBK-217 (U.S. Military Handbook 217), Telcordia (or Bellcore, developed for telecommunications applications), GJB/Z299 (China), FIDES (France), and PRISM (developed by the Reliability Analysis Center, USA, for the U.S. Air Force). Among these, the U.S. Military Handbook 217 is the most widely used handbook for predicting reliability of electronic components [3]. The handbook estimates the reliability of a system based on the failure rates of the individual components which form the system. The limitations of the handbook for designing an electronic assembly have been studied and have been shown to be most effective when considering thermal interactions that found in real-time situations [4]. Handbook estimates are based on constant failure rate assumptions that may not reflect the real life cycle failure trends and take into account the environment and operational loading conditions. The last update to MIL-HDBK-217 was implemented in 1995 in reaction to a contract where the supplier found that the models had no scientific foundation and the results were highly inaccurate [5, 6]. The updated version carries the same deficiencies from its predecessors; however, it is still being used by military and aerospace industries in their reliability and contractual documents.

The electronics industry is transitioning at an unprecedented pace and is developing new technologies with new materials, innovative architecture, and wide application conditions. Traditional handbook-based reliability estimation techniques and field failure data based on reliability estimation have become irrelevant and outdated because they do not incorporate the new technologies and/or provide timely data before the product reaches millions of customers. Physics of failure (PoF) assesses each technology based on its design and considering its application conditions. Unlike traditional techniques, it "models" the failure based on a deeper "failure mechanism" analysis rather than just looking at the external "failure modes."

## 12.2  Life Cycle Aspects and Failure Distributions

In engineering applications, reliability is defined as the probability that a product or system will perform as intended for a specified time, subject to its anticipated life cycle conditions. The stresses that a system or system component experiences during its life cycle has a profound impact on its useful life expectancy. However, the handbooks mentioned above and the statisticians who are concerned only about the number of failures during testing or in the field often do not accurately account for life cycle stresses. They tend to generalize the failures and encompass all

**Fig. 12.1** Typical Weibull
failure distribution curve

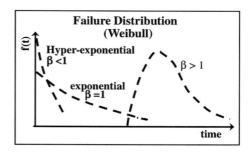

failures due to various causes into a single failure rate for a particular type of component or system. One method to understand reliability is to plot the probability of failure versus time. An example plot, traditionally referred to as Weibull plot, is presented in Fig. 12.1, which shows the Weibull failure function with the shape factor set at three distinct failure classifications. The hyper-exponential region ($\beta < 1$) indicates failures of a product in the early part of its life cycle. This is a characteristic of failures due to manufacturing problems such as quality issues, wrong testing/screening, or human error. The relatively flatter region ($\beta = 1$) constitutes the most useful part of a product's life cycle. The failures in this region are considered to be random because the product is expected to have reached maturity by the time it reaches this region and to be free of any specific assignable causes. The final region ($\beta > 1$) indicates failure due to aging and wear out of the product. A traditional bathtub curve describes the failure rate of a product without providing much information on the specific underlying failure causes. These plots are generally preferred by reliability statisticians for tracking system-level failure data for logistical purposes, i.e., inventory management for planned and corrective maintenance and warranty estimation.

However, this traditional view of reliability does not provide much information on the mode of failure, such as the way a failure precipitates and the underlying failure cause, which might be due to design issues or overstress conditions that were not anticipated during product development. Some of the drawbacks of this view of reliability are:

• The reliability information is obtained either from field failure data or testing data. Both methods are time- and cost-intensive and need hundreds or thousands of parts to be fielded or tested to get the highest possible level of confidence in the data. The highly competitive marketplace does not allow manufacturers the luxury of time to test enough parts to get the reliability data and then carry out corrective actions.

• The reliability data of one type of component cannot be meaningfully adopted for another component with the same design and materials. Similarly, the effect of different environments on the same component type cannot be extracted or extrapolated from data for particular environmental conditions.

- The method can be adversely affected by deficiencies in data collection even after spending significant time and resources on it. Human error, such as missed collection and erroneous judgment of a failure type, can result in inaccurate reporting of failures. Thus, the design improvement process is ineffective if the designers are not addressing the real problem because the data is pointing otherwise.

## 12.3   Physics of Failure-Based Reliability

Physics of failure (PoF) works on the principle that there needs to be a "deeper" analysis of failure than just collecting the failure numbers and "external" observations. As the name suggests, PoF tries to capture the causality of failure by going below the macroscopic observations to the microscopic level of the actual physical processes that culminate in failure. As mentioned in the definition of reliability, the life cycle conditions determine these physical processes, also referred to as the mechanism that makes the product deteriorate over time. A failure mechanism is a physical phenomenon caused by lifecycle stresses that accumulate over a period of time, or it can be a sudden event caused by an "overstress" (or product abuse), which is reflected in the end performance of a product. The degradation or failure in the end application of the product is generally referred to as the "failure/degradation mode." These failure modes can be an observable physical phenomenon such as crack propagation or discoloration, or they can be electrical performance issues such as voltage drops and electrical dysfunctions. PoF provides reliability information that can:

- Create mathematical models that can be used to predict failures for components based on environmental conditions and operating conditions. These models are a set of equations that determine the effect of certain parameters such as temperature, humidity, and voltage on the life of the product over a period of time. They can be used to predict the reliability of a component under different sets of conditions rather than requiring tests for each condition as in the traditional approach.
- Help determine a product's reliability building before it is built rather than after. By identifying the mechanisms that can shorten a product's life, PoF ensures the product can be designed to be robust against those mechanisms.
- Save significant time and resources that would be otherwise spent on numerous testing procedures, redesign, and rebuilding. The advantages are compounded when the components are intended to be used in a diverse set of conditions because all the factors can be included in the model beforehand.
- Reduce costs by eliminating the need for comprehensive testing and better warranty agreements with the knowledge of safe operating conditions of the product.

**Fig. 12.2** Failure distribution
from a PoF perspective

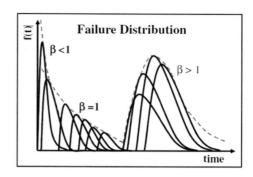

PoF-based reliability engineers conceived the Weibull plot shown in Fig. 12.1 from the perception of failure mechanisms. Each region of the plot, i.e., the early failures phase ($\beta < 1$), the nearly constant failure rate phase ($\beta = 1$), and the wear-out phase ($\beta > 1$), are considered to be a combination of failure distributions for different failure mechanisms. This combination is shown graphically in Fig. 12.2.

Table 12.1 shows that reliability can mean different things to a failure-rate-based reliability engineer versus a PoF-based reliability engineer.

## 12.3.1   Inputs for the PoF Approach

Reliability assessment of a component begins with collecting information about the product and the manufacturing process. The product information includes design specifications (i.e., materials and dimensions) and the intended life cycle conditions (i.e., operational parameters and stresses such as voltage, current, and mechanical load) and environmental conditions (i.e., temperature, humidity, and exposure to any corrosive substances). Product quality depends on the manufacturer's ability to

**Table 12.1** Reliability perspectives

| Statistical reliability | PoF-based reliability |
| --- | --- |
| Product is reliable when the number of failures during a specified period is at an acceptable level | Product is reliable if we are confident that it will not need maintenance for a specified period of time |
| Metric: failure rate, hazard rate or mean time between failures (MTBFs) | Metric: MFOP or FFOP for a given confidence level (maintenance-free or failure-free operating period) |
| Categorizes useful period failures as random failures | Failures are assigned to specific causes |
| Useful for estimating overall reliability of a product group | Useful for focusing on each and every product unit |

produce products consistently and as close as possible to the desired features. This information is generated by the manufacturer and its supply chain. The design and processes are determined based on customer requirements. These customer requirements are collected in the form of performance requirements and reliability requirements. These requirements are translated into specification limits, which in turn drive the design and processes. The amalgamation of manufacturer and customer inputs is used as the basis for further assessment of the component's reliability and in extension the final product's reliability in which the component is intended to be assembled.

### 12.3.2  Failure Modes, Mechanisms, and Effects Analysis (FMMEA)

FMMEA uses the lifecycle profile of a product along with the design information to identify the critical failure mechanisms affecting a product. Failure mechanisms with high criticality and severity determine the operational and environmental stresses that need to be taken into consideration during design. Knowing the causes and consequences of these critical failure mechanisms helps in the product design, qualification, testing, root cause analysis, and prognostics. The FMMEA methodology is shown in Fig. 12.3.

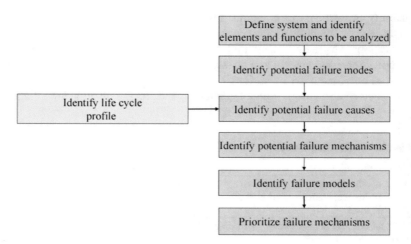

**Fig. 12.3** FMMEA methodology

#### 12.3.2.1  System Definition

The product or system whose reliability is to be predicted is divided into logical elements each of which has associated functions or performance parameters. These functions will be used to define failure criteria for each of the elements. The failure criteria can be either functional (based on what the system "does") or geographic/architectural (based on the location of the elements). The breakdown of elements depends on the design and the lowest level at which information about the part or material can be obtained.

#### 12.3.2.2  Evaluation of the Failure Modes

A failure mode is the way in which a failure is physically observed at each of the demarcated elements or levels. One level's failure mode can be a lower level's failure effect and a higher level's failure cause. Hence, defining the scope of each element in terms of its functionality is essential for avoiding misinterpretation. There can be multiple failure modes for the same element, and all of them must be listed. Some examples of failure modes are cracks, voltage drop, structural damage, and sparks. In cases where information or knowledge about failure modes do not exist, methodologies such as high-stress testing, numerical analysis, experience, and engineering judgment have to be used.

#### 12.3.2.3  Identification of Life Cycle Conditions

Once the potential failure modes are listed, the life cycle conditions of the system are identified. The life cycle conditions are obtained from the life cycle profile a product is anticipated to experience. The life cycle profile (LCP) includes the manufacturing phases, transportation, storage (at the dealer's or manufacturer's location), testing, rework, usage conditions, operational loads (e.g., voltage and current), and maintenance. The LCP should encompass both the occurrences and duration of these conditions. Typical life cycle conditions include loads such as temperature, humidity, pressure, vibration, shock, chemical environments, radiation, contaminants, current, voltage, and power.

#### 12.3.2.4  Assessment of the Failure Causes

The failure cause is the specific process, design, and/or environmental condition that initiated the failure, and whose removal will eliminate the failure. Knowledge of potential failure causes helps identify the failure mechanisms driving the failure modes for a given element. Failure causes are usually identified by brainstorming among the FMMEA group, which reviews the product's LCP and design to come up with failure causes. Typical failure causes include vibration, shock, loss of lubrication, and moisture ingression.

### 12.3.2.5   Identification of the Failure Mechanisms

Failure mechanisms are the processes by which specific combination of physical, electrical, chemical, and mechanical stresses induce failure. A failure mode is not necessarily always associated with a unique failure mechanism. Multiple failure mechanisms can precipitate the same failure mode; for example, both fatigue and overstress can lead to cracks in the solder of components on printed circuit boards (PCBs). JEDEC standard JEP122H [7] provides a list of failure mechanisms that are observed in electronic packages at different levels. This list is not exhaustive, but it contains some of the most studied and understood mechanisms.

### 12.3.2.6   Failure Models

Failure models are mathematical equations that provide a relationship between stress parameters and the life of the component. The fundamental nature of materials to degrade with time forms the basis for all failure models. The models incorporate the effects of environmental factors such as temperature and humidity and mechanical stresses that can contribute to the degradation rate over a period of time. Experiments and field failure data have shown that most distributions of material failures follow either power law, or exponential distributions [8]. The general forms of the models for time to failure ($T_f$) are as follows:

Power law

$$T_f = A * (X^{-n}) \tag{12.1}$$

Exponential

$$T_f = B * \exp(-Y * X) \tag{12.2}$$

where $A$ and $B$ are the material-/device-dependent factors, $X$ represents the stresses/factors that bring about material degradation, $Y$ is the exponential stress parameter, and $n$ is the power law exponent. The Arrhenius equation is a popular failure model. In some cases, a hybrid model that combines the power law and exponential may be needed when multiple stress factors are present.

Failure models are created for a specific failure mechanism such as time-dependent dielectric breakdown, solder fatigue, and hot carrier injection. The life cycle conditions, failure mechanism, and the root cause of the failure determine the model to be chosen as well as the stress parameter to be studied. Models have been developed for various failure mechanisms observed in electronics based on testing and field failure data [8–10]. For mechanisms for which failure models do

**Table 12.2**  Failure mechanisms and models at different package levels

| Failure location | Failure mechanism | Failure model | Stress parameters |
|---|---|---|---|
| Component | Time-dependent dielectric breakdown | Exponential (Fowler–Nordheim) | Electric field in the oxide ($E_{ox}$), temperature ($T$) |
|  | Hot-carrier injection | Power law | Peak current ($I_{sub}$), temperature ($T$) |
| Board trace/ metallization | Corrosion | Exponential | Relative humidity (RH), temperature ($T$) |
|  |  | Power law |  |
|  | Electromigration | Eyring (Black) | Current density ($J$) |
| Interconnects | Fatigue | Nonlinear power law (Coffin–Manson) | Strain range ($\Delta\varepsilon$) |
|  | Stress migration | Exponential model | Tensile stress ($\sigma$), Temperature ($T$) |

not exist or for materials without $A$ and $B$ factors, testing can be conducted on a statistically large sample size to precipitate the specific failure mechanism of interest and obtain the model or the coefficients from the failure data. The failure mechanisms and a few key models that describe them are listed in Table 12.2.

A PoF model relates time to failure with construction and loading conditions. Loading conditions can include environmental factors such as temperature, humidity; electrical parameters such as voltage, current, and power; and mechanical parameters such as vibration and torque. Some of the widely used failure models are discussed below. More models and their relevance can be found in [10].

Electromigration

Electromigration involves migration of metal atoms in interconnects, through which large DC current densities pass. Because reduction in device and node size leads to a rise in current densities, metallizations are prone to failure due to electromigration (EM) of metal ions. The time-to-failure model of EM will be a function of current density and temperature at the location as stress parameters. Some of the additional factors include metal composition, grain structure, grain texture, and interface structure.

The general model for EM is in the form,

$$T_f = P_o \left( J^{(e)} - J^{(e)}_{crit} \right)^{-n} \exp\left( \frac{A}{K_B T} \right) \tag{12.3}$$

where ($T_f$) is the time to failure, $P_o$ is the process-/material-dependent coefficient, $J^{(e)}$ is the applied current density, $J^{(e)}_{crit}$ is the threshold current density that must be exceeded to cause significant EM damage, $n$ is an experimentally determined exponent, $K_B$ is the Boltzmann's constant, $A$ is activation energy of the diffusion process, and $T$ is the temperature in Kelvins. Specific values are proposed for each of the experimental and material coefficients based on numerous experiments.

Time-Dependent Dielectric Breakdown

This degradation involves the phenomenon of leakage current and finally leads to short circuits due to failure of the transistor gates. The degradation mechanism involves creation of charge traps within the gate dielectrics, diminishing the potential barrier. Understanding the trap generation mechanism is the key to evaluating oxide degradation. The model used to predict the failure in dielectric resistance was proposed by Fowler–Nordheim [10]

$$t_{bd} = n.e^{\left[s\left(\frac{t}{V}\right)\left\{1 + \left(\frac{C}{k}\right)\left(\frac{1}{T} - \frac{1}{300}\right)\right\} - \left(\frac{\Delta E}{k}\right)\left(\frac{1}{T} - \frac{1}{300}\right)\right]} \tag{12.4}$$

where $t_{bd}$ is time to dielectric breakdown, $n$ is the ambient temperature pre-exponent value, $s$ is the electric field acceleration parameter, $t$ is the effective oxide thickness, $\Delta E$ is the activation energy, $V$ is the voltage across the oxide, $C$ is an experimentally determined factor, and $T$ is the steady state temperature in Kelvins.

Hot-Carrier Injection (HCI)

Hot carriers refer to electrons that travel from source to drain in a metal–oxide–semiconductor field-effect transistor (MOSFET) device, attaining high kinetic energy due to the surrounding high electric field and breaking through interfaces that are not intended, such as in gate dielectric and forming interfacial states.

The time-to-failure expression developed for HCI is

$$T_f = P_o \left(\frac{I_{gate}}{w}\right)^{-n} \exp\left(\frac{A}{KT}\right) \tag{12.5}$$

where $I_{gate}$ is the peak gate current for $p$-type MOSFETs(however, $n$-type devices use $I_{sub}$, which is the peak substrate current produced by the electrons knocking into the gate-channel interface), $w$ is the width of the transistor, $n$ is the power law exponent, and $P_o$ is the device-dependent parameter.

Fatigue

As it relates to electronics, fatigue is the degradation of materials due to cyclic loading conditions. Failures induced by fatigue can include electrical opens and loss of temperature control. Solder interconnect fatigue failures are arguably the most studied electronic failure mechanism. Solder interconnects are used extensively to electrically and mechanically attach packaged electronic devices to printed wiring boards to create the electronic circuits necessary for achieving the desired product function. Solder is also used within packaged devices to connect semiconductors to heat spreaders, lead frames, and interior patterned connections. In use, solder interconnects are often subjected to cyclic loading due to power cycling or surrounding mechanical disturbances. Over time, these cyclic loads can nucleate and propagate cracks that result in electrical opens such as those depicted in Fig. 12.4.

Due to extensive study, a variety of failure models have been adopted for solder fatigue. While the models are generally power law, they broadly fall into two classes based on the selected stress metric. One class of models uses strain energy, while the other uses strain range. In some cases, the solder interconnect failure models are broken into components to address the time to nucleate a crack and the time to propagate a crack that results in a solder interconnect open, while the majority of models tend to simply relate the time to complete a solder interconnect open. Well-known models include Darveaux's crack initiation and growth model [11] and Engelmaier's solder life model [12]. These and others have their roots in the works of Basquin [13], Coffin [14], Manson [15], and Morrow [16]. While Basquin's work is usually restricted to modeling high-cycle fatigue failures, the others are used in modeling low-cycle fatigue failures.

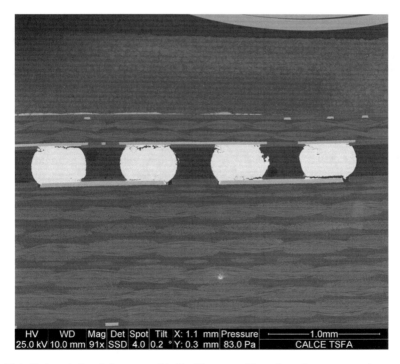

**Fig. 12.4** Fatigue-induced fractures of ball-grid array solder sphere interconnects (courtesy of CALCE)

The distinction between high- and low-cycle fatigue is important in electronic equipment as temperature excursions which normal fall in the low-cycle fatigue region are the focus of the most significant research and study. More importantly, solder materials show a strong difference in fatigue life under large strain range/ energy levels compared to low strain range/energy levels. This dependence is illustrated in Fig. 12.5.

To cover the full range, a model referred to as a generalized or modified Manson–Coffin model has been proposed and used. The model uses strain range as the stress metric and is expressed as

$$\frac{\Delta\varepsilon}{2} = \frac{\sigma'_f}{E}(2N_f)^b + \varepsilon'_f(2N_f)^c \tag{12.6}$$

where $\Delta\varepsilon$ is the strain range, $N_f$ is the cycles to failure, $E$ is the elastic modulus the solder, $\sigma_f$ is the strength coefficient, $b$ is the solder strength exponent, $\varepsilon_f$ is the solder ductility coefficient, and $c$ is the solder ductility exponent.

For temperature cycle-induced solder interconnect fatigue, the strength terms are frequently ignored and the fatigue model is reduced to

$$N_f = \frac{1}{2}\left(\frac{\Delta\varepsilon}{2\varepsilon_f}\right)^{\frac{1}{c}} \tag{12.7}$$

Engelmaier adapted the reduced form of the equation and provided further assumptions for his model, which found its way into an industry test standard, IPC-9701 [17]. Engelmaier's formulation of the fully relaxed shear strain for leadless parts was expressed as

$$\Delta\gamma = \frac{L_d(\alpha_b - \alpha_c)\Delta T}{2h} \tag{12.8}$$

where $L_d$ is the distance of critical terminal from the neutral expansion point, $\alpha_b$ and $\alpha_c$ are the coefficients of temperature expansion for the board and component, $\Delta T$ is the temperature range for the applied temperature cycle, and $h$ is the height of the solder joint. From this formulation, it is clear that a large part would incur a higher cyclic strain and a low solder joint height would also result in a higher cyclic strain.

**Fig. 12.5** Illustration of solder fatigue life (courtesy of CALCE)

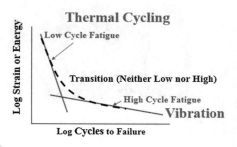

To account for the temperature and hold times that can occur in the field and test, Engelmaier expressed the ductility exponent as a function of temperature and hold time as

$$c = c_0 + c_1 T_{sj} + c_2 \ln\left(1 + \frac{360}{t_h}\right) \tag{12.9}$$

where $c_x$ is the solder material constants, $T_{sj}$ is the mean cycle temperature, and $t_h$ is the cycle hold time. For tin–silver–copper solder, the model constants developed by Osterman and Pecht [18] can be found in Table 12.3.

In addition to electrical opens, solder fatigue can cause loss of thermal performance, which contributes to electromigration, hot carrier, and dielectric breakdown failures. In particular, field-effect transistors (FETs), which are used in power applications, can be particularly susceptible to device failures induced by loss solder bonding due to fatigue.

There are numerous other models that are used for various applications and mechanisms. Some of them are discussed in [10]. Traditional product development methodology involves design, prototype building, testing, and redesign based on test results. And the redesigned products are introduced back into this cycle, making it an expensive iterative process. On the other hand, failure models facilitate a proactive approach where existing knowledge of the materials and their tendency to deteriorate under certain stresses is used to predict the life of the component even before building a prototype. Hence, confidence in a product is improved through this "virtual" qualification process even before the first prototype is made. Testing is later conducted to validate the model estimates rather than find root cause of failure and redesign.

### 12.3.2.7   Prioritization of Failure Mechanisms

As was mentioned in the previous section, a single failure mode can be caused by multiple failure mechanisms. In fact, in real applications, more often than not, a failure is observed as a result of culmination of several failure mechanisms based on the life cycle conditions the product experienced. Based on the storage conditions and the extent of usage, abuse, and mishandling, the environmental stresses as well as operational stress parameters vary. This in turn drives different mechanisms in the same component. Only a handful of stress parameters actually cause these

**Table 12.3** SAC305 thermal fatigue constants [18]

| Constant | Value |
|----------|-------|
| $c_0$ | $-0.347$ |
| $c_1$ | $-1.74 \times 10^{-3}$ |
| $c_2$ | $7.83 \times 10^{-3}$ |
| $\varepsilon_f$ | $0.29$ |

various mechanisms to result in high-risk failures. Hence, identifying these high-risk failure mechanisms can help manufacturers focus their resources on designing to prevent these failures instead of over-designing the product with multiple safety factors.

"High-risk" mechanisms are identified and quantified based on their frequency of occurrence and severity. If certain environmental and operating conditions are non-existent or generate a very low-level load, the failure mechanisms that are exclusively dependent on those environmental and operating conditions are assigned low occurrence. For overstress mechanisms, failure susceptibility is evaluated by conducting a stress analysis to determine if the failure would occur under the given environmental and operating conditions. For wear-out mechanisms, failure susceptibility is evaluated by determining the time to failure (TTF) from the failure models under the given LCP. The levels are assigned based on benchmarking the individual failure mechanism TTF with expected product life, past experience, and engineering judgment. In cases where no failure models are available, the evaluation is based on past experience and engineering judgment. The other criterion for categorizing a failure as high risk is its severity, which is determined based on the effect the failure has on the safety of the user and the mission, and the availability of the product. These effects and their extent are determined case to case based on past data, experience, cost analysis under downtime, and engineering judgment. The ability to detect these failures should also be factored into the prioritization process. The failure mechanisms can be prioritized by calculating the "risk priority number" (RPN) associated with each mechanism, as follows:

$$\text{RPN} = \text{Severity} \times \text{Occurrence} \times \text{Detection} \qquad (12.10)$$

Severity is assigned a number based on the effect of the failure. Fatal failures are given the highest number. Occurrence is assigned a number based on the frequency of occurrence. The tougher the failure is to detect before it occurs, the higher the number assigned to detection. Standards and other methodologies to assign RPNs are discussed in detail in Kapur and Pecht's chapter on FMMEA [9].

## 12.4  Virtual Qualification and Testing

The FMMEA process provides a method for identifying failure mechanisms and failure models and ranking failures in a product. When failure models are available for the identified failure mechanisms of a product, they can be used to estimate the life expectancy of the product. This process can be done while the product is in the design phase, prior to physical prototypes being created. Through the life estimation from the evaluation of the failure models, it can be determined if the product will meet its life requirement. Thus, the reliability of the product can be assured prior to the creation of the physical product. The process of using simulation to determine if a product meets its life requirement is referred to as "virtual qualification."

While virtual qualification can be used to determine if a product will meet its field life requirements, the above-mentioned simulation process can also be used to determine how the product will fare under physical test conditions. When the loading conditions are shifted from life conditions to test conditions, the simulation-based prediction process is referred to as "virtual testing." Virtual testing is important in establishing test expectations as well as in determining test times requirements. When combined with physical testing, the virtual testing approach helps ensure that the virtual qualification process is valid.

As presented above, the FMMEA process is applicable to any type of product whether the product is electronic or mechanical. While electronic products can exhibit a wide range of functions, the building blocks are generally the same. These building blocks include a printed wiring board with patterned copper metallization to provide connects between discrete packaged electronic devices. The packaged electronic devices come in standardized formats designed to protect the device and provide terminations that can be attached to the copper metallization on the printed wiring board. While packaging continues to evolve and new structures are being introduced, failure mechanisms observed in electronic products are being studied and failure models for structures found in electronics are being documented [19, 20]. By collecting the failure models, a simulation environment can be created to model electronic products.

The Simulation Assisted Reliability Assessment (SARA) software developed by the Center for Advanced Life Cycle Engineering at the University of Maryland is an example of a simulation environment for conducting virtual qualification and testing of electronic products [21]. The CALCE SARA software provides a growing collection of failure models for structures found in electronic products. Through the implementation of failure models, geometric, material, and environmental loading parameters that are required input to the failure models have been identified. By assigning design-specific values to these parameters, the time to failure for individual failure mechanisms can be determined by evaluating the associated failure model.

In addition to the computer-implemented individual failure models, the CALCE SARA software provides the ability to create a computer model of printed wiring board assemblies, the common building block of electronic products. Figure 12.6 provides a screen capture of the analysis manager for the CALCE SARA board assembly tool. The computer model for the printed wiring board assembly is informed by the growing collection of failure models. As part of the model creation, the software provides a facility for assigning the life cycle loading conditions to which the assembly is expected to be subject to either in use or in test. Figure 12.7 depicts color-coded lifetimes for components from a printed wiring board assembly life assessment. Since the environmental loading parameters

**Fig. 12.6** CALCE SARA
printed wiring board analysis
modules

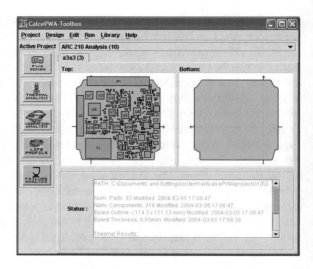

**Fig. 12.7** CALCE SARA
time-to-failure plot

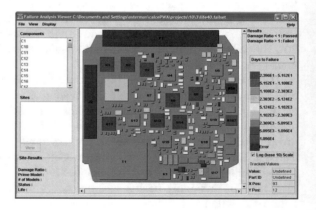

often vary among assembly components due to physical locations, the
CALCE SARA printed wiring board assembly model has been created so that it is
appropriate for assessing assembly components temperatures, as well as displacement and strain levels imposed on assembly components due to mechanical exci-
tation such as vibration and shock. Using the assigned life cycle loading conditions,
the software performs a virtual FMMEA and evaluates the identified failure models
to provide ranked listing for individual times to failure of the assembly components.
In this manner, the CALCE SARA software can be used to perform both virtual
qualification and testing.

## 12.5   Physical Qualification

While FMMEA and virtual qualification, respectively, provide a means for identifying the critical failure mechanisms and predicting time to failure in under-anticipated field conditions, physical testing is still often needed to verify that the physical situation has been accurately captured. Due to the pressure to bring products to market, physical tests on products are often carried out at load conditions that differ from anticipated field conditions. Two common ways of decreasing test time are to increase the load frequency or to increase the load level.

The advantage of elevated test frequency is that the load level stays within anticipated design limits. The ability to use an elevated test frequency requires that the active failure mechanisms are not influenced by the loading frequency. For example, if the failure mechanisms for a particular product are such that the number of on/off cycles is independent of the on/off cycle duration, then an elevating test frequency could be used. In this instance, a product with a design life of 5 years with an anticipated once a day on/off cycle could be tested beyond its anticipated useful life. Under an elevated test frequency, a 5-year life would equate to 1,825 on/off cycles. With a safety factor of two, the test could be designed to turn the product on and off 3,650 times. If a power on/off cycle takes 30 min, twice the product's lifetime can be accounted with 76 days of continuous testing. However, if the product's on/off cycle time requires a full day, then the continuous testing time for one lifetime would be 5 years, which is unlikely to be acceptable from a profitability standpoint.

While elevated load levels can be used to reduce testing time, knowledge of the failure mechanisms that are excited by the load is needed. When the test load condition can be related to the field load condition, the test may be referred to as an accelerated test. For accelerated tests, the ratio of time in the field, $t_{field}$, to time in test, $t_{test}$, is known as the acceleration factor (AF).

$$\text{AF} = \frac{t_{field}}{t_{test}} = \frac{F(L_{field})}{F(L_{test})} \tag{12.11}$$

where $F(x)$ is the failure model for the failure mechanism, $L_{field}$ is the load in field, and $L_{test}$ is the load in test. As previously discussed, failure models relate the time to failure for a specific failure mechanism to geometry, materials, and loading of the failure site. Since the loading may be due a field condition or test condition, failure models can be used to establish the AF for the failure mechanism under test. For example, the AF of a voltage-based exponential failure model can be of the form:

$$\text{AF} = \frac{L_{use}}{L_{test}} = \frac{C.e^{\frac{B}{V_{use}}}}{C.e^{\frac{B}{V_{test}}}} = e^{\left(\frac{B}{V_{use}} - \frac{B}{V_{test}}\right)} \tag{12.12}$$

where $L$ refers to lifetime, $V$ refers to voltage, and $C$ and $B$ are material and/or geometric constants.

As presented in the above equation, it is possible that not all parameters of the failure model are needed to determine the AF. In some cases, acceleration models have been created based on curves created by comparing the same failure site architecture to multiple load levels. One example of an acceleration model is the Norris-Landzberg [22] model. While the model was developed for interconnect failure of parts with solder ball attachments, it has been applied to other package geometries [23]. However, it should not be confused with a failure model because it ignores the package geometry and materials properties.

For complex systems with multiple failure sites, such as an electronic product, there can be a unique AF for each failure site within the product. When designing an accelerated test, it is important to understand what failure mechanisms and sites are being excited. This information is critical in establishing failure criteria, defining the proper monitoring equipment, and relating the time in test to time in the field.

In many instances, accelerated tests are time terminated. Time-terminated tests are tests that are conducted for a fixed period of time. With time-terminated tests, the test time is often set based on the product's field life requirement determined by the application of a selected AF. However, time-terminated tests do not provide confirmation of the failure mechanism and site. A preferred test method is a failure-terminated test. Failure-terminated tests are conducted until a fixed number of failures have occurred. Failure-terminated tests provide the advantage of allowing the failure mechanism and failure site to be confirmed.

In designing an accelerated test, the failure criteria should be clearly defined and active monitoring of the failure conditions should be conducted during the test period. For performance-based failure criteria, active monitoring is needed to identify when the failure condition has occurred as well as to isolate the failure. Active monitoring is important because the failure conditions that occur under test conditions may disappear when the product is not under test conditions. Intermittent failures are a known issue in electronics, and failure to actively monitor during a test may lead to invalid conclusions if a failure is not revealed when monitoring is only applied at the end of a time-terminated test [24].

## 12.6 System-Level Reliability and Standards

The results of accelerated tests can be extrapolated to life under actual usage conditions by multiplying the acceleration factor with the accelerated test life. By this method, component life for different lifecycle conditions can be estimated without having to conduct tests for each scenario. Component reliability of s can also be extended to the system level to estimate a system's performance limits and reliability.

The nature of the interaction between components and their effect on system performance can vary from complex to highly complex based on scale and purpose of the system. With the ever-increasing demand for high functionality in single devices such as cell phones, computers, and even wrist watches, the component-to-component relationship is getting even more complex and critical to the success of the product. In general, system reliability follows the same relationship as the functional relationship. Two functionally interrelated components and their effect on system performance exhibit equivalent reliability relationships and effects on system reliability. However, cases of redundancies have to be considered in a different manner as they have generally an alleviating influence on reliability even if they are not functionally active during a system's service.

PoF-based failure models are usually created for a particular mechanism in a single component. However, PoF principles can be used to predict system-level reliability by extending the effect of failure mechanism to system performance. This is done by combining PoF with other techniques such as reliability block diagram (RBD) to predict system-level reliability. RBD is a graphical method that consists of system-level block diagrams with the definition of each subsystem and equipment failures and repair distributions. RBD has been discussed in detail in the literature [9, 25] and has been computerized and developed into software such as in BlockSim-Reliasoft.

Similar to RBD, fault tree analysis (FTA) is a graphical method that relates the component-level failures to system-level reliability. The difference is that this is a top-down approach, where it starts with the end effect of the system in the top and branches down to different causes of failure associated with different components of the system. Detailed information and instructions on using FTA can be found in various books and journals [9, 26].

JEDEC Solid State Technology Association has published the standard for "Failure Mechanisms and Models for Semiconductor Devices" JEP122H [7]. This publication provides a list of failure mechanisms and their associated activation energies or acceleration factors that may be used in making system failure rate estimations when the only available data is based on tests performed at accelerated stress test conditions. The method to be used is the sum-of-the-failure-rates method. This publication provides guidance in the selection of reliability modeling parameters, namely functional form, apparent thermal activation energy values, and sensitivity to stresses such as power supply voltage, substrate current, current density, gate voltage, relative humidity, temperature cycling range, and mobile ion concentration. The standard states that the failure rate of electronic systems can be estimated by using a single representative "equivalent" apparent thermal activation energy for a given product or product group. The standard adopts the traditional approach of investigating the relationship between a maximum stress failure rate and a system failure rate by choosing a single representative "equivalent" apparent thermal activation energy for a given product or product group.

## 12.7  Prognostics and Health Management

In addition to design and testing, PoF can be instrumental in prognostics and health management of electronic equipment [27]. In this discussion, prognostics is the forecasting failure of a system or component within a system. Health management consists of actions taken to extend the life or service of a system or system component. Where design and test make assumptions about the end-use conditions to which the electronic equipment may be subjected, prognostics can be applied to the individualized systems and has the advantage of being updated or informed by using the actual loading conditions.

With respect to PoF, prognostics can be implemented based on an understanding of the failure mechanisms occurring in the system. Two PoF approaches for prognostics are conducted by life consumption monitoring [28] and "canary" monitoring [29, 30]. Both approaches are presented in this section.

### 12.7.1  Life Consumption Monitoring

The load monitoring prognostic approach [28] uses identified critical failure mechanisms and associated stress drivers which are output from the FMMEA process. Failure models associated with failure mechanisms can be used to develop a life consumption monitoring prognostic. The prognostic is the remaining useful life (RUL) of the system or system component under examination. As previously discussed, failure models predict time to failure based on the architecture of the failure site and the loading condition to which the design is subjected. Based on the failure models, the key loading parameters that induce failure are identified. Example loading parameters could be temperature, relative humidity, and/or mechanical strain. Based on the identified loading parameters, appropriate sensors for monitoring the load parameter can be identified. If the positioning of the sensor in the system does not allow direct monitoring of the identified critical failure site, a load transform may need to be developed that allows the measured load condition to be converted to the load condition for the failure site.

By design, most failure models predict a period of time to failure under a specific load condition and level. When load levels vary, a more sophisticated modeling approach is needed. A common approach for fatigue failure models is the cumulative damage approach introduced by Palmgren [31] and popularized by Miner [32]. The cumulative damage approach defines a numerical metric that ranges from 0 to 1, which is assumed to represent the state of material degradation. For example, the material degradation could be a physical crack and the damage index could be the length of the crack in a bonded structure divided by the overall bond length. In a purely numerical expression, the damage index could be the time under a degrading condition $i$, $t_i^{\mathrm{applied}}$, and the time to failure under the degrading condition $i$, $t_i^{\mathrm{available}}$.

$$D_i = \frac{t_i^{\text{applied}}}{t_i^{\text{available}}} \tag{12.13}$$

If the damage is accumulated linearly, meaning the damage due a load level and application period is independent of the damage state prior to the loading period, the damage to date can be simply the sum of the collected damage or

$$D_{\text{TD}} = \sum_i \frac{t_i^{\text{applied}}}{t_i^{\text{available}}} \tag{12.14}$$

Further, the average damage rate can be expressed based on the ratio of the damage and accumulated time.

$$R = \frac{\sum_i \frac{t_i^{\text{applied}}}{t_i^{\text{available}}}}{\sum_i t_i^{\text{applied}}} \tag{12.15}$$

If the damage level at failure is defined to be $D_{\text{f}}$, then the remaining cycles to failure can be expressed as

$$t_{\text{RUL}} = \frac{(D_{\text{f}} - D_{\text{TD}})}{R} \tag{12.16}$$

A flowchart of the life consumption approach is presented in Fig. 12.8.

In some cases, the failure model may have a time component and the time-based load must be processed to extract load features needed by the failure model. For instance, a fatigue model may only need the cyclic strain range to predict the number of cycles to failure and different strain ranges would result in different cycles to failure. To facilitate data processing for fatigue-based failure, cycle count

**Fig. 12.8** Life consumption monitoring

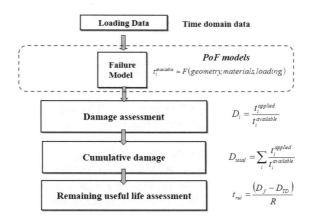

methods have been developed. One of the most well-known methods is the rain flow counting method, which was presented by Downing and Socie [33]. This method was later codified into a standard by ASTM [34]. While simple cycle counting is sufficient for some failure mechanisms, fatigue failure mechanisms that are sensitive to hold time at cyclic peaks require more sophisticated methods. Cluff et al. [35] was the first to address the issue of hold times for fatigue models. Later work on this topic was conducted by Vichare et al. [36].

As can be gleaned from the above discussion, the life consumption monitoring approach can be embedded into a system or performed outside of the system. When properly implemented, the life consumption monitoring approach can be quite effective in maintenance and logistics activities. The strength of the life consumption monitoring approach is that the user has real-time assessment of the health of a system or system component. Specifically, the remaining useful life of the system or system component can be projected based on user-assigned damage rate levels. This feature is critical for asset management and maintenance, as well as failure avoidance.

### 12.7.2   "Canary" Prognostics

As an alternative to a life consumption monitoring approach, a purely embedded approach for PoF-based prognostics is to design a device that provides a signal when the system or a system component has neared its end of life. Such a device is referred to as a "canary." The term "canary" is adopted from the historical use of canary birds in early coal mine operations to detect the presence of hazardous gases.

Dasgupta et al. [37] identified three types of PoF-based canaries: expendable, sensory, and conjugate pairs. Expendable canaries are devices formed with structures similar to the structures in the target system that have been tailored to fail sooner than the structures in the target system. Sensory canaries are monitoring devices that track a measurable change in the target system and that have been correlated to failure. For example, an acoustic sensor that monitors noise immersion from a cooling fan may be used as a sensory canary. Conjugate-pair canaries are sensors that monitor a stress pair whose deviation-established pairing is correlated with time to failure. A number of PoF-based canaries have been presented in the literature [29, 30, 38–49].

To be effective, the canary device has to provide an alert before failure of the target system or system component. The difference between the time of the canary device signal and failure of the target system or system component is referred to as "prognostic distance" (PD). Due to variations in materials, geometry, and loading, it can be expected that the time to failure of the actual system or system component will vary to a certain degree. Variation in canary alert time should also be expected. Figure 12.9 shows the concept of PD as well as various ways of defining it. The subscript assigned to the PD defines the probability percentage of the canary signal occurring followed by the probability of system failure.

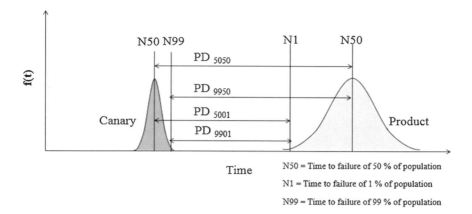

**Fig. 12.9** Prognostic distance

While the PD is important in designing the canary, one must also consider the probability that the canary will provide an alert after failure of the target system or system component. To determine the probability of a canary failing to provide advanced warning of a system failure, the sensing distribution of the canary and the failure distribution of the target system or system component must be known. With this information, the probability that the canary will fail after the target can be expressed as

$$P[t_c \geq t_s] = \int_0^\infty f_s(t) R_c(t) dt \qquad (12.17)$$

where $f_s(t)$ is the probability density function for the systems and $R_c(t)$ is the reliability function for the canary. Here, a $P = 0.5$ would indicate a toss-up over whether the canary alert or target system would fail first. $P < 0.5$ indicates a less than 50% change that the canary will fail after the target system. So, for design, the objective would be to create a canary that would have a very low $P$.

In addition to understanding the distribution of the canary warning signal and the system failure, it is also important to understand how the PD between the canary and the system varies with load level. As an example, consider the varying PDs depicted in Fig. 12.10, which shows that the PD decreases with increasing load levels. As such, the PD demonstrated in test at an elevated load maybe considerably shorter than the PD experienced in actual use. An acceptable PD in test may not be acceptable in the field. Alternatively, a large PD in test may disappear under use conditions.

**Fig. 12.10** Prognostic
distance varying with load
levels

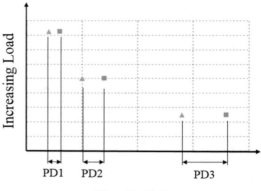

Time To Failure

As can be gleaned from the above discussion, PoF is necessary for implementing canaries. The canary development process is as follows:

1. Identify the critical failure mechanisms of the target system. Since canaries are supposed to provide warning of the target system failure, FMMEA can be used to identify and prioritize the dominant mechanisms that can lead to the host failure.

2. Based on the identified mechanisms, the life-governing parameters such as material properties, physical dimensions, and operational and environmental stresses can be established. Further measurable physical changes that may occur in close proximity to the failure site may be established. Failure models that are available for the identified mechanism can be used to arrive at these parameters and in turn describe their relationship to degradation rates.

3. Based on the information extracted from step 2, canaries can be proposed. For expendable canaries, structures similar to the identified failure site may be proposed. With similar structures, the same failure mechanism is active and the time to failure of the structure can be tailored. The time to failure of a similar structure may be reduced by employing any of, or a combination of, three controlled error-seeding processes. These are geometric error-seeding, material error-seeding, and load error-seeding.

    (a) Geometric error-seeding: As the name indicates, the size, shape, or location of a canary's particular failure site is designed to increase stress at that site. For example, canary solder joints can be designed to have a smaller footprint than normal ones to attain faster fatigue degradation rates under temperature or power cycling.

    (b) Material error-seeding: The composition and microstructure of a canary can be tailored to alter material properties. The material properties can be dielectric constants, dielectric strength, glass-transition temperature, diffusivity, creep resistance, ductility, and fracture toughness. For example, canaries that sense electrochemical migration between conductive traces on a board can be made of traces without solder mask between the traces. Here, effectively, the dielectric property of the separator that aids easier migration of metals across traces is changed.

(c) Load error-seeding: Canaries can be subjected to heavier loads (mechanical, electrical, or chemical) than the functional elements. For example, canaries for electromigration in a solder and die metallization can be subjected to higher current densities than normal.

4. The purpose of error-seeding is to accentuate stresses/governing parameters that affect the life of the canary. Hence, it is vital to determine the appropriate equipment to measure these governing parameters as the canary is employed in the field. This process can ensure the canary works as intended. In addition, in cases of load error-seeding, determining proper means of applying higher loads on canaries while isolating the functional elements from these loads is essential.

5. The prognostic distance for the canary must be established along with the distribution of the canary's sensing time and target system. Failure models identified in the FMMEA process may be used to determine prognostic distance as well as the probability of the canary failing to provide early warning. If not, physical testing should be conducted. Further, the shift in prognostic distance and probability of detection failure should be characterized over the range of expected life cycle loading conditions.

6. If models have been used to establish prognostic distance and failure distributions of the canary and target system, tests should be conducted to verify the validity of the model.

To help understand the canary development process, an example is provided. For this example, a system where the critical failure mechanisms in solder interconnect failure under temperature cycling is considered. Based on an examination of all parts, it is found that the critical, first-to-failure sites are the solder joints of a 192 IO BGA. For this critical BGA, the 3 corner solder spheres for each corner are not part of the functional circuit but are connected so that a resistance network can be designed into the board to identify a solder interconnect failure of one of the 12 corner solder joints. Figure 12.11 depicts the package with the boxes drawn around the non-functional daisy-chainable contacts. Here, the 12 corner contacts can act as the canary if a circuit is constructed to issue a signal when one of the contacts fails by thermal fatigue.

For the canary contacts, the diagonal length for estimating the fatigue life is 16.97 mm. For the functional solder interconnects, the diagonal length is 14.88 mm. For this example, the coefficient of temperature expansion (CTE) for the board is assumed to 14 parts per million per degree Kelvin and the CTE of the component is assumed to be 10 parts per million per degree Kelvin. For a test condition of −55 to 125 °C with a 15 min dwell at both the lower and upper temperature, the fatigue life of the SAC305 solder interconnects of the sacrificial contacts would be estimated to be 2,200 cycles to failure and the fatigue life of the

**Fig. 12.11** 192 IO BGAS
with sacrificial contacts
depicted in red boxes

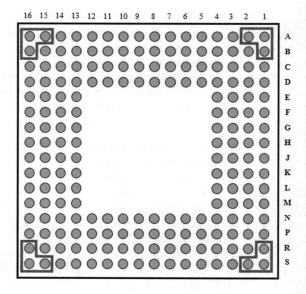

functional interconnects would be estimated to be 3,100 cycles to failure. Based on these estimates, the PD5050 is estimated to be 900 cycles. For an expected use condition of 20–60 °C with a 480 min dwell, the PD5050 changes to 22,800 cycles. As can be observed in this example, the PD5050 increases substantially with a decrease in the temperature range. In addition, if the failure distributions follow a two-parameter Weibull function and the corner connections experience a beta of 8 and the function interconnects experience a beta of 5, the probability of the canary failing after the function interconnects would be 11%. If the beta for the canary corner connection was 5 instead of 9, the probability of the canary failing would increase to 12%.

From the example above, it can be seen that PoF life models are important for identifying critical failure sites. Further, the failure models allow for creating or identifying canary structures. These models can be used to estimate the prognostics distance between the canary device and the target component. Finally, the statistics of the failure of both the canary device and the target component are important for determining the viability of the canary.

## 12.8   Conclusions

In this chapter, physics of failure (PoF) as it relates to electronic products has been introduced. PoF is necessary in electronics due to the rapidly expanding array of devices, packages, materials, and end-use conditions. Traditional methods that rely on reliability data collection simply cannot keep pace with the industry. Further, testing without the insight of PoF may lead to misinterpretation of test results and/or the use of improper tests. The true reliability of a product can be achieved and even estimated only after a thorough, systematic, and conscious examination of every aspect of the life cycle of a product and correlating the life cycle conditions with the design of the product.

PoF deals with the fundamental understanding of the processes that result in product failure. This understanding extends to how geometry and materials create a function and how life cycle loads both intrinsic and extrinsic to the function can degrade and result in loss of the function. This understanding provides a basis for identifying potential failures as well as establishing a ranked order for the failures. It also allows for the creation of mathematical models that can be used to relate time to failure with geometry, materials, and loading conditions.

With PoF, simulation can be used to determine if a planned design can meet lifetime requirements. PoF provides a basis for defining physical tests and interrupting test results. Finally, PoF can be used to establish methods for monitoring the health of a system and avoiding costly loss of service. PoF is a multi-disciplinary practice and requires continuous review of new information. Continuous learning and improvement is the only way to keep pace with the rapid advances occurring within the electronics industry.

# References

1. R.R. Tummala, SOP: What is it and why? A new microsystem-integration technology paradigm-Moore's law for system integration of minimized convergent systems of the next decades. IEEE Trans. Adv. Packag. **27**(2), 241–249 (2004)
2. T. Simonite, Moore's law is dead. Now what? *MIT Technology Review* (2016). https://www.technologyreview.com/s/601441/moores-law-is-dead-now-what/
3. Military Handbook, Reliability prediction of electronic equipment. MILHDBK-217F, Department of Defense, Washington, DC (1991)
4. M. Pecht, W.C. Kang, A critique of MIL-Hdbk-217E reliability prediction methods. IEEE Trans. Reliab. **37**(5), 453–457 (1988)
5. P. Charpenel, P. Cavernes, V. Casanovas, J. Borowski, J.M. Chopin, Comparison between field reliability and new prediction methodology on avionics embedded electronics. Microelectron. Reliab. **38**(6), 1171–1175 (1998)
6. M. Nilsson, Ö. Hallberg, A new reliability prediction model for telecommunication hardware. Microelectron. Reliab. **37**(10), 1429–1432 (1997)
7. JEP122H, *Failure Mechanisms and Models for Semiconductor Devices* (JEDEC Solid State Technology Association, 2016)
8. J.W. McPherson, *Reliability Physics and Engineering* (Springer, New York, 2010), p. 171
9. K.C. Kapur, M. Pecht, *Reliability Engineering* (Wiley, New York, 2014)
10. P. Singh, P. Viswanadham, *Failure Modes and Mechanisms in Electronic Packages* (Springer Science & Business Media, 2012)
11. R. Darveaux, K. Banerji, A. Mawer, G. Dody, Reliability of plastic ball grid array assembly, in *Ball Grid Array Technology*, ed. by J. Lau (McGraw-Hill, New York, 1995)
12. W. Engelmaier, Fatigue life of leadless chip carriers solder joints during power cycling. IEEE Trans. Compon. Hybrids Manuf. Technol. **6**, 232–237 (1983)
13. O.H. Basquin, The exponential law of endurance tests. ASTM Proc. **10**, 625–630 (1910)

14. L.F. Coffin Jr., A study of the effects of cyclic thermal stresses on a ductile metal. Trans. ASME **76**, 931–950 (1954)
15. S.S. Manson, Fatigue: A complex subject-some simple approximations. Exp. Mech. **5**, 193–226 (1965)
16. J. Morrow, Cyclic plastic strain energy and fatigue of metals, in B. Lazan (ed.) *STP43764S Internal Friction, Damping, and Cyclic Plasticity, STP43764S*. ASTM International, West Conshohocken, PA, 1965, pp. 45–87. https://doi.org/10.1520/STP43764S
17. IPC-9701, Performance test methods and qualification requirements for surface mount solder attachments, Northbrook, IL, Jan. 2002
18. M. Osterman, M. Pecht, Strain range fatigue life assessment of lead-free solder interconnects subject to temperature cycle loading. Solder. Surf. Mt. Technol. **19**(2), 12–17 (2007)
19. C. Cohn, C. Harper (eds.), *Failure-Free Integrated Circuit Packages* (McGraw-Hill, New York, 2005)
20. M. Pecht (ed.), *Handbook of Electronic Package Design* (CRC Press, New York, 1991)
21. CALCE SARA, CALCE, www.calce.umd.edu/software. Accessed Nov 2017
22. C. Norris, A.H. Landzberg, Reliability of controlled collapse interconnections. IBM J. Res. Dev. 266–271 (1969)
23. N. Pan, et al., An acceleration model for Sn-Ag-Cu solder joint reliability under various thermal cycle conditions, in *Proc. SMTA*, pp. 876–883 (2005)
24. Q. Haiyu, S. Ganesan, M. Pecht, No-fault-found and intermittent failures in electronic products. Microelectron. Reliab. **48**(5), 663–674 (2008)
25. M. Čepin, Reliability block diagram, in *Assessment of Power System Reliability* (Springer, London, 2011), pp. 119–123
26. C.A. Ericson, Fault tree analysis, in *System Safety Conference*, Orlando, Florida, pp. 1–9 (1999)
27. J. Gu, M. Pecht, Prognostics and health management using physics-of-failure, in *IEEE Reliability and Maintainability Symposium*, 2008, pp. 481–487
28. A. Ramakrishnan, M.G. Pecht, A life consumption monitoring methodology for electronic systems. IEEE Trans. Compon. Packag. Technol. **26**(3), 625–634 (2003)
29. P. Chauhan, S. Mathew, M. Osterman, M. Pecht, In situ interconnect failure prediction using canaries. IEEE Trans. Device Mater. Reliab. **14**(3), 826–832 (2014)
30. S. Mathew, M. Osterman, M. Pecht, A canary device based approach for prognosis of ball grid array packages, in *IEEE Conference on Prognostics and Health Management* (PHM), Denver, CO, 18–22 June 2012
31. A. Palmgren, Die lebensdauer von kugellagern. Veifahrenstechinik **68**, 339–341 (1924). (Berlin)
32. M. Miner, Cumulative damage in fatigue. J. Appl. Mech. **67**, AI59–AI64 (1945)
33. S.D. Downing, D.F. Socie, Simple rain flow counting algorithms. Int. J. Fatigue **4**(1), 31–40 (1982)
34. Standard Practice for Cycle Counting in Fatigue Practices, ASTM Standard E1049-85 (2011)
35. K.D. Cluff, D. Robbins, T. Edwards, B. Barker, Characterizing the commercial avionics thermal environment for field reliability assessment. J. Inst. Environ. Sci. **40**(4), 22–28 (1997)
36. N. Vichare, P. Rodgers, M. Pecht, Methods for binning and density estimation of load parameters for prognostics and health management. Int. J. Perform. Eng. **2**(2), 149–161 (2006)
37. A. Dasgupta, R. Doraiswami, M. Azarian, M. Osterman, S. Mathew, M. Pecht, The use of canaries for adaptive health management of electronic systems, in *ADAPTIVE 2010, IARIA Conference*, Lisbon Portugal, 21–26 Nov. 2010
38. Y.Z. Rosunally, S. Stoyanov, C. Bailey, P. Mason, S. Campbell, G. Monger, I. Bell, Fusion Approach for Prognostics Framework of Heritage Structure. IEEE Trans. Reliab. **60**(1), 3–13 (2011)
39. D.K. Han, M.G. Pecht, D.K. Anand, R. Kavetsky, Energetic material/systems prognostics, in *53rd Annual Reliability & Maintainability Symposium* (RAMS) (2007)

40. H.R. Shea, A. Gasparyan, H.B. Chan, S. Arney, R.E. Frahm, D. López, S. Jin, R. P. McConnell, Effects of electrical leakage currents on MEMS reliability and performance. IEEE Trans. Device Mater. Reliab. **4**(2), 198–207 (2004)
41. Y. Otsuka, T. Sato, T. Yoshiki, T. Hayashida, Multicore energy reduction utilizing canary FF, in *IEEE 2010 International Symposium on Communications and Information Technologies* (ISCIT), Tokyo, pp. 922–927, 26–29 Oct. 2010
42. B.H. Calhoun, A. Chandrakasan, Standby power reduction using dynamic voltage scaling and canary flip-flop structures. IEEE J. Solid State Circuits **39**(9) (2004)
43. J. Wang, A. Hoefler, B.H. Calhoun, An enhanced canary-based system with BIST for standby power reduction. IEEE Trans. Very Large Scale Integr. VLSI Syst. **19**(5), 909–914 (2011)
44. Ridgetop Group Inc., Ridge top Products, www.ridgetopgroup.com/products
45. D. Goodman, B. Vermeire, J. Ralston-Good, R. Graves, A board-level prognostic monitor for MOSFET TDDB, in *IEEE Aerospace Conference*, Mar. 2006
46. C.R. Keese, I. Giaever, A biosensor that monitors cell morphology with electrical field. IEEE Eng. Med. Biol. **13**, 402–408 (1994)
47. M.S. Petrovick, J.D. Harper, F.E. Nargi, E.D. Schwoebel, M.C. Hennessy, T.H. Rider, M.A. Hollis, Rapid sensors for biological-agent identification. Linc. Lab. J. **17**(1), 63–84 (2007). (Special Issue: Chemical and Biological Defense)
48. P. Chauhan, M. Osterman, M. Pecht, Canary approach for monitoring BGA interconnect reliability under temperature cycling, in *The MFPT 2012 Proceedings* (2012)
49. P. Lall, N. Islam, K. Rahim, J. Suhling, S. Gale, Leading indicators-of-failure for prognosis of electronic and MEMS packaging, in *Proceedings of 54th Electronic Components & Technology Conference*, 2004, pp. 1570–1578, Las Vegas, NV, 1–4 June 2004

# Chapter 13
# Prognostics and Health Management

*The effectiveness to be aimed at calls for the application and refinement of all conceivable prognostic techniques for adding to knowledge of the future, including those which can be effectively developed over an ever-wider time scale.*
—Fred Polak (1971) Prognostics, pp. 65–66

## 13.1 Introduction

Nuclear power, with over 430 nuclear power plants (NPPs) operating around the world, is the source of about 17% of the world's electricity. The nuclear industry has arrived at a point where it is dealing with two major issues. First, it must address life extension for legacy units while complying with present-day safety regulations. Second, new systems must be designed with enhanced safety features so that the core damage frequency meets the target of $10^{-6}$ failures per reactor-year or less.

The available literature suggests an increasing role for a risk-based or -informed approach to the design, operation, and regulation of NPPs in order to improve safety [1–3]. Although risk-based or -informed applications are growing for many engineering systems, such as process or chemical plants and aviation systems, NPPs have inherent or specific aspects that need to be addressed. These requirements are met by employing a defense-in-depth approach through efficient and fast-acting mechanisms to address dynamics of nuclear reactions, highly reliable and effective cooling systems to remove the decay heat, maintenance containments by a series of

---

This chapter is a revised and updated version of paper entitled "Role of Prognostics in Support of Integrated Risk-based Engineering in Nuclear Power Plant Safety" by P. V. Varde and Michael Pecht (2002), published in the International Journal of Prognostics and Health Management (ISSN 2153-2648), PHM Society.

P. V. Varde and M. G. Pecht, *Risk-Based Engineering*, Springer Series in Reliability Engineering, https://doi.org/10.1007/978-981-13-0090-5_13

barriers to contact source of radioactivity, and maintenance of emergency measures in general and long-term consequences.

The prognostics approach requires development of models and methods for irradiation-induced degradation. Apart from this, the aspects related to accessibility to reactor core components is a special issue that needs to be addressed while developing applications of prognostics as part of a risk-based or -informed approach to NPPs.

However, the major limitations of the risk-based approach in its present form are: (i) the inability to handle dynamic scenarios (e.g., fault trees and event trees are static in nature), (ii) the uncertainty in prediction of life and reliability of components and systems, (iii) the absence of a well-defined framework for monitoring and tracking system performance, and (iv) the lack of mechanisms to generate input for dynamic PSA models. The prognostics approach has the potential to overcome or reduce these limitations. The prognostics framework uses online monitoring of precursors and feature extraction to predict degradation trends and the life of the component. This feature enables application of dynamic PSA while reducing the uncertainty in life or reliability prediction because the predictions are based on real-time operational and environmental stresses. The failure mode, effects, and criticality analysis (FMECA) performed as part of a prognostics approach provides an effective framework to monitor the precursor parameters. The predictions provide the required input to dynamic PRA models and updates of the risk models in real time. This chapter presents a role for prognostics—a relatively new paradigm— as part of a risk-based approach to extend the present periodic activities of monitoring, surveillance, in-service inspection (ISI), and maintenance to condition-based activities through the application of prognostic methods.

The major elements of prognostics are online monitoring of precursor parameters and the detection of deviation from the reference conditions using prognostic algorithms [4]. Here, the evaluation of remaining useful life (RUL) for the monitored component or system and the use of insights from this evaluation is a crucial part of risk-based or -informed applications [5]. Figure 13.1 shows the major steps in prognostics as part of integrated risk-based engineering (IRBE) applications. The main aim here is to monitor the degradation in a dynamic manner and enable prediction of the failure well in advance so that failure can be avoided altogether, or advance action can be taken to repair or mitigate the consequences associated with the failure.

Traditionally, the nuclear industry has employed online status monitoring of safety and process parameters so that any deviation from the reference operating conditions can be detected in time, and, if necessary, automatic safety actions can be initiated. Various levels of defense also exist in the form of alternate provisions

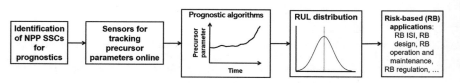

**Fig. 13.1** Simplified representation of prognostics as part of integrated risk-based applications

that provide coping time for systems and equipment should the preceding level of defense fail.

However, there is a need to predict the life and reliability of each level of defense in order to enhance the safety of a plant. The prognostic approach facilitates the health management of systems and components based on the remaining life prediction of components. Even though in the current generation of plants, prognostic principles are used in the form of qualitative reliability and life attributes, the full potential of prognostics has yet to be realized through the formal implementation of a prognostics-based health management program.

The available literature shows that the role of prognostics is growing in many fields of components and systems where safety forms the bottom line, such as in aerospace [6], electronics systems [7–9], telecommunications, and structural systems [10]. Specific engineering applications include prognostics for bearings and gears [11], engine/turbine condition monitoring [12], aircraft engine damage modeling [13], aircraft AC generator model simulation [14], health monitoring of lithium-ion batteries [15], and development of an intelligent approach in support of diagnostics and prognosis [16]. Based on the experience in these fields and the knowledge that has been generated over the years, it can be argued that a prognostics-based approach, as an extension of a condition monitoring approach, is expected to go a long way to address the surveillance and monitoring requirements of new as well as legacy nuclear plants. For legacy plants, life extension programs can be implemented on a sound footing by integrating prognostics and health management (PHM) models to complement the risk-based approach. For new plants, enhanced safety can be achieved by the implementation of prognostics-based health management of systems and components. To realize risk reduction through the prognostic approach, design specifications should ensure that a plant is built with online monitoring capabilities for the identified precursor parameters. This basic setup will focus on online prediction of remaining life and reliability considering the postulated loads and stresses such that risk reduction by detecting failure in advance can be realized. The same approach also applies to legacy plants. The existing sensors and monitoring systems in these plants can be adapted in support of prognostics. For example, the vibration monitoring data on rotating machines can be used for PHM because it is relatively easy to install a vibration monitoring network for existing check valves. However, some specific locations will be challenging to monitor online (e.g., in-core and reactor support and structural components may not be easily accessible). For these systems, the monitoring of derived or secondary parameters may work. For example, the annulus gas monitoring system provides information of leakage, if any, as an online assessment of the integrity of the coolant channel in the existing fleet of pressurized heavy water reactors (PHWRs). Periodic inspection and installation of coupons (e.g., to assess corrosion of subsoil piping) may also provide an effective approach in the absence of online monitoring for legacy plants.

This chapter presents a review of the current approaches to monitoring and surveillance. Likewise, this chapter assesses the prognostic requirements as part of an integrated risk-based approach for legacy and new NPPs. Even though the implementation of prognostics varies depending on the type of components and the objective of the prognostic applications, this chapter proposes a general framework that addresses the basic or broader aspects of various applications. The prognostic performance metrics and other related issues that are relevant to NPPs are also discussed.

## 13.2  A Brief Overview of Surveillance and Condition Monitoring in NPPs

The safety design philosophy for NPPs requires the implementation of defense-in-depth, fail-safe criteria: The design is fault-tolerant to the extent that a single failure event will not adversely affect plant safety. The selection of online process parameters and associated limiting condition settings ensures that all postulated conditions are monitored and timely actions do not compromise safety. Most of the NPPs operating globally belong to first- and second-generation systems. Based on the accumulated operating time logged by NPPs, the average life of the NPPs works out to be over 20 years [17]. In general, NPPs have a design life of more than 40 years. The evidence of aging may manifest in many ways, such as frequent failure of components in process and safety systems and subsequent interruption of plant operation, overall reduction in plant availability, and adverse impacts on the available redundancy or safety margins in safety systems [18, 19]. With effective inspection and maintenance practices, age-induced degradation can be managed, and operational life can be extended. For over 30 years, the U.S. nuclear power industry and the U.S. Nuclear Regulatory Commission (USNRC) have worked together to develop aging management programs that ensure the plants can be operated safely well beyond their original design life [20].

Third-generation plant designs are characterized by the use of inherent safety features such as negative void coefficient of reactivity, incorporation of passive features, the shift from analog to digital plant protection systems, and added redundancy from 2-out-of-3 in second generation to 2-out-of-4 trains and channels (including the control and protection system and improved accident management features in containment). Application of the leak-before-break concept in design and operation has also been associated with new plants. Apart from this, condition monitoring using vibration signatures, current signatures, insulation resistance assessments, temperature trends, acoustic signatures, and other process parameter variations forms part of the diagnostic and, in a limited way, prognostic assessment of third-generation plants, components, and systems. Some examples of condition monitoring include assessment of the health of the fuel by online monitoring of radiation levels, assessment of rotating machine mechanical bearing condition based on online or offline measurement of vibration and temperature, current signature analysis to assess the health of induction motors, electromagnetic interference mapping to assess the effect of magnetic field, pump shaft performance monitoring using eddy current technique, exhaust air temperature and smoke quality monitoring to assess health of the diesel generators, and oil sample analysis of foreign material to assess degradation and wear out of mechanical parts.

There are also examples of built-in-test (BIT) facilities for online diagnostics in systems and control systems. For safety channels, the protection channel will be activated only when there is demand. In these types of systems, the latent fault remains passive and reveals itself only when a channel is required to be activated. For such cases, periodic testing is conducted to reveal a passive fault so that a system is available when there is an actual demand. However, the test interval determines system availability. The safety objective requires testing the channel as frequently as possible to ensure the maximum availability of the channel. For a protection channel, this testing is conducted by incorporating a fine impulse test (FIT) feature. An FIT module sends an electrical pulse of very short duration of around $\sim 2$ ms. This duration is long enough to test electronic cards but short enough to not activate an actuation device such as an electromagnetic relay, because actuation of a 48 VDC relay requires a signal that lasts at least $\sim 40$ ms.

From the structural health monitoring point of view, annulus gas monitoring, where $CO_2$ gas is passed through an annular gap between the pressure tube and a calandria tube, is a good example of condition monitoring [21, 22]. The dew point of the $CO_2$ is monitored at the exit point of the channel to identify any indications of leak. Any increase in dew point from the reference dew point of around –40 °C indicates a possible leak in the annular region from the pressure tube or calandria tube and prompts an analysis of the region. This is an example of an implementation of leak before break strategy in real-time mode. The examples listed above are not exhaustive. They indicate the state of the art in operating NPPs that have condition monitoring provisions and limited features for prognostics. However, this background provides a basis for identifying gap areas for the implementation of a PHM program as part of a risk-based approach.

The first step in implementing a prognostics program for a complex system such as an NPP is to classify the systems, structures, and components (SSCs) into different categories, keeping in view the NPP's design and operation characteristics, which will also determine the type and level of prognostics. The classifications and categorizations performed in this chapter are not comprehensive but rather indicative. The objective here is to present the state of the art of monitoring for these components from the point of assessing the prognostic maturity level for these components. Table 13.1 shows the status of online monitoring, condition monitoring, ISI, and diagnosis and prognosis. This categorization has been primarily carried out keeping in mind the PHWR systems and components and is representative and not exhaustive. The basic idea is to categorize the systems and components, as shown in Table 13.1, keeping in mind the reactor type and prognostic requirements. The following points are drawn from this table with respect to criteria required for the classification of NPP SSCs, the status of various surveillance methods, and existing gaps in the implementation of prognostics.

This classification is required for both new and legacy reactors. Table 13.1 provides a good starting point to generate prognostic specifications for the new design. In each category, the representative components are chosen such that they

**Table 13.1** Categorization of SSCs and status of monitoring, diagnosis, and prognostics in existing NPPs (up to Generation III)

| Component and system type (*representative items*) | M | OFL ISI | OFL CM | OFL D | OFL P | ONL D | ONL P | Remarks |
|---|---|---|---|---|---|---|---|---|
| (a) **Reactor Structures**: Reactor pressure vessel, *coolant channels*, reactor block, reactor vault and its lining, shielding structures, steam generator and associated fittings and penetration and nozzles, ventilation plenum and ducts, etc. | **** | **** | **** | **** | **** | **** | *** | The life prediction for Canada deuterium uranium (CANDU) and PHWR coolant channels [21, 112, 113] |
| (b) **Non-reactor Structures**: *Containment* and civil structures, fuel transfer and storage block, overhead tanks and reservoirs, airlocks, structural support, bridges and jetties, guide and support, etc. | **** | **** | **** | *** | *** | *** | *** | Structural health prediction in the R&D stage [114, 115] |
| (c) **Mechanical Components**: *Pumps and turbines, piping*, valves, heat exchangers (shell and tube and plate type, fueling machine, fans and dampers, hydraulic drives and systems, strainers and filters, bearings, diesel generators, compressors, cranes, traveling water screens, etc.) | **** | **** | **** | **** | ** | *** | *** | State of the art is available on online diagnosis, but prognostics is in the R&D stage [115, 116] |

(continued)

**Table 13.1** (continued)

| Component and system type (*representative items*) | M | OFL ISI | OFL CM | OFL D | OFL P | ONL D | ONL P | Remarks |
|---|---|---|---|---|---|---|---|---|
| **(d) Electrical Power Systems**: Electrical buses and cables, high-voltage transformers, *motors*, breakers and isolators, power relays, motor generators and alternator sets, battery banks, etc. | **** | *** | *** | *** | *** | ** | *** | CM of rotating machines [116] |
| **(e) Power Electronics Systems**: *Uninterrupted power supplies*, convertors, inverters, rectifiers, etc. | *** | ** | ** | ** | ** | ** | * | R&D work on capacitor, insulated-gate bipolar transistors (IGBTs) reported [117, 118] |
| **(f) Microelectronic Systems**: *Digital cards*, integrated circuits, programmable logic controllers, field-programmable gate arrays, interconnects and control cables, control connectors, etc. | **** | *** | **** | *** | ** | **** | *** | Prognostics in the R&D stage [4] |
| **(g) Process Instrumentation**: *Electrical and pneumatic transmitters*, pressure and flow gauges, resistance temperature detectors, thermocouples, impulse tubing, control valve telemetry, solenoids, pH, conductivity meters, etc. | **** | **** | **** | *** | ** | **** | ** | Smart sensors and periodic calibrations [119] |

(continued)

**Table 13.1**   (continued)

| Component and system type (*representative items*) | M | OFL ISI | OFL CM | OFL D | OFL P | ONL D | ONL P | Remarks |
|---|---|---|---|---|---|---|---|---|
| **(h) Nuclear Instruments**: *Fission counters*, ion chambers, etc. | **** | **** | **** | *** | ** | **** | ** | Often saturation characteristics indicate remaining useful life |

*Note* The characterization of the metrics has been performed considering the "representative items" identified in columns as "bold and italics"

*OFL* Offline; *ONL* online; *M* monitoring; *CM* condition monitoring, *D* diagnosis; *P* prognosis; *ISI* in-service inspection

**** Technologies available for NPPs; *** technologies are required for further qualifications and specific applications; ** technology in the R&D domain, feasibility demonstrated; * work initiated
Important: The items shown in this table provide an overview and do not claim, in any way, to provide specifics and guidelines

fulfill one or more of the following criteria: The component allows prognostic implementation (the most challenging); the component allows prognostic, diagnostic, ISI, or condition monitoring implementation; the component represents a typical sample from the category; and for other components in the group, prognostic implementation will be similar to the representative component. The monitoring program has matured for all the categories of components in NPPs. ISI is applicable to mechanical components in general and piping and associated fittings in particular. It may be noted that the capability of ISI in terms of various coverage factors such as detection, location, and isolation remains a subject of research and development (R&D) [23]. The surveillance activities, which include testing and maintenance, performed on electronics channels and electrical power supply systems have also been categorized under the ISI program. Condition monitoring programs for reactor coolant channels, pumps, motors, bearings, reactor containment, fission counters, and transmitters are mature in NPPs.

In general, diagnostics is provided for selected components, such as diesel generators, pumps, and digital cards. For example, the complete protection channel is monitored in an online mode for detecting failure of any card using a BIT or a FIT facility. There are examples of online surveillance and health management programs in NPPs, including coolant channel inspection activities (item (a) in Table 13.1) in a pressurized water reactor or an advanced aging management program for mechanical components. Similarly, the FIT facility for monitoring all the redundant channels (item (f) in Table 13.1) is also an example of a verification and validation tool for the health assessment of electronic parts of protection channels. The saturation characteristics of ion chambers or fission counters provide an online indication of the remaining life of these components. However, regulation and protection channels only have diagnostic features, and R&D is required for the implementation of prognostics for these components. At the component level, condition monitoring of rotating machines using vibration and temperature monitoring and diesel generator sets is arguably a mature health management program, except that it lacks the capability of life prediction. A literature review suggests that for most of the components in NPPs, online prognostics either have not been developed or are still in an R&D stage.

## 13.3 Prognostics and Health Management and Integrated Risk-Based Engineering

Figure 13.2 shows the interrelationship between PHM and IRBM. The deterministic and probabilistic methods can be used to build an integrated risk model of the plant. The integrated risk model provides information about safety issues in general, which is vital for the implementation of prognostics. This includes information on components and modes and sequences of system failures, which are precursors that can form candidate components for the implementation of prognostics. From here the organization gets vital input to focus on a small group of safety-critical items to improve technical specifications. This approach provides a valuable tool for prognostic coverage so that maximum benefits can be realized by investing the available resources.

While probabilistic input is critical for uncertainty characterization in prognostics, the overall effect is to help in the assessment of a safety margin at the reference level. Once a prognostic assessment is performed, it is possible to understand the

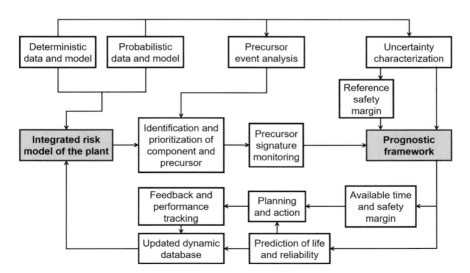

**Fig. 13.2** Interrelationship between PHM and IRBE

safety margin in a dynamic manner. The prognostics strategy complements the present risk-based framework by online performance tracking and generation of feedback for operators as well as regulators. The overall gain from the implementation of a prognostics strategy is that it allows the assessment of safety issues by predicting life, reliability, and prognostic distance in a dynamic manner.

## 13.4  Requirements of Prognostics for IRBE Applications

There are three major areas that need to be strengthened in the risk-based approach. These include dynamic modeling, improved methods for uncertainty characterization, and realistic assessment of safety margins. It can be argued that a risk-based approach is a rational one, in that it combines the plant configuration, including operational logic, and performance parameters through probabilistic reasoning. Hence, this approach, in conjunction with the deterministic approach, is expected to provide more flexibility compared to the traditional approach. Apart from this, performance monitoring and generation of feedback following the implementation of changes and modifications form an integral feature of risk-based engineering. Thus, to develop and implement a prognostics program, the above-mentioned issues need to be addressed in the following manner:

- Develop suitable sensors that can measure precursor parameters of interest;
- Assess reliability and RUL of the monitored systems online, based on identified precursors or degradation characteristics;
- Develop a prognostic algorithm that provides advance remaining life prediction with an adequate time window;
- Characterize uncertainty in the prediction, instead of point estimates of RUL, such that management issues can be addressed in an efficient and effective manner;
- Employ multi-objective algorithms that take into consideration risk, cost, and reduction in radiation dose;
- Include a provision for database, model-based (physics-of-failure), and fusion approaches, keeping in view the varying nature of prognostics programs for a range of components, such as mechanical and electrical systems, electronics, and nuclear components; and
- Include a provision to provide feedback online to track the performance of modifications in a component that has been replaced or has undergone a maintenance procedure or calibration of some instrumentation.

The use of input from prognostics may require modification to the existing risk assessment approach. For example, an existing database in the risk-based approach may have only static reliability data, failure criteria, and maintenance and test schedules. However, when the dynamic aspects are implemented as part of prognostic feedback to risk-based engineering, then there is a need to re-organize the complete database framework such that dynamic inputs and outputs can be managed.

The complex nature of NPP design requires a different set of design metrics that satisfies the requirements of a particular application. There are many areas that require R&D with respect of material degradation, development of special sensors, and suitable algorithms for online feature extraction and analysis. In some situations, the design constraints may make it challenging to implement a prognostics program. Table 13.2 shows the requirements that can be used for the design of a prognostics program for a given application. The applicability of these parameters for legacy or new plants is indicated in the table by * and #, respectively.

## 13.4.1 Plant Stage

The scope of a prognostics program will be governed by such factors as at what stage of the plant the prognostics is being implemented. As mentioned earlier, for the implementation of a prognostics program in new plants, prognostic requirements and specifications should be part of the design strategy. Since the plant has yet to be built, provisions can be made in advance, keeping in view criteria such as safety and availability. Prognostics as part of life extension will have activities focused on select systems, components, and structures. However, the plant's design and operational constraints will dictate the implementation levels. Prognostic requirements for refurbished plants will be similar to a plant whose life has been extended. In a refurbished plant, prognostics is useful particularly for those systems where clear insights into the remaining life of certain components are not available, where the cost of bulk replacement would have been prohibitive, and where it is felt that online prognosis and diagnosis would be useful, such as in coolant channels, piping, and power supply cables.

## 13.4.2 Objective

The objective of prognostics is to view the plant status, determine logistics, and understand the material degradation phenomenon via various prognostic algorithms. For new plants, institutions typically prefer a model-based approach, whereas for legacy plants, where enough data are available, the data-driven approach is preferred. It should be noted that most NPPs either have a Level 1 PRA model with internal initiating events for full-power conditions or a reactor core as the source of radioactivity. These plants have an obvious advantage over using risk modeling to formulate or identify and prioritize a prognostics program. The available literature shows that most prognostics implementations result in improved availability and cost/benefit objectives. There have been applications of prognostics

**Table 13.2** Prognostics design requirements

| Parameter | Applicable levels | | | | |
|---|---|---|---|---|---|
| | 1 | 2 | 3 | 4 | 5 |
| Plant stage (* & #) | Under design new | Operating plant useful life (20–40 years) | Operating plant aged (more than 35 years) | Shutdown under-refurbishing | Operating after refurbishing |
| Objective (* & #) | Safety improvement | Availability improvement | Mission reliability improvement | Follow-up after retrofitting | Reduction in operational cost |
| State-of-the-art enabler (*) | Online monitoring, offline diagnosis | Online monitoring, offline diagnosis, conditioning monitoring | Online monitoring, diagnosis, condition monitoring | Online monitoring, diagnosis, condition monitoring, offline prognosis | Online monitoring, online diagnosis, condition monitoring, online prognosis |
| Subject (* & #) | Microelectronics/digital control channels | Power electronics | Electrical | Structural/mechanical | Interdisciplinary |
| Implementation level (* & #) | Level 1 | Level 2 | Level 3 | Level 4 | – |
| Risk assessment approach (* & #) | Qualitative/deterministic | FMECA | Hazard and operability analysis | System-level reliability modeling, fault tree, event tree | Level 1 PRA |
| Existing maintenance/health management Strategy (*) | Preventive maintenance and scheduled testing | Condition-based test and maintenance | Reliability-centered | Risk-based test and maintenance | Online reconfiguration |
| Stakeholders (* & #) | Design team | Operating organization | Regulators | – | – |
| Implementation approach (* & #) | Model-based | Data-driven | Risk-based | Fusion | – |
| Availability of tools and methods/challenges (* & #) | Prognostic algorithm | Availability of sensors | Degradation models | Feature extraction methods | – |
| Cost benefits (* & #) | In the context of NPPs, cost benefits for a particular level are addressed using safety and availability indications | | | | |

to aircraft health monitoring, where the emphasis has been on safety. There have also been applications of PHM with mission safety as a driving force.

### 13.4.3 State of the Art

The state of the art refers to the current status of monitoring and surveillance methods used in a plant. If a plant is in the design stage, metrics may include the requirements for monitoring and health management. Provisions can then be made throughout the design stage for implementation of a prognostics program. However, in operating plants the existing and new sensors/provisions sensors will determine the level and scope of prognostics. In general, the state of the art in the current generation of plants facilitates online monitoring of important safety parameters, condition monitoring, and surveillance. Such programs are not fully automated, however; diagnosis is performed in offline mode for most systems. There are closed feedback loops wherein corrective actions are automatic. These feedback loops ensure the maintenance of plant parameters within set limits. These metrics help to determine the specifications and the scope of the prognostic requirements. The available literature on NPPs does not appear to provide information about prognostics applications based on online life and reliability prediction.

### 13.4.4 Application-Specific Prognostics

For complex systems such as NPPs, most of the systems require an interdisciplinary approach to implement a prognostics program. However, particular disciplines may require a unique focus. For example, prognostics for reactor protection channels and structural components such as reactor blocks will differ with respect to degradation mechanisms, the time window available, and the monitoring and sensor requirements. Hence, even though the broad framework may remain the same, the specifics will vary by applications.

### 13.4.5 Level of Implementation

The metric for the level of implementation is derived/adopted from the procedure developed by Gu et al., keeping in mind the NPP requirements [24]. In [24], various levels have been presented for the implementation of prognostics of electronic systems. For complex systems such as NPPs, there are different prognostic levels. Level 0 is the component level, which includes items such as fuel assembly, feeder, bearing, motor, control and power cables, alternator, pipeline, battery, relay, microprocessor chip, switch, and electronic cards. Level 1 includes the assembly of components of a particular class, such as mechanical, electrical, or nuclear and associated connections that perform a basic function. Examples include pumps with connected piping to suction and discharge and the suction strainer, a compressor with subcomponents such as coolers and associated connections, diesel generators with support systems, and power supply modules, amplifier modules, and function generators. Level 2 includes those systems that are activated only on demand from the plant control system. They can also be referred to as safety support systems such as Class III electrical power supply, Class II control supply, and Class I power supply systems. Level 3 systems include those systems that are required to be operational when the reactor is in an operational state, including the main coolant system, Class IV power supply systems, feed water systems, regulation systems, and process water systems. The structural systems that are basically passive but require a structural approach for health monitoring, such as the reactor vessel and reactor shielding components, reactor pile, and containment building, are categorized as Level 4. These categories are based on the broad characterization of component functional requirements and their place in a system, whether as an independent unit, sub-block, block, major function, or assembly of functions to deliver an objective function.

## 13.4.6  Risk Assessment Approach

Most NPPs have a Level 1 PRA implemented, considering the internal initiating of events for full-power conditions. Even though risk-based applications require shutdown or a low-power operation PRA, the availability of a full-power PRA can be considered for initiating a PHM implementation program. Apart from this, FMECA forms an integral part of PHM implementation. A comprehensive FMECA program is recommended, keeping in mind the focus of prognostic implementation.

## 13.4.7  Existing Maintenance and Health Management Strategy

This section determines the current maintenance and health management strategy, an important reference for building a PHM program. Typically, most NPPs use preventive maintenance as the major approach for health management. However, condition monitoring, ISI, and scheduled test and maintenance are the general features for health management. The available literature shows that in some NPPs and industrial systems, reliability-centered maintenance, risk-based ISI, and risk-based technical specification optimizations are also used. The available framework is important, as the data generated on the maintenance and health of these systems and pieces of equipment form a fundamental part of the data-driven approach for prognostics. Along with inputs from risk models, these data and insights will help to identify and prioritize the prognostics program.

### 13.4.8   Stakeholders

Although stakeholders are not a metric, they affect which agency is interested in prognostic applications. The designers would like to have a prognostics program for identified systems or as part of a design policy for systems that they feel will determine the life of the plant. The object systems could be in-core components or structures that form an integral part of systems such as reactor vessels, pile blocks, storage pool linings, or containment, or the object systems could be some safety or process system for which they are important to track performance. For operational agency, certain aspects of the plant could affect plant availability, such as performance of the strainer, check valves, and certain pipelines and bearings, which require continuous monitoring and remaining life assessment such that repair and replacement of these components can be scheduled to improve plant availability. Regulatory agencies want to track the performance of a system where the changes have been implemented. Here the role of prognostics is to provide feedback on the RUL or performance monitoring of a system for a specified period of time or for an extended duration to ensure that safety has not been compromised.

### 13.4.9   Approach for Implementation

The approach to prognostics implementation is governed by many factors, including the objective or purpose, the level of detail required, the availability of data, and plant constraints. If a prognostics program requires performance monitoring as part of a risk-informed or -based approach, then the focus will be on monitoring the performance metrics of the system under regulatory review. If a prognostics program is being designed for a new plant and the objective is to strengthen the safety function, then the task should include the prognostic specifications in the design phase and keep provisions not only for online monitoring but also for the implementation and management of the health of the plant. If prognostics is being implemented as part of a life extension strategy, then the focus should be on structural remaining life assessment. As a rule, NPPs require close monitoring of structural health, particularly where safety is the major metric, even if they have not entered the aging phase.

## 13.4.10  Tools and Methods

Prognostic tools and methods are identified when FMECA has been performed, precursors have been identified, and the PoF-based and/or data-driven approaches have been evaluated, keeping in mind the requirements of applications. However, detailed studies, literature searches, or required meetings with consultants may present issues associated with the selection of prognostic algorithms, the availability of sensors, the availability or limitations of degradation models, and the approaches that will be required for feature extraction, deriving useful data and information from a host of complex data, and signatures collected from experiments.

## 13.4.11  Cost/Benefit Studies

The available literature shows that cost/benefit evaluation can be used to demonstrate the net benefit of the implementation of PHM results [25]. In the context of NPPs, benefits need not be in terms of monitory gain; they could be in terms of safety improvement, life extension, or lessening the burden on the operating staff.

## 13.5  Prognostic Framework for Nuclear Plants

Major elements of prognostics implementation include monitoring system performance through noise analysis, detecting changes by trending, understanding and identifying root causes of failure, prognostics, and health management [26]. Even though there will be variation in the tools and methods, the level of accuracy required when prognostics is implemented for mechanical, structural, or electrical systems will remain the same.

## 13.5.1  Existing Setup

Traditionally, online monitoring and maintenance, including surveillance of passive and structural systems as well as maintenance of active components, form an integral part of NPP operations, as shown in Fig. 13.4. The existing approach is shown within the boundary drawn on the left side of Fig. 13.3. The monitoring provisions exist at the component, system, and plant levels. These monitoring provisions are limited to process parameter values and an equipment status display, which indicates various states of reactor operation including transient states. However, condition monitoring and surveillance for many systems is performed in an offline mode as part of plant policy for selected equipment. Even though

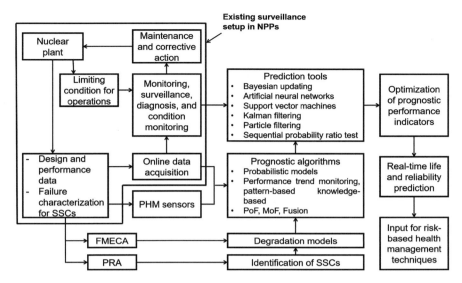

**Fig. 13.3** A prognostic framework for NPPs

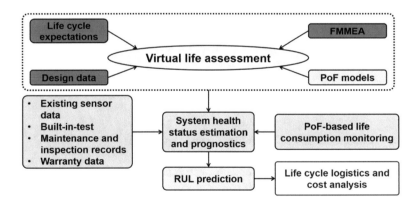

**Fig. 13.4** CALCE's PoF-based approach to prognostics

condition monitoring approaches have matured and are being used in health management, the prognostic quotient in terms of the prediction of remaining life is low. One of the major reasons for this is complexity in terms of material characterization, such as irradiation-induced degradation of core components and structures [15]. Apart from this, the nuclear industry operates on conservative criteria; hence, strict regulations for design and operation dictate that uncertainty in real-time assessment should be as low as possible. However, in the present situation, advances made in other application areas (such as space, aircraft, and civil engineering) can be implemented in NPPs by incorporating adequate provisions for some identified systems, which can provide insight into the application of prognostics in a graded manner as well as into safety-critical systems.

Figure 13.4 shows the framework for PHM for NPPs. The proposed approach, while utilizing the data and information that are available in the traditional approach, envisages development of prognostic sensor systems to monitor the identified precursor parameters. The data available through the sensors are mapped on the prognostic algorithms to track deviation and therefore provide information on incipient faults. Here, the role of intelligent tools such as support vector machine or Bayesian estimation is to predict the prognostic distance such that action can be taken well before the situation results in safety or availability consequences.

## 13.5.2   PHM Approaches

PHM is a multifaceted approach to protect the integrity of systems and avoid unanticipated operational problems leading to mission performance deficiencies, degradation, and adverse effects on mission safety. More specifically, prognostics is the process of predicting a system's RUL. By estimating the progression of a fault given the current degree of degradation, the load history, and the anticipated future operational and environmental conditions, PHM can predict when an object system will no longer perform its intended function within the desired specifications. Health management is the process of decision-making and implementing actions based on the estimate of the state of health derived from health monitoring and expected future use of the systems. Several approaches can be considered to implement PHM for NPPs, and this section mainly discusses the approaches.

### 13.5.2.1   Probabilistic or Reliability-Based Approach

The probabilistic or reliability-based approach is used extensively to predict the life and reliability of mechanical, electrical, or electronic components. Even though this approach is considered an approximate approach, its advantage is that uncertainty characterization comes naturally. The Weibull distribution is often used. Other distributions, such as exponential and log-normal distribution, are also common as a life prediction model [27, 28]. The weakness of this approach is that the predictions are based on the past performance data of equipment or components. This implies that the prediction does not account for variable operational and environmental loads. For example, for a given component in a component database, the failure rate estimations are based on an operational environment where the average temperature and relative humidity are 28 °C and 65%, respectively. The condition for which the failure rate estimation is required to work is a ground benign environment of 22 °C and humidity of 55%. These environmental conditions are bound to affect the failure rate—in this case, the failure rate is reduced. Certain external factors such as vibration and seismic shocks adversely affect the life and performance of a component. If these aspects are not factored into the estimates based on historical data, then the estimates tend to be either optimistic or conservative, depending on the severity levels of the component in the database compared to the component for which failure rates are being estimated. If a given component experiences less vibration and seismic shock than a component with a failure rate estimate based on higher vibration and shocks, then the failure rate estimates will not be accurate. Often these situations are handled by providing uncertainty bounds.

This approach involves prediction of the mean life of a component along with its upper and lower uncertainty bounds. A wide uncertainty bound indicates that the prediction is based on limited data sets, that reliance on such estimates should be lower, and that these estimates should be used as an indicator. In such situations, precursor-monitoring techniques such as vibration or temperature monitoring of the components represent an effective strategy for prognostics. The Bayesian model features probabilistic estimates that form a priori and has data coming from the precursor monitoring that can be used as evidence for updating the strategy for prediction [27]. So, even though the approach is primarily probabilistic, trend monitoring is used to improve the prediction capability.

### 13.5.2.2   Physics-of-Failure (PoF)-Based Approach

The PoF-based approach deals with the application of first principles models to understand the various failure mechanisms and thereby predict the RUL and reliability of components. In other words, this approach is based on the development and application of scientific models that predict component life. Unlike statistical approaches for reliability estimation, past performance data are not required [29]. The predictions are based on the component characteristics, such as material properties, geometrical attributes, and activation energy for applicable degradation processes for given environmental, operational, and environmental stressors. Accelerated life testing is central to the PoF-based approach. PoF enables the identification of dominant failure modes and mechanisms, and thereby precursors for monitoring component health. FMECA forms the cornerstone of this approach to identify and prioritize the applicable degradation mechanisms. Identification of precursors is an important part of this approach. The precursors are the parameters that can be monitored using the available sensors [30]. Precursor monitoring provides advanced information about the underlying degradation mechanism.

The PoF model can be expressed in general form as:

$$t = f(x_1, x_2, x_3, \ldots) \tag{13.1}$$

where $t$ is the median life and $x_i$ are parameters of any given PoF model $f()$. An example of the commonly known PoF models for life prediction includes an Arrhenius model, expressed as follows:

$$t = A\exp\left[\frac{E_a}{kT}\right] \qquad (13.2)$$

where $A$ is a process constant, $E_a$ is the activation energy of the process in eV (electron volts), $k$ is the Boltzmann constant (i.e., $k = 8.617 \times 10^{-5}$ eV/K), and $T$ is temperature in Kelvin.

There are many models available to predict life, such as the Eyring model. These models only recognize temperature as an environmental stress, whereas in real applications, there are many environmental and operational stresses that need to be considered. Likewise, these models fail to account for the geometrical and other design features such as material finishes and materials of the mating parts.

These limitations have led to further research into developing PoF models that consider the various stresses associated with each component in reliability modeling. This became possible with a basic understanding of the physics of degradation of materials under various stresses. Accordingly, accelerated life testing has become central to PoF modeling. Root cause analysis is performed to understand the degradation mechanisms responsible for failure. Figure 13.4 illustrates the framework of a PoF-based approach developed at the Center for Advanced Life Cycle Engineering (CALCE), University of Maryland, USA [31]. Figure 13.4 shows that information about the component, including physical specifications, geometry, construction materials, operating environment (including temperature, humidity, and vibration), and operational stresses (current, voltage, and electric field) forms the main input for modeling.

Failure Modes, Mechanisms, and Effects Analysis (FMMEA)

In PoF-based prognostics approaches, FMMEA has been used for the sake of the identification of PoF models associated with potential failure mechanisms for potential failure modes and the prioritization of failure mechanisms. In fact, FMMEA, developed by CALCE to deal with weak points inherent in conventional failure modes and effects analysis (FMEA) and failure modes, effects, and criticality analysis (FMECA) [32], is based on the understanding of the relationship between product requirements and physical characteristics of the product, the interactions of product materials with loads, and their influence on the product's susceptibility to failure under usage conditions.

A failure mode is referred to as the manner or form by which a product, system, or component fails. Examples of the failure mode include open or short circuits in electronics and cracking or creep in machinery. FMMEA provides a list of all possible failure modes for a target object, where failure modes can be identified with the help of using numerical stress analysis, accelerated life tests, and domain knowledge. In FMMEA, failure modes should be directly observable by visual inspection, electrical measurement, or other tests and measurements. A failure cause (or root cause) is defined as the circumstances during design, manufacture, or use

**Table 13.3** Example of failure mechanisms [120]

|  | Overstress | Wear out |
|---|---|---|
| Mechanical | Delamination, fracture | Fatigue, creep |
| Thermal | Excessive heating of component past glass-transition temperature | Diffusion voiding, creep |
| Electrical | Dielectric breakdown, electrostatic discharge | Electromigration |
| Radiation | Single-expose failure | Embrittlement, charge trapping |
| Chemical | Etching | Dendrites, whiskers, corrosion |

that have led to a failure. In fact, the identification of failure causes can be useful for investigating failure mechanisms, defined as the physical, chemical, thermodynamic, or other processes that result in a failure. In general, failure mechanisms can be categorized as either overstress or wear-out failure mechanisms, examples of which are presented in Table 13.3. In FMMEA, information about life cycle conditions for a target object can be used to eliminate failure mechanisms that may not occur under usage conditions. As aforementioned, the aim of FMMEA is to identify PoF models associated with failure mechanisms of interest for a target object that can be used to determine time to failure or likelihood of occurrence of a failure. For overstress mechanisms, PoF models can offer stress-/strength-based analysis to estimate whether a target object will fail under the given conditions. For wear-out mechanisms, PoF models quantify damage accumulated over a period of time for a target object by using both stress and damage analyses.

## Remaining Useful Life (RUL) Predictions

As briefly mentioned above, PoF models taking various stress parameters and their relationships to materials, geometry, life of a target object as input are used for RUL predictions. For electronic products, many PoF models [33–38] have been developed to describe various failure mechanisms of components, such as printed circuit boards, solder bond pads, and interconnects under various conditions, such as temperature cycling, vibration, humidity, and corrosion. For example, Wong et al. [36] reviewed creep and fatigue models of solder joints and developed a failure model for creep and fatigue. Hendricks et al. [37] presented a summary of the failure models such as electromigration and time-dependent dielectric breakdown.

RUL can be calculated from the beginning of the product life cycle and continues to assess the degradation of the product by monitoring its life cycle environment in order to provide an estimate of the remaining life in the application environment. At each time period, incremental product damage can be calculated from various stresses, which are caused by environmental or operational loads. Then, damage accumulation can be performed for a certain period. At last, the remaining life can be calculated based on the accumulated damage.

Canaries

One of the notable and significant features of the PoF-based approach (see Fig. 13.4) is the development and application of canaries [39]. Expendable devices, such as fuses and canaries, have been a traditional method of protection for structures and electrical power systems. Fuses and circuit breakers are examples of elements used in electronic products to sense excessive current drain and to disconnect power. Fuses within circuits safeguard parts against voltage transients or excessive power dissipation and protect power supplies from shorted parts. For example, thermostats can be used to sense critical temperature-limiting conditions and to shut down the product, or a part of the system, until the temperature returns to normal. In some products, self-checking circuitry can also be incorporated to sense abnormal conditions and to make adjustments to restore normal conditions or to activate switching means to compensate for a malfunction [40].

The word "canary" is derived from one of coal mining's earliest systems for warning of the presence of hazardous gas using the canary bird. Because the canary is more sensitive to hazardous gases than humans, the death or sickening of the canary was an indication to the miners to get out of the shaft. The canary thus provided an effective early warning of catastrophic failure that was easy to interpret. The same approach has been employed in prognostic health monitoring. Canary devices mounted on the actual product can also be used to provide advance warning of failure due to specific wear-out failure mechanisms.

Mishra et al. [41] studied the applicability of semiconductor-level health monitors by using precalibrated cells (circuits) located on the same chip with the actual circuitry. The prognostics cell approach, known as Sentinel Semiconductor™ technology, has been commercialized to provide an early warning sentinel for upcoming device failures [42]. The prognostic cells are available for 0.35, 0.25, and 0.18 μm complementary metal-oxide semiconductor processes; the power consumption is approximately 600 μW. The cell size is typically 800 μm² at the 0.25 μm process size. Currently, prognostic cells are available for semiconductor failure mechanisms such as electrostatic discharge, hot carrier, metal migration, dielectric breakdown, and radiation effects.

A fuse was used in a study on life cycle cost prediction for helicopter avionics [43, 44]. In this study, unscheduled maintenance and fixed-interval scheduled maintenance were compared with maintenance that is led by precursor-to-failure and life consumption monitoring PHM approaches. Optimal safety margins and prognostic distances were determined. In this study, the PHM line replaceable unit (LRU)-dependent model is defined as a fuse. The shape and the width of the HM distribution represent the probability of the monitored structure, indicating the precursor to a failure at a specific time relative to the actual failure time. The parameter to be optimized in the LRU-dependent case is the prognostic distance, which is a measure of how long before system failure the monitored structure is expected to indicate failure.

The time to failure of prognostic canaries can be precalibrated with respect to the time to failure of the actual product. Because of their location, these canaries

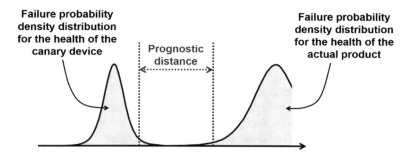

**Fig. 13.5** Failure probability density distributions for canaries and actual products, showing prognostic distance or RUL

contain and experience substantially similar dependencies as does the actual product. The stresses that contribute to degradation of the circuit include voltage, current, temperature, humidity, and radiation. Since the operational stresses are the same, the damage rate is expected to be the same for both circuits. However, the prognostic canary is designed to fail faster through increased stress on the canary structure by means of scaling.

Scaling can be achieved by controlled increase of the stress (e.g., current density) inside the canaries. With the same amount of current passing through both circuits, if the cross-sectional area of the current-carrying paths in the canary is decreased, a higher current density is achieved. Further control of current density can be achieved by increasing the voltage level applied to the canaries. A combination of both of these techniques can also be used. Higher current density leads to higher internal (joule) heating, causing greater stress on the canaries. When a higher-density current passes through the canaries, they are expected to fail faster than the actual circuit [41].

Figure 13.5 shows the failure distribution of the actual product and the canary health monitors. Under the same environmental and operational loading conditions, the canary health monitors wear out faster to indicate the impending failure of the actual product. Canaries can be calibrated to provide sufficient advance warning of failure (prognostic distance) to enable appropriate maintenance and replacement activities. This point can be adjusted to some other early indication level. Multiple trigger points can also be provided using multiple canaries spaced over the bathtub curve.

The extension of this approach to board-level failures was proposed by Anderson and Wilcoxon [45], who created canary components (located on the same printed circuit board) that include the same mechanisms that lead to failure in actual components. Anderson and Wilcoxon identified two prospective failure mechanisms: low cycle fatigue of solder joints, assessed by monitoring solder joints on and within the canary package, and corrosion monitoring using circuits that are susceptible to corrosion. The environmental degradation of these canaries was assessed using accelerated testing, and degradation levels were calibrated and correlated to actual failure levels of the main system. The corrosion test device included an electrical circuitry susceptible to various corrosion-induced mechanisms. Impedance spectroscopy was proposed for identifying changes in the circuits by measuring the magnitude and phase angle of impedance as a function of frequency. The change in impedance characteristics can be correlated to indicate specific degradation mechanisms.

The use of fuses and canaries for PHM presents unanswered questions. For example, if a canary monitoring a circuit is replaced, what is the impact when the product is re-energized? What protective architectures are appropriate for post-repair operations? What maintenance guidance must be documented and followed when fail-safe protective architectures have or have not been included? The canary approach is also difficult to implement in legacy systems because it may require re-qualification of the entire system with the canary module. Also, the integration of fuses and canaries with the host electronic system could be an issue with respect to real estate on semiconductors and boards. Finally, the company must ensure that the additional cost of implementing PHM can be recovered through increased operational and maintenance efficiencies.

The PoF-based approach is particularly suited for assessing electronic component reliability. Even though this approach is in the R&D stage, there are many models available for electronic components. The state of the art in microelectronic reliability shows that more advances have been made for reliability modeling of microelectronic components compared to power electronic components.

## Outputs of the PoF-Based PHM

The PoF-based PHM methodology provides outputs that can be used to provide advance warning of failures; minimize unscheduled maintenance, extend maintenance cycles, and maintain effectiveness through timely repairs; reduce the life cycle cost of equipment by decreasing inspection costs, downtime, and inventory; and improve qualification by assisting in the design and logistical support of fielded and future systems.

Compared with data-driven PHM methods, PoF-based methods have certain advantages in both new and legacy systems. This is mainly because an insufficient amount of historical data will be available for data-driven methods in such systems. Instead, PoF models can still be used when material properties and structure geometries of a target object are available. For legacy systems, PoF-based PHM first utilizes all available information (such as previous loading conditions and maintenance records) to assess the health status of the legacy system. Then it calibrates the health status using individual unit data so that an assessment of an individual legacy system's health can be derived. After that, it uses sensors and prognostic algorithms to update the health status on a continual basis to provide the most up-to-date prognosis of the system [46].

Although PoF-based PHM has advantages, the following issues/challenges need to be properly addressed to improve accuracy of RUL predictions. The first challenge is to account for the accuracy or uncertainty of converting the operational and environmental loading conditions to the local stresses, which are the actual parameters that play a role in most failure models. The second challenge is to ensure the accuracy of the feature extraction procedure for monitored data, such as the counting, amplitude, and duration of thermal cycles under a combination of actual operational and environmental conditions, which is more irregular than the experimental conditions. The third challenge is how to calculate the accumulated damage. Most of the current applications use a linear model (e.g., Miner's rule) to simply add up the damages caused by different types or levels of stresses. The last challenge is to basically ensure the availability and accuracy of PoF models.

### Concerns in PoF-Based PHM

Despite the fact that PoF-based PHM has been a promising option for RUL predictions, the following concerns still exist.

- Multiple and correlated failure mechanisms—One often inherently assumes the existence of one and only one failure mechanism in play at any time for a component or system that is subject to condition monitoring and PHM study. This assumption, however, is idealistic. In practical cases, several failure mechanisms could be coexistent, and some could be independent of one another while others could be correlated to each other, arising due to common driving forces that may govern multiple failure mechanisms. In such cases, the PoF model has to be modified to include the cumulative role of multiple mechanisms that may or may not play an additive role in its impact on the degradation metric being monitored. In fact, it is even a challenge to estimate how many mechanisms are in play in the product using just the sensor data as the input. If multiple mechanisms coexist, they could be due to geometrical proximity of the degradation events or process-induced defects or some sort of a positive feedback between the driving forces. One example in microelectronic devices is oxide breakdown, whereby the current density and temperature tend to have a positive feedback effect. Higher current tunnels through the dielectrics in these devices when the temperature rises, and this increase in charge injection (current) further enhances the localized joule heating (self-heating), resulting in a vicious cycle of temperature and current intensification, eventually culminating in a hard breakdown [47].
- Damage is not always additive—The use of the Miner's rule for accumulating damage is a simplistic assumption as it does not consider the interactions between different driving forces of failure. As explained above, several forces causing degradation can be highly interrelated to each other, as a result of which an additive damage model would be an overly simplistic and optimistic formulation.
- Latent degradation trends—There are certain failure mechanisms wherein the signals pointing to degradation appear to be hidden or latent in the noise existing

in the sensor. In such cases, with no evident degradation trend, application of the PoF models for remaining life estimation is a challenge and may not yield good results. The same holds true for data-driven approaches as well. One example is that of time-dependent dielectric breakdown (TDDB) in thick dielectrics used in power devices [48]. The TDDB event often tends to be sudden and catastrophic, with no presignals of any anomaly in advance. In such cases, we might want to resort to a canary test structure with thinner dielectrics for some early insights into the trend of degradation.

- Extrapolating canary test structure results to the actual device—Most often, the purpose of a canary test structure has been to qualitatively imply that the actual device is nearing its end of life, once the canary has failed. Although this information is useful, it does not allow us to fully leverage the data and results obtained from the canary for any further inference. It would be good to use the degradation and failure data from a canary to quantify the remaining useful life of the actual device under test. However, this requires us to scale-up the failure distribution of the canary device to the actual device using physical theory-based extrapolation rules such as area scaling, voltage/temperature scaling, material property scaling, and several others. In short, the scale-up of canary prognostic results to real device prognosis is a complicated one. It depends on the similarity of the design of the canary test structure to the real device. Sometimes, the canary test structure may experience a completely new and unanticipated failure mode/mechanism due to its scaled-down dimensions, different geometry, or choice of material. In such cases, the data collected cannot be used for any RUL inference for the actual device. The design of the right canary structure therefore requires a lot of attention.

- Failure immunity and incubation time—Some mechanisms tend to be "active" only for a certain set of operating test conditions. In these cases, it is all the more important to account for this failure-immune region implicitly or explicitly in the PoF model, before it is used for failure prognosis for varying operating and environmental load conditions. One such example is the Blech length effect for Cu and Al interconnects [49], whereby for interconnect segments shorter than a critical value, the atomic flux divergence due to electron wind force and back stress gradient nullify each other, resulting in no voiding or hillock formation. Another example is the critical voltage ($V_{critical}$) in dielectrics for progressive breakdown of the percolated defective path [50]. For voltages lower than $V_{critical}$, the positive feedback of percolation current and localized joule heating is insufficient to sustain further evolution and dilation of the percolation path—this is often known as "progressive" or "analog" breakdown. In effect, thinner dielectrics are more immune to this progressive breakdown, which implies that very high currents and destructive breakdown spots are seldom observed in transistors and memory technology that are fabricated with very thin dielectrics (less than 2 nm). Other failure mechanisms tend to have a so-called incubation time, which is a failure-free period wherein the defects have to pre-align themselves in a certain fashion before they can effectively deteriorate the performance or functionality of the device.

- Reliability of the failure model is critical—Prognostic estimations using PoF
  models are only as good as the models themselves. There is often a compromise
  to be made when considering the complexity of a model and its associated
  computational cost for real-time inference and prognosis. The same failure
  mechanism can have different models—some empirical, some phenomenolog-
  ical, and some completely physical. The physical models, which are more
  accurate and effective in describing the physics (most desirable), also tend to be
  more involved in the computational side. Therefore, there is a need to strike a
  balance between efficiency and accuracy, more so in a real-time context wherein
  condition monitoring and prognostic inference have to go hand in hand.

### 13.5.2.3   Data-Driven Approach

This section deals with the data-driven approach (see Fig. 13.6, which illustrates a
general process of the data-driven approach), except that there is a distinction
between various data forms and the input for prognostics. For example, trend
monitoring of operational and environmental parameters through online instrumentation may provide information about some precursors. A pattern comprising
the status of a finite set of alarms as "registered as 1" and "cleared as 0" is another

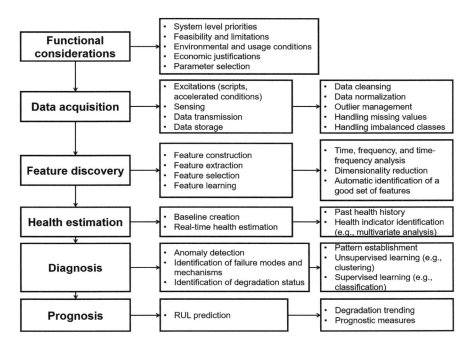

**Fig. 13.6**  A general procedure of a data-driven approach to prognostics

representation of data. A probabilistic distribution of time to failure based on individual components provides time-to-failure estimates of the systems being monitored. Input can be in the form of linguistic variables in place of a numerical value. All these require different approaches.

Likewise, the term "symptom-based approach" can be used to extend the context of input data and information used in prediction, particularly for NPP applications. As mentioned above, often information is not available in the form of a numerical or binary value (0/1 or yes/no). Instead, the information about the model parameters comes from experts in linguistic expressions. This information is not suitable for use as input; however, the information cannot be ignored, as it provides much stronger input for prediction or estimation of RUL. In such instances, treating expert's opinions, which can be considered imprecise information, using fuzzy algorithms can provide one with improved assessment of imprecise parameters [51].

The reason to have provision for some information is that establishing a pattern is important, as often, instead of a single parameter, a pattern can provide more data and information. For example, a comparative value of three parallel components experiencing the same operational and environmental stresses may form a pattern, which may provide an effective mechanism to assess the health of the component and thereby provide an effective input for predicting the remaining life. The only issue is that even this information could be expressed in terms of linguistic variables and will require the fuzzy approach to address the challenge. This background is an obvious reason to formulate the data and information in two ways: trend monitoring using precursor symptoms and a pattern-driven knowledge-based approach.

## Feature Engineering

This section discusses feature engineering, which is fundamental to data-driven PHM applications using machine learning, within the category of feature construction, feature extraction, and feature selection.

### Feature Construction

In general, the types of data collected for the sake of condition monitoring of a target object can be classified as follows: value (e.g., temperature and humidity), waveform (e.g., vibration, current, and voltage), and multi-dimensional (e.g., X-ray image and thermal infrared image) [52]. Among these, waveform data are the most commonly used in data-driven PHM applications. Hence, this section provides a review of methods of feature calculation from waveform data. To create features from waveform data, the following analyses have been widely employed: time-domain analysis, frequency-domain analysis, and time-frequency-domain analysis. Through time-domain analysis, the statistical features presented in Table 13.4 have been commonly used for PHM of machinery [53]. Rather than using statistical features, time series models have been used to create features. For

**Table 13.4**  Examples of time-domain features [53]

| Feature | Equation | Feature | Equation |
|---------|----------|---------|----------|
| Peak ($x_{peak}$) | $max(\lceil x \rceil)$ | Kurtosis | $\frac{1}{N}\sum_{i=1}^{N}\left(\frac{x_i-\bar{x}}{\sigma}\right)^4$ |
| Mean ($\bar{x}$) | $\frac{1}{N}\sum_{i=1}^{N}x_i$ | Crest factor | $\frac{max(|x|)}{x_{rms}}$ |
| Root-mean-square ($x_{rms}$) | $\sqrt{\frac{1}{N}\sum_{i=1}^{N}x_i^2}$ | Impulse factor | $\frac{max(|x|)}{\frac{1}{N}\sum_{i=1}^{N}|x_i|}$ |
| Skewness | $\frac{1}{N}\sum_{i=1}^{N}\left(\frac{x_i-\bar{x}}{\sigma}\right)^3$ | Shape factor | $\frac{x_{rms}}{\frac{1}{N}\sum_{i=1}^{N}|x_i|}$ |

where $x$ is an input waveform-type signal, $N$ is the total number of samples, and $\sigma = \sqrt{\frac{1}{N}\sum_{i=1}^{N}(x_i - \bar{x})^2}$, respectively

instance, coefficients in a fitted autoregressive moving average model were used as indicators for health monitoring [54].

In addition, frequency-domain analysis has been widely used for capturing frequency features. Fourier transform (e.g., fast Fourier transform) is one of the most common frequency analysis techniques, which decomposes a waveform into its constituent frequencies. Besides Fourier transform, other frequency analysis techniques, such as cepstrum analysis [55], high-order spectral analysis [56], and Holo-Hilbert spectral analysis [57], have been employed for PHM. Despite successes using frequency analysis in PHM applications, its inability to deal with non-stationary waveform signals further requires time-frequency-domain analysis. Wavelet analysis has been successfully applied to the creation of features in various applications [58–62]. Other time-frequency analysis techniques used for PHM include spectrogram [63], Wigner–Ville distribution [64–66], and Choi–Williams distribution [67].

*Feature Extraction*

Feature extraction, also known as dimensionality reduction, is the transformation of high-dimensional data into a meaningful representation of reduced dimensionality, which should have a dimensionality that corresponds to the intrinsic dimensionality of the data. Here, the intrinsic dimensionality indicates the minimum number of parameters needed to account for the observed properties of the data [68]. Since dimensionality reduction mitigates the "curse of dimensionality" and other undesired properties of high-dimensional space, it is an important task in the development of data-driven PHM applications.

Although a variety of feature extraction methods have been developed in recent years, this section does not review all the methods. Instead, this section provides an overview of the following feature extraction methods widely used in PHM:

principal component analysis (PCA), kernel PCA, linear discriminant analysis (LDA), kernel LDA, Isomap, and self-organizing map (SOM).

- PCA and kernel PCA—PCA [69] is an unsupervised method to carry out dimensionality reduction by embedding the data into a linear subspace of lower dimensionality. More specifically, PCA reduces the dimensions of a $d$-dimensional data set $X$ by projecting it onto a $k$-dimensional subspace, i.e., $k < d$, in order to increase the computational efficiency while retaining most of the information about the data set. In fact, PCA is a linear projection technique that works well if the data are linearly separable. However, in the case of linearly inseparable data, a nonlinear technique is preferred. The basic idea to deal with linearly inseparable data is to project it onto a higher-dimensional space where it becomes linearly separable with the help of kernel functions [70].

- LDA and kernel LDA—LDA [71] aims at projecting data points onto a lower-dimensional space with good class-separability in order to avoid over-fitting ("curse of dimensionality") and also reduce computational costs. PCA can be described as an "unsupervised" algorithm, since it "ignores" class labels and its goal is to find the directions (the so-called principal components) that maximize the variance in a data set, whereas LDA is "supervised" and computes the directions ("linear discriminants") that will represent the axes that that maximize the separation between multiple classes. Figures 13.7 and 13.8 pictorially illustrate the difference between PCA and LDA. Although it might sound intuitive that LDA is superior to PCA for a multi-class classification task where the class labels are known, this might not always the case. For example, comparisons between accuracies in fault diagnosis applications after using PCA or LDA show that PCA tends to outperform LDA if the number of samples per class is relatively small. Analogous to kernel PCA, kernel LDA, also known as generalized discriminant analysis and kernel Fisher discriminant analysis, uses the kernel trick. That is, using the kernel trick, LDA is implicitly performed in a new feature space, which allows nonlinear mappings to be learned.

- Isomap—Classical scaling[1] has proven to be successful in many PHM applications, but it suffers from the fact that it mainly aims to retain pairwise Euclidean distances, and does not take into account the distribution of the neighboring data points. If the high-dimensional data lie on or near a curved manifold,[2] classical scaling might consider two data points as near points, whereas their distance over the manifold is much larger than the typical inter-point distance. Isomap [72] is a technique that resolves this issue by attempting to preserve pairwise geodesic (or curvilinear) distances between data points,

---

[1]PCA is identical to the traditional technique for multi-dimensional scaling called "classical scaling."

[2]A manifold is a topological space that locally resembles Euclidean space near each point. More precisely, each point of ann-dimensional manifold has a neighborhood that is homeomorphic to the Euclidean space of dimension. In this more precise terminology, a manifold is referred to as ann-manifold.

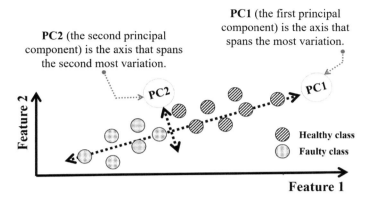

**Fig. 13.7** PCA, where PC1 and PC2 indicate the first and second principal components obtained from PCA

**Fig. 13.8** LDA, where the variables $\mu$ and $s$ indicate the mean and standard deviation obtained from a given class (i.e., health or faulty class), and the objective of LDA is to find a new axis that maximizes the separability

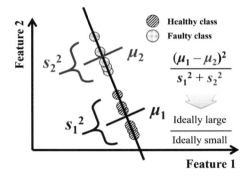

where geodesic distance is the distance between two points measured over the manifold. In Isomap, the geodesic distances between the data points $x_i$ $(i = 1, 2, \ldots, n)$ are computed by constructing a neighborhood graph $G$, in which every data point $x_i$ is connected with its $k$-nearest neighbors $x_{ij}$ $(j = 1, 2, \ldots, k)$ in the data set $X$. The shortest path between two points in the graph forms an estimate of the geodesic distance between these two points and can easily be computed using Dijkstra's or Floyd's shortest-path algorithm [73, 74]. The geodesic distances between all data points in $X$ are computed, thereby forming a pairwise geodesic distance matrix. The low-dimensional representations $y_i$ of the data points $x_i$ in the low-dimensional space $Y$ are computed by applying classical scaling on the resulting pairwise geodesic distance matrix.

- SOM—The SOM [75] consists of a regular, usually two-dimensional (2-D), grid of map units (or neurons). Each unit $i$ is represented by a prototype vector (or weight vector) $w_i = [w_{i1}, w_{i2}, \ldots, w_{id}]$, where $d$ is the dimension of a data point in the data set $X$. The units are connected to adjacent ones by a neighborhood

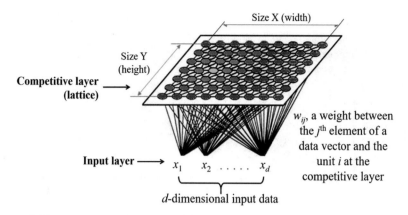

**Fig. 13.9** A standard structure of the SOM

relation. The number of map units, which typically varies from a few dozen up to several thousand, determines the accuracy and generalization capability of the SOM. Figure 13.9 illustrates a standard structure of the SOM.

During training, the SOM forms an elastic net that folds onto the "cloud" formed by the input data. Data points lying near each other in the input space are mapped onto nearby map units. Thus, the SOM can be interpreted as a topology preserving mapping from input space onto the 2-D grid of map units. The SOM is trained iteratively. At each training step, a sample vector $x$ is randomly chosen from the input data set $X$. Distances between $x$ and all the prototype vectors are computed. The best matching unit (BMU), which is denoted here by $b$, is the map unit with prototype closest to $x$

$$b = \min_i \|x - w_i\| \tag{13.3}$$

Next, the weight vectors are updated. The BMU and its topological neighbors are moved closer to the input vector in the input space. The update rule for the prototype vector of unit $i$ is

$$w_i^{(t+1)} = w_i^{(t)} + \alpha^{(t)} h_{bi}^{(t)} \left( x - w_i^{(t)} \right) \tag{13.4}$$

where $t$ is time, $\alpha^{(t)}$ is an adaptation coefficient, and $h_{bi}^{(t)} = \exp\left( \frac{-\|r_b - r_i^2\|}{2\sigma^{(t)2}} \right)$, respectively. Likewise, $r_b$ and $r_i$ are positions of neurons $b$ and $i$ on the SOM grid. Both $\alpha^{(t)}$ and $\sigma^{(t)}$ decrease monotonically with time. There is also a batch version of the algorithm where the adaptation coefficient is not used [75]. The SOM algorithm is applicable to large data sets. The computational complexity scales linearly with the number of data samples, it does not require huge amounts of memory—basically just the prototype vectors and the current training vector—and can be implemented both in a neural, online learning manner as well as parallelized [76].

*Feature Selection*

Feature selection, also called "variable selection" or "attribute selection," is the process of selecting a subset of relevant features for use in model construction. Likewise, feature selection is different from dimensionality reduction. Both methods seek to reduce the number of features in the given data set, but dimensionality reduction methods do so by creating new combinations of features, whereas feature selection methods include and exclude features present in the data without changing them.

Feature selection is usually carried out for the following purposes. First, feature selection improves the performance of a machine learning algorithm. For example, some features are not relevant for a classification problem or they may consist of noise. These features contribute to over-fitting, and thus the classification result can be biased or have undesired variance. Second, feature selection improves model interpretability. After feature selection, some features are discarded, and the model is simplified. Feature selection also ranks the importance of the features, which provides a better understanding of which features contribute most to the model. Third, feature selection reduces resource spending on computation and data acquisition. For example, if sensor data are used as features, feature selection helps to reduce the number of sensors and thus the cost of the sensor system, data acquisition, data storage, and data processing is reduced. Finally, similar to dimensionality reduction, feature selection helps to reduce the risk of the curse of dimensionality.

In general, the following two procedures are performed in the feature selection process, i.e., a number of feature subsets are first formed and then evaluated. Based on the evaluation process, feature selection schemes are basically categorized into filters or wrappers. Filter approaches employ an evaluation strategy that is independent from any classification scheme, while wrapper methods use accuracy estimates for specific classifiers during the assessment of feature subset quality [77]. Accordingly, wrapper methodologies theoretically offer better diagnostic performance for predefined specific classifiers than filter methods. However, filter approaches are computationally efficient since they avoid the accuracy estimation process for a certain classifier. Moreover, different wrapper methods and filter methods have different assumptions on the data. A specific method is selected if its assumption matches the properties of the data. In the following subsections, feature selection is explained using binary classification, which classifies the data into two classes, as the machine learning task. It serves as the basis for the understanding of other machine learning tasks such as multi-class classification and regression.

To achieve high-computational efficiency and diagnostic performance concurrently, recent intelligent fault detection and diagnosis approaches have adopted hybrid feature selection (HFS) schemes that appropriately exploit the advantages of the filter and wrapper methods. Liu et al. [78] presented an HFS approach for the effective identification of various failures in a direct-drive wind turbine. More specifically, the HFS method consists of a global geometric similarity scheme that yields promising feature subsets and a predefined classifier (e.g., support vector machine or general regression neural network) to predict diagnostic performance (or classification accuracies) with these feature subsets. Yang et al. [79] proposed a method to improve diagnostic performance by introducing an HFS framework, which is an unsupervised learning model. This method is effective for bearing fault diagnosis with fewer fault features that are closely related to single and multiple-combined bearing defects.

1.  Feature Selection: Filter Methods

Filter methods select features based on the individual features' properties toward the objective of the specified machine learning task. For example, in binary classification, individual features are evaluated such that any feature provides a certain degree of separation of the data independently. Usually, hypothesis testing is performed in the filter methods. Training data from both classes are prepared first, and then hypothesis testing is performed with a null hypothesis that the data from the two classes are sampled from the same distribution. The hypothesis is tested using the feature under evaluation. If the hypothesis is rejected, the data from the two classes are not regarded as from the same distribution, and it means the feature is able to separate the two classes. The hypothesis testing methods in filter methods can be based on $t$ distribution, $F$ distribution, and K–S distribution. These hypothesis testing methods have different assumptions on the data, and specific methods should be selected accordingly.

- $t$-test Feature Selection

A two-sample $t$-test evaluates if the data from two classes can be separated by their mean values of the selected feature. It assumes the data in any of the two classes are sampled from a Gaussian distribution. When the assumption is satisfied, under the null hypothesis that the data from the two classes have the same mean value, $t$ statistics can be constructed:

$$\text{t} = \frac{\bar{x}_A - \bar{x}_B}{\sqrt{\frac{s_A^2}{n_A} + \frac{s_B^2}{n_B}}} \tag{13.5}$$

where $t$ is the test statistic that follows t distribution under the null hypothesis; $\overline{x_A}$ and $\overline{x_B}$ are the sample means, $S_A$ and $S_B$ are the sample standard deviations, and $n_A$ and $n_B$ are the sample sizes of the Class $A$ and Class $B$, separately. A larger $t$-statistic means the data from the two classes are less likely to have the same mean values.

A value of the $t$ statistic corresponds to a $p$ value on the $t$ distribution, which is available in manuals and software packages. $p$ value is the probability that under the null hypothesis the test statistic takes extreme values. It is the evidence against the hypothesis and determines the statistical significance in the hypothesis testing.

A smaller $p$ value indicates the null hypothesis is less likely to be accepted. Usually, a significance level $\alpha$ is used as a threshold to make decisions on the hypothesis testing. $\alpha$ is the probability that the hypothesis testing rejects the null hypothesis, given that it was true. When $p < \alpha$, the null hypothesis is rejected with a significance of $\alpha$. That is, the means of the data from the two classes are different, and the feature can separate the data. A larger $\alpha$ means it is stricter to accept the null hypothesis, but a larger type I error rate would occur. A commonly used $\alpha$ is 0.05.

For example, supposing there are $m$-dimensional data with two classes of a total of $n$ observations, and the data from each of the class follow a Gaussian distribution, to select features using $t$-test, the $p$ value of the $t$ statistic for each of the $m$ features is calculated. If the significance level $\alpha$ is 0.05, the $p$ values of $k$ features are smaller than 0.05, and the $p$ values of $m-k$ features are larger than 0.05, those $k$ features are selected because they can separate the data by rejecting the null hypothesis that the data from the two classes have the same mean values.

- $F$-test Feature Selection

Sometimes the research is more interested in the separating of different classes by the variance. In this case, a two-sample $F$-test is applied. The $F$-test assumes the data from each of the classes follow a Gaussian distribution. The null hypothesis is that the data from the two classes have the same variance. Under the null hypothesis, the $F$ statistic is:

$$F = \frac{s_A^2}{s_B^2} \tag{13.6}$$

where $F$ is the $F$ statistic that follows $F$ distribution; $S_A$ and $S_B$ are the sample standard deviations of the data from Class $A$ and Class $B$, respectively. A larger $F$ statistic means the data from the two classes are less likely to have the same variance.

Similar to the $t$-test procedure, the $p$ value that corresponds to the $F$ statistic is calculated and compared to a predetermined significance level $\alpha$. If the $p < \alpha$, the null hypothesis is rejected, and the feature is regarded as being able to separate the data by the variance.

- K–S test Feature Selection

$t$-test and $F$-test are used when the interest is on the separation of classes by the mean or the variance. However, there are situations when the data from the classes are not different by the mean values or the variances. For example, the data may have different skewness or kurtosis. Moreover, both $t$-test and $F$-test assume the data follow Gaussian distributions, which cannot be met in a wide range of applications. The two-sample Kolmogorov–Smirnov (K–S) test is an alternative hypothesis testing that avoids the above challenges. The K–S test is a nonparametric method that does not impose any assumptions on the distributions of the data under test. The test statistic, which is called the K–S statistic, is the supreme of the

**Fig. 13.10**  K–S statistic

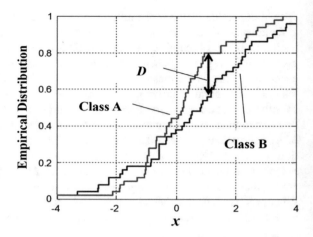

**Fig. 13.11**  A pictorial
schematic of an SVM

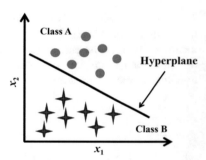

difference between the empirical distributions of the data from the two classes.
A larger K–S statistic means the data are less likely to be sampled from the same
distribution (Fig. 13.10).

$$D = \sup_{x \in \Re} |F_A(x) - F_B(x)| \tag{13.7}$$

where $D$ is the K–S statistic; $F_A$ and $F_B$ are the empirical distributions of the data
from Class $A$ and Class $B$, respectively. The K–S distance is demonstrated in
Fig. 13.11.

The K–S statistic $D$ follows a K–S distribution. Therefore, a $p$ value can be
calculated and compared to a predetermined significance level $\alpha$. If the $p < \alpha$, the
null hypothesis is rejected, and the feature is regarded as being able to separate the
data by rejecting the hypothesis that the data are from the same distribution.

2. Feature Selection: Wrapper Methods

If feature selection is carried out based on the classification performance of a classifier, wrapper feature selection is applied. Wrapper feature selection tries to find a subset of the features that optimize an objective function. Two widely used wrapper selection approaches are forward search and backward search. Forward search starts with an empty subset and repeatedly includes one feature at a time into a subset that minimizes an objective function, which is usually the generalization error from the cross-validation on the selected machine learning algorithm. The procedure stops when the generalization error is smaller than a threshold. Backward search starts with using all features and repeatedly removes features one at a time until a criterion is satisfied. In a complete search, the approach takes $O(n^2)$ calls of the machine learning algorithm, and thus it is impractical for some applications due to the cost of computation. Therefore, heuristic search algorithms have been implemented. For example, simulated annealing, genetic algorithm, and particle swarm optimization have been applied to search the optimized subset of features [80].

3. Filter Feature Selection: Embedded Methods

Embedded feature selection incorporates the feature selection as part of the training process. The basic idea is to rank the features according to their weights assigned by a classifier. A classification algorithm used in feature selection is linear support vector machine (SVM). The hyperplane of a linear SVM is the optimized linear model that maximizes the separation of the data from two classes, as shown in Fig. 13.11 and Eq. (13.8). The features with the largest absolute values of weight contribute most to the separating hyperplane and are the most sensitive to the separation of the classes.

$$w_1x_1 + w_2x_2 + \cdots + w_mx_m + b = 0 \tag{13.8}$$

where $w_m$ is the weight of the $m$th feature $x_m$; $b$ is the constant of the linear model.

To perform the feature selection using linear SVM, all the features are used in the training of SVM, and then the features are ranked according to their absolute values of the weight. The features are selected using certain rules. For example, if the desired classification accuracy is $q$, the linear SVM of the top $k$ features from the ranked features that reaches the accuracy of $q$ is the model with the selected $k$ features. Similarly, feature selection can be achieved based on the weight of features in LDA. In addition, the coefficients in penalized classification and the network pruning of the neural network have also been used in feature selection.

4. Advanced Feature Selection

Based on filter methods, wrapper methods, and embedded methods, some advanced methods have been developed, such as ensemble feature selection, stability feature selection, and hybrid feature selection.

Ensemble feature selection aims to aggregate the power of different feature selection algorithms. In [81], a single feature selection algorithm is run on different subsets of data samples obtained from the bootstrapping method. The results are aggregated to obtain a final feature set. Filter methods to rank the features and various aggregation methods such as ensemble-mean, linear aggregation, weighted aggregation methods are combined to obtain the final feature subset.

Stability feature selection tries to improve the consistency of the feature selection procedure [82]. Stability of a feature selection algorithm can be viewed as the consistency of an algorithm to produce a consistent feature subset when new training samples are added or when some training samples are removed. A strategy to improve the stability is to generate multiple subsets of the training data and use these subsets for feature selection. The features that have the highest frequency to be selected are the features of interest.

## Diagnostics and Prognostics

Diagnostics is not only the detection of an existing failure or degradation condition of an object system or component, but also the identification of the nature of the fault or type of degradation. Expert systems, such as rule-based and fuzzy rule-based systems, have been treated as common diagnosers [83, 84]. However, the increased number of rules to accommodate a number of failure modes and mechanisms limits the use of such systems. Alternatively, various machine learning algorithms have been applied to diagnostics, including $k$-nearest neighbors (kNNs), neural networks, SOMs, $k$-means clustering, and fuzzy $c$-means clustering [85–87].

Prognostics is used to assess the future state of system health and integrate that assessment within a framework of available recourses and operation demand. Recent data-driven prognostic approaches include sensory data fusion, feature engineering, and life prediction using interpolation [88, 89], extrapolation [90], or machine learning [91]. Examples are a similarity-based interpolation approach with a relevance vector machine as a regression technique [88, 92], a similarity-based interpolation with a support vector machine as a regression technique [88, 93], a similarity-based approach with least-squares exponential fitting [88], a Bayesian linear regression with least-squares quadratic fitting, and a recurrent neural network approach [91, 94].

## Machine Learning

Machine learning (ML) algorithms can be classified into broad categories based on whether they are trained with human supervision (supervised, unsupervised, semi-supervised, and reinforcement learning); whether they can learn incrementally on the fly (online versus batch learning); and whether they work by simply comparing new data points to known data points, or instead detect patterns in the training data and build a predictive model (instance-based versus model-based learning). The following subsections explain each category of ML algorithms. ML algorithms can be divided into the following four categories depending on the amount and type of supervision they need while training: supervised, unsupervised, semi-supervised, and reinforcement learning.

In supervised learning, the training data fed to the ML algorithms include the desired solutions, called "labels." Classification is a typical supervised learning task. The diagnostic system is a good example of classification problems. That is, it is trained with many variables or features along with their class (e.g., faulty or healthy),[3] and it must learn how to classify new variables or features. Another typical task is to predict a target numeric value, such as a product's remaining useful life, given a set of features called predictors. This sort of task is called regression. To train the ML algorithms, the training data set must contain predictors with responses (or labels). Note that some regression algorithms can be used for classification, and vice versa. For example, logistic regression [95] is commonly used for classification, as it can output a value that corresponds to the probability of belonging to a given class (e.g., 90% chance of being a healthy product). The supervised ML algorithms widely used for both classification and regression in PHM of electronics involve $k$-nearest neighbor, naïve Bayes classifiers, support vector machines, neural networks, decision trees, random forest, linear regression, and logistic regression.

Unlike supervised learning, the training data set is unlabeled in unsupervised learning. The major tasks using unsupervised learning in PHM are clustering (e.g., $k$-means, fuzzy $c$-means, hierarchical cluster analysis, and SOM) and dimensionality reduction (e.g., PCA, locally linear embedding, and $t$-distributed stochastic neighbor embedding). In fact, clustering has been widely used for anomaly detection, also known as outlier detection, under the assumption that the majority of the instances in the data set are normal and facilitate the detection of anomalies in an unlabeled test data by looking for instances that seem to fit least to the remainder of the data set. In PHM, dimensionality reduction is primarily used for simplifying the data without losing too much information. One way to do this is to merge correlated features into one. For example, a capacitor's capacitance can be highly correlated with its age, so the dimensionality reduction algorithm will merge correlated features representing the capacitor's wear into one feature, which is called "feature extraction." In fact, it is often a good idea to reduce the dimensions of the data set before it is fed to ML algorithms (e.g., supervised ML algorithms for classification). This is mainly because the data set will run much faster, the data will take up less disk and memory space, and in some cases, the data set may also perform better. Likewise, unsupervised learning has been used to output 2-D or 3-D representation of the high-dimensional data that can be easily plotted.

Semi-supervised learning is a class of supervised learning tasks and techniques that make use of unlabeled data for training; the training data set involves a lot of

---

[3]The classification task is further divided into a binary classification task and a multi-class classification task based on the number of classes it addresses. For example, if the diagnostic system is used to identify whether the product is healthy or not, this would be treated as a binary classification task. On the other hand, if the diagnostic system is used to pinpoint multiple failure modes or failure mechanisms of the product, this would be treated as a multi-class classification task.

unlabeled data and a little bit of labeled data. Anomaly detection in a system can be a good example of semi-supervised learning [96]. For example, a system's anomalies can be detected by comparing in situ parameters to monitor against a healthy baseline, which must be known (i.e., labeled) in advance. Likewise, the baseline data set is often composed of a collection of parameters that represent all the possible variations of the healthy operating states of the system. The combination of deep belief networks with unsupervised components called "restricted Boltzmann machines" stacked on top of one another is another example of semi-supervised learning approaches that can be used for health diagnosis [97]. Restricted Boltzmann machines are trained sequentially in an unsupervised manner, and then the whole approach is fine-tuned using supervised learning techniques.

### 13.5.2.4   Fusion Approach

The strength of the PoF-based prognostics approach is its ability to identify the root causes and failure mechanisms that contribute to system failure and predict RUL under different usage loading even before the system is actually in use. However, the weakness of the PoF-based approach is that prognostics is being carried out with the assumption that all components and assemblies are manufactured the same. The strength of the data-driven approach is its ability to track any anomaly that occurs in the system, no matter the mechanism or the cause. The weakness of the data-driven approach is that, without knowledge about what mechanism is causing the anomaly, it is very difficult to set a threshold to link the level of data anomaly to the failure definition.

General experience has been that often one approach may not be adequate to provide the desired results. Hence, the data-driven approach is integrated with the

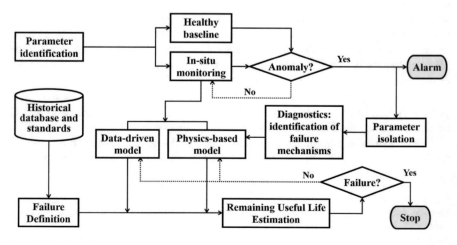

**Fig. 13.12**   A fusion prognostics approach

PoF-based approach (see Fig. 13.12). While the PoF-based approach provides fundamental requirements for prognostic models, the database approach complements the model with a knowledge base that has already been developed for various failure modes.

Cheng and Pecht [98] used the fusion method to predict the RUL of multilayer ceramic capacitors (MLCCs). The fusion method was carried out in 9 steps. The first step was parameter identification. Failure modes, mechanisms, and effects analysis was used to identify potential failure mechanisms and determine the parameters to be monitored. The MLCCs in the case study underwent temperature-humidity-bias testing. Two possible failure mechanisms were identified—silver migration and overall degradation of the dielectric of capacitors. The next steps were parameter monitoring and creation of a healthy baseline from training data, followed by data trend monitoring by the multivariate state estimation technique (MSET) and sequential probability ratio test (SPRT) algorithms. When an anomaly is detected, the RUL should be predicted at that moment. All three parameters contributed to the anomaly. During the anomaly detection, MSET generated residuals for each parameter. The residuals of the three parameters were used to generate a residual vector, which is an indicator of the degradation between the monitored MLCCs and the baseline. When an anomaly is detected, the parameters are isolated, and the failure can be defined based on the potential failure mechanism. If PoF models are available, failure can be defined based on the PoF model and the corresponding parameters, otherwise, it should be based on historical data. The example given in the article has a predicted failure time in the interval of [875,920] h. The predicted RUL from the 839th hour is in the intervals. The actual failure time was at the 962nd hour. The fusion PHM method predicts the failures of this capacitor in advance.

Patil et al. [30] proposed a fusion method for remaining life prediction of an IGBT power module. The proposed method is described as follows: As a first step, parameters to be monitored in the application should be identified. Examples of parameters include the collector-emitter ON voltage, ambient temperature, and module strains. A baseline for the healthy behavior of the parameters is to be created. Identified parameters are continuously monitored in the application and compared with the healthy baseline. When anomalies are detected, the parameters contributing to the anomalies are isolated. Using failure thresholds, methods such as regression analysis can be applied to trend the isolated parameters with time. Further, isolation of parameters causing anomalous behavior helps identify the critical failure mechanisms in operation. In a power module, a drop-in collector-emitter voltage indicates damage in the solder die attach. Trending of the collector-emitter voltage drop to failure thresholds would provide the data-driven RUL estimates, which can then be fused with the PoF estimates. The fusion approach therefore provides an estimate of the RUL of the product based on the combination of information from anomaly detection, parameter isolation, PoF models, and data-driven techniques.

The fusion approach is also beneficial when the available data are inadequate to implement prognostics. To improve the prediction accuracy and precision, it is often necessary to use Bayesian updating to incorporate new data for prediction [28]. Hence, the probabilistic approach is used in conjunction with online precursor trends to update estimates with new data available from online sensors. While the trend monitoring identifies the deviation from normal equipment operation, the probabilistic model with uncertainty bands provides an estimate of the prognostic distance—a performance metric crucial for fixing the deficiency either through repair or replacement. The prognostic distance also prompts the plant manager to plan the maintenance action in advance such that plant availability and safety can be optimized.

## 13.5.3   Uncertainty in PHM

One must acknowledge that prognostics, which is used to predict the future health status of an object system, can be clouded with uncertainty. Hence, dealing with uncertainty is an indispensable task in PHM. In an ideal scenario, it may be possible to predict the future health status of an object system with a significant amount of confidence. Unfortunately, however, this confidence would not be expected in real applications because the future behavior of an object system is uncertain; the models built on the principles of PoF or sensory data may not represent the future behavior of the system, resulting in model errors and uncertainties; and the use of various sensors for data fusion and data pre- or post-processing tools in the development of PHM methods introduces additional uncertainty. As a consequence, since it is obvious that all activities in PHM can render uncertainty [99], it is important to deeply understand the sources of uncertainty that need to be considered in PHM and properly address them to deploy PHM techniques in real applications.

### 13.5.3.1   Sources of Uncertainty

For the sake of facilitating uncertainty quantification and management, sources of uncertainty can be classified into either "aleatory uncertainty" arising due to physical variability or "epistemic uncertainty" arising due to lack of knowledge. However, this classification may be questionable in prognostics because true variability is not present while the health status of an object system is being monitored.

Present Uncertainty

Prior to prognosis, it is of importance to accurately estimate the condition/state of an object system at the time at which RUL prediction needs to be carried out. Typically, since damage (or faults) can be represented in terms of states, state estimation can be equivalent to estimating the extent of damage in the system, which will be accomplished by various filtering-based approaches, such as Kalman filtering and particle filtering. With the help of advanced sensor technologies and improved variants of such filtering techniques, it is practically possible to improve the accuracy of state estimation and reduce uncertainty. However, it is important to know that the system is at a certain state at any time instant and "uncertainty" in this section merely describes the lack of knowledge regarding the true state of the system.

### Future Uncertainty

From the perspective of prognostics, the most significant source of uncertainty is mainly because the future is unknown, e.g., unknown loading, operational, environmental, and usage conditions. Hence, it is highly recommended to address this uncertainty when performing prognosis.

### Modeling Uncertainty

Typically, a degradation model(s) is used for RUL prediction, e.g., a model that represents the response of the system to anticipated loading, operational, environmental, and usage conditions. Likewise, for RUL predictions, the end of life (EOL) is defined using a Boolean threshold function model to indicate whether or not failure has occurred. Such models are jointly used in prognosis. However, the fact that it is practically impossible to develop models that perfectly trend the degradation behavior of the system further introduces "modeling uncertainty," which is referred to as the difference between the predictive and true responses and is due to various factors, such as uncertain model parameters and process noise. Note that "true responses" can neither be known nor measured accurately.

### 13.5.3.2 Uncertainty Treatment in PHM

In PHM, uncertainties have been discussed in terms of uncertainty representation and interpretation, uncertainty quantification, and uncertainty management [100–102].

### Uncertainty Representation and Interpretation

In many practical applications, uncertainty representation and interpretation can be guided by the choice of modeling and simulation frameworks. For uncertainty representation, the following theories have been used: classical set theory, probability theory, fuzzy set theory, plausibility and belief theory, and rough set theory. Among these theories, probability theory has been widely applied to various PHM applications [103]. Likewise, uncertainty is mostly interpreted in two different ways: frequentist (classical) versus subjective (Bayesian). The former interpretation implies that uncertainty exists only when there is natural randomness across multiple nominally identical experiments, whereas the latter enables associating uncertainty even with events that are not random, and such uncertainty is simply reflective of the analyst's belief regarding the occurrence or non-occurrence of such events.

Uncertainty Quantification

The aim of uncertainty quantification is to identify and characterize various sources of uncertainty that can affect RUL predictions. As aforementioned, since common sources of uncertainty in typical PHM applications include unknown future loading, operational, environmental, usage conditions, and modeling errors caused by inaccurate model parameters, process noise, and measurement noise, it is important to address each of these uncertainties separately and quantify them using probabilistic/statistical methods. As an example, the Kalman filter is essentially a Bayesian tool for uncertainty quantification, where the uncertainty in the states is estimated continuously as a function of time, based on measurement data, which is also typically available continuously as a function of time.

Uncertainty Management

In general, uncertainty management refers to various activities that aid in managing uncertainty in PHM applications. For uncertainty management, one needs to think about the following aspects of uncertainty management. Uncertainty management attempts to answer the question: "Is it possible to improve the uncertainty estimates?" The answer to this question lies in identifying which sources of uncertainty are significant contributors to the uncertainty in RUL predictions. Uncertainty management also addresses how uncertainty-related information can be used in the decision-making process. From the perspective of decision-making to facilitate risk mitigation activities, it is highly recommended to fuse the quantified uncertainty into RUL predictions.

## 13.5.4   Performance Metrics

This section primarily reviews performance metrics used in data-driven diagnostics and prognostics in PHM.

### 13.5.4.1 Diagnostic Metrics

From a machine learning point of view, diagnostics, defined as the action of determining the presence, location, and severity of a fault (or faults), can possibly be a binary or multi-class classification task. Accordingly, performance metrics used in ML classification tasks can also be useful for assessing the diagnostic performance in PHM.

To assess the performance of a classification model (or classifier) on a test data set for which the true values (or classes) are known, a confusion matrix constituting true positive (TP), true negative (TN), false positive (FP), and false negative (FN), respectively, is widely used, as presented in Table 13.5.

In Table 13.5, TP is the case that a test instance in the positive class is correctly recognized as the positive class, TN is the case that a test instance in the negative class is correctly identified as the negative class, FP is the case that a test instance belonging to the positive class is incorrectly recognized as the negative class, and FN is the case that a test instance belonging to the negative class is incorrectly assigned to the positive class, respectively.

The common performance measures for diagnostics include accuracy, sensitivity (or recall), and specificity. These measures are computed based on the number of TPs, TNs, FPs, and FNs, defined as

$$\text{Accuracy} = \frac{(\text{TP} + \text{TN})}{(\text{TP} + \text{TN} + \text{FP} + \text{FN})} \tag{13.9}$$

$$\text{Sensitivity (or Recall or True Positive Rate)} = \frac{\text{TP}}{(\text{TP} + \text{FN})} \tag{13.10}$$

$$\text{Specificity} = \frac{\text{TN}}{(\text{TN} + \text{FP})} \tag{13.11}$$

$$\text{Matthews Correlation Coefficient (MCC)}$$
$$= \frac{(\text{TP} \cdot \text{TN} + \text{FP} \cdot \text{FN})}{\sqrt{(\text{TP} + \text{FP}) \cdot (\text{TP} + \text{FN}) \cdot (\text{TN} + \text{FP}) \cdot (\text{TN} + \text{FN})}} \tag{13.12}$$

$$F_\beta = \frac{(\beta^2 + 1) \cdot \text{TP}}{((\beta^2 + 1) \cdot \text{TP} + \text{FP} + \beta \cdot \text{FN})} \tag{13.13}$$

**Table 13.5** A standard form of the confusion matrix for a binary classification problem

|  |  | Predicted | |
|---|---|---|---|
|  |  | Positive | Negative |
| Actual | Positive | True positive (TP) | False negative (FN) |
|  | Negative | False positive (FP) | True negative (TN) |

Accuracy is the proportion of true assessments, either TP or TN, in a population. That is, it measures the degree of diagnostic veracity. However, the problem that can be faced by the accuracy measure is the "accuracy paradox." The accuracy paradox [104] states that a classification model with a given level of accuracy may have greater predictive power than models with higher accuracy. Accordingly, it may be better to avoid the accuracy metric in favor of other metrics such as precision and recall. For example, a well-trained classifier was tested on 100 unseen instances—a total of 80 instances were labeled "healthy" and the remaining 20 instances were labeled "faulty"—and yielded classification accuracy of 80%. At first glance, it seems that the classifier performs well. However, 80% accuracy can be a frustrating result because the classifier may not be able to predict "faulty" instances at all.

Sensitivity measures the proportion of TPs (i.e., the percentage of "healthy" instances that are correctly identified as "healthy"). Accordingly, a classifier with high sensitivity is especially good at detecting a system's health status (not TNs, i.e., a system's faulty status). In Eq. (13.11), specificity measures the proportion of TNs (i.e., the percentage of "faulty" instances that are correctly identified as not "healthy"). More specifically, a classifier with high specificity is good at avoiding false alarms. In summary, both sensitivity and specificity are widely used with accuracy as diagnostic metrics.

To assess classification performance (especially for a binary classification problem), a well-known method is receiver operating characteristics (ROC) analysis using the true positive rate (TPR), also called sensitivity (or recall), against the false positive rate (FPR), where FPR can be measured as:

$$FPR(1\text{-Specificity}) = \frac{FP}{FP + TN} \tag{13.14}$$

In Eq. (13.13), FPR is equivalent to 1-specificity. All possible combinations of TPR and FPR consist of an ROC space. That is, a location of a point in the ROC space can show the trade-off between sensitivity and specificity (i.e., the increase in sensitivity is accompanied by a decrease in specificity). Accordingly, the location of the point in the space can represent whether the (binary) classifier performs accurately or not. As illustrated in Fig. 13.14, if a classifier works perfectly, a point determined by both TPR and FPR would be a coordinate (0, 1), indicating that the classifier achieves a sensitivity of 100% and a specificity of 100%, respectively. If the classifier yields a sensitivity of 50% and a specificity of 50%, a data point can be lain on the diagonal line (see Fig. 13.14) determined by coordinates (0, 0) and (1, 0), respectively. Theoretically, a random guess would give a point along the diagonal line. In Fig. 13.14, an ROC curve can be plotted by employing TPR against FPR for different cut-points, starting from a coordinate (0, 0) and ending at a coordinate (1, 1). More specifically, the x-axis represents FPR, 1-specificity, and the y-axis represents TPR, sensitivity. In the ROC curve, the closer the point on the ROC curve to the ideal coordinate (1, 0), the less accurate a classifier is. In ROC

**Fig. 13.13** Example of an ROC space

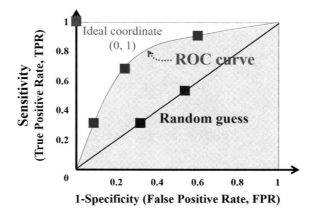

analysis, the area under the ROC curve, also known as AUC, can be calculated to provide a way to measure the accuracy of a classifier (i.e., a binary classifier):

$$\text{AUC} = \int_0^1 \text{ROC}(t)\mathrm{d}t \tag{13.15}$$

where $t$ equals to FPR and $\text{ROC}(t)$ is TPR (see Fig. 13.14). Likewise, the larger the area, the more accurate the classifier is. In practice, if the classifier yields $0.8 \leq \text{AUC} \leq 1$, its classification performance can be said to be good or excellent (Fig. 13.13).

Besides the above-mentioned diagnostic metrics, such as accuracy, sensitivity (or recall and TPR), specificity, and AUC, both Matthews correlation coefficient (MCC) and $F_\beta$ are also useful for evaluating classification performance of a binary classifier, where MCC is a correlation coefficient calculated from all values in the confusion matrix (i.e., TPs, TNs, FPs, and FNs). Additionally, $F_\beta$ is a harmonic mean of recall and precision. Precision is the ratio of TPs to all positives (i.e., TPs and FPs), defined as $\frac{\text{TP}}{\text{TP}+\text{FP}}$. The F-score reaches its best value at 1 and worst at 0. In fact, two commonly used F-scores are the $F_2$ measure (i.e., $\beta = 2$ in Eq. 13.13), which weights recall higher than precision (by placing more emphasis on FNs), and the $F_{0.5}$ measure (i.e., $\beta = 0.5$ in Eq. 13.13), which weights recall lower than precision (by attenuating the influence of FNs).

In PHM, one can often meet multi-class classification problems. For example, the identification of failure modes can be a multi-class classification task because the number of classes (i.e., failure modes) to be classified is greater than 2. The above-mentioned diagnostic metrics can extend to the metrics for multi-class classification, defined as:

$$\text{Average accuracy} = \frac{1}{N_{\text{class}}} \sum_{i=1}^{N_{\text{class}}} \frac{(\text{TP}_i + \text{TN}_i)}{(\text{TP}_i + \text{TN}_i + \text{FP}_i + \text{FN}_i)} \tag{13.16}$$

$$\mu\text{-averaging of sensitivity} = \frac{\sum_{i=1}^{N_{\text{class}}} \text{TP}_i}{\sum_{i=1}^{N_{\text{class}}} (\text{TP}_i + \text{FN}_i)} \tag{13.17}$$

$$M\text{-averaging of sensitivity} = \frac{1}{N_{\text{class}}} \sum_{i=1}^{N_{\text{class}}} \frac{\text{TP}_i}{(\text{TP}_i + \text{FN}_i)} \tag{13.18}$$

$$\mu\text{-averaging of specificity} = \frac{\sum_{i=1}^{N_{\text{class}}} \text{TN}_i}{\sum_{i=1}^{N_{\text{class}}} (\text{TN}_i + \text{FP}_i)} \tag{13.19}$$

$$M\text{-averaging of specificity} = \frac{1}{N_{\text{class}}} \sum_{i=1}^{N_{\text{class}}} \frac{\text{TN}_i}{(\text{TN}_i + \text{FP}_i)} \tag{13.20}$$

$$\mu\text{-averaging of MCC} = \frac{\sum_{i=1}^{N_{\text{class}}} (\text{TP}_i \cdot \text{TN}_i + \text{FP}_i \cdot \text{FN}_i)}{\sum_{i=1}^{N_{\text{class}}} \sqrt{(\text{TP}_i + \text{FP}_i) \cdot (\text{TP}_i + \text{FN}_i) \cdot (\text{TN}_i + \text{FP}_i) \cdot (\text{TN}_i + \text{FN}_i)}} \tag{13.21}$$

$$M\text{-averaging of MCC} = \frac{1}{N_{\text{class}}} \sum_{i=1}^{N_{\text{class}}} \frac{(\text{TP}_i \cdot \text{TN}_i + \text{FP}_i \cdot \text{FN}_i)}{\sqrt{(\text{TP}_i + \text{FP}_i) \cdot (\text{TP}_i + \text{FN}_i) \cdot (\text{TN}_i + \text{FP}_i) \cdot (\text{TN}_i + \text{FN}_i)}} \tag{13.22}$$

$$\mu\text{-averaging of } F_\beta = \frac{\sum_{i=1}^{N_{\text{class}}} (\beta^2 + 1) \cdot \text{TP}_i}{\sum_{i=1}^{N_{\text{class}}} ((\beta^2 + 1) \cdot \text{TP}_i + \text{FP}_i + \beta \cdot \text{FN}_i)} \tag{13.23}$$

$$M\text{-averaging of } F_\beta = \frac{1}{N_{\text{class}}} \sum_{i=1}^{N_{\text{class}}} \frac{(\beta^2 + 1) \cdot \text{TP}_i}{((\beta^2 + 1) \cdot \text{TP}_i + \text{FP}_i + \beta \cdot \text{FN}_i)} \tag{13.24}$$

where, $\text{TP}_i$, $\text{TN}_i$, $\text{FP}_i$, and $\text{FN}_i$ are true positive, true negative, false positive, and false negative obtained for the $i$th class, respectively. Likewise, $N_{\text{class}}$ is the total number of classes that can be specified by the given classification problem. Additionally, the terms "$\mu$-averaging" and "$M$-averaging" are used to indicate micro- and macro-averaging methods, respectively. That is, in the $\mu$-averaging method, one can get statistics by summing up the individual TPs, TNs, FPs, and FNs, whereas the $M$-averaging method simply takes the average of sensitivity, specificity, MCC, and F-score for different classes.

### 13.5.4.2   Prognostic Metrics

Prognostics is defined as the process of estimating an object system's RUL (mostly with a confidence bound) by predicting the progression of a fault given the current degree of degradation, the load history, and the anticipated future operational and environmental conditions. In other words, prognostics predicts when an object system will no longer perform its intended function within the desired specifications. RUL is specified by the length of time from the present time to the estimated time at which the system is expected to no longer perform its intended function. This section reviews a variety of prognostic metrics rather than provide details about prognostic methods.

Figure 13.14a illustrates the times related to a prediction event in the operational life of an object system. Initially, the PHM designer specifies the upper and lower failure thresholds,[4] and the upper and lower off-nominal thresholds for the PHM sensor in the system. In Fig. 13.14a, $t_0$ can be assumed to start at any time (e.g., when the system is turned on), and $t_E$ is the occurrence of the off-nominal event. Off-nominal events occur when the PHM sensor measures an exceedance of the threshold limits specified by the PHM designer. The PHM metrics are initiated when such an event is detected at time $t_D$ by a PHM system. The PHM system then computes a predicted failure time of the part or subsystem with its associated confidence interval. The response time $t_R$ is the amount of time the PHM system uses to produce a predicted time of failure and make a usable prediction at time $t_P$. In Fig. 13.14a, $t_F$ is the actual time that the system fails and the RUL is the time difference between $t_P$ and $t_F$.

Figure 13.14b further shows the end-of-prediction time $t_{EOP}$ to measure prognostic metrics. The common prognostic metrics include mean absolute error (MAE), mean squared error (MSE), and root-mean-squared error (RMSE). The MAE is a quantity used to measure how close the estimated performance degradation trend $\hat{y}$ (or estimates) is to the actual performance degradation trend $y$ (or actual responses), defined by:

$$\text{MAE} = \frac{1}{(t_{EOP} - t_P + 1)} \sum_{t=t_P}^{t_{EOP}} |\hat{y}(t) - y(t)| \qquad (13.25)$$

The MAE is also known as a scale-dependent accuracy measure and therefore cannot be used to make comparisons between series using different scales. Likewise, the MSE, also known as the mean squared deviation, is a measure of the average of the squares of the errors or deviations—that is, the difference between $\hat{y}$ and $y$ is expressed as:

---

[4]The upper and lower failure thresholds can also be specified by standards, historical data, and so forth.

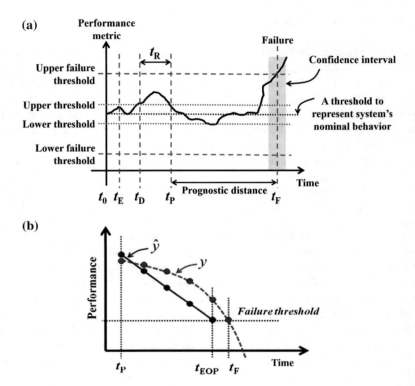

**Fig. 13.14** **a** Milestones on the path to object system failure and **b** the end-of-prediction (EOP) time $t_{\mathrm{EOP}}$ to measure the goodness of fit between the actual performance degradation trend $y$ and estimated degradation trend $\hat{y}$

$$\mathrm{MSE} = \frac{1}{(t_{\mathrm{EOP}} - t_{\mathrm{P}} + 1)} \sum_{t=t_{\mathrm{P}}}^{t_{\mathrm{EOP}}} (\hat{y}(t) - y(t))^2 \tag{13.26}$$

In practice, the MSE is a risk function, corresponding to the expected value of the squared error loss [105]. Although the MSE is widely used in the field, it has the disadvantage of heavily weighting outliers. This is a result of the squaring of each term, which effectively weights large errors more heavily than small ones. This property sometimes has led to the use of alternatives such as the MAE.

The RMSE, also called the root-mean-squared deviation, is a measure of the differences between the values predicted by a prediction model and the values actually observed, defined as:

$$\mathrm{RMSE} = \sqrt{\mathrm{MSE}} \tag{13.27}$$

The RMSE is a good measure of accuracy, but only to compare different prediction errors for a particular variable and not between variables, because it is scale-dependent.

Four more prognostic metrics include prediction horizon, $\alpha - \gamma$ performance, relative accuracy, and convergence [106]. The prediction horizon identifies whether a prediction model can predict within a specified error margin, which can be specified by the parameter $\alpha$ around the actual EOL of an object system. Then the $\alpha - \gamma$ performance further identifies whether the prediction model performs within desired error margins of the actual RUL at any given time instant, where the margins and time instant are specified by the parameters $\alpha$ and $\gamma$, respectively. The relative accuracy is obtained by quantifying the accuracy levels relative to the actual RUL, whereas the convergence quantifies how fast the prediction model converges provided that it meets all the aforementioned prognostic metrics.

## 13.5.5  Verification and Validation of PHM Capabilities

Prior to the development of PHM technologies in real applications, it is crucial that the techniques must undergo verification and validation (V&V); verification is referred to as the process where the stakeholder answers the query, "Are we building it right?", whereas validation indicates the process where stakeholders answer the query, "Are we building the right thing?" Intuitively, verification is a quality control process of evaluating whether or not an object product or system complies with testable constraints imposed by requirements, whereas validation is a quality assurance process of evaluating whether or not an object product or system accomplishes its intended function when fielded in the target application domain.

For verification and validation of PHM capabilities, a standardized set of mathematical metrics for rigorously evaluating the performance and effectiveness of PHM technologies will be necessary. Further, such an evaluation process needs to be capable of assessing PHM technologies with respect to their ability to detect, diagnose, and predict the progression of faults, and available metrics include accuracy, reliability, sensitivities, stability, risk, and economic cost/benefit. More specifically, performance metrics must be designed to be able to quantify how well a technology responds to changes in normal operation in terms of the ability to detect anomalies, isolate a root cause fault or failure mode, or predict the time to a given fault/failure condition. Likewise, prognostic metrics should be capable of evaluating prediction time accuracy in terms of RUL or system degradation levels. Overall performance of a PHM system that consists of diagnostic and prognostic modules may then be estimated as the weighted sum of performance metrics and cost/benefit analysis considerations.

## 13.5.6  Limitations of Prognostic Methods

Even though prognostics has evolved into a relatively new paradigm with applications in areas such as space, aircraft, and structural engineering, the development and deployment of prognostics in NPPs is very limited. In fact, there are certain issues that need to be addressed through R&D efforts.

The major challenges to the implementation of prognostics include sensors and associated networks, PoF and damage models with failure criteria, uncertainty characterization, and organizational frameworks. The availability of sensors in general and the development of an integrated sensor network can be considered one bottleneck in the implementation of prognostics. This is particularly true for electronic components, as this application requires miniaturization of the sensors such that newly developed sensors and networks can fulfill the requirements of an application. Keeping in view the enhanced performance of prognostics for future applications, wireless sensor networks (WSNs), along with utilization of miniaturized sensors such as Pt-100 for online temperature measurement, provide an effective technique for the implementation of prognostics for electronic components [107, 108].

The designers of newly built plants have to take a proactive approach for making prognostic provisions. This requires focused efforts on preparing design specifications based on safety and availability studies that identify not only components and processes that require PHM, but also selecting a PHM approach depending on failure mechanisms. For instance, if prognostic provisions are required for certain in-core components, suitable provisions should be made right before the start of construction activity. For existing plants, plant constraints will dictate the level of prognostics to be implemented. However, when life extension is being explored for new plants, implementation of prognostics tools and methods can provide valuable insights into tracking aging mechanisms as well as help in assessing performance of systems in online mode or at periodic intervals. Keeping in mind advances in WSN applications, the pros and cons of this technology should be evaluated. While there is immense potential for WSN technology, there are some limitations, including: It is a low-speed network, it requires a power supply to the node, it is complex to configure, and the performance of the node is easily affected by surroundings such as walls, microwaves, and large distances between nodes (resulting in signal attenuation) [109].

In the nuclear industry, the principles of defense in depth ensure the implementation of redundant and diverse electronic channels; however, the common cause failure (CCF) aspects require special attention. The effect of any degradation or failure mode induced by material, environmental, or operational parameters needs to be analyzed, particularly for assessing its CCF impact as part of PHM implementation [110]. To develop a PoF model for electronic protection channels, the potential failures due to whisker growth, electromigration-induced shorting of parallel metallization, or coupling of the redundant path due to field effects and solder joint failure require special CCF considerations.

Similarly, development of degradation models for in-core structural components requires not only the monitoring provisions but also considerations of irradiation-induced degradation. The prediction accuracy of these models will require assessment of changes in material property in a dynamic manner [21]. When implemented, these models are expected to provide input for a risk-based approach.

If the PoF and damage models are available, it is often challenging to define the failure criteria and the uncertainty associated with failure definitions. The lack of knowledge related to failure criteria is often addressed by conservative assumptions. Here, the role of prognostics becomes crucial, as the online signal can be used with the available data and models to characterize the incipient failures.

There are two types of uncertainty that need to be addressed in prognostics: aleatory and epistemic. Aleatory uncertainty, which is inherent and cannot be reduced, arises from data and models. Epistemic uncertainty is reduced by acquiring additional knowledge or data. The integrated approach is a typical example of reducing epistemic uncertainty. Reducing uncertainty in PHM becomes more important from the point of estimating prognostic distance. At a higher level, it affects the accuracy of the assessment of the safety margin as part of risk-based applications. Other approaches to modeling or reducing uncertainty involve updating the prior data with new evidence using well-known techniques such as Bayesian updating, Kalman filtering, constrained optimization, and particle filtering.

In spite of these developments, the accuracy of uncertainty assessment is a lingering issue. Other nonparametric methods that are expected to reduce subjectivity in uncertainty assessment are being developed. One method is the imprecise probability-based approach. However, there are limited applications of this approach. Further R&D in this area may provide a new approach to uncertainty modeling and analysis. Because PHM is a resource-intensive application, organizational will to implement and operate a PHM program is a prerequisite. Whether it is for routine health management of components in support of surveillance or life extension studies for new plants, the involvement of not only implementation-level staff but also plant management is an important factor in the success of the PHM approach. The availability of a PoF or damage model is one of the major challenges to the initiation and implementation of a PHM program in a nuclear plant. Even though limited application up to condition monitoring has found wider applications in the nuclear sector, the full potential of prognostics can be realized only after development of a damage model for mechanical and structural engineering components and PoF models for power and microelectronics and electrical components.

## 13.6  Conclusions and Recommendations

Condition monitoring is increasingly being used to support the operation and maintenance of NPPs. The diagnostic and prognostic approaches can be used as part of a risk-based approach, which in turn can support a prognostics program.

Looking at the publications in the areas of mechanical and structural engineering, it can be argued that a prognostics framework for NPPs can be established by adopting the models and methods developed for space, aircraft, and civil engineering systems for core components where radiation-induced degradation may not play much of a role in dictating the RUL. For core components, a limited knowledge base is available that can be used with certain uncertainty bounds.

Prognostics can be applied to new plants by making the complete monitoring and surveillance and maintenance management process more effective through the prediction of fault and degradation trends, such that adequate time is available for recovery and repair actions. That is, the inclusion of prognostics into the plants is important because prognostics improves safety and availability. For legacy plants with constraints imposed by design, layout, or operational limitations, the prognostic approach is expected to be very effective for life extension studies that are carried out as part of aging studies and performance monitoring after the changes and modifications have been incorporated. All of these gains further consolidate the risk-based approach.

Advances in any field and their application to real-life situations are normally judged by the availability of codes and standards. Even though there are many standards and codes for surveillance and condition monitoring, there are few standards for PHM. In this direction, the development of the first IEEE PHM standard for electronic systems is at an advanced stage and appears to be undergoing review [111]. This standard will mark a significant step: It will channel the knowledge base available in advanced laboratories for system applications to industry. As far as the nuclear industry is concerned, this standard will be a clear incentive to develop prognostics for electronics in reactor controls and protection.

The following recommendations are based on the review of the status of existing surveillance and monitoring programs and the potential role that prognostics can play as part of integrated risk-based engineering in NPPs:

- A prognostic approach brings the element of dynamics to the existing risk-based approach. Hence, there is a strong argument in favor of initiating a prognostics-based health management program in NPPs. The current knowledge of prognostics is such that extensive R&D is required to develop accurate prediction of RUL, particularly for power electronic systems and electrical systems.
- The development of PoF models requires elaborate life testing setups and material characterization facilities. Apart from this, the study of irradiation-induced degradation requires research reactor test facilities. The available resources can be networked in a coordinated manner to support this development work. There should be provision in the operating reactor organization to communicate data and insights on failure to a prognostics laboratory, on the one hand, while providing prognostic solutions for real-time issues on the other.
- Even though the prognostic approach for estimating the life and reliability of the components in new and legacy plants remains similar, the emphasis in new

plants is to develop a host of prognostic performance metrics for the identified components, whereas for legacy plants prognostics must support inspection, testing, and condition monitoring. Development efforts should adopt the prognostic systems that have been developed in other fields, such as navigation, aircraft, space, and infrastructure systems, so that the program is more effective in terms of deliverables.

- Nuclear research laboratories generally have a reasonable infrastructure for developing prognostic sensors and associated systems. Hence, early identification of sensor requirements is important for the success of prognostics programs.
- Work should start on the development of codes and standards for PHM for nuclear components and systems.

Keeping in mind the benefits that can be realized through the implementation of a risk-based prognostics program, this chapter emphasizes the significance of R&D on prognostics for complex engineering systems like nuclear power plants. The R&D activities are required because the PHM approach has the potential to benefit existing plants entering the aging zone and new plants, for which a target life of more than 90 years can be met with online prognostics that enable degradation monitoring of critical systems and components.

# References

1. International Atomic Energy Agency (IAEA), "Risk informed in-service inspection of piping systems of nuclear power plants: process, status, issues and development," IAEA Nuclear Energy Series No. NP-T-3.1, 2010
2. International Atomic Energy Agency (IAEA), "Risk-based optimization of technical specifications for operation of nuclear power plants," IAEA-TECDOC-729, 1993
3. A.C. Kadak, T. Matsuo, The nuclear industry's transition to risk-informed regulation and operation in the United States. Rel. Eng. Syst. Safety **92**(5), 609–618 (2007)
4. M. Pecht, *Prognostics and Health Management of Electronics* (Wiley, Hoboken, 2008)
5. J. Coble, M. Humberstone, J. Wes Hines, Adaptive monitoring, fault detection and diagnostics, and prognostics system for the IRIS nuclear plant (pp. 1–10), in *Proceedings of Annual Conference on the Prognostics and Health Management Society,* Portland, OR, USA, 10–14 Oct 2010
6. K.R. Wheeler, T. Kurtoglu, S.D. Poll, A survey of health management user objectives in aerospace systems related to diagnostic and prognostic metrics. Int. J. Prognostics Health Manag. **1**, 1–19 (2010)
7. P.W. Kalgren, M. Baybutt, A. Ginart, C. Minnella, M.J. Roemer, T. Dabney, Application of prognostic health management in digital electronic systems (pp. 1–9), in *Proceedings of 2007 IEEE Aerospace Conference*, Big Sky, MT, USA, 3–10 March 2007
8. J.K. Bhambra, S. Nayagam, I. Jennions, Electronic prognostics and health management of aircraft avionics using digital power converts (pp. 1–2), in *Proceedings of Annual Conference of the Prognostics and Health Management Society,* Montreal, Quebec, Canada, 25–29 September 2011
9. S. Mishra, S. Ganesan, M. Pecht, J. Xie, Life consumption monitoring for electronics prognostics (pp. 3455–3467), in *Proceedings of 2004 IEEE Aerospace Conference*, Big Sky, MT, USA, 6–13 March 2004

10. X. Guan, Y. Liu, R. Jha, A. Saxena, J. Celaya, K. Goebel, Comparison of two probabilistic fatigue damage assessment approaches using prognostic performance metrics. Int. J. Prognostics Health Manag. **2**(1), 1–11 (2011)

11. R. Klein, E. Rudyk, E. Masad, M. Issacharoff, Model based approach for identification of gears and bearings failure modes. Int. J. Prognostics Health Manag. **2**(2), 1–10 (2011)

12. K.-T. Wu, M. Kobayashi, Z. Sun, C.-K. Jen, P. Sammut, Engine oil condition monitoring using high temperature integrated ultrasonic transducers. Int. J. Prognostics Health Manag. **2**(2), 1–7 (2011)

13. A. Saxena, K. Goebel, D. Simon, N. Eklund, Damage propagation modeling for aircraft engine run-to-failure simulation (pp. 1–9), in *Proceedings of 2008 International Conference on Prognostics and Health Management*, Denver, CO, USA, 6–9 October 2008

14. A. Tantawy, X. Koutsoukos, G. Biswas, Aircraft power generators: hybrid modeling and simulation for fault detection. IEEE Trans. Aerosp. Electron. Syst. **48**(1), 552–571 (2012)

15. C. Chen, M. Pecht, Prognostics of lithium-ion batteries using model-based and data-driven methods (pp. 1–6), in *Proceedings of 2012 IEEE Conference on Prognostics and Health Management*, Beijing, China, 23–25 May 2012

16. C. Chen, D. Brown, C. Sconyers, B. Zhang, G. Vachtsevanos, M.E. Orchard, An integrated architecture for fault diagnosis and failure prognosis of complex engineering systems. Expert Syst. Appl. **39**(10), 9031–9040 (2012)

17. L.J. Bond, S.R. Doctor, T.T. Taylor, Proactive management of materials degradation—a review of principles and programs, U.S. Department of Energy, PNNL-17779, 2008

18. International Atomic Energy Agency (IAEA), Living probabilistic safety assessment (LPSA), IAEA-TECDOC-1106, 1999

19. International Atomic Energy Agency (IAEA), Protecting against common cause failures in digital I&C systems of nuclear power plants, IAEA Nuclear Energy Series No. NP-T-1.5, 2009

20. F. Gregor, A. Chockie, Aging management and life extension in the US nuclear power industry, CGI Report Prepared for the Petroleum Safety Authority Norway, 2006

21. International Atomic Energy Agency (IAEA), Assessment and management of ageing of major nuclear power plant components important to safety: CANDU pressure tubes, IAEA-TECDOC-1037, 1998

22. S. Baskaran, Role of NDE in residual life assessment of power plant components, Available: http://www.ndt.net/article/v05n07/baskaran/baskaran.htm. Accessed 8 Nov 2017

23. A. Coppe, R.T. Haftka, N.-H. Kim, C. Bes, A statistical model for estimating probability of crack detection (pp. 1–5), in *Proceedings of 2008 International Conference on Prognostics and Health Management*, Denver, CO, USA, 6–9 October 2008

24. J. Gu, N. Vichare, T. Tracy, M. Pecht, Prognostics implementation methods for electronics (pp. 101–106), in *Proceedings of 2007 Annual Reliability and Maintainability Symposium*, Orlando, FL, USA, 22–25 January 2007

25. S.M. Wood, D.L. Goodman, Return-on-investment (ROI) for electronic prognostics in high reliability telecom applications (pp. 1–3), in *Proceedings of 28th Annual International Telecommunications Energy Conference*, Providence, RI, USA, 10–14 September 2006

26. N.M. Vichare, M. Pecht, Prognostics and health management of electronics. IEEE Trans. Compon. Packag. Manuf. Technol. **29**(1), 222–229 (2006)

27. S.W. Yates, A. Mosleh, A Bayesian approach to reliability demonstration for aerospace systems (pp. 611–617), in *Proceedings of Annual Reliability and Maintainability Symposium*, Newport Beach, CA, USA, 23–26 January 2006

28. M. Modarres, M.P. Kaminskiy, V. Krivtsov, *Reliability Engineering and Risk Analysis: A Practical Guide*, 3rd edn. (CRC Press, Boca Raton, 2016)

29. M. White, J.B. Bernstein, Microelectronics reliability: physics-of-failure based modeling and lifetime evaluation. Jet Propulsion Laboratory Publication 08-5, 2008

30. N. Patil, J. Celaya, D. Das, K. Goebel, M. Pecht, Precursor parameter identification for insulated gate bipolar transistor (IBGT) prognostics. IEEE Trans. Rel. **58**(2), 271–276 (2009)

31. M. Pecht, A. Dasgupta, Physics-of-failure: an approach to reliable product development (pp. 1–4), in *Proceedings of IEEE International Integrated Reliability Workshop*, Lake Tahoe, CA, USA, 22–25 October 1995

32. K.C. Kapur, M. Pecht, Chapter 10. Failure modes, mechanisms, and effects analysis, in *Reliability Engineering* (Wiley, Hoboken, 2014)

33. H. Huang, P.A. Mawby, A lifetime estimation technique for voltage source inverters. IEEE Trans. Power Electron. **28**(8), 4113–4119 (2013)

34. M. Musallam, C. Yin, C. Bailey, C.M. Johnson, Application of coupled electro-thermal and physics-of-failure-based analysis to the design of accelerated life tests for power modules. Microelectron. Rel. **54**(1), 172–181 (2014)

35. L. Yang, P.A. Agyakwa, C.M. Johnson, Physics-of-failure lifetime prediction models for wire bond interconnects in power electronic modules. IEEE Trans. Devices Mater. Rel. **13**(1), 9–17 (2013)

36. E. Wong, W. Drieal, A. Dasgupta, M. Pecht, Creep fatigue models of solder joints: a critical review. Microelectron. Rel. **59**, 1–12 (2016)

37. C. Hendricks, E. George, M. Osterman, M. Pecht, M, Physics-of-failure (PoF) methodology for electronic reliability. Rel. Characterisation Electr. Electron. Syst. 27–42 (2015)

38. Y. Zhou, X. Li, C. Wang, R. Gao, A new creep-fatigue life model of lead-free solder joint. Microelectron. Rel. **55**, 1097–1100 (2015)

39. A. Dasgupta, R. Doraiswami, M. Azarian, M. Osterman, S. Mathew, M. Pecht, The use of "Canaries" for adaptive health management of electronic systems (pp. 176–183), in *Proceedings of 2nd International Conference on Adaptive and Self-adaptive Systems and Applications*, Lisbon, Portugal, 21–26 Nov 2010

40. A. Ramakrishnan, T. Syrus, M. Pecht, Chapter 22. Electronic hardware reliability, in *Avionics Handbook* (CRC Press, Boca Raton, 2000)

41. S. Mishra, M. Pecht, D.L. Goodman, In-situ sensors for product reliability monitoring, in *Proceedings of SPIE 4755, Design, Test, Integration, and Packaging of MEMS/MOEMS*, 19 April 2022

42. Ridgetop Semiconductor-Sentinel SiliconTM Library, Hot Carrier (HC) Prognostic Cell, 2004

43. K.C. Kapur, M. Pecht, *Reliability Engineering* (Wiley, Hoboken, 2014)

44. D. Rizopoulos, *Joint Models for Longitudinal and Time-to-Event Data: With Applications in R* (Chapman and Hall/CRC, UK, 2012)

45. N. Anderson, R. Wilcoxon, Framework for prognostics of electronic systems, in *Proceedings of International Military and Aerospace/Avionics COTS Conference*, Seattle, WA, USA, 3–5 Aug 2004

46. B. Tuchband, N. Vichare, M. Pecht, A method for implementing prognostics to legacy systems, in *Proceedings of IMAPS Military, Aerospace, Space and Homeland Security: Packaging Issues and Applications (MASH)*, Washington, D.C., USA, 6–8 June 2006

47. S. Lombardo, J.H. Stathis, B.P. Linder, K.L. Pey, F. Palumbo, C.H. Tung, Dielectric breakdown mechanisms in gate oxides. J. Appl. Phys. **98**(12, 121301) (2005)

48. N. Raghavan, K.L. Pey, K. Shubhakar, High-κ dielectric breakdown in nanoscale logic devices-scientific insight and technology impact. Microelectron. Rel. **54**(5), 847–860 (2014)

49. E.T. Ogawa, A.J. Bierwag, K.D. Lee, H. Matsuhashi, P.R. Justison, A.N. Ramamurthi, P.S. Ho, V.A. Blaschke, D. Griffiths, A. Nelsen, M. Breen, Direct observation of a critical length effect in dual-damascene Cu/oxide interconnects. Appl. Phys. Lett. **78**(18), 2652–2654 (2011)

50. N. Raghavan, A. Padovani, X. Li, X. Wu, V. Lip Lo, M. Bosman, L. Larcher, K. L. Pey, Resilience of ultra-thin oxynitride films to percolative wear-out and reliability implications for high-κ stacks at low voltage stress. J. Appl. Phys. **114**( 9, 094504) (2013)

51. C. Chen, G. Vachtsevanos, Bearing condition prediction considering uncertainty: an interval type-2 fuzzy neural network approach. Robot. Comput. Integ. Manuf. **28**(4), 509–516 (2012)

52. A.K.S. Jardine, D. Lin, D. Banjevic, A review on machinery diagnostics and prognostics implementing condition-based maintenance. Mech. Syst. Signal Process. **20**(7), 1483–1510 (2006)
53. M. Kang, J. Kim, L.M. Wills, J.-M. Kim, Time-varying and multiresolution envelope analysis and discriminative feature analysis for bearing fault diagnosis. IEEE Trans. Ind. Electron. **62**(12), 7749–7761 (2015)
54. M. Yang, V. Makis, ARX model-based gearbox fault detection and localization under varying load conditions. J. Sound Vib. **329**(24), 5209–5221 (2010)
55. H. Li, H. Zheng, L. Tang, Gear fault diagnosis based on order cepstrum analysis. J. Vib. Shock **25**(5), 65–68 (2006)
56. C.L. Nikias, J.M. Mendel, Signal processing with higher order spectra. IEEE Signal Process. Mag. **10**(3), 10–37 (1993)
57. L. Qu, Y. Chen, J. Liu, The holospectrum: a new FFT based rotor diagnostic method (pp. 196–201), in *Proceedings of 1st International Machinery Monitoring and Diagnostics Conference*, Las Vegas, NV, USA, 11–14 September 1989
58. J. Jin, J. Shi, Feature-preserving data compression of stamping tonnage information using wavelets. Technometrics **41**(4), 327–339 (1999)
59. J. Jin, J. Shi, Automatic feature extraction of waveform signals for in-process diagnostic performance improvement. J. Intell. Manuf. **12**(3), 257–268 (2001)
60. H. Zheng, Z. Li, X. Chen, Gear fault diagnosis based on continuous wavelet transform. Mech. Syst. Signal Process. **16**(2–3), 447–457 (2002)
61. N.G. Nikolaou, I.A. Antoniadis, Rolling element bearing fault diagnosis using wavelet packets. NDT E Int. **35**(3), 197–205 (2002)
62. J. Lin, L. Qu, Feature extraction based on Morlet wavelet and its application for mechanical fault diagnosis. J. Sound Vib. **234**(1), 135–148 (2000)
63. F. Leonard, Phase spectrogram and frequency spectrogram as new diagnostic tools. Mech. Syst. Signal Process. **21**(1), 125–137 (2007)
64. Q. Meng, L. Qu, Rotating machinery fault diagnosis using wigner distribution. Mech. Syst. Signal Process. **5**(3), 155–166 (1991)
65. W.J. Staszewski, K. Worden, G.R. Tomlinson, Time-frequency analysis in gearbox fault detection using the Wigner-Ville distribution and pattern recognition. Mech. Syst. Signal Process. **11**(5), 673–692 (1997)
66. N. Baydar, A. Ball, A comparative study of acoustic and vibration signals in detection of gear failures using Wigner-Ville distribution. Mech. Syst. Signal Process. **15**(6), 1091–1107 (2001)
67. S.U. Lee, D. Robb, C. Besant, The directional Choi-Williams distribution for the analysis of rotor-vibration signals. Mech. Syst. Signal Process. **15**(4), 789–811 (2001)
68. K. Fukunaga, *Introduction to Statistical Pattern Recognition* (Academic Press Professional Inc, San Diego, 1990)
69. S. Wold, K. Esbensen, P. Geladi, Principal component analysis. Chemometrics Intell. Lab. Syst. **2**(1–3), 37–52 (1987)
70. B. Scholkopf, A. Smola, K.-R. Muller, Kernel principal component analysis. Lect. Notes Comput. Sci. **1327**, 583–588 (1997)
71. A.J. Izenman, *"Linear Discriminant Analysis"*, in *Modern Multivariate Statistical Techniques* (Springer, New York, 2013), pp. 237–280
72. J.B. Tenenbaum, Mapping a manifold of perceptual observations (pp. 682–688), in *Proceedings of 1997 Conference on Advances in Neural Information Processing Systems*, Denver, CO, USA, 1–6 December 1997
73. E.W. Dijkstra, A note on two problems in connexion with graphs. Numer. Math. **1**, 269–271 (1959)
74. R.W. Floyd. Algorithm 97: shortest path. Commun. ACM **5**(6), 345 (1962)
75. T. Kohonen, *Self-organizing Maps* (Springer, Berlin, 1995)

76. R.D. Lawrence, G.S. Almasi, H.E. Rushmeier, A scalable parallel algorithm for self-organizing maps with applications to sparse data problems. Data Mining Knowl. Discovery **3**(2), 171–195 (1999)

77. B. Li, P.-L. Zhang, H. Tian, S.-S. Mi, D.-S. Liu, G.-Q. Ren, A new feature extraction and selection scheme for hybrid fault diagnosis of gearbox. Expert Syst. Appl. **38**, 10000–10009 (2011)

78. C. Liu, D. Jiang, W. Yang, Global geometric similarity scheme for feature selection in fault diagnosis. Expert Syst. Appl. **41**, 3585–3595 (2014)

79. Y. Yang, Y. Liao, G. Meng, J. Lee, A hybrid feature selection scheme for unsupervised learning and its application in bearing fault diagnosis. Expert Syst. Appl. **38**, 11311–11320 (2011)

80. J. Yang, V. Honavar, Feature subset selection using a genetic algorithm. IEEE Intell. Syst. Appl. **13**(2), 44–49 (1998)

81. T. Abeel, T. Helleputte, Y. Van de Peer, P. Dupont, Y. Saeys, Robust biomarker identification for cancer diagnosis with ensemble feature selection methods. Bioinformatics **26**(3), 392–398 (2009)

82. N. Meinshausen, P. Bühlmann, Stability selection, J.R. Stat. Soc. Ser. B Stat. Methodol. **72** (4), 417–473 (2010)

83. R. Milne, Strategies for diagnosis. IEEE Trans. Syst. Man, Cybern. **17**(3), 333–339 (1987)

84. S.H. Rich, V. Venkatasubramanian, Model based reasoning in diagnostic expert systems for chemical process plants. Comput. Chem. Eng. **11**(2), 111–122 (1987)

85. V. Venkatasubramanian, R. Rengaswamy, S.N. Kavuri, A review of process fault detection and diagnosis: part II: qualitative models and search strategies. Comput. Chem. Eng. **27**(3), 313–326 (2003)

86. V. Venkatasubramanian, R. Rengaswamy, S.N. Kavuri, K. Yin, A review of process fault detection and diagnosis: part III: process history based methods. Comput. Chem. Eng. **27**(3), 327–346 (2003)

87. V. Venkatasubramanian, R. Rengaswamy, K. Yin, S.N. Kavuri, A review of process fault detection and diagnosis: part I: quantitative model-based methods. Comput. Chem. Eng. **27** (3), 293–311 (2003)

88. T. Wang, J. Yu, D. Siegel, J. Lee, A similarity-based prognostics approach for remaining useful life estimation of engineered systems, in *Proceedings International Conference on Prognostics and Health Management*, Denver, CO, USA, 6–9 Oct 2008

89. E. Zio, F. Di Maio, A Data-driven fuzzy approach for predicting the remaining useful life in dynamic failure scenarios of a nuclear system. Rel. Eng. Syst. Safety **1**, 49–57 (2010)

90. J.B. Coble, J.W. Hines, Prognostic algorithm categorization with PHM challenge application, in *Proceedings of International Conference on Prognostics and Health Management*, Denver, CO, USA, 6–9 Oct 2008

91. F.O. Heimes, Recurrent neural networks for remaining useful life estimation, in *Proceedings of International Conference on Prognostics and Health Management*, Denver, CO, USA, 6–9 Oct 2008

92. M.E. Tipping, Sparse Bayesian learning and the relevance vector machine. J. Mach. Learn. Res. **1**, 211–244 (2001)

93. A.J. Smola, B. Scholkopf, A tutorial on support vector regression. Stats. Comput. **14**(3), 199–222 (2004)

94. M. Ernansky, M. Kakula, U. Benuskova, Organization of the state space of a simple recurrent network before and after training on recursive linguistic structures. Neural Netw. **20**(2), 236–244 (2007)

95. D.W. Hosmer, Jr., S. Lemeshow, R.X. Sturdivant, Chapter 9. Logistic regression models for the analysis of correlated data, in *Applied Logistic Regression*, 3rd ed. (Wiley, Hoboken, 2013)

96. M. Pecht, R. Jaai, A prognostics and health management roadmap for information and electronics-rich systems. Microelectron. Rel. **50**, 317–323 (2010)

97. P. Tamilselvan, P. Wang, Failure diagnosis using deep belief learning based health state classification. Rel. Eng. Syst. Safety **115**, 124–135 (2013)
98. S. Cheng, M. Pecht, A fusion prognostics method for remaining useful life prediction of electronic products, in *Proceedings of IEEE International Conference on Automation Science and Engineering*, Bangalore, India, 22–25 Aug 2009
99. S. Sankararaman, Significance, interpretation, and qualification of uncertainty in prognostics and remaining useful life prediction. Mech. Syst. Signal Process. **52–53**, 228–247 (2015)
100. H. McManus, D. Hastings, A framework for understanding uncertainty and its mitigation and exploitation in complex systems (pp. 1–20), in *Proceedings 15th Annual International Symposium of the International Council on Systems Engineering*, Rochester, NY, USA, 10–15 July 2005
101. M. Orchard, G. Kacprzynski, K. Goebel, B. Saha, G. Vachtsevanos, Advances in uncertainty representation and management for particle filtering applied to prognostics (pp. 1–6), in *Proceedings International Conference on Prognostics and Health Management*, Denver, CO, USA, 6–9 Oct 2008
102. L. Tang, G.J. Kacprzynski, K. Goebel, G. Vachtsevanos, Methodologies for uncertainty management in prognostics (pp. 1–12), in *Proceedings Aerospace Conference*, Big Sky, MT, USA, 7–14 March 2009
103. J.R. Celaya, A. Saxena, K. Goebel, Uncertainty representation and interpretation in model-based prognostics algorithms based on Kalman filter estimation (pp. 23–27), in *Proceedings of Annual Conference of the PHM Society*, Minneapolis, MN, USA, 23–27 September 2012
104. M. Reiner, D.D. Lev, A. Rosen, Theta neurofeedback effects on motor memory consolidation and performance accuracy: an apparent paradox? Neurosci. (2017)
105. E.L. Lehmann, G. Casella, *Theory of Point Estimation*, 2nd edn. (Springer, New York, 1998)
106. A. Saxena, J. Celaya, E. Balaban, K. Goebel, B. Saha, S. Saha, M. Schwabacher, Metrics for evaluating performance of prognostic techniques (pp. 1–17), in *Proceedings of International Conference on Prognostics and Health Management*, Denver, CO, USA, 6–9 Oct 2008
107. D. Puccinelli, M. Haenggi, Wireless sensor networks: applications and challenges of ubiquitous sensing. IEEE Circ. Syst. Mag. **5**(3), 19–31 (2005)
108. R. Lin, Z. Wang, Y. Sun, Wireless sensor networks solutions for real time monitoring of nuclear power plant, in *Proceedings of 5th World Congress on Intelligent Control and Automation*, Hangzhou, China, 15–19 June 2004
109. D. Bhattacharyya, T.-H. Kim, S. Pal, A comparative study of wireless sensor networks and their routing protocols. Sensors **10**(12), 10506–10523 (2010)
110. International Atomic Energy Agency (IAEA), Proactive management of ageing for nuclear power plants, Safety Reports Series No. 62, 2009
111. PHM-Prognostics and Health Management of Electronic Systems, Available: https://standards.ieee.org/develop/wg/PHM.html. 8 Nov 2017
112. N. Dharmaraju, A. Rama Rao, Dynamic analysis of coolant channel and its internals of Indian 540 MWe PHWR reactor. Sci. Technol. Nucl. Install. **2008**(764301), 1–7 (2008)
113. K. Chatterjee, M. Modarres, A probabilistic physics-of-failure approach to prediction of steam generator tube rupture frequency. Nucl. Sci. Eng. **170**(2), 136–150 (2012)
114. A. Andonov, K. Apostolov, M. Kostov, G. Varbanov, Structural health monitoring of VVER-1000 containment structure (pp. 1–8), in *21st International Conference on Structural Mechanics in Reactor Technology (SMiRT 21)*, New Delhi, India, 6–11 Nov 2011
115. J. Coble, P. Ramuhalli, L.J. Bond, B.R. Upadhyaya, Prognostics and health management in nuclear power plants: a review of technologies and applications, U.S. Department of Energy, PNNL-21515, 2012
116. A. Heng, S. Zhang, A.C.C. Tan, J. Mathew, Rotating machinery prognostics: state of the art, challenges and opportunites. Mech. Syst. Signal Process. **23**(3), 724–739 (2009)
117. C. Yin, H. Lu, M. Musallam, C. Bailey, C.M. Johnson, Prognostic reliability analysis of power electronics modules. Int. J. Performability Eng. **6**(5), 513–524 (2010)

118. H. Ye, C. Basaran, D.C. Hopkins, Experimental damage mechanics of micro/power electronics solder joints under electric current stresses. Int. J. Damage Mech. **15**(1), 41–67 (2006)
119. H. M. Hashemian, On-line monitoring and calibration techniques in nuclear power plants (pp. 1–11), in *Proceedings of International Conference on Opportunities and Challenges for Water Cooled Reactors in the 21st Century*, Vienna, Austria, 27–30 Oct 2009
120. I. Snook, J.M. Marshal, R.M. Newman, Physics of failure as an integrated part of design for reliability (pp. 46–54), in *Proceedings of 2003 Annual Reliability and Maintainability Symposium*, 27–30 January 2003

# Chapter 14
# Risk-Informed Decisions

> *Robust decisions are based on sufficient technical evidence and characterization of uncertainties to determine that the selected alternative best reflects the decision-maker's preferences and values, given the state of knowledge at the time of decision, and is considered insensitive to credible modelling perturbations and realistically foreseeable new information.*
>
> NASA, Risk-informed Decisions, 2010 [1, 2]

## 14.1 Introduction

The traditional approach to decision-making is essentially deterministic in nature where the principles of defense in depth along with conservative rules and criteria form higher-level requirements, while the application of single-failure criteria, redundancy, diversity, and quality assurance elements forms the lower-level requirements [2]. The defense in depth is implemented by incorporating multiple barriers and levels of protection such that the probability of severe consequences can be reduced to a very low or negligible level. The objective is to ensure application of five defense-in-depth levels [3, 4] summarized as follows:

*Level 1*: Maintain plant in normal operation, and prevent deviation from normal operation using conservative approach, high quality in construction and operation.
*Level 2*: Control abnormal operation and have provisions to detect failure, should they occur, by having control and surveillance program such that plant is brought back to normal operation.
*Level 3*: In case the accident occurs, then control the accident in design basis envelope by automatic activation of engineered safety features in the plant, for example actuation of emergency core cooling system as a mitigation measure in response to a loss of coolant accident (LOCA) scenario.
*Level 4*: Control the consequences, should the plant enter severe by mitigating the consequences applicable to beyond design basis accidents through severe accident management program. Provision of exclusion zone around an NPP and emergency preparedness should ensure the consequences of the accident.

© Springer Nature Singapore Pte Ltd. 2018  509
P. V. Varde and M. G. Pecht, *Risk-Based Engineering*, Springer Series in Reliability Engineering, https://doi.org/10.1007/978-981-13-0090-5_14

*Level 5*: In the eventuality of release of radioactivity to the environment during an accident condition, provision should exist to mitigate the consequences in public domain through off-site response measure.

The defense-in-depth provision to design and operation of the nuclear plants, as above, ensures that there are multiple means to implement safety functions. The deterministic approach is proven and has worked well for over six decades such that nuclear community can claim to have highest safety standard. However, in spite of this glory, there are three accidents, Three Mile Island in 1979, Chernobyl in 1986, and Fukushima in 2011, which has provided vital lessons on safety, and the safety community in response had introspection that led to new insights, and perhaps development of risk-informed approach has been one of the major outcomes of this introspection. One of the lessons was to have look into the limitations of the deterministic approach. The following summarizes these limitations:

- The notion of safety is qualitative and subjective; particularly, it lacks as an approach to accounting of system performance for providing safety level of the plant in an explicit manner.
- The approach is prescriptive in nature, i.e., the design and operations, governed by strict code, technical specification requirements which at times tend to be over conservative, and does not provide any flexibility to view the issues in terms of benefits that can be availed in light of improved insights into safety and other aspects.
- The deterministic approach is governed by predefined sets of credible initiators where the emphasis is placed on detailing the selected few scenarios.
- The notion of likelihood and consequences is primarily qualitative in nature; hence, it is not very effective in identification and prioritization of the safety issues.
- The notion of safety margins in terms of availability of redundancy, diversity, and single-failure criteria is qualitative with no rational framework to demonstrate that the performance of individual systems forms an integral part of the evaluation.
- Human interactions are an important element in design and operation of the nuclear plant; however, the deterministic approach does not have an effective framework for integrating human performance for demonstrating safety of the plant.

Probabilistic risk assessment (PRA) is a relatively new approach and now considered to be matured enough that it can be used to evaluate safety aspects of plants. In fact, the state of the art of risk-informed decisions makes it possible to use insights from PRA to complement or supplement the deterministic approach. Probabilistic safety assessment (PSA) is an analytical tool that enables risk modeling of the plant by considering various credible postulated accidents and estimating the likelihood and consequences of potential scenarios. PRA provides a systematic and rational-based framework for quantified risk estimation.

When a decision is based on insights from risk assessment of engineering systems, it is called a "risk-based decision." However, when the insights from risk assessment form just one of the inputs in support of decision-making, then this decision-making approach is called a "risk-informed decision."

Even though the risk-informed or risk-based approach is being applied in many areas, including nuclear systems, hydraulic dam safety, financial systems, space systems, process and chemical industries, and health care systems, the discussion in this chapter is primarily limited to nuclear facilities. However, the approach can be adapted to other engineering systems with due considerations to safety characteristics of the particular application.

The risk-informed philosophy has primarily been developed in support of regulatory decision-making. Here, the results of the PRA form one of the inputs along with other insights, such as ensuring adequate safety margins, maintenance of defense in depth, compliance with existing regulation standards/guidelines, and monitoring performance. In this context, it can be argued that the integrated risk-based approach (IRBE) developed in this book forms part of regulatory decision as part of risk-informed approach.

There is a general perception that the traditional approach to decision-making is based on the deterministic approach. However, if we look at the basic approach to address the safety, we can easily see that probabilistic aspects and decision criteria form part of decision-making. Also, there are frequent references where a qualitative notion of uncertainty also forms part of discussions and debate in support of decisions. The only difference is that the notion of probability or likelihood as well as uncertainty is qualitative in deterministic approach, whereas the probabilistic approach provides a framework for quantification of these aspects. Hence, the traditional approach to decision-making can also be argued to be essentially risk-informed in nature. The only difference between the traditional and present approaches to decision-making has been that in traditional approach a qualitative notion of risk was used; however, the flexibility was not there because the plant technical specifications formed the governing considerations. At the implementation level, the present risk-formed approach, as it is perceived in the regulatory domain, has legal or regulatory considerations; i.e., the regulatory framework allows use of insights from quantified results from the PRA studies to enable flexibility in the decision-making process. This flexibility is limited by prescriptive deterministic considerations. It may be noted that probabilistic arguments as part of risk-informed approach must consolidate the provisions of defense in depth and should in no way dilute the defense-in-depth provisions. For example, while the defense-in-depth principle of redundancy requires more than one system to meet the safety function, the PRA methods evaluate the adequacy of the redundancy and provide rational arguments for levels of redundancy that meet the objective criteria by comparison of quantified results with set goals and criteria—again, a quantified metrics. Risk-informed decisions are the way of life for many application areas, be it safety-critical systems such as nuclear plants, processes, or industrial system, in support of designs, operations, and regulations.

This chapter essentially reviews the application of risk-based technology as input to risk-informed decision-making in support of regulatory review [5]. Even though this chapter was written after reviewing the major applications of risk-informed approaches, mainly the IAEA, USNRC, and NASA and NEA approaches, no claim is made to correctness of legal as well as policy depiction as given in their respective national documents. The objective here is to provide academic treatment to the subject.

## 14.2   Generic Steps in Decision-Making

There are two terms in the phrase "risk-informed decision-making," viz. "risk-informed" and "decision-making." In the previous section, the meaning of the term risk-informed was explained, while the rest of the chapter will bring out the activities related to risk modeling/analysis, particularly PRA and insights drawn from this technique for procedural implementation aspects that need to be considered as part of risk-informed decisions. The second term "decision-making" as a field of engineering and management needs to be revisited to have improved appreciation of the process of risk-informed decision-making. The following section reviews in brief the major steps and associated aspects of decision-making [6]. Even though this framework is primarily a means for mathematical problems, if we look at the individual elements of this approach we can easily see that this is a good starting point for going into the details of risk-informed decision-making as the process is analytical in nature.

Figure 14.1 shows a general flowchart depicting major steps in decision-making. The first step in this generic framework is to define the problem, i.e., the statement on the issues that need to be addressed, system boundary and identification of stakeholders; e.g., in the context of risk-informed decision, the definition of the problem could be "Request for a proposal for increasing the ECCS Surveillance test interval time from once per month to two months for X NPP", and the problem formulation statement should be concise and clear. The second step identifies the associated requirements and constraints that form part of the input. This includes the major considerations, such as change in core damage frequency (CDF), cost/benefit analysis, dose reduction, and treatment of uncertainty. Assumptions and constraints associated with the problems often form part of the requirements. The goals could include the achievable targets and the need to satisfy all the proposed alternatives. For example, a qualitative goal could be to increase the test interval without compromising safety or with a minor increase in safety such that the increase can be justified. Normally, there are more than one alternative to arrive at a solution. This might include changes in system configurations and operational and maintenance policies or the level of detail that needs to be used.

The third step is setting the criteria for decision-making; for example, in the context of application of the risk-informed approach, the criteria could be limit on CDF and changes in CDF, dose consumption limit or as low as practicable ALARP

**Fig. 14.1** Procedural chart for the decision-making process [5]

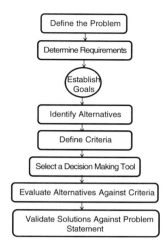

policy and cost/benefit benchmarks. Selection of decision-making tool, the fourth step, is dictated by the approach to problem-solving, the available resources, and requirements of the problem. For example, the major framework could be "risk-informed," while at a lower level it could be the availability of PRA and reliability methods, optimization techniques, and an integration approach to arrive at a conclusion from various indicators. Evaluation of alternatives requires an analytical approach to find a solution for each alternative and compare the same with integrated indicators.

The fifth and final step is validation, which consists of checking for consistency in the solution considering the input for each alternative and ensuring that the objective of the change has been met by monitoring the performance of the system, as applicable to the example considered here.

This process provides a vital point for a discussion on risk-informed decision-making. In the next section, the requirements of risk-informed decision-making as an integrated tool are discussed.

## 14.3 Basic Requirements for Risk-Informed Decisions

Following are the broad requirements for implementation of the risk-informed approach:

- A national regulatory policy for moving from a traditional deterministic decision-making process to use insights from PSA in decision-making.
- A policy framework that brings out the interrelationship of various inputs including traditional deterministic and probabilistic insights in support of decisions.

- Broad regulatory guidelines and identification of activities for implementation of risk-informed approach to regulatory reviews of safety cases submitted by utilities and regulatory activities to oversee the safety of nuclear facilities.
- Regulatory framework on evaluating the quality of a PRA and deterministic safety assessment.
- Availability of safety analysis reports, which includes deterministic and probabilistic studies that meet the required quality standards.
- Development of probabilistic and deterministic requirements/goals, risk criteria, and requirements related to the treatment uncertainties.
- Characterization of systems for which the risk-informed regulation is being developed. For example, the requirements for mission-oriented systems such as space, ships, and nuclear facilities will require a framework that meets their safety goals and objectives.
- Provision for regulatory as well as plant-level monitoring and oversight with clearly defined performance goals such that impact of change can be tracked.
- The success of risk-informed approach may also depend on training of decision.
- Availability of trained staff/agencies to deal with probabilistic and deterministic aspects as a part of risk-informed approach implementation.

Often the requirements are governed by the specific nature of change implementation and the decision attributes that forms basis for the decision/approval. Hence, there is need to develop additional requirements based on regulatory stipulations.

## 14.4   Role of PRA in the Risk-Informed Approach

Even though the available literature shows many models of risk-informed/risk-based applications where the risk insights are used in support of decisions, there are three models for implementation of risk-informed approach that use PRA insights, viz. the IAEA, USNRC, and NASA models. These approaches are similar in principle, but differ in their framework and implementation. This difference can be seen in the context of their application and the expectation from the respective approach. The details about these approaches will be discussed in the next section; however, keeping in view the scope of this book, i.e., application to nuclear facilities, we will review the role of PRA in the risk-informed approach.

PRA plays a key role in supporting the risk-informed approach as follows:

- Insights from PRA are just one of the inputs in risk-informed decision-making along with engineering, operational, and safety inputs.
- PRA provides an integrated and effective rational-based framework/model of the plant/facility.

- Once a computerized model of PRA is available, then this provides an effective mechanism for simulation for various options that need to be evaluated in support of decisions.
- PRA provides an integrated framework for risk assessment toward providing the quantified statement of safety at the plant level and unavailability/reliability at the system level.
- The quantified PRA criteria/goals and the estimation of as-built and operated plant safety (i.e., CDF/large early release frequency (LERF) estimates) provide an elegant means of comparing the alternatives/options available and thereby using this information in support of decisions.
- The quantified estimates of uncertainty at the plant as well at system level propagated from basic components through subsystems provide an effective mechanism for understanding uncertainty, which otherwise is regarded as imprecise information in the traditional deterministic approach.
- The living PRA model of the plant can easily be used for plant as-built and operated conditions of the plant. This enables an efficient use of PRA information for risk-informed application.
- Sensitivity analysis forms an integral part of change evaluation. The impact of the proposed change can be evaluated in terms of change in safety indicators like CDF or LERF at plant level and system unavailability at system level.

It may be noted that many applications may require a full-scope Level 1 PRA study; hence, the scope of the available study will determine whether the study qualifies to be used in support of risk-informed decisions. For full-scope Level 1 PRA apart from the internal events, the scope should include considerations of low-power and shutdown and external events. In case the decisions are related to emergency conditions, then the scope of PRA might include Level 2 and Level 3 PRA also.

## 14.5    Acceptability of PRA as Part of Risk-Informed Applications

Acceptability of PRA can be seen in the context of the inherent limitations of PRA process or as a tool and certain requirements that the PRA needs to fulfill, such as PRA quality, scope, level of detail, uncertainty, and sensitivity aspects. The following section deals with major considerations in respect of evaluating the acceptability of PRA and its results toward risk-informing the decisions.

## 14.5.1   Limitations of PRA

Even though PRA is now considered mature enough to be used in support of decision-making, the results of PRA should be seen in the context of the inherent limitations this approach suffers from. In fact, these limitations become all the more relevant when PRA insights are used as risk-based applications because the decisions are based exclusively on the insights drawn from PRA. The major limitations of PRA are as follows:

- *Completeness*: There is no way to ensure that the scenario or postulations or list of postulated initiating events is comprehensive and complete. The Fukushima Accident in Japan in 2011 has brought in focus considerations of combination of initiating events that need to be considered in the PRA model. Hence, keeping view the available literature and plant-specific experience the list of initiating events should be as complete as possible.
- *Adequacy of Data and Model*: The quality and adequacy of data have great bearing on the results of the study. Also, the selection of models is crucial and affects the estimates. The models must be validated such that they reflect the as-built plant conditions and operational policies and practices.
- *Capability to Address Dynamic Scenario*: The fault tree and event tree form the cornerstone of PRA modeling which in the traditional form is incapable of handling the dynamic scenario. However, the limited-scope application of the Markov model and the recent trend to use dynamic fault tree (DFT) and dynamic event tree (DET) has shown promising results to create dynamic scenario in PRA. Their use is limited and is confined to the R&D domain.
- *Challenges Posed by New Systems*: One of the vital challenges is the availability of software modeling in digital systems. There is no consensus for an acceptable approach to software reliability quantification for safety-critical applications. Apart from this, incorporation of new technologies such as field-programmable gate arrays (FPGA) and other systems which have their own failure modes (e.g., single event upsets (SEU) for FPGA) poses added challenges.
- *Human Factor Modeling*: Even though a host of techniques are available for human factor modeling in PRA, further consensus is required to have improved understanding toward reducing uncertainty in human reliability estimates.
- *Modeling of Common Cause Failures*: Even though there are many approaches based on empirical models available for common cause failure (CCF) modeling, there is a need to develop advanced methods based on physics-of-failure approaches. At present, the approach is to ensure that adequate defenses are available to address CCF for redundant and diverse systems.
- Apart from safety aspects, system security aspects need to be addressed in PRA study; particularly, the system vulnerability against potential CCF needs special considerations.

Apart from these limitations, the availability of resources in terms of trained manpower, limited knowledge of new components, and lack of understanding of degradation mechanisms also affects the results of PRA studies.

## 14.5.2 PRA Requirements for Specific Applications

The PRA requirements in respect of given application need to be reviewed for the following considerations:

(a) *Quality Requirements*: For given PSA study to qualify as a candidate for risk-informed application, it must conform to the available national or international standards. For example, one of the frameworks for quality conformance is IAEA-TECDOC-1511 [6], which defines that general attributes are further detailed into special attributes for each of the major element of PSA. ASME-std 2002 [7] is a standard that sets out quality specifications for PRA application in nuclear power plants. The IAEA and ASME documents/standard provide an effective mechanism for ensuring quality of PRA for a given application.

- *Scope of PRA Study*: A full-scope Level 1 PRA is recommended for developing a risk-informed application. A full-scope PRA includes internal and external events (e.g., seismic, flood, impact of external objects); low-power and shutdown PSA; reactor core and reactor block as a source of radioactivity; consideration of human factors in the plant model; and performance of uncertainty and sensitivity analyses. However, the present situation is that most of the nuclear facilities have PRA with internal initiating events and reactor core as the source of radiation where uncertainty analysis and human factor modeling are also performed to estimate the CDF. While developing a risk-based application using these limited-scope studies, suitable margins should be kept for non-availability of the other modes and associated consequences. For example, for a decision related to actions for shutdown of the plant, it should be factored that there can also be contribution from the shutdown mode of the plant.
- *Level of Details*: The level of detail determines the applicability of PRA. For example, for a risk-informed maintenance management application for diesel generator (DG) sets, the PRA fault tree model should not be terminated to DGs as basic events, but should go down to the level of various DG subsystems and further to the basic components in the subsystems to evaluate contributors to DG failures. This detailed model will enable development of a risk-informed maintenance management system for emergency DG systems.
- *Input Data Assumptions*: The PRA developed using plant-specific data forms the ideal case for risk-informed applications. However, it is not possible to develop a PRA using plant-specific data only, and the generic data are used for many reasons in a PRA study. In such cases, uncertainty and sensitivity analysis should be carried out to see the impact of assumptions and uncertainty in the data.

USNRC Draft Regulatory Guide DG-1122, "An Approach for Determining the Technical Adequacy of Probabilistic Risk Assessment Results for Risk-Informed Activities" [8], provides a comprehensive framework to evaluate the applicability of a PRA for risk-informed applications.

## 14.6  Overview of Risk-Informed Developments

The available literature shows that there is noticeable interest in application of risk-informed approach employing PRA insights. The scope of this chapter is limited to review of the published technical documents, guides, and research publications in support of risk-informed approaches of IAEA, USNRC, NASA, and NEA.

### 14.6.1  IAEA RIDM Development Status

As an international nuclear safety body, the IAEA spearheaded development of standards and technical documents on application of PSA to nuclear power plants. Even though most of the documents were specific to NPPs, a good number of documents were also developed for research reactors and other nuclear facilities. The IAEA Safety Standard on safety assessment for facilities and activities encouraged application of PRA through the *requirement 15* on deterministic and probabilistic approach which states that "both deterministic and probabilistic approaches" shall be included in the safety analysis as these two approaches complement each other [8]. As can be seen, this policy statement of IAEA brought deterministic and probabilistic given equal weightage without any prejudice to deterministic or probabilistic. This was a major shift from earlier policy where the role of PRA was either complementary or supplementary in nature. This was in a way boost for application of risk-informed approach. In the recent times, revision and publication of PRA Level 1 and 2 safety guide to safety standard, SSG-3 for Level 1 PRA [9] and SSG-4 for Level 2 PRA [10] in 2010 provided the needed confidence in member states to apply PRA approach to risk-informed decisions. IAEA through TECDOC-1209 publication inspired the member states to have risk management approach to improve the nuclear power plant performance, in 2001 [11]. IAEA-TECDOC-1436 published in 2005 provides the status of risk-informed regulation for nuclear facilities and provides an overview and status of risk-informed regulation in member states [13]. A report by the IAEA's International Nuclear Safety Group, INSAG-25, proposes a framework, principles, and key elements for an integrated risk-informed decision-making process [8]. The objective of this report is not only to propose the framework, but also promote a common understanding among the international community on how the risk-informed concept can be used in decision-making. Even though INSAG-25 [14] is targeted to NPPs, it can be adopted for other nuclear facilities and non-nuclear applications. This approach is discussed in detail in the next section. IAEA has been actively promoting, apart from safety, the nuclear security in member states. The IAEA Nuclear Security Series No. 24-G makes the case for implementation of risk-informed approach for nuclear security measures for nuclear and other radioactive material [15].

## 14.6.2  USNRC Development Program

The USNRC Policy Statement in 1995 on the use of PRA insights in regulatory matters can be seen as initiation of adoption of PRA insights in risk-informed decision-making in nuclear plants [16]. In 1995, the NRC Guide 1.174, entitled "An approach for using PRA in risk-informed decisions on plant specific changes to the Licensing Basis," provided a regulatory framework for the risk-informed approach to nuclear power plants (NPPs) in the USA [17]. In 1998, a Chapter on General Guidance on use of PRA in plant specific risk informed decision making formed part of NUREG 0800 [18]. Later, three application-specific USNRC Guides to the risk-informed approach were developed, namely 1.175 for in-service testing [19], 1.176 for graded quality assurance [20], 1.177 for technical specification [21], and 1.178 for in-service inspection of piping [22]. It was considered important that a PRA should meet the minimum quality standards to qualify for risk-informed applications. In this context, a draft regulatory guide DG-1122, published in November 2002, entitled "An Approach for determining the technical adequacy of PRA results for risk-informed activities" came out with an acceptable approach for determining the quality of PRA into or in part for specific applications [23]. Another milestone in the history of PSA quality assurance was through the American Society for Mechanical Engineers (ASME) standard ASME RA-S-2002, in April 2002, on PRA for nuclear plant applications.

The USNRC Guide 1.174 provides guidelines on application of risk-informed plant-specific changes to licensing basis decision-making [17]. Figure 14.2 shows USNRC's principles of risk-informed decision-making. This approach is based on the use of insights from traditional deterministic philosophy complemented by insights from PRA toward evaluation of risk for changes to the licensing basis. Each of the elements should be considered in integrated risk-informed decision-making. For details, refer to the USNRC Guide 1.174 [17].

Figure 14.3 shows the major elements of the USNRC approach to risk-informed decision-making. As mentioned in the previous section, the first element is "defining the change" that is proposed by the stockholder that means a utility (e.g., nuclear plant operator). This change could be either a modification, retrofitting, a change in operational and maintenance policy, or a specific change in the technical specifications. Traditional as well as probabilistic analysis is carried out in support of the decision. The next step is to bring out the implementation program, including performance monitoring for the change as well as the other systems it is interfacing

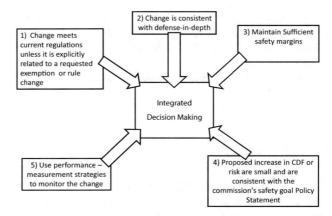

**Fig. 14.2** USNRC principles for integrated risk-informed decision-making [17]

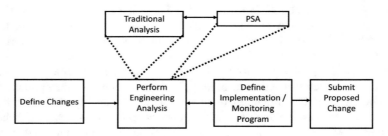

**Fig. 14.3** Elements of USNRC's approach to risk-informed decision-making [17]

with. The final element involves submission of the proposal for the change for regulatory considerations.

Figure 14.4 shows the risk metrics that depict the acceptance criteria in terms of CDF versus changes in CDF ($\Delta$CDF). As can be seen, the flexibility for a change reduces as the CDF value moves toward the right side. It may be noted that the boundary as shown in the figure appears to be crisp and distinct, but for the actual interpretation there is a gradient while moving from one region to another. If $\Delta$CDF is less than $10^{-6}$/r-y, then an even higher reference CDF can be accepted. Similar acceptance criteria have been formulated using LERF in USNRC Guide 1.174.

Modeling, representation, interpretation of uncertainties in PRA model is critical from the point of risk-informed applications. Probabilistic nature of PRA model accounts for aleatory uncertainty while the lack of data and adequacy of model that represent the data introduces epistemic uncertainties. The NUREG-1855 provides guidance on how to treat uncertainties in PRA as a part of risk-informed approach implementation [24].

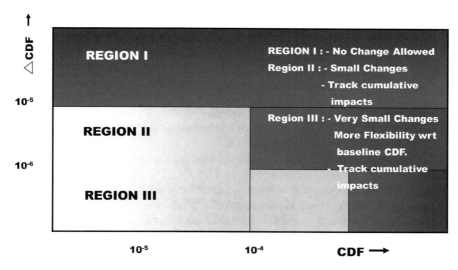

**Fig. 14.4**   Representation of USNRC acceptance criteria [17]

## 14.6.3   NASA Risk-Informed Development Program

The NASA PRA procedure guide published in 2002 provides the needed risk assessment document in support of risk-informed application to the space program [25]. Later NASA published document 'Agency Risk Management Procedural Requirements (NPR 8000.4A) technical document' [26] in 2008 which introduced risk-informed decision-making (RIDM) as a complementary process to the existing continuous risk management program of NASA. In 2010, a risk-informed decision-making handbook brought out the procedural element employed through the performance requirements for higher levels as well as lower levels to cater to the safety- and mission-critical system modeling requirements of NASA [2]. This handbook addressed the risk-informed decision-making component of risk management program of NASA.

The available literature shows that even though the fundamental approach for NASA and USNRC risk-informed program in terms of broad framework and approach for integration of various performance indicators is similar, there are differences in terms of application because the orientation is to model mission-oriented safety-critical systems, and this leads to differences of procedural elements. In NASA's risk-informed development program, unlike the USNRC's, interaction/deliberation process between various stakeholder takes place on continuous basis. The major distinguishing features of NASA's risk-informed decision-making process can be summarized as follows:

- The approach caters to mission-oriented safety-critical systems such as space and aviation systems.
- NASA approach has been designed to cater to organizational settings where top-level objectives flow down progressively to more detailed performance requirements, such as safety, technical, cost/benefit, and schedule objectives. Lower-level performance requirements should be satisfied to arrive at top-level objective functions, e.g., mission success.
- In NASA RIDM, the process of risk-informed development interaction takes place according to established protocol.
- For risk assessment, the PRA approach is preferable at the quantitative level or qualitative level.
- The complete process of risk-informed decision-making has been formulated in three major steps, viz. (a) identification of decision alternatives, (b) risk analysis of decision alternatives, and (c) risk-informed alternative selection.
- One of the noticeable features of this approach is the application of a trade tree to compile decision alternatives and selection of the alternative by inducing the system attributes.
- The technical basis is produced as input for the deliberative process by propagation of uncertainty for each performance measure.
- The final step comprises of deliberation and selection of an alternative and documentation of the rationales that back the final decision.

### 14.6.4 NEA RIDM Development Status

The OECD/Nuclear Energy Agency report entitled "Nuclear Regulatory Decision Making" presents that to a large extent the decision-making process is deterministic in nature; however, it notes that "Most regulators find that assessing the safety significance of an issue can be improved through the use of PRA insights" [27]. This report also endorses that even in the traditional approach safety regulations were risk-informed in nature, whereas to a large extent the safety insight was qualitative in nature. The report discusses that PRA methodology has matured and found widespread usage in OECD countries and is generally accepted by regulatory bodies. It is realized that PRA, like any other methodologies, has limitations; e.g., it cannot model the safety culture of the plant. It is reported that within the spectrum of OECD countries there is a general consensus that PRA, if properly used, can be an effective tool for supporting regulatory decision-making. There are some guidelines that the PRA should be of high quality, the operator and regulator should be knowledgeable of PRA methodology and its limitations, the PRA information should not be used to replace the defense in depth, and the PRA results should be judiciously interpreted and used with the considerations of uncertainties.

### 14.6.5   Related Literature on Risk-Informed Decision-Making

International Risk Governance Council (IRGC) published a white paper on risk governance through an integrated approach [28]. In this context, Serbanescu and Vetere propose the role of WPI—risk-informed decision-making (IRDM) in an integrated risk governance framework where, apart from IRDM, the other major elements are "precaution and risk reduction principle" and "risk deliberation" [29]. Zio and Pedroni while reviewing the risk-informed approach of USNRC and NASA introduce the general concept, definitions, and issues related to RIDM process [30]. They argue in favor of implementation of RIDM for systems where decision-making requires dealing with complexity in terms of solution space that poses multiple objective functions coupled with high level of uncertainty. Pulkkinen and Simola describe an expert panel methodology for supporting risk-informed decision-making where the panel achieves a balanced utilization of data, and information from several discipline, including PRA, in support of decision-making [31]. They also summarized application of expert panel methodology for a pilot study on risk-informed in-service inspection program. Further, Simola and Pukkinnine performed a prestudy in 2004 to review the status of risk-informed decision-making in Swedish and Finnish nuclear power plants and made recommendations like development of risk-informed procedure on decision analytic, requirement of documentation format, organizational mandate for PSA team and definition of goals and criteria and finally arguing the demonstration of risk-informed methodology by revisiting or developing a in-service inspection activities [32]. When we discuss risk-informed decision-making, there are two questions that need to be evaluated, viz. how a decision would affect in short as well as in long term in terms of consequences and what makes a good decision. Ersdal and Aven present their research on risk-informed decision-making and its ethical basis to explore the second question that deals with moral and norm [33]. The authors conclude and opined that awareness of the ethical element of a risk-informed decision is an important aspect.

## 14.7   Integrated Decision-Making

As mentioned earlier, the requirement 15 of the IAEA Safety Standard entitled Safety Assessment for Facilities and Activities states that "Both deterministic and probabilistic approaches shall be included in the safety analysis. Item 4.53 of this document further states that deterministic and probabilistic approaches have been shown to complement one another and can be used together to provide input into an integrated decision-making process. Item 4.55 further discusses the role of PRA and benefits of applying PRA in enhanced understanding of risk. One major point emerges here is that the notion of "risk" comprised of deterministic and probabilistic elements, where the deterministic element provides the qualitative and the probabilistic element and provides the quantitative argument in support of decisions. This argument is in line with the IRBE approach.

The development work of IAEA-integrated risk-informed decision-making, more commonly referred to as IRIDM, has been described in IAEA-TECDOC-1436. This document describes the integrated risk-informed decision-making process. The inputs to this decision-making process are based on probabilistic, deterministic, mandatory regulatory, and legal and technical specification requirements. Integration of these parameters is performed using suitable "weights" that reflect the importance given to these parameters. This document provides guidelines on the use of risk-informed decision-making by the regulatory body for safety issues at nuclear plants, also referred to as risk-informed decision-making, and how risk information is used to support/optimize or prioritize various activities of regulatory bodies, referred to as risk-informed regulation. Details of these two activities have been elaborated in the subject TECDOC.

Discussion on the probabilistic criteria typically related to CDF and LERF is one of the notable features of this document. Figure 14.5 reproduces the relevant text from this document.

---

*Probabilistic Criteria*

*A possible framework for the definition of probabilistic Criteria was given in INSAG [35]. This defines a "threshold of tolerability" above which the level of risk would be intolerable and a "design target" below which the risk would be broadly acceptable. Between these two levels there is region where the risk would only be acceptable if all reasonable achievable measures have been taken to reduce it.*

*Based on the current experience with nuclear power plants design and operation, numerical values are proposed that could be achieved by current and future designs. For the CDF, the objective is $10^{-4}$ per reactor-year for <u>existing</u> plants and $10^{-5}$ per reactor-year for <u>future plants.</u> For a large release of radioactive material the objective is $10^{-5}$ per reactor-year for <u>existing</u> plant and $10^{-6}$ per reactor-year for <u>future</u> plants. In the USA, acceptance criteria for addressing changes in the design or operation of a plant that would lead to a change in the risk (CDF and LERF) are as follows:*

- *Changes that lead to a reduction in the risk would normally be allowed;*
- *Changes that lead to small increase in the risk ($<10^{-6}$ to $10^{-5}$ per reactor-year for CDF and $10^{-7}$ per reactor-year for LERF would normally be allowed  unless the overall risk is high ($>10^{-4}$ per reactor-year for CDF or $> 10^{-5}$ per reactor-year for LERF) in which case the focus would need to be on finding ways to reduce the risk;*
- *Changes that lead to a moderate increase in the risk ( in the range $10^{-6}$ to $10^{-5}$ per reactor-year for CDF or $10^{-7}$ to $10^{-6}$ per reactor-year for LERF) would normally be allowed only if it can be shown that the overall risk is small (that is CDF $< 10^{-4}$ to $10^{-5}$ per reactor-year and LERF $< 10^{-5}$ per reactor-year for CDF);*
- *Changes that would lead to a large increase in the risk ($>10^{-5}$ per reactor-year for CDF or $10^{-6}$ per reactor-year for LERF) would not be allowed.*

*These Guidelines are intended for comparison with the results obtained by using a full scope Level 2 PSA to determine the change in the CDF or LERF for the proposed change in the design or operation of the plant.*

---

**Fig. 14.5**  Acceptance probabilistic criteria reproduced from IAEA-TECDOC [13]

The IAEA document in collaboration with USNRC and other organizations where risk-informed approach was initiated brings out the technical documents [13] for implementing this approach in member states. This section provides the salient features of the integrated decision-making process proposed by IAEA, which combines the insights from the deterministic approach and probabilistic analysis with other requirements, viz. legal, regulatory, cost/benefit, in making risk-informed decisions.

It may be noted that the phrases "existing plants" and "future plants" refer to the plant operating in the year when the NUREG and the USNRC Guides were written. If this is so, then the applicability of these criteria should be seen in this light. The criteria shown in Fig. 14.5 in the bullet have been derived from the risk metrics of USNRC metrics shown in Fig. 14.4. These criteria are vital because they enable application of PRA insights into risk-informed decision-making.

The following discussion provides a brief commentary and salient observations on the elements of IAEA's risk-informed decision approach shown in Fig. 14.6:

- *Issues to Be Considered*: In the regulatory decision-making framework, there are two major agencies/stakeholders, viz. the licensee or nuclear plant operator and the regulatory body that makes the final decision. The issue or the problem and available options/alternatives that need to be considered are defined by the

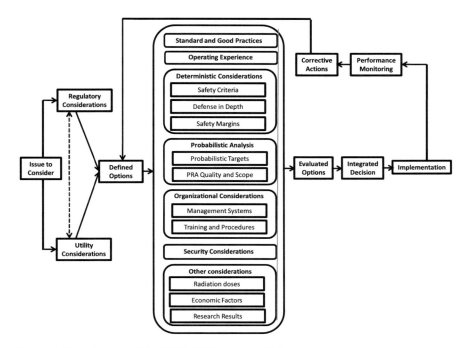

**Fig. 14.6**  Key elements of the IAEA IRIDM process [14]

licensee for reference to the regulatory body in the form of a proposal. This issue is subjected to regulatory as well as utility considerations.

- *Regulatory Considerations*: The primary line of regulatory considerations deals with identification of risk issues. Other considerations such as the adequacy/ suitability of the proposal form added steps.
- *Utility Considerations*: Utility considerations include argument on effect of this proposal on safety and the benefits that can be achieved either in terms of safety improvement or cost/benefit analysis.
- *Regulatory and Utility Considerations*: These considerations together enable formulating options that are available to resolve the issue.

In the following steps, the options are subjected to evaluations against standards, criteria, and organizational and other considerations:

1. All standards and good practices as applicable for a given option or alternative should be followed. For example, national and international standards need to be followed. One of the important documents, i.e., plant technical specifications, needs to be strictly followed. In case the proposal is concerning relaxation on plant technical specification, then this should be seen that basic safety standards of the plant are not compromised.
2. *Operating Experience*: The evaluation of options in light of past operating experience on the facility being considered and other facilities is carried out. In this context, a good database on regulatory decisions at the facility, national, and international level helps in evaluation of operational aspects of the options. IAEA-TECDOC further details utilization of operating experience feedback in safety matters.
3. *Deterministic Considerations*: Each option is evaluated for maintenance of defense in depth and safety margin. Further considerations of criteria such as redundancy, diversity, and single-failure criteria form an integral part of the deterministic considerations. In case the options require reduction of safety margins, i.e., requiring less conservatism than stipulated in deterministic approach, risk evaluation should be done to evaluate the feasibility of the options.
4. *Probabilistic Analysis*: Probabilistic risk analysis forms the cornerstone of risk-informed applications. However, for implementation of the risk-informed approach, probabilistic criteria are required for judging the gap area in risk by comparing the results of PRA with the set criteria as discussed in previous section. The scope of the PRA study and quality criteria needs to be evaluated using the available national or international standards, such as IAEA or ASME standards.
5. *Organizational Considerations*: The experience in implementation of risk-informed decisions is that it requires motivation and consensus at the national/regulatory level. Availability of the trained human resources particularly at the executive level is important to appreciate the potential role of the risk-informed approach toward addressing safety issues while keeping plants

availability interests in place. An organizational setup for lines of authority, communications, and documentation maintenance forms crucial steps.

6. *Security Considerations*: This is a new element of the risk-informed approach of IAEA compared to other parallel approaches. PRA studies have been important for not only safety but also security considerations as it evaluates potential reactor accidents. It is interesting to note that the IAEA has pioneered development of guidelines on security considerations through its Technical Safety Series [15]. For details, the document can be referred.

## 14.8 Final Remarks and Conclusions

This chapter has been developed considering the prevalent definition of risk-informed decision, viz. decisions where PRA input forms just one input along with other deterministic indicators. If we look at the core idea of IRBE where it is being proposed that the notion of risk can be qualitative based on deterministic considerations and quantitative from probabilistic considerations. In short, IRBE argues that risk component does not come only from probabilistic approach but also from deterministic approach. Hence, the IRBE approach would replace "risk" without going into the debate of probabilistic or deterministic. However, all through this chapter the authors have used the term PRA and deterministic approach considering the existing definition (PRA is just one among other input) for two reasons: (a) This chapter provides an overview of the state of the art in existing risk-informed approaches, where probabilistic and deterministic have been considered as two explicit entity, and (b) even if we have the existing notion of "risk-informed," it is vital that the readers get to know the state of the art in risk-informed applications.

This chapter reviews the technological, procedural, documentation-related, administrative, and even ethical aspects of risk-informed decision-making. It has been observed that risk-informed approach in decision-making is actively being considered for nuclear power plant applications. The factors that led to the growth of risk-informed decision-making are efforts made by international organizations (IAEA, NEA) and national regulatory bodies, like USNRC, NKS, STUK, and many other regulatory bodies who encourage use of probabilistic risk assessment in support of decisions. It can be argued, based on the reviews, that implementation of risk-informed approach in any country can start with risk-informed ISI as this is one application that is being considered for not only nuclear plants, but chemical, process industries.

Even though the risk-informed decision-making is increasingly being used, there are some limitations which are coming in the way of its accelerated growth. These issues can be summarized as follows: (a) For implementation of risk-informed approach, a country-specific policy is required that may encourage the regulatory and utility organization to draft guidelines on the subject, (b) clearly defined goals and criteria including ways to represent uncertainty in results of the study is a must, (c) regulatory framework to qualify a PRA study for developing risk-informed application as in the absence of quality criteria/attributes one not sure, whether to use a PRA to build a risk-informed application, (d) administrative provision that provides for a line of communication and authority to review, (e) scope and objective of deliberative framework to address risk-informed proposals, and (f) ethical framework that provides incentive for right decisions; for example, finally the public safety and interest in long as well as short run should form the basis of decision.

# References

1. National Aeronautical Space Administration, *Risk-Informed Decision Making Handbook, NASA/SP-2010-576* (NASA, 2010)
2. National Aeronautical and Space Administration, *NASA Risk-Informed Decision Making Handbook, NASA/SP-2010/Ver.1.0* (NASA, Washington DC, 2010)
3. International Atomic Energy Agency, *Basic Safety Principles for Nuclear Power Plants-INSAG-12* (IAEA, Vienna, 1999)
4. International Atomic Energy Agency, *Defence in Depth in Nuclear Safety—INSAG-10* (IAEA, Vienna, 1996)
5. J. Fulop, *Introduction to Decision Making* (Computer and Automation Institute, Hungarian Institute, Hungarian Academy of Science, Hungary, BDEI-3 workshop Washington, 2005-citeseerx.ist.psu.edu (INSAG-25)
6. International Atomic Energy Agency, *Determining the Quality of Probabilistic Safety Assessment (PSA) for Applications in Nuclear Power Plants, IAEA-TECDOC-1511* (IAEA, Vienna, 2006)
7. American Society for Mechanical Engineers, *Standard for Probabilistic Risk Assessment for Nuclear Power Plant Applications, ASME-RA-S-2002* (ASME, 2002)
8. United States Nuclear Regulatory Commission, *An Approach for Determining Technical Adequacy of Probabilistic Risk Assessment Results for Risk-Informed Activities* (USNRC, Washington DC, 2002)
9. International Atomic Energy Agency, *Safety Assessment for Facilities and Activities, GSR-Part 4* (IAEA, Vienna, 2009)
10. International Atomic Energy Agency, *Development and Application of Level 1 Probabilistic Safety Assessment for Nuclear Power Plants, SSG-3* (IAEA, Vienna, 2010)
11. International Atomic Energy Agency, *Development and Application of Level 2 Probabilistic Safety Assessment for Nuclear Power Plants, SSG-4* (IAEA, Vienna, 2010)
12. International Atomic Energy Agency, *Risk Management: A Tool for Improving Nuclear Power Plant Performance, IAEA-TECDOC-1209* (IAEA, Vienna, 2001)
13. International Atomic Energy Agency, *Risk Informed Regulation of Nuclear Facilities: Overview of the Current Status, IAEA-TECDOC-1436* (IAEA, Vienna, 2005)
14. International Atomic Energy Agency, *A Framework for an Integrated Risk Informed Decision Making Process* (IAEA, Vienna, 2011)
15. International Atomic Energy Agency, *Risk Informed Approach for Nuclear Security Measures for Nuclear and Other Radioactive Material Out of Regulatory Control, Security Series 24G* (IAEA, Vienna, 2015)
16. United States Nuclear Regulatory Commission, *Use of Probabilistic Risk Assessment Methods in Activities: Final Policy Statement, 60-FR-42622* (USNRC, 1995)

17. United States Nuclear Regulatory Commission, *An Approach for Using Probabilistic Risk Assessment in Risk-Informed Decisions on Plant Specific Changes to Licensing Basis, Regulatory Guide 1.174 revision 1* (USNRC, Washington DC, 2002)
18. United States Nuclear Regulatory Commission, *Use of Probabilistic Risk Assessment in Plant Specific, Risk-Informed Decision Making: General Guidelines—Chapter 19 of the Standard Review Plan, NUREG-0800* (USNRC, Washington DC, 1998)
19. United States Nuclear Regulatory Commission, *An Approach for Using Probabilistic Risk Assessment in Risk-Informed Decisions on In-Service Testing, Regulatory Guide 1.175* (USNRC, Washington DC, 2002)
20. United States Nuclear Regulatory Commission, *An Approach for Using Probabilistic Risk Assessment in Risk-Informed Decisions on Graded Quality Assurance, Regulatory Guide 1.176* (USNRC, Washington DC, 1998)
21. United States Nuclear Regulatory Commission, *An Approach for Using Probabilistic Risk Assessment in Risk-Informed Decision Making: Technical Specification, Regulatory Guide 1.177* (USNRC, Washington DC, 1998)
22. United States Nuclear Regulatory Commission, *An Approach for Using Plant-Specific Risk-Informed Decision Making for In-Service Inspection of Piping, Regulatory guide 1.178* (USNRC, Washington DC, 2003)
23. United States Nuclear Regulatory Commission, *An Approach for Determining the Technical Adequacy of PRA Results for Risk-Informed Activities, Draft Regulatory Guide DG-1122* (USNRC, Washington DC, 2002)
24. United States Nuclear Regulatory Commissions, *Guidance on the Treatment of Uncertainties Associated with PRAs in Risk-Informed Decision Making, NUREG-1855*, vol. I (USNRC, Washington DC, 2009)
25. National Aeronautical and Space Administration, *Probabilistic Risk Assessment Procedures Guide for NASA Managers and Practitioners* (Washington DC, 2002)
26. NPR 8000.4A, NASA Procedural Requirements: Agency Risk Management Procedural Requirements (16 Dec 2008)
27. Nuclear Energy Agency, *OECD/NEA—Nuclear Regulatory Decision Making, 5356* (OECD/NEA, Paris, 2005)
28. International Risk Governance Council, *White Paper on Risk Governance—Towards an Integrated Approach, 7-9, IRGC, Chemin de baleexert Chatelaine* (Geneva, 2006)
29. D. Serbanescu, A.L. Vetere Arellano, *WPI—Risk-Informed Decision Making (RIDM): Proposal Contract No. FP6-036720—Comparison of Approaches to Risk Governance. Sixth Framework Programme: Citizens and Governance in a Knowledge Based Society* (2007)
30. E. Zio, N. Pedroni, *Risk-Informed Decision Making Process, FonCSI-2012-10* (Foundation for Industrial Safety Culture, Toulouse, France, 2012)
31. U. Pulkkinen, K. Simola, *An Expert Panel Approach to Support Risk-Informed Decision Making STUK-YTO-TR-172* (STUK—Radiation and Nuclear Safety Authority, Helsinki, 2000)
32. K. Simola, U. Pulkkinen, *Risk Informed Decision Making—A Pre-study, NKS-93* (NKS—Nordic Nuclear Safety Research, Denmark, 2004). ISBN 87-7893-151-7
33. G. Ersdal, T. Aven, Risk Informed Decision-Making and Its Ethical Basis. Reliab Eng Syst Saf **93**, 197–205 (2008)

# Chapter 15
# Risk-Based/Risk-Informed Applications

*Any sufficiently advanced technology is indistinguishable from Magic.*

Aurthue C. Clark, Picturequotes.com

## 15.1 Introduction

This chapter presents the developmental work related to application of risk-based/risk-informed methodology to various aspects of design and operation of complex safety-critical engineering systems by discussing case studies on nuclear power plants (NPPs). There are efforts at the international level to identify areas for integrated risk management during design, operations, and regulation as part of the strategies for competitive NPPs [1]. The objectives, motivation, approach/methodology, and conclusions for each case study are discussed. Some of these applications have been used in support of regulation while others have been performed as part of R&D work to address real-time conditions. The following Sect. 15.2 deals with discussion on the general areas of PRA/risk-based applications, and Section 15.3 presents the experience on development of risk-based/risk-informed applications through specific case studies. Out of seven case studies, six deals with demonstration of risk-based approach to nuclear plant, and one case study deals with aging assessment of an irradiator plant. The point being made here is that the risk-based approach can suitably be adapted for other areas, such as process and chemical plants, aviation and space, and rail and transport.

## 15.2 Risk-Informed/Risk-Based Application Areas

The factors that enable risk-based/risk-informed applications are the development of PRA methodology and a general acceptance of this tool to address real-time scenarios; the availability of international documents/standards to evaluate the quality of given applications; and the availability of regulatory frameworks to check

© Springer Nature Singapore Pte Ltd. 2018
P. V. Varde and M. G. Pecht, *Risk-Based Engineering*, Springer Series in Reliability Engineering, https://doi.org/10.1007/978-981-13-0090-5_15

compliance levels of given applications. For example, ASME/ANS standard and IAEA-TECDOC-1511 provide the needed guidance to fine tune not only PRA quality but also given applications [2, 3]. Implementation of risk-informed applications requires a national regulatory approach, which includes acceptance criteria on reliability and risk target/goals, a policy on the use of PRA results and insights, and detailed guidance on various application categories. For example, while on the one hand, IAEA documents such as IAEA-TECDOC-729 [4], IAEA-TECDOC-1200 [5], and IAEA-INSAG-25 [6] provide general guidance, some documents provide national guidance on specific aspects, such as USNRC Guide 1.174 [7] and USNRC Guide 1.178 [8], on the other.

The major risk-informed/risk-based applications can be summarized as follows [5]:

- Integrated evaluation of plant design, from concept to operation and aging management;
- Surveillance test interval and allowable outage time evaluation as part of technical specification optimization;
- Management of operational configuration using risk monitoring or applications such as operator advisor systems;
- Identification and prioritization of systems, structures, and components (SSCs) that support safety categorization;
- Plant life assessment and life extension studies;
- Evaluation of criteria and assumptions in the design basis report or safety analysis report;
- Evaluation of retrofits or changes in configurations or operational and maintenance policies;
- Accident sequence precursor analysis;
- Planning for postulated emergency scenarios.

The risk-informed/risk-based approach can be employed in many situations including ones where no other approach provides adequate solutions, evaluation of financial risks, prioritization of maintenance programs, and evaluation of plant emergency operating procedures. The following section provides a brief overview of some applications.

## 15.3   Risk-Informed/Risk-Based Case Studies

The applications range from a rather broad objective to demonstrate overall safety of the plant employing the IRBE technique in specific areas, viz. (a) integrated design evaluation, (b) surveillance test interval optimization as part of technical specification evaluation, (c) plant life extension as part of re-licensing, (d) risk monitor in support of design and operational safety management, (e) risk-based in-service inspection, (f) risk-based operator support system development, and

(g) safety margin re-evaluation. This section deals with R&D work performed on development of these applications. The applications presented here were developed over the past ten years, while the approach presented in this book was developed based on the experience of application of PRA methodology in general, and development of specific applications in particular; hence, one should not expect all the elements of the IRBE methodology in the case studies presented here. The other aspects of this chapter are that all the case studies will be briefly presented giving their salient features while for details the link or reference of the published material/report has been given. This is done to limit the size of this chapter at the same time provide essential features of the application.

## 15.3.1   Case Study 1: Integrated Design Evaluation

The evaluation of a plant's design should start at the conception stage or when the design basis report is prepared. The risk-based approach can effectively provide the component configuration at the system level, and it is easier to implement changes during the initial design stages rather than at later stages. The risk-based approach becomes prohibitive in certain situations once the plant is built and begins operations.

This case study deals with the PRA performed for Apsara, a new research reactor being developed at Bhabha Atomic Research Centre (BARC). The PRA model for the plant was developed when the reactor was in the conceptual stage. Even though the major design features were similar to the old Apsara 1 $MW_{th}$ reactor design that operated for over 50 years, the design team started with a fresh design for the new Apsara 2 $MW_{th}$ reactor.

A Level 1 PRA was performed in three stages: (a) the conceptual stage when the broad configuration and requirements were being worked out, (b) the preliminary design stage when the design is being finalized, and (c) the final design stage. The journey from conceptual stage to final design was marked by more design clarity, more information on process variables, and integration of process and safety systems to represent the model of the plant.

The following major steps were followed for the design evaluation:

1. *Project report*: A project report was developed that envisaged that the Apsara risk model will be created in three stages, as mentioned above. The project report consisted of the objective, scope, methodology and organizational structure, documentation requirements, quality assurance program, required resources, selection of PRA team, and project schedule.
2. *Plant familiarization*: The advantage in this project was that the design was an extension of an old plant, and the PRA modeling was also being performed for the same group involved in the operation of the old plant; hence, the information gathering process did not consume many resources.
3. *Creation of plant model*: The Level 1 PRA procedure discussed in Chap. 5 was followed, keeping in view the project objectives. To ensure that the list of

initiating event was as complete as possible, a detailed screening was performed considering (a) the operating experience available from the old plant, (b) experience on other reactor facilities located in the same site [9], (c) generic sources list [10, 11], and (d) initiating event (IE) list on existing other research and power reactors. The Level 1 PRA model of the plant was created as this reactor. Being a low-power, low-enthalpy, and pool-type designs that can be termed an inherently safe design, Level 2 PRA and Level 3 PRA did not form part of the project required. The complete PRA modeling was performed in an M/s Isograph Fault Tree+ environment [12].

4. *Quantification of the model and estimation of CDF*: The failure data used in this project were based on plant-specific experience, generic data [13], and data from other research reactors in BARC [9]. The system unavailability and CDF estimates were evaluated.

5. *Design evaluations*: Evaluations were performed keeping in view the CDF goal of $1 * 10^{-5}$/year [14] for all operational states, all the sources of radiation and internal, as well as external hazards. This demonstrated safety of the plant as the estimated CDF was found to be commensurate with the system unavailability and CDF targets.

6. *Recommendations*: Some of the major recommendations of the analysis included (a) the technical specification conditions should be based on the surveillance test interval and allowable outage time from the risk model of the plant as it reflects both deterministic and probabilistic aspects of the design; (b) the shutoff rod and control rod configurations should be two-fourth rods even though they have independent banks; and (c) the natural circulation valve (NCV) has high safety importance, and a further study should be performed to understand its failure mechanisms.

7. *Actions*: (a) The surveillance test interval and allowable outage time in the deterministic approach were either based on qualitative arguments involving engineering judgment, experience, or vendor recommendations, and they did not have a sound rationale to systematically base on risk model of the plant. The test intervals and allowable outage times used in Level 1 PRA model of the Apsara reactor were based on sound rationale and had a quantitative basis to support the arguments recommended by this study. (b) The shutoff rod configuration suggested by risk analysis evaluation was accepted and implemented in the design. (c) Further, modeling for the NCV was carried out that involved failure model and effect analysis (FMEA) to understand the critical mechanism, and it was identified that the valve pin wear was one of the mechanisms that determine life and reliability of the valve.

8. *Validation and verification*: Wear modeling for the NCV pin was carried out, and the model was validated with the testing data collected on the NCV. It was concluded that the lifetime demand on the NCV, i.e., of the order of 10,000 open and closed operations, will not lead to jamming of the pin due to wear and is expected not to lead to failure in open/closed conditions.

The observations made in steps 7 and 8 bring out the potential of the risk-based approach where it is proposed that deterministic and probabilistic components are integral to achieving higher safety. Experience gained from the PRA modeling for the research reactor resulted in insights not only for Apsara but for two other research reactors, Cirus and Apsara [9].

## 15.3.2  Case Study 2: Surveillance Test Interval Optimization

The operating experience suggests that the surveillance test interval of the emergency core cooling system (ECCS) for Dhruva reactor located at BARC, Mumbai, a 100 MW$_{th}$ research reactor, should be increased from the present three months to four or preferably six months. Periodic testing is integral part of surveillance program for safety systems that remain on standby and poised to start on demand for catering to safety function, like core cooling in emergency mode during a loss of coolant accident. The objective of testing is to precipitate the latent failures which remain dormant in a safety system till a demand is put to start and operate the safety system. The ECCS is required during an unlikely event of a loss of coolant accident (LOCA), which the plant will not experience even during its lifetime.

Although Dhruva's primary coolant system operates at low temperatures and moderate pressures, provisions have been made for a situation arising out of significant loss of the primary coolant due to a double-ended rupture of the system's largest pipe. An engineered safety feature in the form of an ECCS has been provided to mitigate the consequences of significant loss of coolant inventory (Fig. 15.1). The primary coolant system caters to core cooling requirement by operating main coolant system during normal operation while through auxiliary cooling pump (Aux) during unavailability of main coolant pumps. The heat from the coolant is extracted in heat exchangers. The reactor power is controlled by varying moderator level in reactor vessel. The coolant and the moderator systems are designed to be interconnected. The primary coolant and moderator system equipment are located inside the shielded rooms. Floor drains are provided in each room, which are interconnected through a network of piping and lead to the four sequence dump tanks, T-1 to T-4 under gravity flow. Any loss of coolant from the main coolant system will be automatically replenished by heavy water inventory from the moderator system. A common manifold from this tank provides suction to the emergency core cooling pumps P-1 and P-2, which pump back the heavy water to the coolant system through two independent lines. One of the lines injects water into the system at the core outlet through injection valves MV-3198 and MV-3199, while the other line injects water at the core inlet through MV-3101 and MV-3104. Each pipeline is provided with rupture disks, RD-1 in line 1 and Rd-2 in line 2, which act as a passive barrier between the heavy water system and the ECCS during normal condition and facilitates ECCS injection during LOCA scenario. The emergency core cooling system actuates based on demand generated from LOCA signal when the leakage rate is more than 75 m$^3$/h. When the leakage rate from the

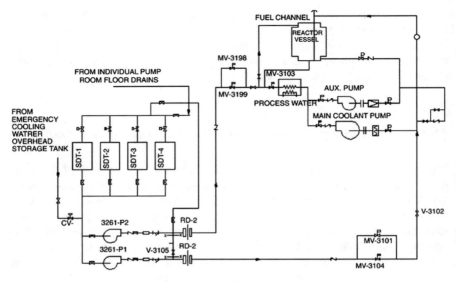

**Fig. 15.1** Emergency core cooling system of Dhruva reactor [16]

main coolant system is <75 m³/h, the coolant system will not void for at least 30 min because the lost inventory is replenished from the reactor vessel heavy water inventory and the auxiliary coolant pumps provide core cooling. Provision has been made to actuate the ECCS manually, if required, even for leakage rates less than 75 m³/h. The system can be maintained in a re-circulating mode to provide long-term cooling. Provisions for manual injection of demineralized light water have also been made for the situation when the leaking out coolant inventory is unable to reach the underground tanks or when the ECCS pumps are not able to pump the coolant back into the core. The system is subjected to quarterly testing to ensure its reliable status [15].

The PRA model of the plant is available, and the CDF of the plant is in the range of $10^{-5}$/reactor-year, which meets the target/CDF goal criteria. The modeling of the ECCS forms part of the PRA model for the LOCA category of initiating events.

The challenge associated with evaluation surveillance test interval is that it requires optimization of the surveillance test interval against multiple objective functions, which include minimization of system unavailability (such that either there is no increase in CDF or even if there is, then it should be in an acceptable range, say, $\leq 1\%$), dose consumption, and aging due to mechanical wear out. The second issue is there is consensus on a well-accepted methodology for aging-related degradation due to testing of standby components. At qualitative level, it can be argued that relatively more frequent testing means enhanced wear and tear of components that accelerate aging. The third issue is related to considerations of ECCS requirements during the shutdown state of the plant because the system is depressurized the decay heat is a fraction of full power. However, the argument against this requirement was that there are maintenance activities performed so any error in maintenance may lead to loss of inventory even though it can be arrested as maintenance team can take recovery action to mitigate the situation. However, owing to these complexities R&D is required to address this problem.

For the purpose of this case study, it was assumed that the factors discussed above may not affect safety in significant way. Hence, optimization was performed ignoring aging-related aspects risk from maintenance activities performed during shutdown and depressurized coolant state of the plant. The salient feature of this study is as follows:

- The objective of this study was to establish a procedure for surveillance test interval optimization for a multi-objective function problem.
- A Level 1+ PRA model, with limited-scope Level 2 and Level 3 PRA considering LOCA scenario with internal events, full-power operations, and reactor core as the source of radiation provided the basic framework for this study.
- This modeling was performed with the multi-objective functions, viz. system-level, plant-level goals, and dose consumption.
- Plant-specific data, at least for the components in the ECCS that have high safety importance, such as injection pumps, valves, and actuation circuits where no redundancy exists, were used.
- Human reliability analysis was performed employing THERP and HCR model while human reliability data were collected from plant-specific records on surveillance program.
- Modeling was performed considering the requirements of ECCS during normal as well as LOCA condition.
- It was assumed that familiarization or training factors remain unaffected if there is a minor change in surveillance program.
- Optimization of the test interval employing an intelligent approach, here a real-valued genetic algorithm (GA), which has provision to address requirements of multi-objective functions and given constraints, was employed.
- Uncertainty and sensitivity analyses were performed for assessing impact of parameters and assumptions.

This application included risk modeling, application of intelligent approach for optimization, human error evaluation for every scenario, validation of assumptions and therefore needed extensive R&D to arrive at some possible solutions. For details of modeling, interested readers can refer a paper "real-valued GA approach to optimization of ECCS test intervals" [17].

### 15.3.3  Case Study 3: Life Extension in Support of Relicensing

This case study deals with life extension studies for ISOMED, a 30-year-old Co-60 irradiation plant in Trombay, Mumbai, India [18]. The irradiator uses a radiation source and facilitates irradiation of products, such as food products for preservation purposes and sterilization of medical products. After a preset limit of irradiation, the product is dispatched for end application/use. Figure 15.2 shows an overview of the ISOMED facility, viz. the Co-60 source frame, its raise and lower chains, and cooling pipes along with conveyor belt that moves the product around the source are located in the irradiation cell. When not in use, the source frame remains lowered in the concrete pit which also provides required shielding for the source frame. The product irradiation occurs with source raise condition out of the pit with on conveyor bent. Figure 15.2 shows (a) the cross section of the cell, depicting Co-60 source frame, source pit, labyrinth for product entry and exit, product boxes and conveyor and ventilation air duct for cell, (b) the line diagram for product movement showing 4 pass of product movement on each side of the Co-60 source frame, and (c) the cross section of the ISOMED cell and associated structures and components.

A risk-informed/risk-based approach was used to evaluate the remaining life of the vital plant SSCs. This is perhaps the first instance of employing PRA modeling for a non-reactor nuclear system. This work needed a formulation of a set of initiating events, which required data from irradiator accidents worldwide, including operating experience and insights, and failure data from plant-specific sources, particularly irradiator-specific systems, such as cobalt source system design and operational data, conveyor belt system movement system, and locking systems and interlocks. The postulated initiating events (PIEs) are as follows:

- *PIEs for plant operation*: normal power supply failure;
- normal plant trip—conveyor stoppage due to loading misalignment outside the cell;
- source solenoid stuck open;
- snapping of wire rope of source frame;
- entanglement of product carrier box with source frame;
- failure of conveyor;
- source not going down due to mechanical failure;
- cell ventilation system failure;
- unauthorized personnel entry into cell through labyrinth;
- unauthorized personnel entry into cell through auxiliary conveyor room door;
- fire in plant.

*PIEs for plant shutdown/startup*

- Person trapped inside the cell;
- Source hydraulic drive failed to cutoff when source has reached top position;
- Failure of source cooling water system when source is positioned in pit.

*External events*

- Flooding inside the plant and cell;
- Seismic consequences.

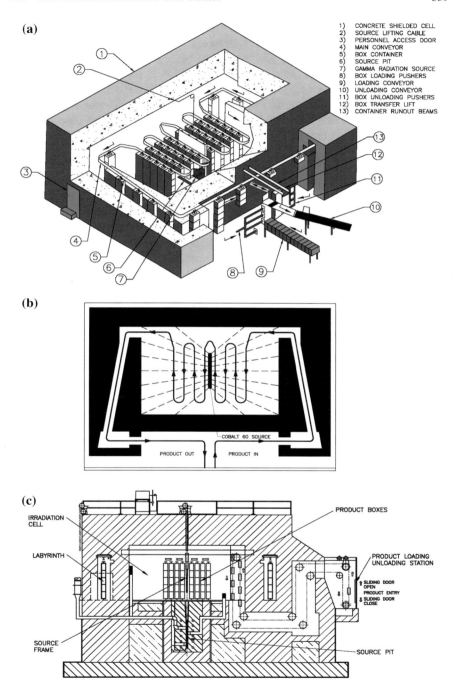

**Fig. 15.2** ISOMED plant **a** cross-sectional view of the irradiator cell showing the irradiator cell location of source in the center and movement of the conveyor in and out of the cell, **b** a simplified sketch showing the product irradiation and entry and exit, **c** cross section of the irradiation cell in ISOMED

The salient features of this approach were PRA modeling ant to get the risk insights of the plant, life testing, and structural testing of the components to assess the remaining life of the plant, and the use of other performance indicators apart from risk. These indicators include dose consumption in the last 30 years, number of occasions and maximum dose received by occupational workers, plant availability and capacity factors, and incident trends (increasing/decreasing/constant). A deterministic procedure was also used to evaluate the potential consequences, such as ozone level in the plant, contribution of human error to the present system, and reduction potential due to implementation of the design changes and operational policies.

The major steps in this methodology are as follows: (1) collection of data and information on design description, operation, and maintenance of the plant; (2) review of existing operational and maintenance policies, operating manuals, emergency procedures, and plant incidents; (3) identification of the PIEs for the plant; (4) identification and prioritization of SSCs in general and safety interlocks in particular, using a probabilistic approach, considered to be critical for the plant safety; (5) development of a probabilistic model for a scenario leading to overexposure of the plant personnel; (6) assessment of the health of critical components and structures in view of aging-related degradation; (7) deterministic analysis for some scenarios identified through engineering judgment and a probabilistic approach; (8) assessment of the plant components categorized as repairable, replaceable, and neither repairable nor replaceable; (9) assessment of the remaining life of a selected category of the components; and (10) prioritizing the critical components/systems/structures that determine the remaining life of the plant (Fig. 15.3).

Major activities identified, prioritized, and performed in support of this life extension project include (a) remaining life prediction of electronic contactors, relays, power, and control cables, (b) requalification of mechanical structures like Co-60 source driving shaft and source frame, (c) root cause analysis of frequent misalignment of source frame, (d) replacement of source cooling system, (e) performance of ventilation system and ozone hazard assessment, and (f) upgradation of electrical power supply systems. While the study was in progress, work on implementing 8-out-of-11 recommendations was completed, and the report that was submitted to regulators had discussed these jobs and the safety improvements.

Risk from the ISOMED plant estimated as likelihood of severe/fatal consequences that have potential for death due to dose of more than 5 Sv is as low as $5.1 * 10^{-6}$/year. This result demonstrates that the plant has adequate safety provision. The International Commission of Radiological Protection (ICRP) has stipulated as 4% per Sv, as cancer death risk to occupational worker. Considering this observation and likelihood estimate for overexposure that was estimated in this study a risk profile has been generated for ISOMED plant for lower dose range. From the graph shown in Fig. 15.4, it can be seen that for probability of cancer-related death for the overexposure case is $\sim 1.0 * 10^{-5}$/year for an overexposure of 0.1 Sv.

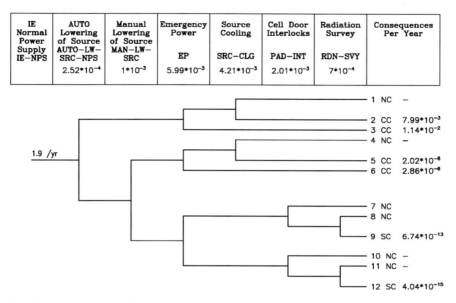

| IE<br>Normal<br>Power<br>Supply<br>IE−NPS | AUTO<br>Lowering<br>of Source<br>AUTO−LW−<br>SRC−NPS | Manual<br>Lowering<br>of Source<br>MAN−LW−<br>SRC | Emergency<br>Power<br><br>EP | Source<br>Cooling<br><br>SRC−CLG | Cell Door<br>Interlocks<br><br>PAD−INT | Radiation<br>Survey<br><br>RDN−SVY | Consequences<br>Per Year |
|---|---|---|---|---|---|---|---|
| | $2.52*10^{-4}$ | $1*10^{-3}$ | $5.99*10^{-3}$ | $4.21*10^{-3}$ | $2.01*10^{-3}$ | $7*10^{-4}$ | |

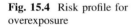

Fig. 15.3 Event tree for off-site power failure initiating event for ISOMAD facility

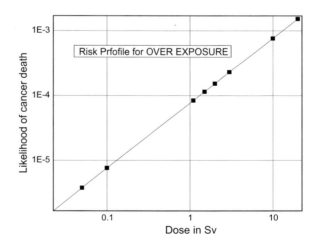

Fig. 15.4 Risk profile for overexposure

This study is unique in the sense that it was a demonstration of a risk-based approach to irradiators. It formed part of a regulatory review for life extension and re-licensing of the plant for the next ten years. Interested reader may refer a paper on "Development of an integrated methodology in support of remaining life assessment of Co-60 based gamma irradiation facilities" by Varde et al. [19] for details this study.

## 15.3.4   Case Study 4: Risk Monitor

A risk monitor is a plant-specific real-time analysis tool used to determine the instantaneous risk based on the actual status of the systems and components [20]. At any instant of time, it provides the risk estimates for the given set of equipment conditions, i.e., it reflects the current plant status. The model used by the risk monitor is based on, and is consistent with, the updated probabilistic safety assessment model for the facility. The design of the risk monitor requires an assessment of the level of compliance to defense in depth for ensuring competitive strategies where safety is not comprised beyond an acceptable level [21]. The first risk monitor was implemented in 1998, since then the deployment of risk monitors in NPP control rooms has grown rapidly to over 110 in 2004 and is expected to reach over 150 soon [22]. In a way, this trend is a testimony to the growing application of risk-based applications in the nuclear power industry worldwide.

This section presents the developmental work performed on risk monitors considering research reactors as the target application. The risk monitor is known as a risk-based operations and maintenance management system (RBOMMS). The developmental work was performed keeping in view of the future major procedural requirements as follows: (a) development of a project document for this task, which provided the project objective, scope, data updating and analysis, evaluation of initiating events (keeping in view the scope, uncertainty and sensitivity analyses, and generation of minimal cut-set equations), and identifying areas of implementation and limitations of the tool; (b) development of software specifications; (c) evaluation of the tool for evaluating defense-in-depth compliance; (d) organizational and administrative arrangement required for implementation of the tool; (e) stages for qualification, e.g., laboratory trials, training; (f) training of the staff; (g) regulatory clearances; and (h) implementation of the tool for training/control room requirements.

For computerization requirements, plant-specific Level 1+ PRA study was updated and modified for developing a risk monitor. One of the major changes was the conversion of all the fixed values of failure probability of standby components into dynamic models by using the failure rate and test intervals for periodically tested components, and failure and mean time to repair (MTTR) for continuously operating components in process systems. This updating enabled evaluation of surveillance test intervals and allowable outage time estimation in a dynamic manner such that instantaneous CDF value can be obtained (Fig. 15.5).

The implementation of risk-based operational management requires a reference plant condition and corresponding CDF where along with median/mean value uncertainty bounds are also forms an input. As shown in Fig. 15.6 for the reference plant, the distribution of CDF obtained employing Monte Carlo analysis provided the reference median CDF and associated 5% (lower bound) and 95% (upper bound). The graphic user interface (GUI) was developed such that apart from the CDF chart, the editor facilitates performance of "what-if" analysis.

The decision metrics are implemented at two levels:

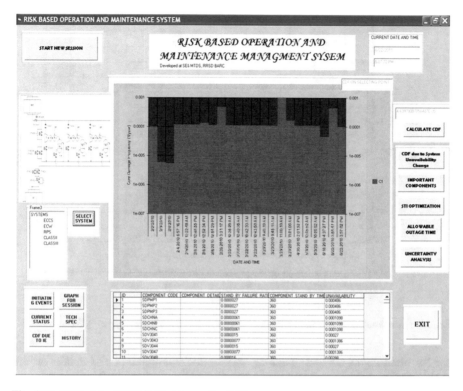

**Fig. 15.5**   GUI of the RBOMMS risk monitor [23]

- No increase in core damage is permitted (in the absence of availability of $\Delta$CDF targets).
- Increase in CDF as permitted by international/regulatory criteria, i.e., $\leq 1\%$ of base CDF or as permitted by the risk metrics.

The GUI of the RBOMMS is shown in Fig. 15.5.

The risk-informed asset management methodology developed making use of the risk monitor mainly involves resource allocation and shutdown scheduling planning. The developed risk monitor has all the important elements for risk-based management, such as CDF calculation, generation of risk profile graph for past configurations, estimation of system unavailability and IE contribution to CDF, importance and uncertainty analysis, store login sessions, comparison of risk-based surveillance test intervals with traditional surveillance test intervals, and scope of technical specifications. This system can also facilitate the shutdown maintenance planning and scheduling. In view of these features, the risk monitor enables plant operators and managers to evaluate the plant risks and problems associated with plant configuration and maintenance activities. Hence, the risk monitor will be of

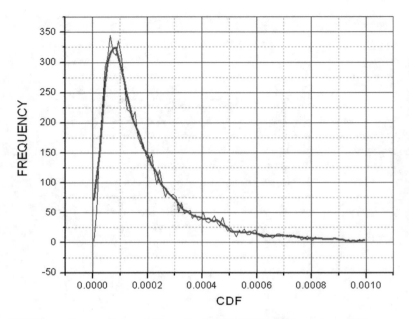

**Fig. 15.6** Uncertainty prediction facilitated by RBOMMS for a safety system [23]

use for plant managers and operating staff in decision-making. The details on this project can be found in a publication by Agarwal and Varde [23].

### 15.3.5   Case Study 5: Risk-Based In-Service Inspection

There is a growing interest in application of risk-based in-service inspection (RB-ISI) in nuclear and other industries to balance the safety and performance of the sustained business model. The phrase "in-service inspection" refers to inspection of SSCs undertaken over the operating lifetime by or on behalf of the operating organization for the purpose of identifying age-related degradation or conditions that, if not addressed, might lead to their failure [20]. The acronyms "RB-ISI" or "RI-ISI" are used interchangeably; nevertheless, the methodology is the same, namely the risk model of the plant forms the core of the inspection program. Risk-based inspection involves the planning of an inspection on the basis of the information obtained from a risk analysis of the equipment. RB-ISI was first introduced by the US Nuclear Regulatory Commission to emphasize the link but not a direct correlation between risk and inspection. If risk-based inspection is understood to be inspection planned based on information obtained about the risk, then the two terms are synonymous [24]. The field of RB-ISI can be argued to be matured than any other RB applications based on the observation that there are standards and guides or technical documents available to structure and develop qualification criteria for an RB-ISI program, for example, Electric Power Research Institute (EPRI)'s methodology [25], Westinghouse Owners Group's Methodology [26], USNRC Regulatory Guide 1.178 [27], the American Petroleum Institute Guide API 581 [28], IAEA Nuclear Energy Series No. NP-T-3.1 [29], and the European Commission's framework document [30] on implementation of RB/RI methodology.

The above discussions pertain to the application of an RB-ISI program for NPPs. However, there is no reported literature available on application of RB-ISI to research reactors perhaps because research reactors have a power level in watts to a couple of hundreds in the $MW_{th}$ range. So it is conceivable that a lower-power and low-enthalpy research reactor may not require ISI; however, when the power level is higher, then the question arises whether ISI is required or not. To this end, item 1.8 of IAEA standard SSR-3 [31] clarifies that "Research reactors with power levels in excess of several tens of megawatts, fast reactors and reactors using experimental devices such as high pressure and temperature loops and cold and hot neutron sources may require the application of supplementary measures or even the application of requirements for power reactors and or additional measures."

This case study was performed for Dhruva, a $100\ MW_{th}$ research reactor in BARC, Mumbai, India [9]. The IAEA guidelines in SSR-3 was the motivation of having an ISI program for this reactor, while the case study discussed here was the subsequent effort that dealt with development of risk-based ISI for the Dhruva reactor. So based on the available literature, it can be argued that this application of risk-based ISI is perhaps the first case for a research reactor.

During the operating life of a nuclear reactor, its components are exposed to influences whose individual or combined effect cannot be fully predicted for the operating life of the plant with the accuracy level desirable for nuclear safety. The most important influences are stress, temperature, irradiation, hydrogen absorption, corrosive attack, vibration, and fretting, which depend upon exposure time and operating history. These influences may result in changes in material properties such as embrittlement, fatigue, formation and/or growth of defects/flaws, and aging.

ISI involves periodic examination of components of a nuclear reactor during its lifetime to assess the defects. Current ISI programs are based on past experience and engineering judgment through deterministic analysis. Over conservatism in these methodologies can result in significant expenditure and reactor outage time. In such cases, implementation of RB-ISI provides the advantages that include, and the ISI program is primarily based on likelihood and consequences which are quantified parameters (unlike in traditional approach where these parameters are qualitative and tends to be arbitrary in nature), and this is a significant aspect as quantification enables identification and prioritization of SSCs based on its safety importance. This results into a more effective inspection strategy by focusing on inspection efforts on safety significant SSCs and expected to reduce overall inspection efforts and resources. Even while ensuring higher level of safety the major advantage includes, e.g., enhanced plant performance, lower, lower radiation exposure budget for ISI program.

As a pilot study, an RB-ISI program was developed for Dhruva research reactor. The three loops of the primary coolant system recirculate heavy water and remove the heat generated during nuclear fission reaction. An ISI document that was developed based on traditional deterministic criteria and operation and maintenance insight formed the reference or base document for this RB-ISI development. This section presents the RI-ISI methodology applied on these three loops of heavy water and associated equipment.

In most of countries, RB-ISI is being carried out by adopting the methodology developed by either Westinghouse Owners Group (WOG) or Electric Power Research Institute (EPRI) [5]. WOG proposed important measures for RB-ISI categorization, which were reported in ASME code case 577 [16]. RAW and RRW are two important measures normally employed for associating safety significance of SSCs considering the failure frequency of the SSC and the impact of the failure of the SSC on the overall core damage frequency. EPRI proposed applying risk metrics to categorize the different pipe segments into high, medium, and low importance based on the degradation mechanism and consequences of its failure. In the risk metrics, based on the plant service data, a basis has been established for ranking pipe segment rupture potential as high, medium, and low simply by understanding the type of degradation mechanism present. Consequence can be quantified through the estimation of conditional core damage probability (CCDP) and then categorized into three or four levels.

**Development of RB-ISI for Dhruva**

As a proactive approach in-service inspection program was developed for Dhruva reactor in 1996. This program was based on deterministic approach where the notion of likelihood and consequence was qualitative in nature. The primary coolant system and secondary coolant and emergency core cooling system formed the scope of this document. Keeping in view the success of RI/RB-ISI for nuclear power plant R&D work was initiated for developing a RB-ISI program for this 100 $MW_{th}$ research reactor.

Major motivation was to (a) optimize the existing ISI program employing prioritization approach based on safety, (b) use the insights from the two campaigns such that the ISI resources are utilized in an effective manner, (c) use the safety insights available from the Level 1+ PRA document to structure the program based on quantitative estimates of likelihood and consequence, and (d) to critically evaluate the safety margins and use available margins without diluting the defense-in-depth provisions.

Figure 15.7 shows a flowchart that depicts the approach for risk-based approach for the research reactor Dhruva. The major steps in the proposed risk-based ISI are in line with EPRI framework [32], like starting from defining the scope and objective of the RB-ISI program identifying the system and candidate pipe segment and performance of failure mode, mechanism, and effect analysis (FMMEA) for each segment. The insight from this analysis forms input for degradation assessment and finally for evaluating the likelihood of failure based on plant-specific data. Often due to limited data, it is required to update the plant-specific data with generic failure data or some generic procedure. The Thomas model is used for estimating the failure frequency estimation [33]. Thomas model provides the estimates the frequency of piping failure by modifying the reference frequency by considering the physical parameters of the pipe like diameter, thickness, number of weld and factors like service period. The approach for estimating the frequency of pipe failure employing Thomas model is given in Annexure 1 of the chapter.

In this project, Bayesian updating was employed to get the final estimate of likelihood of the pipe segment failure [34]. These estimates also form input for the Level 1 PRA study to revise/update the PRA model. This updated model enables evaluation of conditional core damage probability (CCDP) estimates. The consequence evaluation is further modified considering the plant-specific factors, e.g., postulated leal size, leaky zone area, pressure in the given segment, etc. Often it is required to give credit to factors like transient time for the affected system, alternate mode of operation that favors significant reduction in leakage.

A two-dimensional risk metrics are developed based on likelihood and consequences as shown in Fig. 15.8. For creating the risk metrics, the leak/rupture frequency is categorized to high, medium, low and very low. Similarly, the consequences are also created at four levels as high, medium, low, and negligible. The likelihood and consequences so obtained are binned into one of the four safety significant categories, viz. high, medium, low, and negligible. The resultant risk metrics are shown in Fig. 15.8.

The unique feature of this approach is that the proposed risk metrics, shown in Fig. 15.8, can again be modified based on the availability and reliability of mitigating features. The strength of mitigating features can be assessed based on the severity of the accident scenario and the plant characteristics. For example, research reactors are low-pressure and low-enthalpy systems by thermal-hydraulic characteristics, and further pressure and temperature can be brought down by tripping the

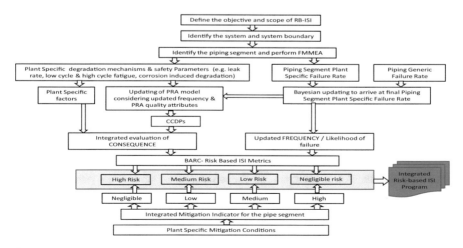

**Fig. 15.7** Flowchart depicting risk-based in-service inspection as part of IRBE approach

| Consequence | Very Low / Insignificant | Low | Medium | High |
|---|---|---|---|---|
| High (>10⁻⁴) → | Low | Medium | High -3 | High -1 |
| Medium | Low | Low | Medium | High - 2 |
| Low | Low | Low | Low | Medium |
| Very Low | Very low | Low | | |

Likelihood frequency yr

**Fig. 15.8** Risk metrics

coolant circulations which provides scope for readjusting the safety significant categories. Hence, the availability of mitigating features in terms of hardware or human action or equipment location consideration may enable the basic metrics shown in figure. In case, the organization involved in performing RB-ISI does not want to consider the strong dependency of mitigation features, and the risk metrics proposed in Fig. 15.8 are final.

The major insights obtained from the development work are (a) the quantitative indicators of likelihood and consequences provided an effective mechanism for identifying safety significant piping segments and support structures, (b) RB approach provided an overall mechanism for prioritization of pipe segments in all the category, (c) this R&D enabled development of a rule-based approach for characterization and classification criteria linking with plant-specific conditions, and the mitigation metrics provided an effective mechanism to account for plant-specific conditions to re-allocate or shift pipe segments from one category to another category based on rationales.

### 15.3.6   Case Study 6: Risk-Based Operator Support System

The available literature shows that human error been one of the major contributors to the accidents. An assessment of the contribution of human performance in nuclear plants to risk operational events was conducted and shows that human error is significant contributor to accidents [35, 36]. Extensive work is being done to reduce the chances of human error through many techniques and management approaches, for example, designing the systems to reduce operator intervention, improving ergonomics, using the insights from risk assessment studies, validating psychological factors in the operational environment, reshaping training programs based on the insights particularly during accident conditions, balancing the approach to automation, and qualifying staff by multi-tier training processes, including physical health assessment of the operating staff [37].

The design of the advanced reactor system appears to be a promising option. For example, in advanced nuclear reactors, operator intervention during emergency conditions is delayed for an extended period ranging from hours to a couple of days by incorporating safety provisions, like provision of a passive shutdown core cooling system during and extended station blackout scenario [38]. However, these provisions require robust systems, for example, for the case of passive shutdown cooling system, it is required to demonstrate that there is an adequate assurance on reliability of shutdown cooling considering the uncertainty associated with the thermal-hydraulic scenario and modeling parameters. The validation of this option poses a challenge in terms of acceptable uncertainty. Therefore, even if the results from advanced systems are optimistic, the issue in the current context is that existing plants and new plants being designed may not have features such as passive and inherently safe designs. Hence, the existing system and new systems without passive features, by and large, require tools and methods that reduce the chances of human error. While the work on other factors, as mentioned above, is being performed, the need for an operator support system is one option where R&D efforts need to be focused.

Deployment of an operator support system for normal and emergency conditions is one of the options to reduce human error and improve safety. The issue with operational conditions is that the advice to the operator should be timely, for example, during an emergency scenario, the diagnostics and evaluation of the plant conditions should be performed in such a manner that there is an adequate time window available for the operator to bring the plant back to a normal condition [39].

However, the plant transient from normal condition to deviation from normal condition is usually associated with flooding of signals, both audio and visual, such that it might pose a challenge to the operator to perform the correct diagnosis and later perform the required action. For example, the loss of off-site power scenario leads to tripping of the reactor, followed by primary pumps, and other electrical loads that flood the control room with audio/visual signals which need to be understood by the operating staff, and such scenarios might overload the operator's cognition, particularly if the scenario is infrequent. Although operators are trained to handle such scenarios, there are scenarios, such as LOCA, where it will be useful if an operator advisory system is available that presents the operator with the evolving status of the plant and further advises on the possible line of actions to restore the plant condition to normal status. This is possibly the reason for the growing interest in developing operator support systems for plant emergency conditions [40].

This section presents the R&D efforts to develop a risk-based operator support system [41]. The system is referred to as a risk-based system because the knowledge representation in this system employs insight and information derived from the PRA model of the plant. The objective is to address a control room scenario where the operational environment is divided into normal operation, deviation from normal operation, and off-normal/accident conditions. Particularly for accident conditions, the operator support system should be capable of handling a scenario that starts with an initiating event in the form of transient and later requires diagnosis followed by actions for system stabilization. For example, a loss of off-site power (LOOP) initiating event requires reactor shutdown because the coolant pump cannot continue operating. The evolving scenario requires automatic starting of reactor shutdown cooling to and resumption of power supply from emergency source (Class III power), i.e., on-site diesel generator set such that essential loads in the plant start and cater to safety functions of the plant. However, in case one or more diesel generators do not start or not adequate to cater to the essential power supply requirements then the management of power supply from station Class I (battery) and Class II (uninterrupted power source) provides the limited coping time to avoid the plant entering station blackout scenario.

Such complex situations put cognitive load on the operator for timely assessments of overall status of the plant systems and for taking necessary action such that plant safety is ensured. Here the role of a computerized operator support system where the signals on status of the equipment, and their status is processed and diagnosis is performed employing intelligent tools to advise the operator such that possibility of human error is reduced or eliminated. Hence, the operator support systems should exhibit intelligent processing of knowledge, such as having memory, parallel processing for transient conditions, diagnostic capabilities in terms of "if-then" rules, and finally the capability for inference with required confidence.

The operator advisory system discussed here is referred to as operator advisory system (OPAD) and has an architecture where the transient identification is performed by an artificial neural network (ANN) system, whereas the rule-based system conducts the plant deviation/failure diagnostics. The ANN has been trained for 14 plant conditions, which include LOCA, LOOP, and further degradation to station blackout and other scenarios that pose a threat to safety function such as shutdown cooling and protection system failure. Each scenario has been constructed using 42 plant parameters that include reactor trip, alarm, and analog process parameters. Each plant condition has been represented by a unique vector. Each vector has either binary, i.e., 0 (signal "normal") or 1 (parameter registered), and the analog parameters are represented by a normalized value of the parameter (current value of parameter divided by range of the instrument) (Fig 15.9).

The ANN was trained for these 14 plant conditions. For identification of each state a unique $14 \times 14$ matrix was presented at the output layer of the ANN. Each row identifies one reactor state. This matrix facilitates accommodation of a new reactor state for which the ANN was not trained. For example, row 1 is associated with the first reactor state, i.e., the reactor status is "normal." In case there is one more condition identified that corresponds to normal operation that can be accommodated in row 1 with one addition "1" in any of the other positions, other than the first column (having "1") which is unique to state-1. This reference output matrix has "1" moving diagonally when we move from row 1 to 14.

Once the reactor conditions are identified, the ANN passes control to a knowledge-based module to process the input using if-then logic. These rules have been derived from the PRA model of the plant (Fig. 15.10).

The knowledge base representation in the expert shell is structured by two entities—frames and rules. The frames provide the semantic arguments for rule-based systems. The system was tested by performing recall tests first with 14 transients and subsequent diagnostics, and it was found that it identifies each transient with the target root mean square (RMS) error. Further, the recall tests were performed with new or unlearned patterns designed to mimic the patterns the system may face during real-time conditions (e.g., the patterns with 20% noise, missing nodes in the recall pattern, and fuzzy data), and it was found that the uncertainty was reflected in the final identification, as happens with normal decisions made by an operator in the control room environment, but the final results met the test objective. Varde et al. [41] present the details of this R&D and the results of the recall tests. The risk monitor along with the operator support system is expected to complement each other in support of operational decisions during normal operations as well as during emergency conditions.

| ST-1 | 1.0 0.0 0.0 0.0 0.0 0.0 0.0 0.0 0.0 0.0 0.0 0.0 0.0 0.0 |
|---|---|
| ST-2 | 0.0 1.0 0.0 0.0 0.0 0.0 0.0 0.0 0.0 0.0 0.0 0.0 0.0 0.0 |
| ST-3 | 0.0 0.0 1.0 0.0 0.0 0.0 0.0 0.0 0.0 0.0 0.0 0.0 0.0 0.0 |
| ST-4 | 0.0 0.0 0.0 1.0 0.0 0.0 0.0 0.0 0.0 0.0 0.0 0.0 0.0 0.0 |
| ST-5 | 0.0 0.0 0.0 0.0 1.0 0.0 0.0 0.0 0.0 0.0 0.0 0.0 0.0 0.0 |
| ST-6 | 0.0 0.0 0.0 0.0 0.0 1.0 0.0 0.0 0.0 0.0 0.0 0.0 0.0 0.0 |
| ST-7 | 0.0 0.0 0.0 0.0 0.0 0.0 1.0 0.0 0.0 0.0 0.0 0.0 0.0 0.0 |
| ST-8 | 0.0 0.0 0.0 0.0 0.0 0.0 0.0 1.0 0.0 0.0 0.0 0.0 0.0 0.0 |
| ST-9 | 0.0 0.0 0.0 0.0 0.0 0.0 0.0 0.0 1.0 0.0 0.0 0.0 0.0 0.0 |
| ST-10 | 0.0 0.0 0.0 0.0 0.0 0.0 0.0 0.0 0.0 1.0 0.0 0.0 0.0 0.0 |
| ST-11 | 0.0 0.0 0.0 0.0 0.0 0.0 0.0 0.0 0.0 0.0 1.0 0.0 0.0 0.0 |
| ST-12 | 0.0 0.0 0.0 0.0 0.0 0.0 0.0 0.0 0.0 0.0 0.0 1.0 0.0 0.0 |
| ST-13 | 0.0 0.0 0.0 0.0 0.0 0.0 0.0 0.0 0.0 0.0 0.0 0.0 1.0 0.0 |
| ST-14 | 0.0 0.0 0.0 0.0 0.0 0.0 0.0 0.0 0.0 0.0 0.0 0.0 0.0 1.0 |

**Fig. 15.9** Reactor state/transient identification matrix

**Fig. 15.10** Procedure to
extract knowledge from a
fault tree is depicted by a
simplified fault tree followed
by if-then rules derived from
the fault tree structure. The
event tree approach was used
for event progression
representation

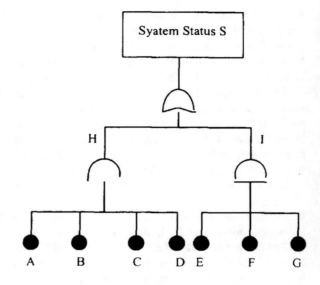

Rule representation in the OPAD for this fault
tree is as follows

| # r1 | # r2 |
|------|------|
| IF | IF |
| Condition A  OR | Condition E  AND |
| Condition B  OR | Condition F  AND |
| Condition C  OR | Condition G |
| Condition D | THEN |
| THEN | Condition I |
| Condition H | |

# r3
IF
Condition H  OR
Condition I
THEN
Condition S

### 15.3.7   Case Study 7: Risk-Based Approach to Re-assess
###              the Enhancement in Safety Margin

This case study is basically an insight while performing Level 1 PSA for the
reference reactor. In this reactor, the gravity insertion of 9 shutoff rods along with
dumping of moderator on demand trips/shuts down the reactor. The shutoff rod
insertion is fast while the dumping action is relatively slow. Hence, the

deterministic approach does not take credit for availability of moderator dumping achieved by opening of six valves (three dumps and three control valves).

While performing and even after the PSA project completed, as part of regulatory review, there were questions whether the credit for a slow-acting shutdown system should be taken for demonstrating safety of the plant. Objective of this case study is to demonstrate of the safety of the reactor crediting moderator dumping as secondary shutdown system for loss of off-site power (LOOP) failure, loss of regulating accidents (LORA), and minor loss of coolant accident (Min-LOCA).

The reactor power regulation is achieved by varying the moderator level with constant inflow and variable outflow. The arrangement for achieving this level control is shown in Fig. 15.11. Three independent channels of instrumentation are provided for every reactor/process parameter. When any parameter exceeds its preset value, reactor trip signal is generated on two-out-of-three co-incidence logic. The instrumentation with adequate redundancy covers monitoring and recording of all important reactor/process parameters and provides audiovisual alarm annunciations to make plant personnel aware about the operational health of each system/subsystem and its various components at all times. Primary shutdown system of the reactor is composed of nine cadmium shutoff rods. Fast shutdown of the reactor is achieved by gravity insertion of all the nine rods into the core. Minimum reactivity worth of all nine rods is about 90 mk. The backup shutdown system (BSS) is

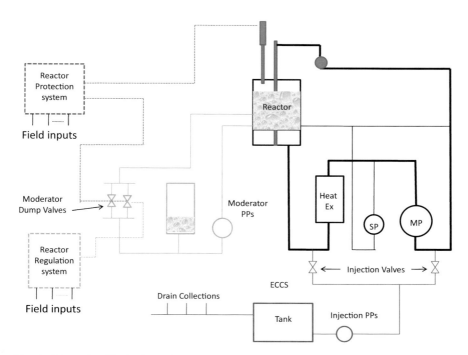

**Fig. 15.11** A simplified schematic showing reactor and coolant system along with protection system (with shutoff rod and moderator dumping system) and emergency core system

dumping of the heavy water moderator from the reactor vessel to the dump tank in order to bring down moderator level in the reactor vessel to a predetermined level. Dumping of the heavy water moderator is achieved by opening three dump and three control valves. Dump valves, which remain fully closed during reactor operation, are air to close and spring to open, thus providing fail-safe feature. The dump valves are provided in parallel with the control valves, which control moderator outflow from the vessel. All the six valves fly open on a completed reactor trip signal.

### 15.3.7.1   Risk Targets/Goals

The risk metrics chosen for this study have been given in Table 15.1. As can be seen, the target goals at plant level as core damage frequency (CDF) and large early release frequency (LERF) and risk to the member of public are given along with uncertainty bounds. These targets and goals are in line with internationally accepted practices. Further, the system-level targets for shutdown system, shutdown cooling system, and emergency core cooling system have also been given. The safety demonstration follows the compliance to these targets.

### 15.3.7.2   The Approach and Results

Even the major feature for this case study involved implementation of IRBE approach where along with deterministic analysis, probabilistic risk assessment, uncertainty characterization, consideration of human factor employing CQB approach, and validation through performance data formed the considerations. The event tree analysis for LOCA, LORA, and LORI in Level 1 PSA along with system reliability analysis was revisited to model the plant with and without crediting the

**Table 15.1**  Target risk metrics considered for this case study

| Risk metrics—parameter | Target/goals | Uncertainty bound |
|---|---|---|
| Core damage frequency (/year) | $1.0 \times 10^{-5}$ | 95% Bound: $5 \times 10^{-5}$ |
| Large early release frequency (/year) | $1.0 \times 10^{-6}$ | 95% Bound: $5 \times 10^{-6}$ |
| Risk to member of public (/year) | $1.0 \times 10^{-7}$ | 95% Bound: $5 \times 10^{-7}$ |
| System-level unavailability | | |
| Reactor having single shutdown system | $1.0 \times 10^{-5}$ | $3.0 \times 10^{-5}$ |
| Reactor with two shutdown systems | $1.0 \times 10^{-3}$ | $3.0 \times 10^{-5}$ |
| Primary shutdown system | $1.0 \times 10^{-3}$ | $3.0 \times 10^{-3}$ |
| Secondary shutdown system | $1.0 \times 10^{-4}$ | $3.0 \times 10^{-4}$ |
| Shutdown cooling system | | |
| Emergency core cooling system | $1.0 \times 10^{-3}$ | $3.0 \times 10^{-4}$ |
| Emergency power supply system | $1.0 \times 10^{-3}$ | $3.0 \times 10^{-4}$ |

moderator dumping. The accident sequence so generated provided basis for comparing these two options. The results of the accident sequence analysis showed that by crediting moderator dumping CDF value is well below the target of $10^{-5}$/reactor-year.

The deterministic analysis considering neutronic and thermal-hydraulic studies were also performed to demonstrate the plant safety. The flow coast down study in support of loss of off-site power (LOOP) was performed to ensure that all the fuel critical parameters, like fuel central line temperature, fuel clad interface temperature, clad surface temperature, and coolant temperatures are well within limit for a case when reactor is tripping by moderator dumping alone. Similarly, LORA condition also could be demonstrated that the above-mentioned parameters are within limit. For the case of loss of coolant accident condition (LOCA) the deterministic study was performed for major LOCA and minor LOCA. It could be demonstrated that for minor LOCA condition, the thermal hydraulics as well as fuel parameters are well within limit even with reactor power as low as 100 kW. All the deterministic reactor physics and thermal-hydraulic calculations were again performed using the best-estimated codes, neutronic and thermal-hydraulic coupled codes [42, 43], and it was found that physical core parameters remained well within the limits for LOOP, LORA, and minor LOCA cases. It was found that by taking credit of moderator dumping along with shutoff rod system, the CDF can be reduced by a factor of 10, which was very encouraging. Details can be found in Varde et al. [44]. For major LOCA conditions further structural analysis is required to validate that probability of large rupture that calls for double-ended rupture in the largest diameter pipe in coolant system is $<10^{-7}$/reactor-year.

For details of this study, readers are requested to refer article by Varde et al. entitled risk-based approach to design evaluation—re-assessment of shutdown safety margin [45].

## 15.4  Concluding Remarks

This chapter presents the potential application of IRBE approach to nuclear plants and presents case studies dealing with application of RB/RI approach for range of issues that include integrated design evaluation to operational applications like risk monitor, test interval optimization, in-service inspection, and finally risk-based operator support system development for control room applications. In fact, one case study deals with application of RI/RB approach to life extension study in support of re-licensing requirement for a 30-year-old Co-60 irradiator. Some of these case studies can be termed as novel application like RB-ISI program development for a research reactor as the tools and methods, including the RI framework that are available for NPPs cannot be directly applied to research reactors due to their specific characteristics, like these systems have typical low-pressure and low-enthalpy conditions compared to NPPs. In fact for risk-based ISI framework proposed in this case study is a modified form of the existing approaches available in the literature. One more case study deals with using available safety margins in the plant to demonstrate the improved safety of the plant. The point made here is that risk-based approach provides a rational framework not only to evaluate the margin but also provides a way to use the excess margin like available redundancy or diversity to demonstrate improved safety of the plant.

The objective of presenting these case studies here is not to present the specific details but provide an overview. Again all the elements of IRBE, like condition monitoring, human factor, uncertainty evaluation, were not presented here to limit the size of the chapter.

# Annexure 1: Thomas Model for Estimation of Pipe Failure Frequency/Likelihood

This model proposes a reference/base frequency obtained from operating experience database on piping systems. This reference frequency is modified by employing the piping segment physical attributes, like piping length, diameter, thickness, and other parameters like number of welds and weld affected zone.

Pipe failure probability ($\lambda_L$) is estimated by modifying the base failure frequency ($\lambda_{Base}$) with factors related to design parameters and service data. The following correlation model suggested by Thomas [46] was used:

$$\lambda_L = \lambda_{Base} \times Q_E \times F \times B$$

where

$\lambda_L$     is the pipe leakage frequency;
$\lambda_{Base}$   is the generic pipe leakage frequency suggested value is $1 \times 10^{-9}$ to $1 \times 10^{-7}$/year;
$F$      is the plant's age factor;
$B$      is the design learning curve; and
$Q_E$     is a factor representing construction and physical features of the pipe segment.

$$Q_E = Q_P + (A Q_w)$$

$$Q_P = L(D/t^2)$$

$$Q_W = N \times 1.75(D/t)$$

$A$ is a weld penalty factor, accounting for vulnerability of weld joint for leakage compared to piping base material. Thomas suggested a constant 50 for this parameter. $L$, $D$, and $t$ are length, diameter, and thickness of the pipe while $N$ is number of weld joint in the pipe. The factor of 1.75 accounts for the width of the weld affected zone. diameter of the is the length

Finally, as suggested by Thomas, $\lambda_L$ was multiplied with conditional probability values $P(C|L)$ taken from the SKI database to estimate catastrophic rupture frequency $\lambda_C$, i.e.,

$$\lambda_C = \lambda_L \times P(C|L)$$

In the proposed risk-based approach, the failure estimate obtained, as above, were used as prior in Bayesian model and the failure frequency obtained from the statistical analysis was used as plant-specific data as evidence to arrive at final failure frequency of pipe segment.

## Annexure 2: Procedure for Conditional Core Damage Probability (CCDP) Evaluation

The consequence evaluation group is organized into two basic impact groups: the initiating event (IE) group and the loss of mitigating ability group. This estimation requires a Level 1 PRA model of the plant. A rupture in the primary coolant water system pressure boundary has potential of LOCA condition. CCDP can be directly obtained from the PSA results by dividing the CDF due to the specific IE ($f_{IE}$) by the frequency of that IE ($\lambda_{IE}$).

$$CCDP = f_{IE}/\lambda_{IE}$$

In the "loss of mitigating ability" group, the event describes the pipe failures in the safety system. The safety system can be in two configurations, standby and demand. While in standby configuration, the failure may not result in an IE, but degrade the mitigating capabilities. After failure is discovered, the plant enters the allowed outage time (AOT). In a consequence evaluation, AOT is referred to as the exposure time.

$$CCDP_i = [CDF_{(\lambda i=1)} - CDF_{(base)}] \times T_E$$

where $CDF_{(\lambda i=1)}$ is the CDF given the piping segment failure in a given safety system; $CDF_{(base)}$ is the base CDF; $\lambda_i$ is the pipe break frequency; and $T_E$ is the exposure time (detection time + AOT).

In the demand configuration, failure occurs when the system/train operation is required by an independent demand. Here, instead of exposure time, the time since the last demand is considered, which is the test interval.

$$CCDP_i = [CDF_{(\lambda i=1)} - CDF_{(base)}] \times T_t$$

where $CDF_{(\lambda i=1)}$ is the CDF given the piping segment failure frequency set to 1, i.e., the piping segment is assumed to be failed; $CDF_{(base)}$ is the base CDF; $\lambda_i$ is the pipe break frequency; and $T_t$ is the mean time between tests or demands.

The measure of risk due to pipe break is:

$$CDF_i = \lambda_i \times CCDP_i$$

# References

1. International Atomic Energy Agency, *IAEA-TECDOC-1123, Strategies for Competitive Nuclear Power Plants* (IAEA, Vienna)
2. ANSI/ASME/ANS, *Standard for Level 1/Large Early Release Frequency Probabilistic Risk Assessment for Nuclear Power Plant Applications, RA-S-2008 (RA-Sa-2009/RA Sb 2013)* (ASME, USA, 2009)
3. IAEA, *Determining the Quality of Probabilistic Safety Assessment IAEA-TECDOC-1511* (IAEA, Vienna, 2006)
4. IAEA, *Risk Based Optimization of Technical Specifications for Operation of Nuclear Power Plants, IAEA-TECDOC-729* (IAEA, Vienna, 1993)
5. International Atomic Energy Agency, *Applications of Probabilistic Safety Assessment (PSA) for Nuclear Power Plants* (IAEA, Vienna, 2001)
6. International Atomic Energy Agency, *A Framework for an Integrated Risk Informed Decision Making Process INSAG-25* (IAEA, Vienna, 2011)
7. USNRC, *As Approach for Using Probabilistic Risk Assessment in Risk-Informed Decisions on Plant Specific Changes to the Licensing Basis, Regulatory Guide 1.174, Revision 2* (USNRC, Washington DC, 2011)
8. USNRC, *An Approach for Plant-Specific Risk-Informed Decision Making for Inservice Inspection of Piping, Regulatory Guide 1.178* (USNRC, Washington D.C., 2003)
9. S. Kumar, N. Joshi, V. Mishra, P. Varde, Experiences and insights in development of probabilistic safety assessment of research reactors in BARC, in *ICRESH-ARMS-2015, Current Trends in Reliability, Availability, Maintainability and Safety* (Lulea, Sweden, 2015)
10. D. Hirate, G. Sabundjian, E. Cabrel, Preliminary Study of Probabilistic Safety Assessment Level 1 for the IEA-R1 Research Reactor of the IPEN/CNEN, in *2007 International Nuclear Atlantic Conference—INAC-2007* (Santos, Brazil, 2007)
11. International Atomic Energy Agency, *Defining Initiating Events for Purposes of Probabilistic Safety Assessment, IAEA-TECDOC-719* (IAEA, Vienna, 1993)
12. ISOGRAPH, *Isograph Fault Tree+ Module Version 10* (ISOGRAPH, UK, 2012)
13. International Atomic Energy Agency, *Generic Component Reliability Data for Research Reactor PSA, IAEA-TECDOC-930* (IAEA, Vienna, 1997)
14. International Atomic Energy Agency, *Basic Safety Principles for Nuclear Power Plants 75-INSAG-3 Rev. 1INSAG-12* (IAEA, Vienna, 1988)
15. S. Agarwal, C. Karhadkar, A. Zope, K. Singh, Dhruva: main design features, operational experience and utilization. Nucl. Eng. Des. **236**, 747–757 (2006)
16. P. Varde et.al., *Level 1+ Probabilistic Safety Assessment of Dhruva Reactor* (BARC, Mumbai, 2002)
17. S. Chowdhury, P. Varde, Surveillance test interval optimization for nuclear plants using multi objective real parameter genetic algorithms. Int. J. Reliab. Qual. Saf. Eng. **18**(2), 159 (2011)
18. P.V. Varde, N. Joshi, N. Bhamra, V. Latey, A. Shrivastava, V. Mishra, L. Bandi, *Safety Re-assessment Study for ISOMED* (BARC, Mumbai, 2007)

19. P.V. Varde, N. Joshi, M. Agarwal, A. Shrivastava, L. Bandi, A. Kohli, Development of an integrated methodology in support of remaining life assessment of Co-60 based gamma irradiation facilities. Nucl. Eng. Des. **241**, 328–338 (2011)
20. International Atomic Energy Agency, *IAEA Safety Glossary, 2016 Revision (IAEA, 2016)*, https://www-ns.iaea.org/downloads/standards/glossary/iaea-safety-glossary-rev2016.pdf. Accessed 4 May 2017
21. H. Yoshikawa, M. Yang, M. Hashim, M. Lind, Z. Zhang, Design of risk monitor for nuclear reactor plants, ed. by H Yoshikawa, Z. Zhang, in *Progress of Nuclear Safety for Symbiosis and Sustainability: Advanced Digital Instrumentations, Control and Information Systems for Nuclear Power Plants* (Springer, 2014)
22. Nuclear Energy Agency, *Risk Monitors—the State of the Art in their Development and Use at NPPs, NEA/CSNI/R(2004)20* (OECD Nuclear Energy Agency, France, 2004)
23. M. Agarwal, P. Varde, Development of risk-informed approach for asset management—a case study on nuclear plant (VII-3A), in *Structural Mechanics in Nuclear Technology Conference 2010—SMiRT-21* (New Delhi, 2011)
24. Health & Safety Executive, *Best Practice for Risk Based Inspection as a Part of Plant Integrity Management, Contract Research Report 363/2001* (H&SE, Norwich, UK, 2001)
25. Electric Power Research Institute, *Revised Risk Informed in Service Inspection Methodology, EPRI-TR-112657, Rev. B-A* (EPRI, 1999)
26. J. Brown et.al., *Westinghouse Owners Group Risk Informed Regulation Efforts—Option 2 nd 3*, https://inis.iaea.org/search/searchsinglerecord.aspx?recordsFor=SingleRecord&RN=39077116
27. U.S. Nuclear Regulatory Commission, *RG 1.178, Revision 1, An Approach for Plant Specific Risk-Informed Decision Making for Inservice Inspection of Piping* (USNRC, Washington DC, 2003)
28. American Petroleum Institute, *API RP 581—Risk Based Inspection Technology* (API, 2016)
29. International Atomic Energy Agency, *Risk Informed In-Service Inspection of Piping Systems of Nuclear Power Plant: Process, Status, Issues and Development, Nuclear Energy Series No NP-T-3.1* (IAEA, Vienna, 2010)
30. Europran Commission—Institute of Energy, *European Framework Document for Risk Informed in Service Inspection, ENIQ report no. 23* (EC Institute of Energy, The Netherlands, 2005)
31. International Atomic Energy Agency, *Safety of Research Reactors, Specific Safety Requirements No SSR 3* (IAEA, Vienna, 2016)
32. K. Fleming et.al., *Piping System Reliability and Failure Rate Estimation Models for Use in Risk-Informed In-Service Inspection Applications, EPRI TR110161* (ERIN Engineering and research Inc. for EPRI, 1998)
33. H.M. Thomas, *Pipe and vessel failure probability.* Reliab. Eng. **2**, 83–124 (1981)
34. M. Modarres, M. Kaminskiy, V. Krivtsov, *Reliability Engineering and Risk Analysis* (CRC Press, FL, 2010)
35. D. Gertman, B. Hallbert, M. Parrish, M. Sattision, D. Brownson, J. Tortorelli, *Review of Findings for Human Error Contribution to Risk in Operating Events, INEEL/EXT-01-01166/NUREG* (USNRC, Washington DC, 2001)
36. E. Swaton, V. Neboyan, L. Lederman, *Human Factors in the Operation of Nuclear Power Plants—Improving the Way Man and Machines Work Together* (IAEA, Vienna, 1987)
37. D. Petersen, *Human Error Reduction and Safety Management*, 3rd edn (Wiley, 1996)
38. A. Sinha, R. Nayak, Role of passive systems in advanced reactors. Prog. Nucl. Energy **49**(6), 486–498 (2007)
39. S. Seong, P. Lee, Design of an integrated operator support system for advanced NPP MCRs: issues and perspectives, in *Progress of Nuclear Safety for Symbiosis and Sustainability: Advanced Digital Instrumentation, 11* (Japan, 2014)
40. A. Anokhin, A. Kalinushkin, V. Gorbaev, V. Sivokon, NPP Operator Support Systems: State of Art and Development prospects. *Izvestia Vysshikh Uchebnykh Zawedeniy. Yadernaya Energetika* (2016), pp. 5–16

41. P. Varde, A. Verma, S. Sankar, An operator support system for research reactor operations and fault diagnosis through a connectionist framework and PSA based knowledge based systems. Reliab. Eng. Syst. Saf. **60**(1), 53–69 (1998)
42. T. Singh, P. Varde, *Thermal Hydraulic Coupled Code Development and Its Validation* (2015)
43. J. Kumar, T. Majumdar, T. Singh, P. Varde, A review of neutronics and thermal hydraulics coupled codes SAC-RIT and RITAC. SRESA's Int. J. Life Cycle Reliab. Saf. Eng. **4**(4), 44–53 (2015)
44. T. Singh, P. Varde, Risk-based approach to use the available provisions to re-assess the enhancement in safety margin. Life Cycle Reliab. Saf. Eng. SRESA J. **6**(3) (2017)
45. P. Varde, T. Singh, T. Muzumdar, S. Kumar, *Risk-Based Approach for Design Evaluation, BARC/Ext/004* (BARC, Mumbai, 2017)
46. H. Thomas, Pipe and vessel failure probability. Reliab. Eng. **2**, 83–124 (1981)

# Annexure

See Table A.1.

**Table A.1** Cumulative standard normal distribution

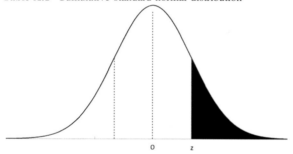

| z | 0.00 | 0.01 | 0.02 | 0.03 | 0.04 | 0.05 | 0.06 | 0.07 | 0.08 | 0.09 |
|---|------|------|------|------|------|------|------|------|------|------|
| 0.0 | 0.5000 | 0.4960 | 0.4920 | 0.4880 | 0.4840 | 0.4801 | 0.4761 | 0.4721 | 0.4681 | 0.4641 |
| 0.1 | 0.4602 | 0.4562 | 0.4522 | 0.4483 | 0.4443 | 0.4404 | 0.4364 | 0.4325 | 0.4286 | 0.4247 |
| 0.2 | 0.4207 | 0.4168 | 0.4129 | 0.4090 | 0.4052 | 0.4013 | 0.3974 | 0.3936 | 0.3897 | 0.3859 |
| 0.3 | 0.3821 | 0.3783 | 0.3745 | 0.3707 | 0.3669 | 0.3632 | 0.3594 | 0.3557 | 0.3520 | 0.3483 |
| 0.4 | 0.3446 | 0.3409 | 0.3372 | 0.3336 | 0.3300 | 0.3264 | 0.3228 | 0.3192 | 0.3156 | 0.3121 |
| 0.5 | 0.3085 | 0.3050 | 0.3015 | 0.2981 | 0.2946 | 0.2912 | 0.2877 | 0.2843 | 0.2810 | 0.2776 |
| 0.6 | 0.2743 | 0.2709 | 0.2676 | 0.2643 | 0.2611 | 0.2578 | 0.2546 | 0.2514 | 0.2483 | 0.2451 |
| 0.7 | 0.2420 | 0.2389 | 0.2358 | 0.2327 | 0.2296 | 0.2266 | 0.2236 | 0.2206 | 0.2177 | 0.2148 |
| 0.8 | 0.2119 | 0.2090 | 0.2061 | 0.2033 | 0.2005 | 0.1977 | 0.1949 | 0.1922 | 0.1894 | 0.1867 |
| 0.9 | 0.1841 | 0.1814 | 0.1788 | 0.1762 | 0.1736 | 0.1711 | 0.1685 | 0.1660 | 0.1635 | 0.1611 |
| 1.0 | 0.1587 | 0.1562 | 0.1539 | 0.1515 | 0.1492 | 0.1469 | 0.1446 | 0.1423 | 0.1401 | 0.1379 |

(continued)

© Springer Nature Singapore Pte Ltd. 2018
P. V. Varde and M. G. Pecht, *Risk-Based Engineering*, Springer Series in Reliability
Engineering, https://doi.org/10.1007/978-981-13-0090-5

**Table A.1** (continued)

| z | 0.00 | 0.01 | 0.02 | 0.03 | 0.04 | 0.05 | 0.06 | 0.07 | 0.08 | 0.09 |
|---|------|------|------|------|------|------|------|------|------|------|
| 1.1 | 0.1357 | 0.1335 | 0.1314 | 0.1292 | 0.1271 | 0.1251 | 0.1230 | 0.1210 | 0.1190 | 0.1170 |
| 1.2 | 0.1151 | 0.1131 | 0.1112 | 0.1093 | 0.1075 | 0.1056 | 0.1038 | 0.1020 | 0.1003 | 0.0985 |
| 1.3 | 0.0968 | 0.0951 | 0.0934 | 0.0918 | 0.0901 | 0.0885 | 0.0869 | 0.0853 | 0.0838 | 0.0823 |
| 1.4 | 0.0808 | 0.0793 | 0.0778 | 0.0764 | 0.0749 | 0.0735 | 0.0721 | 0.0708 | 0.0694 | 0.0681 |
| 1.5 | 0.0668 | 0.0655 | 0.0643 | 0.0630 | 0.0618 | 0.0606 | 0.0594 | 0.0582 | 0.0571 | 0.0559 |
| 1.6 | 0.0548 | 0.0537 | 0.0526 | 0.0516 | 0.0505 | 0.0495 | 0.0485 | 0.0475 | 0.0465 | 0.0455 |
| 1.7 | 0.0446 | 0.0436 | 0.0427 | 0.0418 | 0.0409 | 0.0401 | 0.0392 | 0.0384 | 0.0375 | 0.0367 |
| 1.8 | 0.0359 | 0.0351 | 0.0344 | 0.0336 | 0.0329 | 0.0322 | 0.0314 | 0.0307 | 0.0301 | 0.0294 |
| 1.9 | 0.0287 | 0.0281 | 0.0274 | 0.0268 | 0.0262 | 0.0256 | 0.0250 | 0.0244 | 0.0239 | 0.0233 |
| 2.0 | 0.0228 | 0.0222 | 0.0217 | 0.0212 | 0.0207 | 0.0202 | 0.0197 | 0.0192 | 0.0188 | 0.0183 |
| 2.1 | 0.0179 | 0.0174 | 0.0170 | 0.0166 | 0.0162 | 0.0158 | 0.0154 | 0.0150 | 0.0146 | 0.0143 |
| 2.2 | 0.0139 | 0.0136 | 0.0132 | 0.0129 | 0.0125 | 0.0122 | 0.0119 | 0.0116 | 0.0113 | 0.0110 |
| 2.3 | 0.0107 | 0.0104 | 0.0102 | 0.0099 | 0.0096 | 0.0094 | 0.0091 | 0.0089 | 0.0087 | 0.0084 |
| 2.4 | 0.0082 | 0.0080 | 0.0078 | 0.0075 | 0.0073 | 0.0071 | 0.0069 | 0.0068 | 0.0066 | 0.0064 |
| 2.5 | 0.0062 | 0.0060 | 0.0059 | 0.0057 | 0.0055 | 0.0054 | 0.0052 | 0.0051 | 0.0049 | 0.0048 |
| 2.6 | 0.0047 | 0.0045 | 0.0044 | 0.0043 | 0.0041 | 0.0040 | 0.0039 | 0.0038 | 0.0037 | 0.0036 |
| 2.7 | 0.0035 | 0.0034 | 0.0033 | 0.0032 | 0.0031 | 0.0030 | 0.0029 | 0.0028 | 0.0027 | 0.0026 |
| 2.8 | 0.0026 | 0.0025 | 0.0024 | 0.0023 | 0.0023 | 0.0022 | 0.0021 | 0.0021 | 0.0020 | 0.0019 |
| 2.9 | 0.0019 | 0.0018 | 0.0018 | 0.0017 | 0.0016 | 0.0016 | 0.0015 | 0.0015 | 0.0014 | 0.0014 |
| 3.0 | 0.0013 | 0.0013 | 0.0013 | 0.0012 | 0.0012 | 0.0011 | 0.0011 | 0.0011 | 0.0010 | 0.0010 |
| 3.1 | 0.0010 | 0.0009 | 0.0009 | 0.0009 | 0.0008 | 0.0008 | 0.0008 | 0.0008 | 0.0007 | 0.0007 |
| 3.2 | 0.0007 | 0.0007 | 0.0006 | 0.0006 | 0.0006 | 0.0006 | 0.0006 | 0.0005 | 0.0005 | 0.0005 |
| 3.3 | 0.0005 | 0.0005 | 0.0005 | 0.0004 | 0.0004 | 0.0004 | 0.0004 | 0.0004 | 0.0004 | 0.0003 |
| 3.4 | 0.0003 | 0.0003 | 0.0003 | 0.0003 | 0.0003 | 0.0003 | 0.0003 | 0.0003 | 0.0003 | 0.0002 |
| 3.5 | 0.0002 | 0.0002 | 0.0002 | 0.0002 | 0.0002 | 0.0002 | 0.0002 | 0.0002 | 0.0002 | 0.0002 |

| Chi-square distribution | | | | | | | | | |
|---|---|---|---|---|---|---|---|---|---|
| Degrees of freedom | $\alpha$ | | | | | | | | |
| | 0.995 | 0.990 | 0.975 | 0.900 | 0.100 | 0.050 | 0.025 | 0.010 | 0.005 |
| 1 | 0.0000 | 0.0002 | 0.0010 | 0.0158 | 2.7055 | 3.8415 | 5.0239 | 6.6349 | 7.8794 |
| 2 | 0.0100 | 0.0201 | 0.0506 | 0.2107 | 4.6052 | 5.9915 | 7.3778 | 9.2103 | 10.5966 |
| 3 | 0.0717 | 0.1148 | 0.2158 | 0.5844 | 6.2514 | 7.8147 | 9.3484 | 11.3449 | 12.8382 |
| 4 | 0.2070 | 0.2971 | 0.4844 | 1.0636 | 7.7794 | 9.4877 | 11.1433 | 13.2767 | 14.8603 |
| 5 | 0.4117 | 0.5543 | 0.8312 | 1.6103 | 9.2364 | 11.0705 | 12.8325 | 15.0863 | 16.7496 |
| 6 | 0.6757 | 0.8721 | 1.2373 | 2.2041 | 10.6446 | 12.5916 | 14.4494 | 16.8119 | 18.5476 |
| 7 | 0.9893 | 1.2390 | 1.6899 | 2.8331 | 12.0170 | 14.0671 | 16.0128 | 18.4753 | 20.2777 |
| 8 | 1.3444 | 1.6465 | 2.1797 | 3.4895 | 13.3616 | 15.5073 | 17.5345 | 20.0902 | 21.9550 |
| 9 | 1.7349 | 2.0879 | 2.7004 | 4.1682 | 14.6837 | 16.9190 | 19.0228 | 21.6660 | 23.5894 |

(continued)

(continued)

Chi-square distribution

| Degrees of freedom | α | | | | | | | | |
|---|---|---|---|---|---|---|---|---|---|
| | 0.995 | 0.990 | 0.975 | 0.900 | 0.100 | 0.050 | 0.025 | 0.010 | 0.005 |
| 10 | 2.1559 | 2.5582 | 3.2470 | 4.8652 | 15.9872 | 18.3070 | 20.4832 | 23.2093 | 25.1882 |
| 11 | 2.6032 | 3.0535 | 3.8157 | 5.5778 | 17.2750 | 19.6751 | 21.9200 | 24.7250 | 26.7568 |
| 12 | 3.0738 | 3.5706 | 4.4038 | 6.3038 | 18.5493 | 21.0261 | 23.3367 | 26.2170 | 28.2995 |
| 13 | 3.5650 | 4.1069 | 5.0088 | 7.0415 | 19.8119 | 22.3620 | 24.7356 | 27.6882 | 29.8195 |
| 14 | 4.0747 | 4.6604 | 5.6287 | 7.7895 | 21.0641 | 23.6848 | 26.1189 | 29.1412 | 31.3193 |
| 15 | 4.6009 | 5.2293 | 6.2621 | 8.5468 | 22.3071 | 24.9958 | 27.4884 | 30.5779 | 32.8013 |
| 16 | 5.1422 | 5.8122 | 6.9077 | 9.3122 | 23.5418 | 26.2962 | 28.8454 | 31.9999 | 34.2672 |
| 17 | 5.6972 | 6.4078 | 7.5642 | 10.0852 | 24.7690 | 27.5871 | 30.1910 | 33.4087 | 35.7185 |
| 18 | 6.2648 | 7.0149 | 8.2307 | 10.8649 | 25.9894 | 28.8693 | 31.5264 | 34.8053 | 37.1565 |
| 19 | 6.8440 | 7.6327 | 8.9065 | 11.6509 | 27.2036 | 30.1435 | 32.8523 | 36.1909 | 38.5823 |
| 20 | 7.4338 | 8.2604 | 9.5908 | 12.4426 | 28.4120 | 31.4104 | 34.1696 | 37.5662 | 39.9968 |
| 21 | 8.0337 | 8.8972 | 10.2829 | 13.2396 | 29.6151 | 32.6706 | 35.4789 | 38.9322 | 41.4011 |
| 22 | 8.6427 | 9.5425 | 10.9823 | 14.0415 | 30.8133 | 33.9244 | 36.7807 | 40.2894 | 42.7957 |
| 23 | 9.2604 | 10.1957 | 11.6886 | 14.8480 | 32.0069 | 35.1725 | 38.0756 | 41.6384 | 44.1813 |
| 24 | 9.8862 | 10.8564 | 12.4012 | 15.6587 | 33.1962 | 36.4150 | 39.3641 | 42.9798 | 45.5585 |
| 25 | 10.5197 | 11.5240 | 13.1197 | 16.4734 | 34.3816 | 37.6525 | 40.6465 | 44.3141 | 46.9279 |
| 26 | 11.1602 | 12.1981 | 13.8439 | 17.2919 | 35.5632 | 38.8851 | 41.9232 | 45.6417 | 48.2899 |
| 27 | 11.8076 | 12.8785 | 14.5734 | 18.1139 | 36.7412 | 40.1133 | 43.1945 | 46.9629 | 49.6449 |
| 28 | 12.4613 | 13.5647 | 15.3079 | 18.9392 | 37.9159 | 41.3371 | 44.4608 | 48.2782 | 50.9934 |
| 29 | 13.1211 | 14.2565 | 16.0471 | 19.7677 | 39.0875 | 42.5570 | 45.7223 | 49.5879 | 52.3356 |
| 30 | 13.7867 | 14.9535 | 16.7908 | 20.5992 | 40.2560 | 43.7730 | 46.9792 | 50.8922 | 53.6720 |
| 31 | 14.4578 | 15.6555 | 17.5387 | 21.4336 | 41.4217 | 44.9853 | 48.2319 | 52.1914 | 55.0027 |
| 32 | 15.1340 | 16.3622 | 18.2908 | 22.2706 | 42.5847 | 46.1943 | 49.4804 | 53.4858 | 56.3281 |
| 33 | 15.8153 | 17.0735 | 19.0467 | 23.1102 | 43.7452 | 47.3999 | 50.7251 | 54.7755 | 57.6484 |
| 34 | 16.5013 | 17.7891 | 19.8063 | 23.9523 | 44.9032 | 48.6024 | 51.9660 | 56.0609 | 58.9639 |
| 35 | 17.1918 | 18.5089 | 20.5694 | 24.7967 | 46.0588 | 49.8018 | 53.2033 | 57.3421 | 60.2748 |
| 36 | 17.8867 | 19.2327 | 21.3359 | 25.6433 | 47.2122 | 50.9985 | 54.4373 | 58.6192 | 61.5812 |
| 37 | 18.5858 | 19.9602 | 22.1056 | 26.4921 | 48.3634 | 52.1923 | 55.6680 | 59.8925 | 62.8833 |
| 38 | 19.2889 | 20.6914 | 22.8785 | 27.3430 | 49.5126 | 53.3835 | 56.8955 | 61.1621 | 64.1814 |
| 39 | 19.9959 | 21.4262 | 23.6543 | 28.1958 | 50.6598 | 54.5722 | 58.1201 | 62.4281 | 65.4756 |
| 40 | 20.7065 | 22.1643 | 24.4330 | 29.0505 | 51.8051 | 55.7585 | 59.3417 | 63.6907 | 66.7660 |
| 41 | 21.4208 | 22.9056 | 25.2145 | 29.9071 | 52.9485 | 56.9424 | 60.5606 | 64.9501 | 68.0527 |
| 42 | 22.1385 | 23.6501 | 25.9987 | 30.7654 | 54.0902 | 58.1240 | 61.7768 | 66.2062 | 69.3360 |
| 43 | 22.8595 | 24.3976 | 26.7854 | 31.6255 | 55.2302 | 59.3035 | 62.9904 | 67.4593 | 70.6159 |
| 44 | 23.5837 | 25.1480 | 27.5746 | 32.4871 | 56.3685 | 60.4809 | 64.2015 | 68.7095 | 71.8926 |
| 45 | 24.3110 | 25.9013 | 28.3662 | 33.3504 | 57.5053 | 61.6562 | 65.4102 | 69.9568 | 73.1661 |
| 46 | 25.0413 | 26.6572 | 29.1601 | 34.2152 | 58.6405 | 62.8296 | 66.6165 | 71.2014 | 74.4365 |
| 47 | 25.7746 | 27.4158 | 29.9562 | 35.0814 | 59.7743 | 64.0011 | 67.8206 | 72.4433 | 75.7041 |
| 48 | 26.5106 | 28.1770 | 30.7545 | 35.9491 | 60.9066 | 65.1708 | 69.0226 | 73.6826 | 76.9688 |

(continued)

(continued)

Chi-square distribution

| Degrees of freedom | $\alpha$ | | | | | | | | |
|---|---|---|---|---|---|---|---|---|---|
| | 0.995 | 0.990 | 0.975 | 0.900 | 0.100 | 0.050 | 0.025 | 0.010 | 0.005 |
| 49 | 27.2493 | 28.9406 | 31.5549 | 36.8182 | 62.0375 | 66.3386 | 70.2224 | 74.9195 | 78.2307 |
| 50 | 27.9907 | 29.7067 | 32.3574 | 37.6886 | 63.1671 | 67.5048 | 71.4202 | 76.1539 | 79.4900 |
| 51 | 28.7347 | 30.4750 | 33.1618 | 38.5604 | 64.2954 | 68.6693 | 72.6160 | 77.3860 | 80.7467 |
| 52 | 29.4812 | 31.2457 | 33.9681 | 39.4334 | 65.4224 | 69.8322 | 73.8099 | 78.6158 | 82.0008 |
| 53 | 30.2300 | 32.0185 | 34.7763 | 40.3076 | 66.5482 | 70.9935 | 75.0019 | 79.8433 | 83.2526 |
| 54 | 30.9813 | 32.7934 | 35.5863 | 41.1830 | 67.6728 | 72.1532 | 76.1920 | 81.0688 | 84.5019 |
| 55 | 31.7348 | 33.5705 | 36.3981 | 42.0596 | 68.7962 | 73.3115 | 77.3805 | 82.2921 | 85.7490 |
| 56 | 32.4905 | 34.3495 | 37.2116 | 42.9373 | 69.9185 | 74.4683 | 78.5672 | 83.5134 | 86.9938 |
| 57 | 33.2484 | 35.1305 | 38.0267 | 43.8161 | 71.0397 | 75.6237 | 79.7522 | 84.7328 | 88.2364 |
| 58 | 34.0084 | 35.9135 | 38.8435 | 44.6960 | 72.1598 | 76.7778 | 80.9356 | 85.9502 | 89.4769 |
| 59 | 34.7704 | 36.6982 | 39.6619 | 45.5770 | 73.2789 | 77.9305 | 82.1174 | 87.1657 | 90.7153 |
| 60 | 35.5345 | 37.4849 | 40.4817 | 46.4589 | 74.3970 | 79.0819 | 83.2977 | 88.3794 | 91.9517 |
| 61 | 36.3005 | 38.2732 | 41.3031 | 47.3418 | 75.5141 | 80.2321 | 84.4764 | 89.5913 | 93.1861 |
| 62 | 37.0684 | 39.0633 | 42.1260 | 48.2257 | 76.6302 | 81.3810 | 85.6537 | 90.8015 | 94.4187 |
| 63 | 37.8382 | 39.8551 | 42.9503 | 49.1105 | 77.7454 | 82.5287 | 86.8296 | 92.0100 | 95.6493 |
| 64 | 38.6098 | 40.6486 | 43.7760 | 49.9963 | 78.8596 | 83.6753 | 88.0041 | 93.2169 | 96.8781 |
| 65 | 39.3831 | 41.4436 | 44.6030 | 50.8829 | 79.9730 | 84.8206 | 89.1771 | 94.4221 | 98.1051 |
| 66 | 40.1582 | 42.2402 | 45.4314 | 51.7705 | 81.0855 | 85.9649 | 90.3489 | 95.6257 | 99.3304 |
| 67 | 40.9350 | 43.0384 | 46.2610 | 52.6588 | 82.1971 | 87.1081 | 91.5194 | 96.8278 | 100.5540 |
| 68 | 41.7135 | 43.8380 | 47.0920 | 53.5481 | 83.3079 | 88.2502 | 92.6885 | 98.0284 | 101.7759 |
| 69 | 42.4935 | 44.6392 | 47.9242 | 54.4381 | 84.4179 | 89.3912 | 93.8565 | 99.2275 | 102.9962 |
| 70 | 43.2752 | 45.4417 | 48.7576 | 55.3289 | 85.5270 | 90.5312 | 95.0232 | 100.4252 | 104.2149 |
| 71 | 44.0584 | 46.2457 | 49.5922 | 56.2206 | 86.6354 | 91.6702 | 96.1887 | 101.6214 | 105.4320 |
| 72 | 44.8431 | 47.0510 | 50.4279 | 57.1129 | 87.7430 | 92.8083 | 97.3531 | 102.8163 | 106.6476 |
| 73 | 45.6293 | 47.8577 | 51.2648 | 58.0061 | 88.8499 | 93.9453 | 98.5163 | 104.0098 | 107.8617 |
| 74 | 46.4170 | 48.6657 | 52.1028 | 58.9000 | 89.9560 | 95.0815 | 99.6783 | 105.2020 | 109.0744 |
| 75 | 47.2060 | 49.4750 | 52.9419 | 59.7946 | 91.0615 | 96.2167 | 100.8393 | 106.3929 | 110.2856 |
| 76 | 47.9965 | 50.2856 | 53.7821 | 60.6899 | 92.1662 | 97.3510 | 101.9993 | 107.5825 | 111.4954 |
| 77 | 48.7884 | 51.0974 | 54.6234 | 61.5858 | 93.2702 | 98.4844 | 103.1581 | 108.7709 | 112.7038 |
| 78 | 49.5816 | 51.9104 | 55.4656 | 62.4825 | 94.3735 | 99.6169 | 104.3159 | 109.9581 | 113.9109 |
| 79 | 50.3761 | 52.7247 | 56.3089 | 63.3799 | 95.4762 | 100.7486 | 105.4728 | 111.1440 | 115.1166 |
| 80 | 51.1719 | 53.5401 | 57.1532 | 64.2778 | 96.5782 | 101.8795 | 106.6286 | 112.3288 | 116.3211 |
| 81 | 51.9690 | 54.3566 | 57.9984 | 65.1765 | 97.6796 | 103.0095 | 107.7834 | 113.5124 | 117.5242 |
| 82 | 52.7674 | 55.1743 | 58.8446 | 66.0757 | 98.7803 | 104.1387 | 108.9373 | 114.6949 | 118.7261 |
| 83 | 53.5669 | 55.9931 | 59.6918 | 66.9756 | 99.8805 | 105.2672 | 110.0902 | 115.8763 | 119.9268 |
| 84 | 54.3677 | 56.8130 | 60.5398 | 67.8761 | 100.9800 | 106.3948 | 111.2423 | 117.0565 | 121.1263 |
| 85 | 55.1696 | 57.6339 | 61.3888 | 68.7772 | 102.0789 | 107.5217 | 112.3934 | 118.2357 | 122.3246 |
| 86 | 55.9727 | 58.4559 | 62.2386 | 69.6788 | 103.1773 | 108.6479 | 113.5436 | 119.4139 | 123.5217 |
| 87 | 56.7769 | 59.2790 | 63.0894 | 70.5810 | 104.2750 | 109.7733 | 114.6929 | 120.5910 | 124.7177 |

(continued)

(continued)

Chi-square distribution

| Degrees of freedom | α | | | | | | | | |
|---|---|---|---|---|---|---|---|---|---|
| | 0.995 | 0.990 | 0.975 | 0.900 | 0.100 | 0.050 | 0.025 | 0.010 | 0.005 |
| 88 | 57.5823 | 60.1030 | 63.9409 | 71.4838 | 105.3722 | 110.8980 | 115.8414 | 121.7671 | 125.9125 |
| 89 | 58.3888 | 60.9281 | 64.7934 | 72.3872 | 106.4689 | 112.0220 | 116.9891 | 122.9422 | 127.1063 |
| 90 | 59.1963 | 61.7541 | 65.6466 | 73.2911 | 107.5650 | 113.1453 | 118.1359 | 124.1163 | 128.2989 |
| 91 | 60.0049 | 62.5811 | 66.5007 | 74.1955 | 108.6606 | 114.2679 | 119.2819 | 125.2895 | 129.4905 |
| 92 | 60.8146 | 63.4090 | 67.3556 | 75.1005 | 109.7556 | 115.3898 | 120.4271 | 126.4617 | 130.6811 |
| 94 | 62.4370 | 65.0677 | 69.0677 | 76.9119 | 111.9442 | 117.6317 | 122.7151 | 128.8032 | 133.0591 |
| 95 | 63.2496 | 65.8984 | 69.9249 | 77.8184 | 113.0377 | 118.7516 | 123.8580 | 129.9727 | 134.2465 |
| 96 | 64.0633 | 66.7299 | 70.7828 | 78.7254 | 114.1307 | 119.8709 | 125.0001 | 131.1412 | 135.4330 |
| 97 | 64.8780 | 67.5624 | 71.6415 | 79.6329 | 115.2232 | 120.9896 | 126.1414 | 132.3089 | 136.6186 |
| 98 | 65.6936 | 68.3957 | 72.5009 | 80.5408 | 116.3153 | 122.1077 | 127.2821 | 133.4757 | 137.8032 |
| 99 | 66.5101 | 69.2299 | 73.3611 | 81.4493 | 117.4069 | 123.2252 | 128.4220 | 134.6416 | 138.9868 |
| 100 | 67.3276 | 70.0649 | 74.2219 | 82.3581 | 118.4980 | 124.3421 | 129.5612 | 135.8067 | 140.1695 |
| 101 | 68.1459 | 70.9007 | 75.0835 | 83.2675 | 119.5887 | 125.4584 | 130.6997 | 136.9710 | 141.3513 |
| 102 | 68.9652 | 71.7374 | 75.9457 | 84.1773 | 120.6789 | 126.5741 | 131.8375 | 138.1345 | 142.5322 |
| 103 | 69.7853 | 72.5748 | 76.8086 | 85.0875 | 121.7686 | 127.6893 | 132.9747 | 139.2971 | 143.7122 |
| 104 | 70.6064 | 73.4130 | 77.6722 | 85.9982 | 122.8580 | 128.8039 | 134.1112 | 140.4590 | 144.8913 |
| 105 | 71.4282 | 74.2520 | 78.5364 | 86.9093 | 123.9469 | 129.9180 | 135.2470 | 141.6201 | 146.0696 |
| 106 | 72.2509 | 75.0918 | 79.4013 | 87.8208 | 125.0354 | 131.0315 | 136.3822 | 142.7804 | 147.2470 |
| 107 | 73.0745 | 75.9323 | 80.2668 | 88.7327 | 126.1234 | 132.1444 | 137.5167 | 143.9400 | 148.4236 |
| 108 | 73.8989 | 76.7736 | 81.1329 | 89.6451 | 127.2111 | 133.2569 | 138.6506 | 145.0988 | 149.5994 |
| 109 | 74.7241 | 77.6156 | 81.9997 | 90.5579 | 128.2983 | 134.3688 | 139.7839 | 146.2569 | 150.7743 |
| 110 | 75.5500 | 78.4583 | 82.8671 | 91.4710 | 129.3851 | 135.4802 | 140.9166 | 147.4143 | 151.9485 |
| 111 | 76.3768 | 79.3017 | 83.7350 | 92.3846 | 130.4716 | 136.5911 | 142.0486 | 148.5710 | 153.1218 |
| 112 | 77.2044 | 80.1459 | 84.6036 | 93.2986 | 131.5576 | 137.7015 | 143.1801 | 149.7269 | 154.2944 |
| 113 | 78.0327 | 80.9907 | 85.4728 | 94.2129 | 132.6433 | 138.8114 | 144.3110 | 150.8822 | 155.4662 |
| 114 | 78.8618 | 81.8362 | 86.3425 | 95.1276 | 133.7286 | 139.9208 | 145.4413 | 152.0367 | 156.6373 |
| 115 | 79.6916 | 82.6824 | 87.2128 | 96.0427 | 134.8135 | 141.0297 | 146.5711 | 153.1906 | 157.8076 |
| 116 | 80.5221 | 83.5293 | 88.0837 | 96.9582 | 135.8980 | 142.1382 | 147.7002 | 154.3438 | 158.9771 |
| 117 | 81.3534 | 84.3768 | 88.9551 | 97.8740 | 136.9822 | 143.2461 | 148.8288 | 155.4964 | 160.1460 |
| 118 | 82.1854 | 85.2250 | 89.8271 | 98.7902 | 138.0660 | 144.3537 | 149.9569 | 156.6483 | 161.3141 |
| 119 | 83.0182 | 86.0738 | 90.6996 | 99.7067 | 139.1495 | 145.4607 | 151.0844 | 157.7995 | 162.4815 |
| 120 | 83.8516 | 86.9233 | 91.5726 | 100.6236 | 140.2326 | 146.5674 | 152.2114 | 158.9502 | 163.6482 |

Critical values of the Kolmogorov–Smirnov statistic $D_n$ (α)

| n | α | | | | |
|---|---|---|---|---|---|
| | 0.20 | 0.15 | 0.10 | 0.05 | 0.01 |
| 1 | 0.900 | 0.925 | 0.950 | 0.975 | 0.995 |
| 2 | 0.684 | 0.725 | 0.776 | 0.842 | 0.929 |
| 3 | 0.565 | 0.597 | 0.636 | 0.708 | 0.829 |
| 4 | 0.493 | 0.525 | 0.565 | 0.624 | 0.734 |

(continued)

(continued)

| n | α | | | | |
|---|---|---|---|---|---|
| | 0.20 | 0.15 | 0.10 | 0.05 | 0.01 |
| 5 | 0.447 | 0.474 | 0.510 | 0.563 | 0.669 |
| 6 | 0.410 | 0.436 | 0.468 | 0.519 | 0.617 |
| 7 | 0.381 | 0.405 | 0.436 | 0.483 | 0.576 |
| 8 | 0.358 | 0.380 | 0.410 | 0.454 | 0.542 |
| 9 | 0.339 | 0.360 | 0.387 | 0.430 | 0.513 |
| 10 | 0.323 | 0.342 | 0.369 | 0.409 | 0.489 |
| 11 | 0.308 | 0.326 | 0.352 | 0.391 | 0.468 |
| 12 | 0.296 | 0.313 | 0.338 | 0.375 | 0.449 |
| 13 | 0.285 | 0.302 | 0.325 | 0.361 | 0.432 |
| 14 | 0.275 | 0.292 | 0.314 | 0.349 | 0.418 |
| 15 | 0.266 | 0.283 | 0.304 | 0.338 | 0.404 |
| 16 | 0.258 | 0.274 | 0.295 | 0.327 | 0.392 |
| 17 | 0.250 | 0.266 | 0.286 | 0.318 | 0.381 |
| 18 | 0.244 | 0.259 | 0.279 | 0.309 | 0.371 |
| 19 | 0.237 | 0.252 | 0.271 | 0.301 | 0.361 |
| 20 | 0.232 | 0.246 | 0.265 | 0.294 | 0.352 |
| 21 | 0.226 | 0.249 | 0.259 | 0.287 | 0.344 |
| 22 | 0.221 | 0.243 | 0.253 | 0.281 | 0.337 |
| 23 | 0.216 | 0.238 | 0.247 | 0.275 | 0.330 |
| 24 | 0.212 | 0.233 | 0.242 | 0.269 | 0.323 |
| 25 | 0.208 | 0.228 | 0.238 | 0.264 | 0.317 |
| 26 | 0.204 | 0.224 | 0.233 | 0.259 | 0.311 |
| 27 | 0.200 | 0.219 | 0.229 | 0.254 | 0.305 |
| 28 | 0.197 | 0.215 | 0.225 | 0.250 | 0.300 |
| 29 | 0.193 | 0.212 | 0.221 | 0.246 | 0.295 |
| 30 | 0.190 | 0.208 | 0.218 | 0.242 | 0.290 |
| 31 | 0.187 | 0.205 | 0.214 | 0.238 | 0.285 |
| 32 | 0.184 | 0.201 | 0.211 | 0.234 | 0.281 |
| 33 | 0.182 | 0.198 | 0.208 | 0.231 | 0.277 |
| 34 | 0.179 | 0.195 | 0.205 | 0.227 | 0.273 |
| 35 | 0.177 | 0.193 | 0.202 | 0.224 | 0.269 |
| 36 | 0.174 | 0.190 | 0.199 | 0.221 | 0.265 |
| 37 | 0.172 | 0.187 | 0.196 | 0.218 | 0.262 |
| 38 | 0.170 | 0.185 | 0.194 | 0.215 | 0.258 |
| 39 | 0.168 | 0.182 | 0.191 | 0.213 | 0.255 |
| 40 | 0.165 | 0.180 | 0.189 | 0.210 | 0.252 |
| >40 | $1.07/\sqrt{n}$ | $1.14/\sqrt{n}$ | $1.22/\sqrt{n}$ | $1.36/\sqrt{n}$ | $1.63/\sqrt{n}$ |

Critical values of the Kolmogorov–Smirnov statistic $D_n$ ($\alpha$)

$n$: Number of trials

95th percentile values for the $F$ distribution

| $f2$ | $f1$ | | | | | | | | | | | | | | | | | | |
|---|---|---|---|---|---|---|---|---|---|---|---|---|---|---|---|---|---|---|---|
| | 1 | 2 | 3 | 4 | 5 | 6 | 7 | 8 | 9 | 10 | 12 | 15 | 20 | 24 | 30 | 40 | 60 | 120 | Inf |
| 1 | 161 | 200 | 216 | 225 | 230 | 234 | 237 | 239 | 241 | 242 | 244 | 246 | 248 | 249 | 250 | 251 | 252 | 253 | 254 |
| 2 | 18.51 | 19.00 | 19.16 | 19.25 | 19.30 | 19.33 | 19.35 | 19.37 | 19.38 | 19.40 | 19.41 | 19.43 | 19.45 | 19.45 | 19.46 | 19.47 | 19.48 | 19.49 | 19.50 |
| 3 | 10.13 | 9.55 | 9.28 | 9.12 | 9.01 | 8.94 | 8.89 | 8.85 | 8.81 | 8.79 | 8.74 | 8.70 | 8.66 | 8.64 | 8.62 | 8.59 | 8.57 | 8.55 | 8.53 |
| 4 | 7.71 | 6.94 | 6.59 | 6.39 | 6.26 | 6.16 | 6.09 | 6.04 | 6.00 | 5.96 | 5.91 | 5.86 | 5.80 | 5.77 | 5.75 | 5.72 | 5.69 | 5.66 | 5.63 |
| 5 | 6.61 | 5.79 | 5.41 | 5.19 | 5.05 | 4.95 | 4.88 | 4.82 | 4.77 | 4.74 | 4.68 | 4.62 | 4.56 | 4.53 | 4.50 | 4.46 | 4.43 | 4.40 | 4.36 |
| 6 | 5.99 | 5.14 | 4.76 | 4.53 | 4.39 | 4.28 | 4.21 | 4.15 | 4.10 | 4.06 | 4.00 | 3.94 | 3.87 | 3.84 | 3.81 | 3.77 | 3.74 | 3.70 | 3.67 |
| 7 | 5.59 | 4.74 | 4.35 | 4.12 | 3.97 | 3.87 | 3.79 | 3.73 | 3.68 | 3.64 | 3.57 | 3.51 | 3.44 | 3.41 | 3.38 | 3.34 | 3.30 | 3.27 | 3.23 |
| 8 | 5.32 | 4.46 | 4.07 | 3.84 | 3.69 | 3.58 | 3.50 | 3.44 | 3.39 | 3.35 | 3.28 | 3.22 | 3.15 | 3.12 | 3.08 | 3.04 | 3.01 | 2.97 | 2.93 |
| 9 | 5.12 | 4.26 | 3.86 | 3.63 | 3.48 | 3.37 | 3.29 | 3.23 | 3.18 | 3.14 | 3.07 | 3.01 | 2.94 | 2.90 | 2.86 | 2.83 | 2.79 | 2.75 | 2.71 |
| 10 | 4.96 | 4.10 | 3.71 | 3.48 | 3.33 | 3.22 | 3.14 | 3.07 | 3.02 | 2.98 | 2.91 | 2.85 | 2.77 | 2.74 | 2.70 | 2.66 | 2.62 | 2.58 | 2.54 |
| 11 | 4.84 | 3.98 | 3.59 | 3.36 | 3.20 | 3.09 | 3.01 | 2.95 | 2.90 | 2.85 | 2.79 | 2.72 | 2.65 | 2.61 | 2.57 | 2.53 | 2.49 | 2.45 | 2.40 |
| 12 | 4.75 | 3.89 | 3.49 | 3.26 | 3.11 | 3.00 | 2.91 | 2.85 | 2.80 | 2.75 | 2.69 | 2.62 | 2.54 | 2.51 | 2.47 | 2.43 | 2.38 | 2.34 | 2.30 |
| 13 | 4.67 | 3.81 | 3.41 | 3.18 | 3.03 | 2.92 | 2.83 | 2.77 | 2.71 | 2.67 | 2.60 | 2.53 | 2.46 | 2.42 | 2.38 | 2.34 | 2.30 | 2.25 | 2.21 |
| 14 | 4.60 | 3.74 | 3.34 | 3.11 | 2.96 | 2.85 | 2.76 | 2.70 | 2.65 | 2.60 | 2.53 | 2.46 | 2.39 | 2.35 | 2.31 | 2.27 | 2.22 | 2.18 | 2.13 |
| 15 | 4.54 | 3.68 | 3.29 | 3.06 | 2.90 | 2.79 | 2.71 | 2.64 | 2.59 | 2.54 | 2.48 | 2.40 | 2.33 | 2.29 | 2.25 | 2.20 | 2.16 | 2.11 | 2.07 |
| 16 | 4.49 | 3.63 | 3.24 | 3.01 | 2.85 | 2.74 | 2.66 | 2.59 | 2.54 | 2.49 | 2.42 | 2.35 | 2.28 | 2.24 | 2.19 | 2.15 | 2.11 | 2.06 | 2.01 |
| 17 | 4.45 | 3.59 | 3.20 | 2.96 | 2.81 | 2.70 | 2.61 | 2.55 | 2.49 | 2.45 | 2.38 | 2.31 | 2.23 | 2.19 | 2.15 | 2.10 | 2.06 | 2.01 | 1.96 |
| 18 | 4.41 | 3.55 | 3.16 | 2.93 | 2.77 | 2.66 | 2.58 | 2.51 | 2.46 | 2.41 | 2.34 | 2.27 | 2.19 | 2.15 | 2.11 | 2.06 | 2.02 | 1.97 | 1.92 |
| 19 | 4.38 | 3.52 | 3.13 | 2.90 | 2.74 | 2.63 | 2.54 | 2.48 | 2.42 | 2.38 | 2.31 | 2.23 | 2.16 | 2.11 | 2.07 | 2.03 | 1.98 | 1.93 | 1.88 |
| 20 | 4.35 | 3.49 | 3.10 | 2.87 | 2.71 | 2.60 | 2.51 | 2.45 | 2.39 | 2.35 | 2.28 | 2.20 | 2.12 | 2.08 | 2.04 | 1.99 | 1.95 | 1.90 | 1.84 |
| 21 | 4.32 | 3.47 | 3.07 | 2.84 | 2.68 | 2.57 | 2.49 | 2.42 | 2.37 | 2.32 | 2.25 | 2.18 | 2.10 | 2.05 | 2.01 | 1.96 | 1.92 | 1.87 | 1.81 |
| 22 | 4.30 | 3.44 | 3.05 | 2.82 | 2.66 | 2.55 | 2.46 | 2.40 | 2.34 | 2.30 | 2.23 | 2.15 | 2.07 | 2.03 | 1.98 | 1.94 | 1.89 | 1.84 | 1.78 |
| 23 | 4.28 | 3.42 | 3.03 | 2.80 | 2.64 | 2.53 | 2.44 | 2.37 | 2.32 | 2.27 | 2.20 | 2.13 | 2.05 | 2.01 | 1.96 | 1.91 | 1.86 | 1.81 | 1.76 |

(continued)

(continued)

95th percentile values for the $F$ distribution

| $f2$ \ $f1$ | 1 | 2 | 3 | 4 | 5 | 6 | 7 | 8 | 9 | 10 | 12 | 15 | 20 | 24 | 30 | 40 | 60 | 120 | Inf |
|---|---|---|---|---|---|---|---|---|---|---|---|---|---|---|---|---|---|---|---|
| 24 | 4.26 | 3.40 | 3.01 | 2.78 | 2.62 | 2.51 | 2.42 | 2.36 | 2.30 | 2.25 | 2.18 | 2.11 | 2.03 | 1.98 | 1.94 | 1.89 | 1.84 | 1.79 | 1.73 |
| 25 | 4.24 | 3.39 | 2.99 | 2.76 | 2.60 | 2.49 | 2.40 | 2.34 | 2.28 | 2.24 | 2.16 | 2.09 | 2.01 | 1.96 | 1.92 | 1.87 | 1.82 | 1.77 | 1.71 |
| 26 | 4.23 | 3.37 | 2.98 | 2.74 | 2.59 | 2.47 | 2.39 | 2.32 | 2.27 | 2.22 | 2.15 | 2.07 | 1.99 | 1.95 | 1.90 | 1.85 | 1.80 | 1.75 | 1.69 |
| 27 | 4.21 | 3.35 | 2.96 | 2.73 | 2.57 | 2.46 | 2.37 | 2.31 | 2.25 | 2.20 | 2.13 | 2.06 | 1.97 | 1.93 | 1.88 | 1.84 | 1.79 | 1.73 | 1.67 |
| 28 | 4.20 | 3.34 | 2.95 | 2.71 | 2.56 | 2.45 | 2.36 | 2.29 | 2.24 | 2.19 | 2.12 | 2.04 | 1.96 | 1.91 | 1.87 | 1.82 | 1.77 | 1.71 | 1.65 |
| 29 | 4.18 | 3.33 | 2.93 | 2.70 | 2.55 | 2.43 | 2.35 | 2.28 | 2.22 | 2.18 | 2.10 | 2.03 | 1.94 | 1.90 | 1.85 | 1.81 | 1.75 | 1.70 | 1.64 |
| 30 | 4.17 | 3.32 | 2.92 | 2.69 | 2.53 | 2.42 | 2.33 | 2.27 | 2.21 | 2.16 | 2.09 | 2.01 | 1.93 | 1.89 | 1.84 | 1.79 | 1.74 | 1.68 | 1.62 |
| 40 | 4.08 | 3.23 | 2.84 | 2.61 | 2.45 | 2.34 | 2.25 | 2.18 | 2.12 | 2.08 | 2.00 | 1.92 | 1.84 | 1.79 | 1.74 | 1.69 | 1.64 | 1.58 | 1.51 |
| 60 | 4.00 | 3.15 | 2.76 | 2.53 | 2.37 | 2.25 | 2.17 | 2.10 | 2.04 | 1.99 | 1.92 | 1.84 | 1.75 | 1.70 | 1.65 | 1.59 | 1.53 | 1.47 | 1.39 |
| 120 | 3.92 | 3.07 | 2.68 | 2.45 | 2.29 | 2.18 | 2.09 | 2.02 | 1.96 | 1.91 | 1.83 | 1.75 | 1.66 | 1.61 | 1.55 | 1.50 | 1.43 | 1.35 | 1.25 |
| Inf | 3.84 | 3.00 | 2.60 | 2.37 | 2.21 | 2.10 | 2.01 | 1.94 | 1.88 | 1.83 | 1.75 | 1.67 | 1.57 | 1.52 | 1.46 | 1.39 | 1.32 | 1.22 | 1.00 |

$f1$: Numerator degrees of freedom
$f2$: Denominator degrees of freedom

Printed in the United States
By Bookmasters